ENGINEERING
ECONOMY

PRENTICE-HALL INTERNATIONAL SERIES
IN INDUSTRIAL AND SYSTEMS ENGINEERING

W. J. Fabrycky and J. H. Mize, Editors

BLANCHARD *Logistics Engineering and Management*
BROWN *Systems Analysis and Design for Safety*
FABRYCKY, GHARE, AND TORGERSEN *Industrial Operations Research*
FRANCIS AND WHITE *Facility Layout and Location: An Analytical Approach*
GOTTFRIED AND WEISMAN *Introduction to Optimization Theory*
KIRKPATRICK *Introductory Statistics and Probability for Engineering,
Science, and Technology*
MIZE, WHITE, AND BROOKS *Operations Planning and Control*
OSTWALD *Cost Estimating for Engineering and Management*
SIVAZLIAN AND STANFEL *Analysis of Systems in Operations Research*
SIVAZLIAN AND STANFEL *Optimization Techniques in Operations Research*
THUESEN, FABRYCKY, AND THUESEN *Engineering Economy, 5th edition*
WHITEHOUSE *Systems Analysis and Design Using Network Techniques*

ENGINEERING ECONOMY
5th edition

H. G. THUESEN

Oklahoma State University

W. J. FABRYCKY

Virginia Polytechnic Institute and State University

G. J. THUESEN

Georgia Institute of Technology

PRENTICE-HALL, INC., *Englewood Cliffs, New Jersey* 07632

Library of Congress Cataloging in Publication Data

THUESEN, HOLGER GEORGE. 1898–1973
   Engineering economy.

   (Prentice-Hall international series in industrial and systems engineering)
   Bibliography: p.
   Includes index.
   1. Engineering economy. I. Fabrycky, Wolter J.,
joint author. II. Thuesen, G. J.,
joint author. III. Title.
TA177.4.T48   1977   338.4'7'62   76-26887
ISBN 0-13-277491-7

© 1977 by Prentice-Hall, Inc.
Englewood Cliffs, New Jersey 07632

10 9 8 7

Printed in the United States of America

PRENTICE-HALL INTERNATIONAL, INC., *London*
PRENTICE-HALL OF AUSTRALIA PTY. LIMITED, *Sydney*
PRENTICE-HALL OF CANADA, LTD., *Toronto*
PRENTICE-HALL OF INDIA PRIVATE LIMITED, *New Delhi*
PRENTICE-HALL OF JAPAN, INC., *Tokyo*
PRENTICE-HALL OF SOUTHEAST ASIA PTE. LTD., *Singapore*
WHITEHALL BOOKS LIMITED, *Wellington, New Zealand*

CONTENTS

PREFACE

Engineers are confronted with two important interconnected environments, the physical and the economic. Their success in altering the physical environment to produce products and services depends upon a knowledge of physical laws. However, the worth of these products and services lies in their utility measured in economic terms. There are numerous examples of structures, machines, processes, and systems that exhibit excellent physical design but have little economic merit.

Engineering Economy deals with the concepts and techniques of analysis useful in evaluating the worth of systems, products, and services in relation to their cost. In this text, we emphasize that the essential prerequisite of successful engineering application is economic feasibility. Our objective is to help the reader to grasp the significance of the economic aspects of engineering and to become proficient in the evaluation of engineering proposals in terms of worth and cost.

The engineering approach to problem solution has advanced and broadened to the extent that success often depends upon the ability to deal both with economic aspects and with physical aspects of the problem. Down through history, the limiting factor has been predominantly physical, but with the development of science and technology, goods and services that may not have utility have become physically possible. Fortunately, the engineer can readily extend his inherent ability for analysis to embrace economic factors. To aid him in so doing is a primary aim of this book.

A secondary aim of this book is to acquaint the engineer with operations and operational feasibility. An understanding of the organizational and personnel factors upon which the success of engineering activity depends is essential to the solution of most complex problems. Economic factors in the operation of systems and equipment can no longer be left to chance but must be considered in the design process. A basic understanding of mathematical modeling of operations is becoming more important as operational systems attract the attention of a larger number of engineers.

Being accustomed to the use of facts and being proficient in computation, engineers should accept the responsibility for making an economic interpretation of their work. It is much easier for the engineer to master the fundamental concepts of economic analysis necessary to bridge the gap between physical and economic aspects of engineering application than it is for the person who is not technically trained to acquire the necessary technical background.

We have had no difficulty in teaching this material to engineering sophomores as well as to upper division students in management, economics, and the physical sciences who wish an introduction to engineering from the economic point of view. College algebra is the only mathematical background required.

This text contains more than enough material for a three semester hour course. Some material will have to be skipped for a course of shorter duration. This may be easily done, since the foundation topics are concentrated in the first ten chapters.

Those familiar with earlier editions of this text will note that we have retained the basic conceptual approach with considerable emphasis on examples. Also retained is the functional factor designation system which was originated by the late H. G. Thuesen. This factor designation system has proven its worth in the teaching of compound interest applications. Symbols in the system have been changed as suggested by the ad hoc standardization committee of the Engineering Economy Division of the American Society for Engineering Education.

It is our pleasure to acknowledge the very useful comments offered by many students who were exposed to this material in class. Specific thanks are due Mr. I. D. Moon and Mr. C. S. Park, both of whom assisted in the development of certain sections.

Finally, the expert editorial and typing assistance provided by Mrs. Elizabeth D. Mellichamp and Ms. Kaye L. Watkins is gratefully acknowledged.

W. J. FABRYCKY
G. J. THUESEN

ENGINEERING
ECONOMY

part one

INTRODUCTION TO
ENGINEERING ECONOMY

ENGINEERING AND
ENGINEERING ECONOMY

Engineering activities of analysis and design are not an end in themselves but are a means for satisfying human wants. Thus, engineering has two aspects. One aspect concerns itself with the materials and forces of nature; the other is concerned with the needs of mankind. Because we live in a resource constrained world, engineering must be closely associated with economics. It has become absolutely essential that engineering proposals be evaluated in terms of worth and cost before they are undertaken. In this chapter and throughout the text, we emphasize that the essential prerequisite of successful engineering application is economic feasibility.

1.1. Engineering and Science

Engineering is not a science but is an application of science. It is an art composed of the skill and ingenuity in adapting knowledge to the uses of humanity. As expressed in a definition adopted by the Engineers' Council for Professional Development: "Engineering is the profession in which a

1

knowledge of the mathematical and natural sciences gained by study, experience, and practice is applied with judgment to develop ways to utilize, economically, the materials and forces of nature for the benefit of mankind." In this, as in most other accepted definitions, the applied nature of engineering activity is emphasized.

The purpose of the scientist is to add to mankind's accumulated body of systematic knowledge and to discover universal laws of behavior. The purpose of the engineer is to apply this knowledge to particular situations to produce products and services. To the engineer, knowledge is not an end in itself but is the raw material from which he fashions structures, systems, and processes. Thus, engineering involves the determination of the combination of materials, forces, and human factors that will yield a desired result. Engineering activities are rarely carried out for the satisfaction that may be derived from them directly. With few exceptions, their use is confined to satisfying human wants.

Modern civilization depends to a large degree upon engineering. Most products used to facilitate work, communication, and transportation and to furnish sustenance, shelter, and even health are directly or indirectly a result of engineering activity. Engineering has also been instrumental in providing leisure time for pursuing and enjoying culture. Through the development of the printing process, television, and rapid transportation, engineering has provided the means for both cultural and economic improvement of the human race. In addition, engineering has become an essential ingredient in national survival as is evident from the considerable portion of engineering talent that is employed in national defense.

Science is the foundation upon which the engineer builds toward the advancement of mankind. With the continued development of science and the widespread application of engineering, the standard of living may be expected to improve and further increase the demand for those things that contribute to people's love for the comfortable and beautiful. The fact that these human wants may be expected to engage the attention of engineers to an increasing extent is, in part, the basis for the movement to incorporate humanistic and social considerations in engineering curricula. An understanding of these fields is essential as engineers seek solutions to the complex socio-technological problems of today.

1.2. The Bi-Environmental Nature of Engineering

In dealing with the physical environment engineers have a body of physical laws upon which to base their reasoning. Such laws as Boyle's law, Ohm's law, and Newton's laws of motion were developed primarily by collecting and comparing numerous similar instances and by the use of an inductive process. These laws may then be applied by deduction to specific instances. They are supplemented by many formulas and known facts, all of which enable the engineer to come to conclusions about the physical environment that match the facts within narrow limits. Much is known with certainty about the physical environment.

Much less, particularly of a quantitative nature, is known about the economic environment. Since economics is involved with the actions of people, it is apparent that economic laws must be based upon their behavior. Economic laws can be no more exact than the description of the behavior of people acting singly and collectively.

Want satisfaction in the economic environment and engineering proposals in the physical environment are linked by the production or the construction process. Figure 1.1 illustrates the relationship between engineering proposals, production or construction, and want satisfaction.

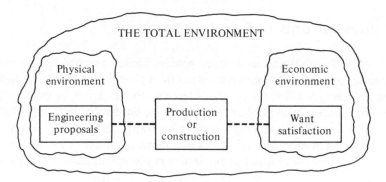

Figure 1.1. The engineering environment.

The usual function of engineering is to manipulate the elements of one environment, the physical, to create utility in a second environment, the economic. However, engineers sometimes have a tendency to disregard economic feasibility and are often appalled in practice by the necessity for meeting situations in which action must be based on estimates and judgment. Yet the modern engineering graduate is increasingly finding himself in positions in which his responsibility is extended to include economic factors.

There are those, and some are engineers, who feel that engineers should restrict themselves to the consideration of the physical and leave the economic and humanistic aspects of engineering to others; some would not even consider these aspects as coming under engineering. This viewpoint may arise in part because those who take pleasure in discovering and applying the well-ordered certainties of the physical environment find it difficult to adjust their thinking to consider the complexities of the economic environment.

Engineers can readily extend their inherent ability of analysis to become proficient in the analysis of the economic aspects of engineering application. Furthermore, the engineer who aspires to a creative position in engineering will find proficiency in economic analysis helpful. The large percentage of engineers who will eventually be engaged in managerial activities will find such proficiency a necessity.

Initiative for the use of engineering rests, for the most part, upon those who will concern themselves with social and economic consequences. Therefore, to maintain the initiative, engineers must operate successfully in both the physical and economic segments of the total environment. It is the objective of *engineering economy* to prepare engineers to cope effectively with the bi-environmental nature of engineering application.

1.3. Physical and Economic Efficiency

Both individuals and enterprises possess limited resources. This makes it necessary to produce the greatest output for a given input, that is, to operate at high efficiency. Thus, the search is not merely for a fair or good opportunity for the employment of limited resources, but for the best opportunity.

Man is continually seeking to satisfy his wants. In so doing, he gives up certain utilities in order to gain others that he values more. This is essentially an economic process in which the objective is the maximization of economic efficiency.

Engineering is primarily a producer activity that comes into being to satisfy human wants. Its objective is to get the greatest end result per unit of resource expenditure. This is essentially a physical process with the objective being the maximization of physical efficiency.

The objective of engineering application is to get the greatest end result per unit of resource input. This statement is an expression of physical efficiency which may be stated as

$$\text{efficiency (physical)} = \frac{\text{output}}{\text{input}}.$$

If interpreted broadly enough, this statement measures the success of engineering activity in the physical environment. However, the engineer must be concerned with two levels of efficiency. On the first level is physical efficiency expressed as outputs divided by inputs of such physical units as Btu's, kilowatts, and foot-pounds. When such physical units are involved, efficiency will always be less than unity or less than 100%.

On the second level are economic efficiencies. These are expressed in terms of economic units of output divided by economic units of input, each expressed in terms of a medium of exchange such as money. Economic efficiency may be stated as

$$\text{efficiency (economic)} = \frac{\text{worth}}{\text{cost}}.$$

It is well known that physical efficiencies over 100% are not possible. However, economic efficiencies can exceed 100% and must do so for economic undertakings to be successful.

Physical efficiency is related to economic efficiency. For example, a power plant may be profitable in economic terms even though its physical efficiency in converting units of energy in coal to electrical energy may be relatively low. As an example, in the conversion of energy in a certain plant, assume that the physical efficiency is only 32%. Assuming that output Btu's in the form of electrical energy have an economic value of $6.80 per million and that input Btu's in the form of coal have an economic value of $1.34 per million, then

$$\text{efficiency (economic)} = \frac{\text{Btu output} \times \text{value of electricity}}{\text{Btu input} \times \text{value of coal}}$$

$$= 0.32 \times \frac{\$6.80}{\$1.34} = 162\%.$$

Since physical processes are of necessity carried out at efficiencies of less than 100% and economic ventures are feasible only if they attain efficiencies of greater than 100%, it is clear that the economic worth per unit of physical output must always be greater than the economic cost per unit of physical input in feasible economic ventures. Consequently, economic efficiency must depend more upon the worth and cost per unit of physical outputs and inputs than upon physical efficiency. Physical efficiency is always significant, but only to the extent that it contributes to economic efficiency.

In the final evaluation of most ventures, even in those in which engineering plays a leading role, economic efficiencies must take precedence over physical efficiencies. This is because the function of engineering is to create utility in the economic environment by means of altering the elements of the physical environment.

1.4. The Engineering Process

Engineering activities dealing with the physical environment take place to meet the needs of mankind which arise in an economic setting. The engineering process employed from the time a particular need is recognized until the time when the need is satisfied may be divided into a number of phases. These are discussed in this section.

Determination of objectives. One important facet of the engineering process involves the search for new objectives for engineering application; to learn what people want that can be supplied by engineering. In the field of invention, success is not a direct result of the construction of a new device; rather, it is dependent upon the capability of the invention to satisfy human wants. Thus, market surveys seek to learn what the desires of people are.

Automobile manufacturers make surveys to learn what mechanical, comfort, and style features people want in transportation. Highway commissions make traffic counts to learn what construction programs will be of greatest use. Considerations of physical and economic feasibility come only after what is wanted has been determined.

The things that people want may be the result of logical considerations, but more often they are the result of emotional drives. There appears to be no logical reason why one prefers a certain make of car, a certain type of work, or a certain style of clothes. The bare necessities needed to maintain physical existence, in terms of calories of nourishment, clothing, and shelter are limited and may be determined with a fair degree of certainty. But the wants that stem from emotional drives seem to be unlimited.

Economic limitations are continually changing with the needs and wants of people. Physical limitations are continually being pushed back through science and engineering. In consequence, new openings revealing new opportunities are continually developing. For each successful venture an opening through the barrier of economic and physical limitations has been found.

The facet of the engineering process that seeks to learn of human wants requires not only a knowledge of the limitations of engineering capability but also a general knowledge of sociology, psychology, political science, economics, literature and other fields related to the understanding of human nature. A knowledge of these fields is recognized to be useful or essential in most branches of modern engineering.

Identification of strategic factors. The factors that stand in the way of attaining objectives are known as *limiting factors*. An important element of the engineering process is the identification of the limiting factors restricting accomplishment of a desired objective. Once the limiting factors have been identified they are examined to see which ones may be operated on with success. Thus, each of the limiting factors should be examined in order to locate *strategic factors*; that is, those factors which, if altered, will remove limitations restricting the success of an undertaking.

The understanding that results from the delineation of limiting factors and their further consideration to arrive at the strategic factors often stimulates ideas for improvements. There is obviously no point in operating upon some factors. Consider for example a situation in which a truck driver is hampered because he had difficulty in loading a heavy box. Three factors are involved: the pull of gravity, the mass of the box, and the strength of the man. Not much success would be expected from an attempt to lessen the pull of gravity. Nor is it likely that it is feasible to reduce the mass of the box. A stronger man might be secured, but it seems more logical to consider overcoming the need for strength by devices to supplement the strength of the man. This analysis leads to consideration of devices that might circumvent the limiting factor of strength.

The identification of strategic factors is important, for it makes possible the concentration of effort on those areas in which success is obtainable. This may require inventive ability, or the ability to put known things together in new combinations, and is distinctly creative in character. The means that will achieve the desired objective may consist of a procedure, a technical process, or a mechanical, organizational, or managerial change. Strategic factors limiting success may be circumvented by operating on engineering, human, and economic aspects individually and jointly.

Determination of means. The determination of means is subordinate to the identification of strategic factors, just as the identification of strategic factors is necessarily subordinate to the determination of objectives. Strategic factors may be altered in many different ways. Each possibility must be evaluated to determine which will be most successful in terms of overall economy. Engineers are well equipped by training and experience to determine means for altering the physical environment. If the means devised to overcome strategic factors come within the field of engineering they may be termed engineering proposals.

It may be presumed that a knowledge of facts in a field is a necessity for creativeness in that field. Thus, for example, it appears that a person who is proficient in the science of combustion and machine design is more likely to contrive an energy saving internal-combustion engine than a person who has little or no such knowledge. It also appears that knowledge of costs and people's desires as well as of engineering is necessary to conceive of opportunities for profit that involve engineering.

Engineers in any capacity have the opportunity to be creative by considering human and economic factors in their work. The machine designer may design tools or machines that will require a minimum of maintenance and that can be operated with less fatigue and greater safety. The highway designer may consider durability, cost, and safety. Engineers in any capacity can see to it that both private and public projects are planned, built, and operated in accordance with good engineering practice. Engineering is an expanding profession. People have great confidence in the integrity and ability of engineers. Perhaps nothing will enhance the image of the individual engineer and the profession of engineering more than the acceptance of responsibilities which go beyond the determination of means.

Evaluation of engineering proposals. It is usually possible to accomplish a desired result by several means, each of which is feasible from the technical aspects of engineering application. The most desirable of the several proposals is the one that can be performed at the least cost. The evaluation of engineering proposals in terms of comparative cost is an important facet of the engineering process and an essential ingredient in the satisfaction of wants with maximum economic efficiency. Although engineering alternatives are most often evaluated to determine which is most desirable economically, exploratory evaluations are also made to determine if any likely engineering proposal can be formulated to reach a goal profitably.

A wide range of factors may be considered in evaluating the worth and cost of engineering proposals. When investment is required, the time value of money must be considered. Where machinery and plants are employed, depreciation becomes an important factor. Most proposals involve organized effort, thus making labor costs an important consideration. Material is an important ingredient which may lead to market analyses and a study of procurement policy. Risks of a physical and economic nature may be involved and must be evaluated. Where the accepted engineering proposal is successful a net income will be derived, thus making the consideration of income tax necessary. This text offers specific instruction in methods of analysis pertaining to each of these factors and general instruction regarding the engineering process as a whole.

Assistance in decision making. Engineering is concerned with action to be taken in the future. Therefore, an important facet of the engineering process is to improve the certainty of decision with respect to the want satisfying objective of engineering application. Correct decisions can offset many operating handicaps. On the other hand, incorrect decisions may and often do hamper all subsequent action. No matter how expertly a bad decision is carried out, results will be at best unimpressive and at worst disastrous.

To make a decision is to select a course of action from several. A correct decision is the selection of that course of action which will result in an outcome more desirable than would have resulted from any other selection. Decision rests upon the possibility of choice; that is, on the fact that there are alternatives from which to choose. The engineer acting in a creative capacity proceeds on the thought that there is a most desirable solution if he can find it.

The logical determination and evaluation of alternatives in tangible terms have long been recognized as an integral facet of the engineering process. The success of engineers in dealing with this element of application is responsible, in part, for the large percentage of engineers who are either directly or indirectly engaged in decision-making activities.

1.5. A Plan for Engineering Economy Studies

The engineering process described in the previous section involves a creative element based on the employment of engineering economy studies. These economy studies can be made haphazardly or they can be made on the basis of a logical plan. This plan involves a creative step, a definition step, a conversion step, and a decision step.

The creative step. Engineers, whether engaged in research, design, construction, production, operations, or management activity must be concerned with the efficient use of limited resources. When known opportunities fail to hold sufficient promise for the employment of resources, more promising opportunities are sought. People with vision are those who accept the premise that better opportunities exist than are known to them. This view accompanied by initiative leads to exploratory activities aimed at finding the better opportunities. Exploration, research, investigation, and similar activities are creative. In such activities steps are taken into the unknown to find new possibilities, which may then be evaluated to determine if they are superior to those that are known.

Opportunities are not made; they are discovered. The person who concludes that there is no better way makes a self-fulfilling prophecy. When the belief is held that there is no better way, a search for one will not be made, and a better way will not be discovered. Any situation embraces groups of facts of which some may be known and others unknown. The material out of which new opportunities for profit are to be fashioned are the facts as they exist.

Some successful ideas are dependent upon the discovery of new facts. New facts may become known through research effort or by accident. Research is effort consciously directed to the learning of new facts. In pure research, facts are sought without regard for their specific usefulness, on the premise that a stockpile of knowledge will in some way contribute to man's welfare. Much of man's progress rests without doubt upon facts discovered from efforts to satisfy curiosity.

Both new facts and new combinations of facts may be consciously sought. The creative aspect of engineering economy consists in finding new facts and new combinations of facts out of which may be fashioned opportunities to provide profitable service through the application of engineering.

Aside from the often quoted statement that "inspiration is 90% perspiration," there are few guides to creativeness. It appears that both conscious application and inspiration may contribute to creativeness. Some people seem to be endowed with marked aptitudes for conceiving new ideas.

The creative step is believed to be of first importance in economy studies. It is directly related to the delineation and selection of objectives that are without doubt the most important functions of engineering economy and certainly the first steps toward success in any field of endeavor. Since the mental processes involved are in large measure illogical, this step must be approached with considerable alertness and curiosity and a willingness to consider new ideas and unconventional patterns of thought.

The definition step. The definition step consists of defining alternatives that originated in the creative step, or which have been selected for comparison in some other way. A complete and all-inclusive alternative rarely emerges in its final state. It begins as a hazy, but interesting idea. The attention of the individual or group is then directed to analysis and synthesis and the result is a definite proposal. In its final form, an alternative should consist of a complete description of its objectives and its requirements in terms of inputs and outputs.

Both different ends and different methods are embraced by the term alternative. All proposed alternatives are not necessarily attainable. Alternatives are frequently proposed for analysis even though there seems to be little likelihood that they will prove feasible. This is done on the thought that it is better to consider many unprofitable alternatives than to overlook one that is profitable. Alternatives that are not considered cannot be adopted, no matter how desirable they may prove to be.

In addition to the alternatives formally set up for evaluation, another alternative is always present. This is the alternative of making no decision on the formal alternative being considered. The decision not to decide may be a result of either active consideration or passive failure to act; it is usually motivated by the thought that there will be opportunities in the future which will prove more profitable than any known at present.

In the first stage of the definition step, the engineer's aim should be to delineate each alternative on the basis of its major and subordinate physical units and activities. The purpose of this stage is to insure that all factors of each alternative and no others will be considered in evaluating it.

The second stage of the definitive step consists of enumerating the prospective items of output and input of each alternative, in quantitative physical terms as far as possible and then in qualitative terms. Though qualitative items cannot be expressed numerically they may often be of major importance. They should be listed carefully so that they may be considered in the final evaluation.

The conversion step. In order to compare alternatives properly, it is important that they be converted to a common measure. The common denominator usually chosen for economic comparison is value expressed in terms of money.

The first phase of the conversion step is to convert the prospective output and input items enumerated in the definition step into receipts and disbursements at specified dates. This phase consists essentially of appraising the unit value of each item of output or input and determining their total amounts by computation. On completion, each alternative should be expressed in terms of definite cash flows occurring at specified dates in the future, plus an enumeration of qualitative considerations that it has been impossible to reduce to monetary terms. For such items the term "irreducibles" is often employed. This term seems to be superior to the term "intangibles" in this connection, for many items described in qualitative terms are clearly known even though not expressible in numbers.

The second phase of the conversion step consists of placing the estimated future cash flows for all alternatives on a comparable basis, considering the time value of money. This involves employment of the techniques presented in future chapters. Selection of the particular technique depends upon the situation and its appropriateness is a matter of judgment. Consideration of inherent inaccuracies in estimates of the future outputs and inputs may be considered part of the conversion step and should not be overlooked.

The final phase of the conversion step is to communicate the essential aspects of the economy study, together with an enumeration of irreducibles, so that they may be considered by those responsible for making the decision. Responsibility for the acceptance or rejection of an engineering proposal is exercised more often than not by persons who have not been concerned with the technical phases of the proposal. Also, the persons who control acceptance are likely to lack understanding of technical matters.

A proposal should be explained in terms that will best interpret its significance to those who will control its acceptance. The aim of a presentation should be to take persons concerned with a proposal on an excursion into the future to experience what will happen if the proposal is accepted or rejected. For example, suppose that a proposal for a new pollution control system is to be presented. Since those who must decide if it should be adopted rarely have the time and background to go into and appreciate all the technical details involved, the significance of these details in terms of economic results must be made clear. Of interest to those in a position to decide will be such things as the present outlay required, capital-recovery period, flexibility of the system in event of production volume changes, effect upon product price and quality, and difficulties of financing. Cost and other data should be broken down and presented so that attention may be easily focused upon pertinent aspects of the proposal. Diagrams, graphs, pictures, and even models should be used where these devices will contribute to understanding.

The decision step. On completion of the conversion step, quantitative and qualitative outputs and inputs for each alternative form the basis for comparison and decision. Quantitative input may be deducted from quantitative output to obtain quantitative profit, or the ratio of quantative profit to quantitative input may be found. Each of these measures is then supplemented by what qualitative consideration may have been enumerated.

Decisions between alternatives should be made on the basis of their differences. Thus, all identical factors can be canceled out for the comparison of any two or more alternatives at any step in an economy study. In this process great care must be exercised that factors canceled as being identical are actually of the same significance. Unless it is very clear that factors considered for cancellation are identical, it is best to carry them through the first stage of the decision step. This may entail a greater amount of computation and other paper work, but the slight added complexity and loss of time is ordinarily insignificant in comparison with the value to be derived.

When a diligent search uncovers insufficient information to reason the outcome of a course of action, the problem is to render as accurate a decision as the lack of facts permits. In such situations there is a decided tendency, on the part of many, to make little logical use of the data that are available in coming to a conclusion, on the thought that since some pretty rough estimating has to be done on some elements of the situation, the estimate might as well embrace the entire situation. But an alternative may usually be subdivided into parts, and the available data are often adequate for a complete or nearly complete evaluation of several of the parts. The segregation of the known and unknown parts is in itself additional knowledge. Also, the unknown parts, when subdivided, frequently are recognized as being similar to parts previously encountered and thus become known.

After a situation has been carefully analyzed and the possible outcomes have been evaluated as accurately as possible, a decision must be made. Even after all the data that can be brought to bear on a situation have been considered, some areas of uncertainty may be expected to remain. If a decision is to be made, these areas of uncertainty must be bridged by consideration of nonquantitative data or, in other words, by the evaluation of irreducibles. Some call the type of evaluation involved in the consideration of irreducibles *intuition*; others call it *hunch* or *judgment*.

When complete knowledge of all facts concerned and their relationships exists, reason can supplant judgment and predictions become a certainty. Judgment tends to be qualitative. Reason is both qualitative and quantitative. Judgment is at best an informal consideration and weighing of facts; at its worst it is merely wishful thinking. Judgment appears to be an informal process for considering information, past experience, and feeling in relation to a problem. No matter how sketchy factual knowledge of a situation may be, some sort of a conclusion can always be drawn in regard to it by judgment.

Whatever it be called, it is inescapable that this type of thinking or, perhaps better, this type of feeling, must always be the final part in coming to a decision about the future. There is no other way if action is to be taken. There appears to be a marked difference in people's abilities to come to sound conclusions when some facts relative to a situation are missing. Perhaps much more attention should be devoted to developing sound judgment, for those who possess it are richly rewarded. But as effective as intuition, hunch, or judgment may sometimes be, this type of thinking should be reserved for those areas where facts on which to base a decision are missing.

An important aim of engineering economy studies is to gather and analyze the facts so that reason may be used to the fullest extent in arriving at a decision. In this way judgment can be reserved for parts of the situation where factual knowledge is absent. This idea is embraced in the statement, "Figure as far as you can, then add judgment."

1.6. Engineering Economy and the Engineer

Economy, the attainment of an objective at low cost in terms of resource input, has always been associated with engineering. During much of history the limiting factor has been predominantly physical. Thus, a great innovation, the wheel, awaited invention, not because it was useless or costly, but because the mind of man could not synthesize it earlier. But, with the development of science, things have become physically possible that people are interested in only slightly or not at all. Thus, a new type of communication system may be perfectly feasible from the physical standpoint but may enjoy limited use because of its first cost or cost of operation.

Engineers are becoming increasingly aware of the fact that many sound proposals fail because those who might have benefited from them did not understand their significance. A prospective user of a good or service is primarily interested in its worth and cost. The person who lacks an understanding of engineering may find it difficult or even impossible to grasp the technical aspects of a proposal sufficiently to arrive at a measure of its economic desirability. The uncertainty so engendered may easily cause loss of confidence and a decision to discontinue consideration of the proposal.

Since economic factors are the strategic consideration in most engineering activities, engineering practice may be either responsive or creative. If the engineer takes the attitude that he should restrict himself to the physical, he is likely to find that the initiative for the application of engineering has passed on to those who will consider economic and social factors.

The engineer who acts in a responsive manner acts on the initiative of others. The end product of his work has been envisioned by another.

Although this position leaves him relatively free from criticism, he gains this freedom at the expense of professional recognition and prestige. In many ways, he is more of a technician than a professional man. Responsive engineering is, therefore, a direct hindrance to the development of the engineering profession.

The creative engineer, on the other hand, not only seeks to overcome physical limitations, but also initiates, proposes, and accepts responsibility for the success of projects involving human and economic factors. The general acceptance by engineers of the responsibility for seeing that engineering proposals are both technically and economically sound, and for interpreting proposals in terms of worth and cost, may be expected to promote confidence in engineering as a profession.

QUESTIONS

1. Contrast the role of the engineer and the scientist.

2. What is the important difference between a physical law and an economic law?

3. Give several examples of products that would be technically feasible but that would possess little economic merit.

4. Why should the engineer concern himself with both the physical and the economic aspects of the total environment?

5. Explain how physical efficiency below 100% may be converted to economic efficiency above 100%.

6. Why must economic efficiency take precedence over physical efficiency?

7. Name the essential activities in the engineering process.

8. What is involved in seeking new objectives for engineering application?

9. What is a limiting factor? A strategic factor?

10. Give reasons why engineers are particularly well equipped to determine means for the attainment of an objective.

11. How may engineers assist in decision making?

12. As compared with the economic aspect of engineering application, give reasons why the physical aspect is decreasing in relative importance.

13. Discuss the potential benefits of the plan for engineering economy studies.

14. Apply the plan for engineering economy studies to an engineering problem of your choice.

15. How is the creative step related to the satisfaction of human wants?

16. List the methods for discovering means to more profitably employ resources.

17. Why is it not possible to consider all possible alternatives?

18. Discuss the nature of judgment and explain why it is applicable to more situations than reason.

19. Explain why decisions must be based on differences occurring in the future.

20. Why are decisions relative to the future based upon estimates instead of upon the facts that will apply?

21. Why is judgment always necessary to come to a decision relative to an outcome in the future?

22. Why should the economic interpretation of engineering proposals be made by engineers?

SOME FUNDAMENTAL
ECONOMIC CONCEPTS

Concepts are crystallized thoughts which have withstood the test of time. They are usually qualitative in nature and are not necessarily universal in application. Economic concepts, if carefully related to fact, may be useful in suggesting solutions to problems in engineering economy. Those given in this chapter are by no means exhaustive. Many others will be found throughout the text to support the quantitative material presented. The ability to arrive at sound decisions through engineering economic analysis is dependent jointly upon a sound conceptual understanding and the ability to handle the quantitative aspects of the problem. In this chapter we give special attention to some basic economic concepts useful in engineering economy studies.

2.1. Economics Deals With the Behavior of People

The engineer is often concerned by the lack of certainty associated with the economic aspect of engineering. However, it must be recognized that economic considerations embrace many of the subtleties and complexities

2

characteristic of people. Economics deals with the behavior of people individually and collectively, particularly as their behavior relates to the satisfaction of their wants.

The wants of people are motivated largely by emotional drives and tensions and to a lesser extent by logical reasoning processes. A part of human wants can be satisfied by physical goods and services such as food, clothing, shelter, transportation, communication, entertainment, medical care, educational opportunities, and personal services; but man is rarely satisfied by physical things alone. In food, sufficient calories to meet his physical needs will rarely satisfy. He will want the food he eats to satisfy his energy needs and also his emotional needs. In consequence we find people concerned with the flavor of food, its consistency, the china and silverware with which it is served, the person or persons who serve it, the people in whose company it is eaten, and the "atmosphere" of the room in which it is served. Similarly, there are many desires associated with clothing and shelter, in addition to those required merely to meet physical needs.

Anyone who has a part in satisfying human wants must accept the uncertain action of people as a factor with which he must deal, even though he finds such action unexplainable. Much or little progress has been made in discovering knowledge on which to base predictions of human reactions, depending upon one's viewpoint. The idea that human reactions will some-day be well-enough understood to be predictable is accepted by many people; but in spite of the fact that this has been the objective of the thinkers of the world since the beginning of time, it appears the progress in psychology has been meager compared to the rapid progress made in the physical sciences. But, in spite of the fact that human reactions can be neither predicted nor explained, they must be considered by those who are concerned · with satisfying human wants.

2.2. Concepts of Value and Utility

The term *value*, like most other widely used terms, has a variety of meanings. In economics, value designates the worth that a person attaches to an object or a service. Thus, the value of an object is not inherent in the object but is inherent in the regard that a person or people have for it. Value should not be confused with the cost or the price of an object in engineering economy studies. There may be little or no relation between the value a person ascribes to an article and the cost of providing it or the price that is demanded for it.

The general economic meaning of the term *utility* is the power to satisfy human wants. The utility that an object has for an individual is determined by him. Thus, the utility of an object, like its value, is not inherent in the object itself but is inherent in the regard that a person has for it. Utility and value in the sense used here are closely related. The utility that an object has for a person is the satisfaction he derives from it. Value is an appraisal of utility in terms of a medium of exchange.

In ordinary circumstances a large variety of goods and services is available to an individual. The utility that available items may have in the mind of a prospective user may be expected to be such that his desire for them will range from abhorrence, through indifference, to intense desire. His evaluation of the utility of various items is not ordinarily constant but may be expected to change with time. Each person also possesses either goods or services that he may render. These have the utility for the person himself that he regards them to have. These same goods and possible services may also be desired by others, who may ascribe to them very different utilities. The possibility for exchange exists when each of two persons possesses utilities desired by the other.

If the supernatural is excluded from consideration all that has utility is physically manifested. This statement is readily accepted in regard to physical objects that have utility, such as an automobile, a tractor, a house, or a steak dinner. But this statement is equally true in regard to the more intangible things. Music, which is regarded as pleasing to a person, is manifested to him as air waves which strike his ears. Pictures are manifested as light waves. Even friendship is realized only through the five senses and must, therefore, have its physical aspects. It follows that utilities must be created by changing the physical environment.

For example, the consumer utility of raw steak can be increased by altering its physical condition by an appropriate application of heat. In the area of producer utilities the machining of a bar of steel to produce a shaft for a rolling mill is an example of creating utility by manipulation of the physical environment. The purpose of much engineering effort is to determine how the physical environment may be altered to create the most utility for the least cost in terms of the utilities that must be given up.

2.3. Consumer and Producer Goods

Two classes of goods are recognized by economists: consumer goods and producer goods. Consumer goods are products and services that directly satisfy human wants. Examples of consumer goods are television sets, houses, shoes, books, orchestras, and health services. Producer goods also satisfy human wants but do so indirectly as a part of the production or the construction process. Broadly speaking, the ultimate end of all engineering activity is to supply goods and services that people may consume to satisfy their desires and needs.

Producer goods are, in the long run, used as a means to an end; namely, that of producing goods and services for human consumption. Examples of this class of goods are lathes, bulldozers, ships, and railroad cars. Producer goods are an intermediate step in man's effort to supply his wants. They are not desired for themselves, but because they may be instrumental in producing something that man can consume.

Once the kind and amount of consumer goods to be produced has been determined, the kinds and amounts of facilities and producer goods necessary to produce them may be approached objectively. The energy, ash, and other contents of coal, for instance, can be determined very accurately and are the basis of evaluating the utility of the coal. The extent to which producer utility may be considered by logical processes is limited only by factual knowledge and the ability to reason.

Utility of consumer goods. A person will consider two kinds of utility. One kind embraces the utility of goods and services that he intends to consume personally for the satisfaction he gets out of them. Thus, it seems reasonable to believe that the utility a person ascribes to goods and services that are consumed directly is in large measure a result of subjective, non-logical mental processes. That this is so may be inferred from the fact that sellers of consumer goods apparently find emotional appeals more effective than factual information. Early automobile advertising took the form of objective information related to design and performance, but more recent practice stresses such subjective aspects as comfort, beauty, and prestige values.

Some kinds of human wants are much more predictable than others. The demand for food, clothing, and shelter, which are needed for bare physical existence, is much more stable and predictable than the demand for those items that satisfy man's emotional needs. The amount of foodstuffs needed for existence is ascertainable within reasonable limits in terms of calories of energy, and the clothing and shelter requirement may be fairly accurately determined from climate data. But once man is assured of physical existence he reaches out for satisfactions related to his being a person rather than merely to his being a physical organism.

An analysis of advertising and sales practices used in selling consumer goods will reveal that they appeal primarily to the senses rather than to reason and perhaps rightly so. If the enjoyment of consumer goods stems almost exclusively from how one feels about them rather than what one reasons about them, it seems logical to make sales presentation on the basis on which customers ascribe utility.

Utility of producer goods. The second kind of utility that an object or service may have for a person is its utility as a means to an end. Producer goods are not consumed for the satisfaction that can be directly derived from them but as a means of producing consumer goods usually by facilitating the alteration of the physical environment.

Although the utility of consumer goods is primarily determined subjectively, the utility of producer goods as a means to an end may be, and usually is, in large measure considered objectively. In this connection consider the satisfaction of the human want for harmonic sounds as in music. Suppose it has been decided that the desire for music can be met by 100,000 stereophonic records. Then the organization of the artists, the technicians, and the equipment necessary to produce the records becomes predominantly objective in character. The amount of material that must be compounded and processed to form one record is calculable to a high degree of accuracy. If a concern has been making records for some time, it will know the various operations that are to be performed and the unit times for performing them. From these data, the kind and amount of producer service, the amount and kind of labor, and the number of various types of machines are determinable within rather narrow limits. Whereas the determination of the kinds and amounts of consumer goods needed at any one time may depend upon the most subjective of human consideration, the problems associated with their production are quite objective by comparison.

2.4. Economic Aspects of Exchange

Economy of exchange occurs when utilities are exchanged by two or more people. In this connection, a utility means anything that a person may receive in an exchange that has any value whatsoever to him; for example, a lathe, a dozen pencils, a meal, music, or a friendly gesture.

A buyer will purchase an article when he has money available and when he believes that the article has equal or greater utility for him than the amount required to purchase it. Conversely, a seller will sell an article when he believes that the amount of money to be received for the article has greater utility than the article has for him. Thus, an exchange will not be effected unless at the time of exchange both parties believe that they will benefit. Exchanges are made when they are thought to result in mutual benefit. This is possible because the objects of exchange are not valued equally by the parties to the exchange.

As an illustration of the economy of exchange consider the following example. Two workmen, upon opening their lunch boxes, discover that one contains a piece of apple pie and the other a piece of cherry pie. Suppose further that Mr. A evaluates his apple pie to have 10 units of utility for him and the cherry pie to have 30 units. Suppose, also that Mr. B evaluates his cherry pie to have 20 units of utility for him and the apple pie 40 units. If Mr. A consumes his apple pie and Mr. B consumes his cherry pie, the utility realized in the system is 10 plus 20, or 30 units. But, if the two workmen exchange pieces of pie, the resulting utility will be increased to 30 plus 40, or 70 units. The utility in the system can be increased by 70 less 30, or 40 units by exchange.

Economy of exchange is possible because consumer utilities are evaluated by the consumer almost entirely, if not entirely, by subjective consideration. Thus, if the workmen in the example believe that the exchange has resulted in a gain of net satisfaction to them, no one can deny it. On the other hand, at the time of exchange, unless each person valued what he had to give less than what he was about to receive, an exchange could not take place. Thus, we conclude that an exchange of consumer utilities results in a gain for both parties because the utilities are subjectively evaluated by the participating parties. The reason that people can be found who will subjectively evaluate consumer utilities so that an exchange will permit each to gain is that people have different needs by virtue of their history and their environment.

Assuming that each party to an exchange of producer utilities correctly evaluates the objects of exchange in relation to his situation, what makes it possible for each person to gain? The answer is that the participants are in different economic environments. The fact may be illustrated by an example of a merchant who buys lawn mowers from a manufacturer. For example, at a certain volume of activity the manufacturer finds that he can produce and distribute mowers at a total cost of $70 per unit and that the merchant buys a number of the mowers at a price of $80 each. The merchant then finds that by expending an average of $40 per unit in selling effort, he can sell a number of the mowers to homeowners at $160 each. Both of the participants profit by the exchange. The reason that the manufacturer profits is that his environment is such that he can sell to the merchant for $80 a number of mowers that he cannot sell elsewhere at a higher price, and that he can manufacture lawn mowers for $70 each. The reason that the merchant profits is that his environment is such that he can sell mowers at $160 each by applying $40 of selling effort upon a mower of certain characteristics which he can buy for $80 each from the manufacturer in question, but not for less elsewhere.

The questions may be asked, why doesn't the manufacturer enter the merchandising field and thus increase his profit or why doesn't the merchant enter the manufacturing field? The answer to these questions is that neither the manufacturer nor the merchant can do so unless each changes his environment. The merchant, for example, lacks physical plant equipment and an organization of engineers and workmen competent to build lawn mowers. Also, he may be unable to secure credit necessary to engage in manufacturing, although he may easily secure credit in greater amounts for merchandising activities. It is quite possible that he cannot alter his environment so that he can build mowers for less than $80. Similar reasoning applies to the manufacturer. Exchange consists essentially of physical activity designed to transfer the control of things from one person to another. Thus, even in exchange, utility is created by altering the physical environment.

Mutual benefit in exchange. Each party in an exchange should seek to give something that has little utility for him but that will have great utility for the receiver. In this manner each exchange can result in the greatest gain for each party. Nearly everyone has been a party to such a favorable exchange. When a car becomes stuck in snow, only a slight push may be required to dislodge it. The slight effort involved in the dislodging push may have very little utility for the person giving it, so little that he expects no more compensation than a friendly nod. On the other hand, it might have very great utility for the person whose car was dislodged, so great that he may offer a substantial tip. The aim of much sales and other research is to find products that not only will have great utility for the buyer but that can be supplied at a low cost; that is, have low utility for the seller. The difference between the utility that a specific good or service has for the buyer and the utility it has for the seller represents the profit or net benefit that is available for division between buyer and seller, and may be called the range of mutual benefit in exchange.

The factors that may determine a price within the range of mutual benefit at which exchange will take place are infinite in variety. They may be either subjective or objective. A person seeking to sell may be expected to make two evaluations: the minimum amount he will accept and the maximum amount a prospective buyer can be induced to pay by persuasion. The latter estimate may be based upon mere conjecture or upon a detailed analysis of buyers' subjective and objective situation. In bargaining it is usually advantageous to obscure one's situation. Thus, sellers will ordinarily refrain from revealing the costs of the things they are seeking to sell or from referring a buyer to a competitor who is willing to sell at a lower price.

Persuasion in exchange. It is not uncommon for an equipment salesman to call on a prospective customer, describe and explain a piece of equipment, state its price, offer it for sale, and have his offer rejected. This is concrete evidence that the machine does not possess sufficient utility at the moment to induce the prospective customer to buy it. In such a situation, the salesman may be able to induce the prospect to listen to further sales talk, during which the prospect may decide to buy on the basis of the original offer. This is concrete evidence that the machine now possesses sufficient utility to induce the prospective customer to buy. Since there was no change in the machine or the price at which it was offered, there must have been a change in the customer's attitude or regard for the machine. The pertinent fact to grasp is that a proposition at first undesirable now has become desirable as a result of a change in the customer, not in the proposition.

What brought about the change? A number of reasons could be advanced. Usually it would be said that the salesman persuaded the customer to buy, in other words that the salesman induced the customer to believe something, namely, that the machine had sufficient utility to warrant its purchase. There are many aspects to persuasion. It may amount only to calling attention to the availability of an item. A person cannot purchase an item he does not know exists. A part of a salesman's function is to call attention to the things he has to sell.

It is observable that persuasive ability is much in demand, is often of inestimable beneficial consequences to all concerned, and is usually richly rewarded. Persuasion as it applies to the sale of goods is of economic importance to industry. A manufacturer must dispose of the goods he produces. He can increase the salability of his products by building into them greater customer appeal in terms of greater usefulness, greater durability, or greater beauty, or he may elect to accompany his products to market with greater persuasive effort in the form of advertising and sales promotion. Either plan will require expenditure and both are subject to the law of diminishing returns. It is a study in economy to determine what levels of perfection of product and sales effort will be most profitable.

It is interesting to note, in this connection, that the costs of some things that increase the sale of a product may be more than outweighed by the economies that result. Thus, qualities that make a product more desired, or sales promotion, may have two beneficial effects. One of these is that it may be feasible to increase the price asked for the product and the other is that the volume of product sold may increase to the extent that a lower manufacturing cost per unit may result.

Whatever the approach, factual or emotional, persuasion consists of taking a person on an excursion into the future in an attempt to show and convince him what will happen if he acts in accordance with a proposal. The purpose of engineering economy analysis is to estimate, on as factual a basis as possible, what the economic consequences of a desision will be. It is therefore a useful technique in persuasion.

2.5. The Economy of Organization

Man has accumulated an extensive body of physical laws to aid him in dealing with the physical aspect of his environment. With every advance in the physical sciences new relationships are discovered which provide further understanding of physical phenomena. Although much less is known about organized activity and human factors, this important area is now receiving considerable attention.

Organizations consist of the coordinated effort of individuals and have existed since mankind's earliest beginnings. Through organizations, man can either attain ends that cannot be achieved by individual effort or he can attain certain ends more economically. Thus, a critical element in the development of the forces and materials of nature is human effort acting through organization. In fact, organization is considered by many to be mankind's most important innovation.

Labor saving through organization. Organized effort is often a means to economy in accomplishment through labor saving. Suppose, for example, that two men adjacent to each other are each confronted with the task of lifting a box onto a loading platform, and that each of the two boxes is too heavy for one man to lift but not too heavy for two men to lift. Assume that the only practical way for one man to accomplish his task is to obtain a hand winch with which the task can be accomplished in 30 minutes' time. If there is no coordination of effort the cost of getting the two boxes onto the platform will be 60 man-minutes.

Suppose that the two men had coordinated their efforts to lift the two boxes in turn and that the time consumed was one minute per box. The two tasks would have been accomplished at the expense of 4 man-minutes, or about 7% as much as if there had been no coordination of effort.

Coordination of human effort is so effective a means of labor saving that it may be economical to pay for effort to bring about coordination of effort. In the example above, effort directed to bring about coordination would result in a net labor saving of 56 man-minutes of effort.

For further illustration of the creativeness of coordination of human effort, suppose that a water well would have a value of $100 to each of 100 families in a village of a certain undeveloped country. The head of each of the families recognizes this and on inquiry finds that a well will cost $1,000. Each family head, being oblivious of the opportunities for coordination, abandons the well drilling project as unprofitable. If an entrepreneur could bring about a coordination of effort in the village, the net benefit might be as follows: (100 families $\times$ $100) $-$ $1,000 = $9,000. Thus, entrepreneurship is a worthwhile and necessary activity in most situations involving organized activity.

The efficiency of organization. Organizations are a means for creating and exchanging utilities. The fact that they are creative may be inferred from the following example. A person who is employed by an industrial organization may be presumed to value the wages and other benefits he gets more highly than the efforts he contributes to gain them. The person who sells material to the organization must value them less than the money he receives for them or he would not sell. The same may be said of the seller of equipment. Similarly, a person who loans money to an organization will in the long run receive more in return than he advances or he will cease to loan money. The customer who comes with money in hand to exchange for the products of the organization may be expected to part with his money only if he values it less than he values the products he can get for it. This situation is illustrated in Figure 2.1.

In order that the organization illustrated be successful, not only must the total of the satisfactions exceed the total of contributions, but also each contributor's satisfactions must exceed his contribution as he evaluates them. In other words, each contributor must realize his aspirations to a satisfactory degree or he ceases to contribute. Organizations are essentially devices to which people contribute what they desire less to gain what they desire more. Unless people receive more than they put into an organization they withdraw from it. For an organization to endure, its efficiency—output divided by input—must exceed unity.

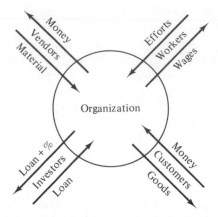

Figure 2.1. Organization as a mechanism for exchange.

The function of a manager is to maintain a system for pooling the activities of people so that each person, including the manager himself, gains more than he contributes and so that individual contributors do not believe that superior alternatives are open to them elsewhere. Those who can successfully manage possess a valuable talent which is usually richly rewarded.

2.6. Classifications of Cost

The ultimate objective of engineering application is the satisfaction of human wants. But human wants are not satisfied without cost. Alternative engineering proposals will differ in regard to the costs they involve relative to the objective of want satisfaction. The engineering proposal resulting in least cost will be considered best if its end result is identical to that of competing proposals.

A number of cost classifications have come into use to serve as a basis for economic analysis. As concepts, these classifications are useful in calling to mind the source and effect of costs that will have a bearing on the end result of a proposal. This section will define and discuss these cost classifications.

First cost. By definition, *first cost* is considered to involve the cost of getting an activity started. The chief advantage in recognizing this classification is that it calls attention to a group of costs associated with the initiation of a new activity that might not otherwise be given proper consideration. Ordinarily, this classification is limited to those costs which occur only once for any given activity.

There are degrees of action in a great many fields below which the effect of the action is insignificant. For example, a small movement of a gear in a gear train will merely take up the backlash without moving an adjacent gear. As an analogy, it must be recognized that many activities that otherwise may be profitable cannot be undertaken for the reason that their associated first cost represents a level of input above that which can be met. Many engineering proposals that are otherwise sound are not initiated because the first cost involved is beyond the reach of the controlling organization. Other engineering proposals meet with failure after initiation because it was found that the threshold input required was not successfully met because of financial limitations.

As an example, consider a proposed mining operation involving the extraction of an estimated 1,200,000 tons of ore from a mountain. Engineering Proposal A involves the construction of a tunnel beginning at an existing railroad spur and terminating within the mountain at the ore deposit. The ore will be removed by a gravity conveyor within the tunnel. The first cost of the tunnel and installed conveyor is estimated to be $1,380,000. The conveyor will cost $0.40 per ton to operate and will have a salvage value of $180,000.

Engineering Proposal B involves the removal of a quantity of over-burden so the ore deposit will be exposed for loading into trucks. The first cost of removing the overburden and improving an existing road is estimated to be $220,000. Trucks with drivers will be leased to haul the ore to the spur at a cost of $1.30 per ton of ore removed. All other costs for the two proposals are equal. Neglecting interest on investment, the cost difference is calculated as follows:

Proposal A $1,380,000 $-$ $180,000 $+$ $0.40(1,200,000) $=$ $1,680,000

Proposal B $220,000 $+$ $1.30(1,200,000) $=$ $1,780,000

Suppose that further analysis indicates that the mining operation will result in sufficient income to yield a profit regardless of the proposal chosen, but that financial limitations will not allow the successful completion of Proposal A because of its high first cost. In such a case, the mining operation might be initiated by accepting Proposal B even though it will result in a cost of $100,000 more than A. Although A appears to be best, it would result in failure because of the inability to meet its first cost.

Fixed and variable cost. *Fixed cost* is ordinarily defined as that group of costs involved in a going activity whose total will remain relatively constant throughout the range of operational activity. The concept of fixed cost has a wide application. For example, certain losses in the operation of an engine are in some measure independent of its output of power. Among its fixed costs, in terms of energy for a given speed and load, are those for the power to drive the fan, the valve mechanism, and the oil and fuel pumps. Almost any task involves preparation independent of its extent. Thus, to paint a small area may require as much effort for the cleaning of a brush as to paint a large area. Similarly, manufacturing involves fixed costs that are independent of the volume of output.

Fixed costs arise from making preparation for the future. A machine is purchased now in order that labor costs may be reduced in the future. Materials which may never be needed are purchased in large quantities and stored at much expense and with some risk in order that idleness of production facilities and men may be avoided. Research is carried on with no immediate benefit in view in the hope that it will pay in the long run. The investments that give rise to fixed cost are made in the present in the hope that they will be recovered with a profit as a result of reductions in variable costs or of increases in income.

Fixed costs are made up of such cost items as depreciation, maintenance, taxes, insurance, lease rentals, and interest on invested capital, sales programs, certain administrative expenses, and research. It will be observed that these arise from the decisions of the past and in general are not subject to rapid change. Volume of operational activity, on the other hand, may fluctuate widely and rapidly. As a result, fixed costs per unit may easily get out of hand. It is probable that this is the cause of more unsuccessful activity than any other, for few have the foresight or luck to make commitments in the present which will fit requirements of the future even reasonably well. Since fixed costs cannot be changed readily, consideration must be focused upon maintaining a satisfactory volume and character of activity.

In a practical situation, fixed costs are only relatively fixed and their total may be expected to increase somewhat with increased activity. The increase will probably not follow a smooth curve but will vary in accordance with the characteristics of the enterprise.

Consider a plant of several units that has been shut down or is operating at zero volume. No heat, light, janitor, and many other services will have been required. Many of these services must be reinstated if the plant is to operate at all, and if reinstated only on a minimum basis it is probable that these services will be adequate for quite a range of activity. Further increases in activity will require expenditures for other services that cannot be provided to just the extent needed. Thus, what are termed "fixed costs" in business may be expected to increase in some stepped pattern with an increase in activity.

Variable cost is ordinarily defined as that group of costs which vary in some relationship to the level of operational activity. For example, the consumption of fuel by an engine may be expected to be proportional to its output of power and the amount of paint used may be expected to be proportional to the area painted. In manufacturing, the amount of material needed per unit of product may be expected to remain constant and, therefore, the material cost will vary directly with the number of units produced. In general, all costs such as direct labor, direct material, direct power, and the like, which can readily be allocated to each unit of product, are considered to constitute the variable costs and the balance of the costs of the enterprise are regarded as fixed costs.

Variable expense may also be expected to increase in a stepped pattern. To increase production beyond a certain extent another machine may be added. Even though its full capacity may not be utilized, it may be necessary to employ a full crew of men to operate it. Also, an increase in productivity may be expected to result in the use of materials in greater quantities, and thus in their purchase at a lower cost per unit due to quantity discounts and volume handling.

The practices followed in designating fixed and variable costs are usually at variance with the strict interpretation of these terms. Analysis in which they are a factor must recognize this fact or the results may be grossly misleading.

Incremental and marginal cost. The terms incremental cost and marginal cost refer to essentially the same concept. The word increment means increase and an increment cost means an increase in cost. Usually, reference is made to an increase of cost in relation to some other factor, thus resulting in such expressions as increment cost per ton, increment cost per gallon, or increment cost per unit of production. The term marginal cost refers specifically to an increment of output whose cost is barely covered by the return derived from it.

Figure 2.2 illustrates the nature of fixed and variable cost as a function of output in units. The incremental cost of producing 10 units between outputs of 60 and 70 units per year is illustrated to be $8. Thus, the average incremental cost of these 10 units may be computed as $\Delta \text{cost}/\Delta \text{output} = \$8/10 = \$0.80$ per unit.

Incremental costs are difficult to determine. In the example above, a curve was given which enabled the increment cost to be precisely determined as $8 for 10 units. As a practical matter in actual situations it is ordinarily difficult to determine increment cost. There is no general approach to the problem, but each case must be analyzed on the basis of the facts that apply to it at the time and the future period involved. Increment costs can be overestimated or underestimated, and either error may be costly. The overestimation of increment costs may obscure a profit possibility; on the other hand, if they are underestimated, an activity may be undertaken which will result in

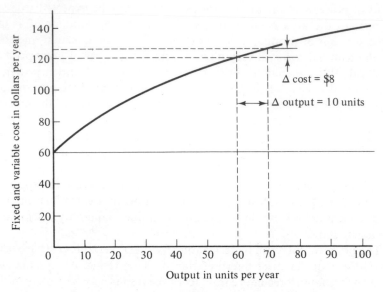

Figure 2.2. Fixed, variable, and incremental cost.

a loss. Thus, accurate information is necessary if sound decisions are to be made.

Sunk cost. When making engineering economic analyses the objective of the decision maker is to choose that course of action which is expected to result in the most favorable future benefits. Because it is only the future consequences of investment alternatives that can be affected by current decisions, an important principle in economic studies is to disregard cost incurred in the past. A past cost or *sunk cost* is one that cannot be altered by future action and is therefore irrelevant in engineering economy studies.

Although the principle that sunk costs should be ignored seems reasonable it is quite difficult for many people to apply. For example, suppose you had purchased 1,000 shares of stock at a price of $50 per share two years ago and now it is worth $15 per share. In all probability there are other stocks available that have a better future than the stock you presently possess. Many people react to this type of situation by holding on to their present stock until they can recover their losses. By not selling your stock you do not have to openly admit your losses and thereby admit your failure in judgment. However, it should be clear that because of better opportunities elsewhere it is certainly sound decision making to acknowledge your losses by selling the present stock and to use what money remains more productively from now into the future. It is the emotional involvement with past or sunk costs that makes it difficult for such costs to be ignored in practice.

2.7. Price Is Determined by Supply and Demand

In a free enterprise system, the price of goods and services is ultimately determined by supply and demand. Cost only enters the picture as a factor in determining profit or loss.

Typical demand and supply curves are illustrated in Figure 2.3. The demand curve shows the relationship between the quantity of a product that people are willing to buy and the price of the product. The supply curve

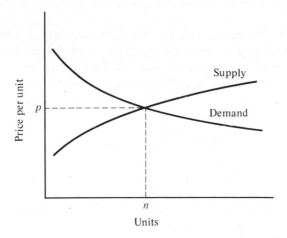

Figure 2.3. Typical supply and demand curves.

shows the relationship between the quantity of a product that vendors will offer for sale and the price of the product. Therefore, the intersection of these curves determines the price at which exchange will take place. The quantity exchanged is equal to both quantity of supply and quantity of demand. In the illustration, the number of units that will be exchanged is n and the price at which exchange will take place is p.

As a concept, the law of supply and demand is important in engineering economy studies since proposed ventures frequently involve action that will increase the supply of a product or influence its demand. The effect of such action upon the price at which the product can be sold is an important factor to be considered in evaluating the desirability of the venture.

The elasticity of demand. Consumer goods and services may be classified as being either necessities or luxuries. The classification is relative, for that which is considered to be a necessity by one person may be considered a luxury by others. The classification depends upon the individual's economic and social position.

People will decrease their consumption of luxury goods at a faster rate than their consumption of necessities for a given increase in price. The extent to which price changes affect demand is measured by a concept called the *elasticity of demand*. If the product of volume and price is constant, the elasticity of demand is unitary. Under this condition, the total amount which will be spent for the product will be constant regardless of the selling price. A drop in price would cause an increase in demand. Thus, the demand for a product is *elastic* when a decrease in price results in a greater than proportionate increase in sales. On the other hand, if a decrease in selling price produces a less than proportionate increase in sales, the demand is said to be *inelastic*.

Obviously, luxuries have a much greater elasticity of demand than do necessities. A change in demand will have a much greater effect on the price of necessities than on the price of luxuries. This relationship is important in evaluating the probable effect of alternate engineering proposals.

2.8. The Law of Diminishing Returns

The term *law of diminishing returns* was originally used to designate the relation of input of fertilizers to land and the output of crops. It is a special application of the law of diminishing productivity, which may be stated as follows: *The amount of product obtained in a productive process varies with the way the agents of production are combined. If only one agent is varied, the product per unit of this agent may increase to a maximum amount, after which the product per unit may be expected to diminish but not necessarily proportionately.*

The law of diminishing returns is directly applicable in a great number of engineering situations where it manifests itself in terms of providing a service at least cost. In a given situation one element may be increased and for a time costs will be decreased to a minimum point only to rise again if the element is further increased. This may be illustrated by a situation in which economy depends upon a balance between losses associated with machine down time and the cost associated with maintaining a crew to undertake preventative maintenance and repair to minimize such losses. Suppose that the failure of a machine will result in a loss of $10 per hour during the period that it is inoperative or "down" for repairs. Past experience shows that a typical repair for this machine can be made by one man in 9 hours, by two men in 6 hours, by three men in 4 hours, by four men in 3.5

hours, and by five men in 4 hours. The wage rate of repair men is taken to be $6 per hour. The costs incident to the failure of the machine are given in Table 2.1.

Table 2.1. COSTS INCIDENT TO REPAIR AND DOWN TIME

Number of men in repair crew	1	2	3	4	5
Hours required to make repair	9	6	4	3.5	4
Man-hours required to make repair	9	12	12	14	20
Labor cost of making repair at $6 per hour	$54	$72	$72	$84	$120
Down-time cost at $10 per hour for machine	$90	$60	$40	$35	$40
Resultant of labor cost of making repair and down-time loss of machine	$144	$132	$112	$119	$160

The economy associated with the loss of use of the machine suggests that the repair crew should be the size that will result in minimum down time. A person concerned only with the down-time loss of a machine would select a four-man crew, which would hold the down-time loss to $35 per repair. The economy of the repair activity points to a crew of one man so that the repair may be made most efficiently and at least cost. A person primarily interested in the efficient utilization of repair men and low repair costs would have his men work singly. Usually it is the overall economy that is desired. In this case the greatest overall economy occurs when a crew of three men is used resulting in a total cost of $112.

Most situations have many elements, and when the optimum whole is sought, the optimum of elements must often be compromised. Long bridge spans reduce the cost of piers but will increase the cost of superstructure. Price can be increased but sales may fall so that profit is less. Managements are continually confronted with finding the balance between supervisory and nonsupervisory workers.

The chief value of the concept of the law of diminishing returns in engineering economy situations is that it produces an awareness that output does not necessarily increase in a straight-line relationship with an increase in input of an agent of production. The solution of many problems in economy centers around adjusting the amounts of agents of productivity to produce maximum output per unit of input. This end is frequently expressed as an effort to find the input ratio that will result in least cost per unit of output. Such expressions as horsepower-hour output per pound of fuel, miles per gallon of fuel, sales per dollar of advertising, units of output per fatal accident, and defects per 100 units of product are associated with analyses in search of a maximum output per unit of input.

2.9. Considering Advantages and Disadvantages

Any contemplated action has advantages and disadvantages. A person who has enthusiasm for a proposal, particularly its originator, is likely to give undue favorable weight to its advantages and to minimize its disadvantages. But experience has taught most people to be skeptical of new things. The opposition to a new proposal may tend to stress its disadvantages and disregard its advantages.

Certain advantages favor a new activity. Since it is independent of previous commitments, it may be implemented in large measure on the basis of present and future consideration. Little consideration need be taken of the past except to profit by previous mistakes. Contrast this with the situation of railroads, whose future development is for all practical purposes limited by the present standard distance between rails. This gauge is not the best for either economy or comfort for higher-speed trains, but to change it is not feasible.

A new activity may enjoy a tremendous temporary advantage from being first in the field. Its launching is analogous to the surprise attack in war, where traditionally a small force overwhelms a much larger force. Where the new enterprise is based upon a patented device, initiative of discovery results in the advantage of monopoly to a greater or lesser extent. But even without patents the discovery and the vigorous prosecution of an economic opportunity often result in a tremendous initial gain before rivals can come into the field.

A serious disadvantage faced in launching many new enterprises is that they often require an input of money and effort for long periods of time prior to realization of income from them. Many new ventures of ultimate promise fail because funds and enthusiasm are exhausted before adequate returns can be established. Unless strong financial backing without returns for long periods of time is available, many otherwise profitable undertakings are not feasible. Inadequate financing is recognized as one of the most important causes of failure of new enterprises.

In any pioneering activity lack of experience often results in the arising of unforeseen contingencies for which provision has not been made. Lack of experience or ability in regard to financing, organization, marketing, production, and other important activities is an important cause of failure in otherwise promising opportunities. Thus, in launching a new enterprise competent personnel is of great importance. Incompetence of management ranks along with inadequate financing as a cause of failure of new enterprises.

Many new enterprises fail because the opportunities they are designed to exploit are inadequate for success. Although the opportunities for profit should be ascertained prior to deciding upon a proposal if possible, many opportunities cannot be evaluated even reasonably well except by trial. When this is the case, the uncertainty of the enterprise should be realized so that steps may be taken to meet the outcome, whatever it may be. If the outlook for the new activity seems promising, immediate steps should be taken to exploit the opportunity with vigor. If it seems unpromising, steps should be taken to withdraw from the venture with a minimum loss of money and effort.

The point to realize is that every proposal has its advantages and disadvantages whether these exist in the proposal itself or in people's regard for it. The sponsor of a proposal should school himself to consider advantages and disadvantages as impersonally as possible, because the economic advantage of a proposal depends upon its advantages and disadvantages in comparison with those of other alternatives.

QUESTIONS AND PROBLEMS

1. Define value; utility.
2. Explain how utilities are created.
3. Describe the two classes of goods recognized by economists.
4. Contrast the utility of consumer goods with the utility of producer goods.
5. Why is it that the utility of consumer goods is determined subjectively whereas the utility of producer goods is usually determined objectively?
6. Explain the economy of exchange.
7. Why is it possible for both parties to profit by an exchange?
8. An automobile parts manufacturer located in City A dispatches a shipment of finished parts to its warehouse in City B each day. The distance is 200 miles and the cost per mile is estimated to be $0.18 plus $6.20 per hour for the driver. The truck makes the return trip without payload. A trucking company located in City B has a contract to haul finished castings to City A and has established its cost per mile to be $0.20 plus $5.80 per hour for the driver. The return trip is made without payload. The round-trip time for the trucking company and the manufacturer is 9 hours.

   The parts manufacturer has offered to subcontract the hauling of the machined castings. It is estimated that the task will cause an increase in expenses equivalent to an additional 20 miles and an increased round-trip time of 1 hour and 45 minutes.

(a) What is the minimum that the manufacturing company can charge for the casting haul?

(b) What is the maximum that the trucking company can pay for subcontracting the haul?

9. Describe the role of human action in the transformation of physical resources into goods and services.

10. Explain why it is usually profitable to expend resources to bring about coordination of human effort in an organization.

11. Relate organizational efficiency to the concept of economic efficiency and contrast both with physical efficiency.

12. What is the function of a manager?

13. A group of 60 employees, whose average salary is $9,900 per year, is directed by 5 foremen, whose average salary is $14,000 per year. It is suggested that the number of foremen be increased to 6.

(a) What % increase in average output per man must result for the two plans to incur equal cost?

(b) If the suggested plan is adopted, what will be the % increase in average time that can be spent in supervision per worker?

14. A foreman receiving a salary of $15,600 per year supervises the work of 18 men whose average annual salary is $11,200 per year. It is proposed that the employment of a second foreman will improve the total situation.

(a) Assuming that the 2 foremen would have to spend 2 hours each day coordinating their work, what will be the average increase in the time spent in supervision per day per man?

(b) What average % increase in output is necessary to justify the employment of the second foreman?

15. Define first cost and explain how it may be a limiting factor in successful engineering activity.

16. Discuss the difference between fixed cost and variable cost.

17. List some difficulties associated with classifying a cost as either fixed or variable.

18. The total cost of manufacturing 260 units of a product is $3,200. If 340 units are manufactured, the total cost will be $3,800.

(a) What is the average manufacturing cost for the first 260 units?

(b) What is the variable cost per unit?

(c) What is the total fixed cost?

(d) What is the average fixed cost per unit for the first 260 units?

19. Define sunk cost and explain why it should not be considered in engineering economic analysis.

20. Explain the relationship between price and supply; price and demand.

21. Why is the law of diminishing returns a useful concept in economy studies?

22. A manufacturer purchases ball bearings in cases containing 1,000 bearings each. Some bearings from each case are inspected as part of the acceptance procedure. The cost of inspecting one bearing is $0.20 and the loss resulting from the acceptance of a defective bearing is $2. From past experience, the number of defective bearings that will be accepted if 100, 200, 300, and 400 bearings are inspected from each case is, respectively, 45, 30, 25, and 20. Determine the most economical number of bearings to inspect from each case.

ELEMENTARY EXAMPLES
OF ECONOMY STUDIES

There are many situations in which alternate designs, materials, methods, and personnel will provide identical results and will involve expenditures occurring in a short time period considered to be the present. Since the results of the alternatives will be identical, the immediate cost of each is a measure of its comparative economy. Economic evaluations that fall into this category do not require consideration of the time value of money. In this chapter some elementary examples are presented to illustrate the point that there are many simple economy studies which can be made where the present difference is the basis for decision. Additional concepts useful in engineering economy are also presented.

3.1. Design and Economy

The results of design effort, as manifested in plans and specifications, crystallize the final form of the product to be produced or the structure to be constructed as to its physical form, material, and production or construction requirements. For this reason design affords many opportunities for economy.

3

Design, no matter how poorly done, is predicated on the thought that the effort devoted to it will be outweighed by the results.

To design is to project and evaluate ideas for attaining an objective. Suppose that a person employed in a machine design department has been assigned the task of designing a machine for a special purpose. His first step will be to project, literally to invent, new combinations of materials, machine elements, forces, and kinematic motions that are believed may meet the purposes of the machine. His next step will be to evaluate these combinations.

There may be many bases of evaluation: certainty of operation, operator attention required, safety of operator, rate of operation, power required, maintenance required, and so forth. On the whole, these bases of evaluation all relate to economy in one way or another. In fact, economy is inherent in design.

One important aspect of design is that it makes it possible to try out a great many ideas on paper more or less conclusively. The cost of these trials may be negligible when compared with the actual accomplishment. A decision to accept a design is a decision to assume the advantages and disadvantages associated with it and to discard other designs that may have been considered. The following examples will illustrate several aspects of economy as it relates to design.

Elimination of overdesign. In the design of a certain building requiring 180 rafters it was found that 2-inch by 8-inch by 16-foot pieces, 16 inches center-to-center, are just sufficient to meet the contemplated load. It was also found that the sheathing to span rafters placed 24 inches center-to-center was just adequate for strength and stiffness. Consequently, the sheathing will be stronger and stiffer than necessary for the 16-inch spacing. The excess is called *overdesign* since it has no functional value and may be disadvantageous because of its extra weight.

A new design is contemplated using 2-inch by 10-inch cross-section rafters. On the principle that the bending moment resisted by a rafter is in accordance with $bd^2/6$, the ratio of the load-carrying capacity of the 2-inch by 10-inch rafter to the 2-inch by 8-inch rafter is 1.56 to 1.00.

As far as load-carrying capacity is concerned, the 2-inch by 10-inch rafters may be spaced 16 inches $\times$ 1.56, or 25 inches. In accordance with practical standards, the rafters are spaced 24 inches center-to-center. The analysis required for 2-inch by 8-inch by 16-foot rafters, 16-inches center-to-center, is as follows:

Number of rafters required 180

Overdesign of rafters... 0%

Overdesign of sheathing, $\dfrac{24 \text{ in.} - 16 \text{ in.}}{24 \text{ in.}}$ 33%

Cost of rafters @ $0.20 per board foot,

$\dfrac{180 \text{ rafters} \times 2 \text{ in.} \times 8 \text{ in.} \times 16 \text{ ft}}{12 \text{ in.}} \times \$0.20$ \$768

For 2-inch by 10-inch by 16-foot rafters, 24 inches center-to-center, the required calculations are:

Number of rafters required, $\dfrac{180 \times 16 \text{ in.}}{24 \text{ in.}}$ 120

Overdesign of rafters, $\dfrac{25 \text{ in.} - 24 \text{ in.}}{24 \text{ in.}}$ 4%

Overdesign of sheathing 0%

Cost of rafters @ $0.20 per board foot,

$\dfrac{120 \text{ rafters} \times 2 \text{ in.} \times 10 \text{ in.} \times 16 \text{ ft}}{12 \text{ in.}} \times \$0.20$ \$640

In this example, the design with 2-inch by 10-inch rafters is not only adequate, but it is less expensive by $128. Additional savings will result from having to handle fewer pieces.

Designing for economy of production. A product may exhibit excellent functional design but may be poorly designed from the standpoint of producibility. The drawings and specifications establishing the design also establish a minimum manufacturing cost regardless of efforts to reduce labor, material, and other costs.

The effect of design on production cost is recognized by designers in many fields. Plastics have replaced metal in children's toys, reducing production costs and increasing safety. In fastening, quick attachments have replaced nuts and bolts in some applications making assembly easier while increasing the ease of disassembly for inspection and repair. Specifying certain quick setting finishes makes possible the production of consumer goods with a minimum need for curing equipment.

Designers concerned with economy must take into consideration the availability of production or construction equipment as a prerequisite to the choice of methods. For example, there is little point in specifying a granular finish for the ceiling of a certain public building if the equipment to apply such a finish cannot be made available economically. There are many instances in which designs must be adapted to account for the availability of both machines and the skills of people.

Economy in the design of producer goods. In the design of equipment for carrying on manufacturing, the designer may become too engrossed in the mechanical features and give insufficient attention to overall economy. As an example, assume that a jig is to be designed for adjusting 100,000 assemblies for a special production order. Two designs are presented and are evaluated as follows:

	Jig A	*Jig B*
Estimated life	100,000 pieces	100,000 pieces
Cost	$1,600	$800
Hourly rate of class of labor required	$4.80	$5.90
Operator hourly output rate	62	54

The cost of adjusting the 100,000 assemblies with Jig A is:

$$\text{Labor cost, } \frac{100{,}000 \text{ pc.} \times \$4.80 \text{ per hr.}}{62 \text{ pc. per hr}} \quad \dots\dots\dots\dots\dots\dots\dots \quad \$\ 7{,}742$$

$$\text{Cost of Jig A} \dots\dots\dots\dots\dots\dots\dots\dots\dots\dots\dots\dots\dots\dots\dots \quad \underline{\$\ 1{,}600}$$

$$\$\ 9{,}342$$

The cost of adjusting the 100,000 assemblies with Jig B is:

Labor cost, $\dfrac{100,000 \text{ pc.} \times \$5.90 \text{ per hr.}}{54 \text{ pc. per hr}}$ $10,926

Cost of Jig B .... $ 800

$11,726

The net advantage of selecting Jig A over Jig B is $2,384. This economic advantage is inherent in the design of Jig A. Its characteristics were such that it could be operated more rapidly with less operator skill.

In the industrial literature, many examples may be found in which economy of production was effected through design of equipment, equipment layout, and plant. In such cases, it is apparent that the economic advantage resulted from wise choice among alternative designs.

Design for economy of maintenance. Drawings and specifications establishing the design of a product, structure, or complex system should exhibit consideration for maintainability. A commercial product or a defense system is not available to serve its intended purpose if it is down for maintenance. Because of the high cost of maintenance manpower and the associated costs of test equipment, spare parts, transportation, and other logistic elements engineers must consider maintainability during design to a greater degree than at present.

The challenge is simple. Designers have the opportunity to reduce the need for maintenance by striving for a product or system that will not fail in use. By achieving high reliability the needed item will be available when needed and, in addition, burdensome costs associated with and resulting from maintenance will be avoided. Of course, high reliability is not achieved without cost.

The ultimate goal of maintainability design is the economic reduction of the support requirements for a system after it goes into operation and the facilitation of whatever remaining maintenance will be required. This goal is logically the concern of the design engineer, for it is not likely that the disruption of service and the burden of maintenance will be reduced unless engineers assume the responsibility for this neglected area. The need is critical, for it appears that both commercial and the military products, structures, and systems may soon consume more economic resources for maintenance and related services than their first cost.

Design for economy of shipping. The shipping costs of products can often be reduced by proper design. Some products are designed so they can be easily assembled upon receipt by the customer. Others are designed so they may be easily packaged for shipment. Weight is often reduced in design so that shipping costs are minimized.

Often it may be advantageous to design so that the product can be nested to save space in shipment. In the shipment of pipe to Saudi Arabia this principle was used. The design called for pipe approximately 30 inches in diameter. To facilitate shipment by water, where bulk freight rates were in effect, half of the pipe was made 30 inches in diameter and the other half was made 31 inches in diameter to make nesting possible. As a result, shipping costs were reduced by several million dollars.

Economy of interchangeable design. Interchangeability in design is an extension of the mathematical axiom that things equal to the same thing are equal to each other. Components of an assembly that are to be interchangeable with components of other assemblies must possess specific tolerance limits.

As an example of the economy inherent in interchangeability, consider an engine re-manufacturing process. If the parts were not originally designed to be interchangeable with like parts of other engines of the same model, it would be necessary to keep all parts of each engine together during re-manufacturing. The needed coordination, tagging, and cost of locating stray parts are eliminated through initial design that allows interchangeability. In this example interchangeability contributes to maintainability.

3.2. Economy of Material Selection

A designer has a choice of specifying either aluminum or grey iron castings for an intricate housing for an instrument to be permanently mounted in a power plant. He has ascertained that either metal will serve equally well. The aluminum casting will weigh 0.8 pound and the grey iron casting 2.2 pounds. His analysis of the cost of providing each type of casting follows:

	Grey Iron	*Aluminum*
Cost of casting delivered to factory	$0.76	$0.92
Cost of machining.	0.63	0.52
	$1.39	$1.44

On the basis that either material will serve equally well and provide an identical service, the grey iron casting will be selected because its immediate or present cost is the lower of the two.

In the above example, in which it is specified that the instrument is to be mounted on a power plant wall, differences in weight were not considered to be a factor. However, for an instrument for airplane service the difference in weight might be the deciding economic factor. In airplane service the lighter instrument casting would have an economic value in lessened fuel consumption or greater pay load and would thus have greater utility.

In determinations of economy, care must always be exercised to see that alternatives provide identical services. "Sales appeal" of one type of casting over the other may easily render their worths nonidentical. Even for power plant use it might have been recognized that the lighter metal had "sales appeal" over the heavier metal.

If concrete value can be given to a quality, this difference in service can be taken into account to render the services of the two alternatives identical. For instance, the quality of lightness might have service value to the manufacturer in that it would reduce the delivery cost of his product to the consumer. If this had applied in the example of the instrument casting, the two types of castings could have been placed on an identical service basis for comparison as follows:

	Grey Iron	Aluminum
Cost of casting delivered to factory	$0.76	$0.92
Cost of machining	0.63	0.52
Average additional cost to deliver grey-iron-casting-equipped instruments to customer	0.26	0.00
	$1.65	$1.44

Where two or more materials may serve a purpose equally well from a functional standpoint, the relationship of their cost, availability, and processing cost should be considered in determining which is chosen. Brass, for example, is often found to be less costly for parts than cold rolled steel because it can be machined at a higher rate, in spite of its greater weight per unit volume and greater cost per pound. Aluminum, which is easily machinable and in addition has a low specific weight, is being used in increasing amounts as a replacement for steel, cast iron, and other metals whose cost per pound is considerably less. Because of the ease with which they can be processed, plastics have proved to be an economy in many applications as a replacement for materials of less cost per pound.

In some cases the decision to substitute one material for another will result in an entirely different sequence of processing. For instance, a change from grey iron to zinc alloy castings will require marked change in equipment. To determine the comparative economic desirability of two materials, it is necessary to make a detailed study of the costs that arise when each is used. In some cases this may involve the economy of disposing of present equipment and acquiring new equipment.

3.3. Standardization and Simplification

A standard is a specification. Products designated as *standardized* conform to a previously accepted specification. Standardization, the conduct of activities in accordance with previously determined specifications, appears to be a modern development, particularly as it relates to the production of goods. Both products and the procedures by which they are made are subject to standardization.

Simplification is a name for the practice of examining a line of products for the purpose of eliminating useless variety in style, color, dimensions, size range, and the like. Its practice results in setting up what might be termed *most useful specifications* for a line of products. Thus, simplification is closely related to standardization. Simplification may be thought of as being the practice of eliminating undesirable standards.

There are a number of reasons, related to economy, why standardization has become a widespread practice. One factor is undoubtedly that human actions are characterized by variation and individualism whereas machine action is characterized by repetition of identical patterns. Thus, it is relatively easy to build machines that operate and produce goods in accordance with predetermined specifications. As the burden of production has been shifted from human effort and skills to machines, it has been necessary, at least to the extent of machine processes, to standardize.

The chief drive to standardization is probably specialization. It is recognized as characteristic that the skill and speed of performance are increased and the effort of performance is decreased as a person repetitiously performs a specified task. It is also known that some people have greater aptitude for doing some things than others; more people may be found with one skill than several. It is rarity that places triple-threat men in demand in football.

Specialization is not economically practical unless it can be engaged in for some time. It is rarely profitable to build a machine for a total lifetime production of a single part. Nor would it be profitable to set up an assembly line of a number of specialized people to assemble a single unit of product.

Standardization is a method of increasing the number of units of one kind for a given total production. Suppose that a pottery has had an annual output of 1,000 nonstandardized vases made in accordance with 40 different patterns, each in 5 different sizes. Thus, the output involves 200 different specifications and an average annual output of 5 units per specification. If it were decided to standardize on one vase of one size, the annual output per specification would be increased from 5 to 1,000 units. Thus, it appears that the economy of standardization lies, in large measure, in the increased opportunity for specialization afforded by increased volume of one kind of product or activity.

Since standards ordinarily have extended usefulness in terms of time and volume, it is often economical to take great pains to perfect them. A standard is in reality a predetermined decision to act in accordance with a certain specification in certain situations. Even though a standard may be developed only after a considerable expenditure, it will probably result in a lower cost per decision than if each specification problem were decided as it arose without benefit of a standard. Suppose that a certain situation necessitating decision arises in a concern 1,000 times per year and that the cost of making each decision individually, owing to confusion, error, and loss of time, amounts to $10. A standard procedure that could reduce this cost to $1 per decision, for instance, would justify an expenditure of $9,000 (or more, if applicable for more than one year).

One valuable feature of standards is that they greatly facilitate communication. Involved directions, for instance, may be given simply by pointing out the standards that should be applied. This is of tremendous value in coordinating the activities of large groups of people as in modern industry.

3.4. The Selection of Personnel

Viewed objectively, subordinates are merely their leader's means for reaching ends. Of the several productive factors used in industry, none has such a variety of characteristics as personnel. Characteristics of both body and mind are of concern. One individual may accomplish several times as much of a task requiring physical strength, dexterity, or acuteness of vision as another. The range of mental proficiencies for given tasks is even wider.

The physical and mental endowments of peoples appear to be subject only to limited change by training and leadership. For this reason it is important that personnel be selected who have inherent characteristics as nearly compatible as possible with the work they are to perform and the position they are to occupy. Contrasted with the meticulous care with which materials are selected, the selection of personnel in most organizations is given but superficial attention.

The range of human capacities. The ratio of the poorest to the best performance time is often as much as 1 to 3 in common operations. It is apparent that the cost of labor input per unit of output might be materially reduced by selecting only employees capable of better than average performance. In many cases superior employees receive little or no greater compensation than less capable employees. This is particularly true when performance is measured with difficulty. Superior managerial ability, for example, may go unrecognized and therefore unrewarded for long periods of time because it cannot be measured. But even when superior ability is proportionally compensated for, savings may still result.

Superior ability that remains unused is not an advantage and may often be a disadvantage because of the resulting frustration and dissatisfaction of the employee. Generally speaking, the work assignment should require the workman's highest skill.

For many tasks, tests can be devised for selection of prospective employees with superior ability. Representative of these are a battery of three tests for the selection of mail distributors for the Postal Service. It was found, from extensive experimental trials, that of those making the highest 25% of scores on these tests, over 93% would be above average in proficiency. The economic desirability of having most employees above average in ability for the task under consideration (even though compensation be in proportion to ability, which it rarely is) can hardly be overestimated.

In this relatively new field, results of the installation of a program of selection practices are most difficult to estimate. Since the input cost of a program can usually be estimated with reasonable accuracy, it may be helpful in arriving at a decision to calculate the benefits necessary to justify a contemplated selection program.

Suppose that a certain plant employs 40 operators. The average output is 28 pieces per hour, the average spoilage is 1.7%, and each rejected piece results in a loss of $0.60. Records reveal that average spoilage for the best 20 operators is 1.1%, whereas that for the poorest 20 operators is 2.3%. A consultant agrees to prepare a set of tests and administer them for a year for $1,800. Administration subsequent to the first year is expected to cost $3 per operator hired or $140 per year based upon the estimated turnover.

Let P equal the percentage reduction in spoilage during the first year to justify expenditure of $1,800. Then $P \times 40$ operators $\times 28$ pieces per hr. $\times 2,000$ hr. per year $\times \$0.60$ per reject $= \$1,800$ from which $P = 0.00134$ or 0.134%.

This percentage reduction in spoilage seems attainable in view of the fact that a program that will reject applicants of less than average ability in relation to spoilage will result in an average reduction of $1.7 - 1.1$ or 0.6%. Moreover, it is reasonable to expect other benefits from the program.

By tending to place people in work for which they are best fitted, sound selection is generally beneficial. Even those who are rejected will generally be benefited; for, if continually rejected for work for which they are unfitted, they must eventually find a job that will permit them to work at their highest skill.

The economy of proficiency. Value of proficiency is not necessarily directly proportional to degree of proficiency. Thus, a baseball player whose batting average is 0.346 will ordinarily command more than twice as much salary as one whose average is 0.173, if they are equal in other respects. In some activities the compensation of those with unusually high proficiencies is extremely high, but those with ordinary abilities can find no market for their services. Consider acting; the motion picture industries pay fabulous salaries for those whose box office appeal is high when they undoubtedly could cast all their productions with volunteer actors of fair ability without cost.

The cost of human effort is a considerable portion of the total cost of carrying on nearly all production and construction activities. People are employed with the idea of earning a profit on the skills they possess. In general, the higher a person's proficiency in any skill, be it manual dexterity, creativeness, leadership, inventiveness, or physical ability, the greater his value to his employer. Consider a machine operation for which a workman is paid $6 per hour for producing 10 pieces per hour on a machine whose rate is $6 per hour. In this example the cost per piece is equal to

$$\frac{\$6 + \$6}{10} = \$1.20.$$

If a second worker of less proficiency completes 9 pieces per hour, his relative worth to his employer (as compared with that of the first workman) is calculated as follows: Let W equal the hourly pay of the second workman to result in a cost per piece of $1.20. Thus,

$$\frac{W + \$6}{9} = \$1.20$$

and

$$W = \$4.80.$$

It will be noted in this example that a 10% reduction in the ability of the workman to turn out pieces results in his services being worth 80% as much per hour to his employer as the services of the first workman, all other things being equal.

If the second workman also received $6 per hour, the cost per piece of the 9 pieces he turns out in an hour will be as follows:

$$\text{cost per piece} = \frac{\$6 + \$6}{9} = \$1.33.$$

It is conceivable that the resulting cost per piece may be so high that the employer cannot profit if he must pay the second workman $6 per hour.

The economy of specialization. To specialize is to restrict a person or thing to a particular activity, place, time, or situation. Specialization in this sense results when there is a division of labor, as for example, when operations in a process performed by a single person are divided among a number of persons. It is clear that machines can also be specialized in this sense, that is, specialized as to activity.

Persons or things can also be specialized as to use, place, or time. Thus, a person whose assignments require that he report to work in Atlanta is specialized as to place. A passenger train that operates on a schedule is specialized as to time and place; so is a person who works in one place from nine to five o'clock, five days a week.

Specialization is of interest in relation to economy studies because it is often a means whereby the cost of accomplishing a given result can be reduced. In any design or planning effort the desirability of specialization should be considered as a routine matter. It is widely recognized that specialization, particularly as it relates to people, may result in improved performance.

Specialization usually requires preplanning and specialized arrangements, the cost and maintenance of which may more than offset the advantage gained. The gains of a specialization of one kind may require specialization of another that is too costly. For instance, it might be desirable to have two men perform one each of two activities that must be performed one after the other, unless the new arrangement would also require the men to be specialized as to time. The latter specialization might easily result in one man's being idle while the other works.

Specialization is subject to the law of diminishing returns. It is apparent that it would not be economical to specialize to such an extent, for example, that one truck driver made only right turns and another made only left turns.

Generalization, the reverse of specialization, may also be a factor in economy. Thus, in machine design it is often desirable to have one part serve several purposes; the crankcase of an engine, for instance, serves as a base for the engine, a container for oil, and a support for the crankshaft and cylinder. The term "all-around man" is evidence of generalization with respect to personnel.

The economy of dependability. There is an economy of each of the innumerable personal qualities, but few transcend dependability in importance. Dependability in relation to organized action means behaving as agreed upon or as expected. For example, the chairman of an industrial committee has called a meeting of 12 persons but arrives 15 minutes after the specified time. If the average salary rate of committee members is $8 per hour the lack of dependability of the chairman may have resulted in a loss of time of $12 \times 0.25 \times \$8$, or $24. This direct loss points out the value of dependability, but it probably is not as great as the frustration and loss of confidence that the lack of dependability engenders in the persons affected by it.

Dependability rests upon ability, consideration for others, willingness to cooperate, and honesty. Lack of dependability in one person, regardless of the reason for it, invariably results in the need for increased vigilance and effort on the part of others. People in supervisory positions are keenly aware of the costs of deviation from dependability, and they thus prize dependability highly.

Willful undependability and dishonesty are particularly difficult to cope with because even infrequent dishonesty leads to the necessity for a supervisor to be continually prepared against the entire range of defections that a dishonest person may cause. Thus, for many positions, even extremely limited dishonesty may outweigh the honest services a person may render.

3.5. Economy of Resource Input

In order for organizations to remain operational, they must receive both tangible and intangible inputs. Incentives are inputs of a particular type directed to the person for his individual benefit. Other resource inputs are directed to the person for his use as a contributor to the organization. To reach a decision relative to the economic desirability of a proposed resource input, a relationship between the cost of the input and the value of the expected output should be established in terms of a comparable measure such as money.

Evaluating a tangible resource input. The cost of a tangible resource input such as equipment for use by an individual in his work can usually be obtained without great difficulty. But the value of output is often very difficult, if not impossible, to obtain. In such case, an analysis is necessary to determine the increment of output that must be obtained for an increment of resource input.

As an example suppose that the desirability of providing an electronic technician with automated test equipment is being considered. Under the present set-up, the annual cost of wages and fringe benefits is $14,800. Under the proposed arrangement, the annual cost would be increased by the equivalent annual cost of the test equipment. Suppose that this increase is $800 and that the technician must increase his productivity in direct proportion to the increase in cost of his wages, fringe benefits, and equipment.

With this assumption, the increase in productivity must be at least

$$\frac{(14{,}800 + \$800) - \$14{,}800}{\$14{,}800} = 0.054 \text{ or } 5.4\%.$$

The decision is now reduced to consideration of the estimated increase in productivity against the minimum increase needed to pay for the automated test equipment.

Evaluating an intangible resource input. All discernible human activity is of a physical nature. Man achieves his ends by manipulation of his physical environment. This is as true of the administrator as it is of a machinist, for whatever thoughts the former may have, he cannot transmit them except as they are manifested in physical ways which can be perceived by others.

The manipulation of the physical environment necessary to produce a desired result can be learned only as the result of experience or of instruction. Experience is often very costly, but the knowledge that a single person has gained may often be transmitted through the processes of instruction at relatively little cost. For example, the ingredients and the method of heat-treating a useful alloy may be learned only after years of costly experimentation, but the knowledge gained may be taught and become useful to other persons at relatively little cost.

Instruction is an economical method of transmitting useful knowledge and therefore an economical means of increasing the abilities of persons to achieve ends not otherwise attainable or at less cost. Through instruction, the best knowledge of the most capable persons can be made available to all. Thus, instruction is a resource input which can be of great value to an organization.

Industrial and business or organizations are making increasing use of instructional programs to improve the competence of employees. The worth of such programs is often difficult to evaluate. However, the cost of an instructional program can usually be estimated with a fair degree of accuracy. Based on the cost, the increase in the outputs of participants necessary to justify a program of instruction can be calculated.

Consider the following example involving instruction as an intangible resource input. The program of instruction is to be 40 hours in duration for a class of 10 engineers at a cost of $40 per hour plus salary costs of participants. Assume that the average salary of participants is $16,500 for a 40-hour, 50-week year and that the value of the instruction will be applicable for one year only. Under these assumptions, the average increase in productivity per engineer during the year to justify the instruction is

$$\frac{\left[10(\$16,500) + \frac{10(\$16,500)}{50} + 40(\$40)\right] - 10(\$16,500)}{10(\$16,500)} =$$

$$\frac{\$169,900 - \$165,000}{\$165,000} = 0.03 \text{ or } 3\%.$$

The decision regarding the intangible input now hinges on the estimated output increase compared with the minimum needed to break even. It should be recognized in the decision that the required productivity increase of 3% is based on a payout in one year when it is likely that the value of instruction will exist for several years.

3.6. Considering Qualitative and Quantitative Knowledge

Qualitative knowledge is knowledge of the attributes of the thing under consideration. It is descriptive and tells how a thing is constituted by naming its distinctive characteristics. In the expression of qualitative knowledge such statements are used as: Repair cost of the truck will be low; many units of this item will be sold this year; this method was less expensive than that method; the turbine had much power and used little fuel. Such expressions cannot be precisely evaluated and have comparatively little value in economy studies.

Quantitative knowledge is knowledge of the amount or extent of the attributes of a thing in terms that are capable of being counted. It is information that is capable of being expressed in numbers. Representative of quantitative knowledge are such statements as: The repair cost of the truck amounted to $358 per year; this year 6,200,000 units of this item were sold; this method costs $0.07 per piece but that method costs $0.09 per piece; the turbine developed 286 horsepower and used 0.48 pound of fuel per horsepower-hour.

It appears that qualitative knowledge is primarily evaluated by one's feeling for it and one's judgment. Different people are likely to vary widely in their interpretation of the significance of qualitative ideas. Quantitative knowledge, on the other hand, is precise and may be evaluated in large measure by the processes of reasoning. Also, most people may be expected to attach about the same significance to quantitative statements pertaining to fields with which they are familiar. This is an important characteristic because it facilitates communication of ideas.

A comparison of the effectiveness of the two kinds of knowledge may be gained from a consideration of two statements: (1) The turbine had much power and used little fuel. (2) The turbine developed 286 horsepower and used 0.48 pounds of fuel per horsepower-hour. Let each statement be considered in conjunction with the knowledge that a machine for which a prime mover is sought will require a torque of 420 foot-pounds at 3,200 revolutions per minute, that the cost of fuel is $0.046 per pound, and that

$$\text{h.p.} = \frac{\text{torque} \times 2\pi \times \text{r.p.m.}}{33,000}.$$

If the economy of using the turbine as the prime mover for the machine were being considered, two questions would be pertinent: (1) Is the turbine powerful enough? (2) What will be the fuel cost per hour? It will be noted that qualitative statements cannot be used to arrive at the desired answers. The quantitative information can be used as follows:

$$\text{required h.p.} = \frac{420 \times 2\pi \times 3,200}{33,000} = 256.$$

This answers the first question. The answer to the second question may be approximated as follows:

$$\text{fuel cost per hour} = 256 \times 0.48 \times \frac{256}{286} \times \$0.046 = \$5.06.$$

One important aspect of quantitative data is that they can often be combined with other quantitative data by logical processes to produce new quantitative knowledge as was illustrated above. This important characteristic suggests that quantitative knowledge should be exhausted in economy studies before considering qualitative knowledge.

3.7. The Appropriate Combination of Elements

There are many instances in which economy rests on the appropriateness of a combination of several elements. Consider the machining of a metal shaft on a lathe. The job will be accomplished economically if the feed and speed of the tool are appropriate in the light of the metal being machined, the character of the tool, the rigidity of the tool holder, the power available, and the strength of the machine components. Each of these elements must be appropriate for the total purpose if the total purpose is to be pursued economically.

A great many machines can be built from such common machine elements as levers, cams, columns, beams, ratchets, gears, and pulleys. However, the economic worth of the resulting machines depends largely upon the appropriateness of the combination of the several machine elements rather than upon the excellence of the individual elements. In a similar manner, the success of any economic venture results from the effectiveness of the combination of elements employed rather than upon the success of the separate elements.

For example, a well-designed office building might be constructed on a well-located site for the purpose of economic gain by lease of office space. But if the building is too small to yield a return sufficient for the purpose as a whole, the income derived might only cover the cost of the building and not the cost of the expensive site. In this case, the building was not matched to the site in a manner that would contribute to the economic success of the entire undertaking.

Much excellent work produces little result because it is directed to perfecting details rather than to perfecting their joint effect. One objective of engineering economy is to prepare the engineer to cope with social and economic elements as well as with physical factors so that his ability will be improved to consider the joint effect of these elements in situations involving engineering application.

PROBLEMS

1. The flooring for an area 12 feet by 15 feet has been designed for a floor joist spacing of 18 inches center-to-center. The joists may span either the 12-foot or the 15-foot dimension. The joists are to support a uniform loading of 120 pounds per square foot including the weight of the flooring. Space restrictions limit the depth of the joists to 10 inches. Joists are to be selected from standard size lumber; that is, the width is available in increments of 1 inch and the length in increments of 2 feet. The allowable stress in the joist is 1,400 psi and the cost of the lumber is $190 per 1,000 board feet.

   The stress in a beam of rectangular cross section is given by the expression $S = 6l^2w/8bd^2$, where S = stress in psi; l = span of joist in inches; w = load on beam in pounds per inch of length; b = width of beam in inches; d = depth of beam in inches. Determine the most economical span for the joists and find the saving over the alternate design.

2. It has been decided by a previous design that 4-inch by 4-inch rough pine shores will be used to support certain concrete forms. The contractor will need 175 shores, each 10 feet long and each capable of supporting a load of 5,000 pounds induced by the forms and concrete.

   The maximum safe load on a shore is given by the expression $P = 1,000 (1 - g/80b)bh$, where P = maximum safe load in pounds; g = height of shore in inches; b = width of shore in inches; h = depth of shore in inches. It has been decided that the 4-inch width of the shores will be maintained in any redesign consideration. If rough pine lumber costs $140 per 1,000 board feet, what total amount can be saved by the elimination of overdesign?

3. The chief engineer in charge of refinery operations is not satisfied with the preliminary design for storage tanks to be used as part of a plant expansion program. The engineer who submitted the design was called in and asked to reconsider the overall dimensions in the light of an article in *The Chemical Engineer*, titled "How to Size Future Process Vessels."

   The original design submitted called for 4 tanks 5.2 meters in diameter and 7.0 meters high. From a graph in the article, the engineer found that the present ratio of height to diameter of 1.35 is 111% of the minimum cost and that the minimum cost for a tank occurred when the ratio of height to diameter was 4 to 1. The cost for the tank design as originally submitted was estimated to be $28,000. What are the optimum tank dimensions if the volume remains the same as for the original design? What total saving may be expected through redesign?

4. Two alternate designs are under consideration for a tapered fastening pin. Either design will serve equally well and will involve the same material and manufacturing cost except for the lathe and grinder operations.

Design A will require 16 hours of lathe time and 4.5 hours of grinder time per 1,000 units. Design B will require 7 hours of lathe time and 12 hours of grinder time per 1,000 units. The operating cost of the lathe including labor is $16 per hour. The operating cost of the grinder including labor is $12 per hour. Which design should be adopted if 95,000 units are required per year and what is the saving over the alternate design?

5. In the design of buildings to be constructed in Northern Greenland, the designer is considering the type of window frame to specify. Either steel or aluminum window frames will satisfy design criteria. Because of the remote location of the building site and the lack of building materials in Greenland, the window frames will be purchased in the United States and transported a distance of 3,500 miles to the site. The price of window frames of the type required is $42 each for steel frames and $56 each for aluminum frames. The weight of steel window frames is 150 pounds each and the weight of aluminum window frames is 58 pounds each. The transportation shipping rate is $0.015 per pound per 100 miles. Which design should be specified and what is the economic advantage of the selection?

6. An architectural engineer is considering means for saving design and construction cost for a new suburban development involving 100 new homes. The approach of using a limited number of house designs repeated throughout the area is eliminated because of the resulting reduction in sales appeal produced by an area of "look alike" houses. Individual designs for each house would increase design and construction costs to such an extent that sales would be difficult.

The engineer decides that one means of decreasing costs without making houses look alike is to use interchangeable foundation and floor structure designs. Estimates show that the average cost of design for a foundation is $260 and that floor structure design averages $180 per design; the average cost of construction of a foundation is $3,500 and the average cost of floor structure construction is $2,600. The use of basic designs repeated throughout the housing area will save the total costs of design after the first 10 houses and will save 6% of the costs of construction of foundations and floor structures after the first 10 houses. How much can the sale price of each house be reduced without a loss of profit by using this system of interchangeable design for foundations and floor structure?

7. The volume of the raw material required for a metal part is 1.2 cubic centimeters. Its finished volume is 0.62 cubic centimeters. The machining time per piece is 0.246 minute for steel and 0.144 minute for brass. The cost of the specified steel is $0.51 per kilogram and the value of steel scrap is negligible. The cost of the specified brass is $1.22 per kilogram and the value of brass scrap is $0.20 per kilogram. The hourly cost of the required machine and operator is $8.90. The weight of brass and steel is 14.02 and 12.84 grams per cubic centimeter, respectively. Determine the comparative costs per piece for steel and brass parts.

8. In the design of a jet engine part, the designer has a choice of specifying either an aluminum alloy casting or a steel casting. Either material will provide equal service, but the aluminum casting will weigh 1.20 pounds as compared with 1.35 pounds for the steel casting.

   The aluminum can be cast for $2.40 per pound and the steel can be cast for $1.30 per pound. The cost of machining per unit is $4.80 for the aluminum and $5.35 for the steel. Every pound of excess weight is assessed a penalty of $40 due to increased fuel consumption. Which material should be specified and what is the economic advantage of the selection per unit?

9. Either aluminum alloy or stainless steel will serve equally well in a certain corrosive environment. Aluminum alloy has a yield strength of 20,000 pounds per square inch and stainless steel has a yield strength of 33,000 pounds per square inch. The aluminum alloy will cost $0.68 per pound and the stainless steel will cost $0.94 per pound. The specific gravities of aluminum alloy and stainless steel are respectively 2.79 and 7.77. If selection is based upon yield strength, which material will be more economical?

10. In a certain manufacturing activity, 40 employees are engaged in identical activities. The average output of the group as a whole is 46.4 units per hour. The average output of the less productive half is 40.2 satisfactory and 1.4 unsatisfactory units per hour and the average for the more productive half is 52.6 satisfactory and 0.8 unsatisfactory units per hour. The employees work on a straight piecework plan and receive $3.30 per hundred satisfactory units. The firm sustains a loss of $0.07 for each unsatisfactory unit. One machine is required for each employee. Each machine has an annual fixed cost of $320 and a variable cost of $0.18 per hour. Supervision and other overhead costs are estimated at $720 per employee per year. The average employee works 1,900 hours per year. How much could be paid annually for a selection, training, and transfer program which would result in raising the average productivity of the entire group to 52.6 satisfactory and 0.8 unsatisfactory units per hour?

11. A foreman supervises the work of Mr. B and 8 other men. The foreman states that "Mr. A requires twice as much and Mr. B requires half as much of my time as the average of my men." Mr. A's output is 8 units per day and Mr. B's output is 7 units per day. On the basis of equal cost per unit, what monthly salary is justified for Mr. B if the foreman receives $1,100 per month and Mr. A receives $820 per month?

12. A foreman is in charge of a construction crew of 8 men and takes great pride in the amount of work he personally performs on the job. Observation shows that the accomplishment of the men is impaired by lack of direction as a result of

the foreman's active participation in the work to be done. The foreman receives $6.50 and his men $5.60 per hour. If the foreman does one-third as much work as the average of his men would do if they were properly directed, what loss in the effectiveness of the crew will just be compensated for by the actual work performed by the foreman?

13. A machine tool operator produces 46 units per hour of which an average of 6 are defective. He is paid on a straight piecework basis at the rate of $5 per 100 satisfactory units. The firm sustains a loss of $0.02 per defective unit produced. The machine used in the process has a total operating cost of $3 per hour which includes depreciation, return on investment, and overhead. The operator works 1,800 hours per year. If a training program is initiated, it is anticipated that the production rate will increase to 52 units per hour only 2 of which will be defective.

   (a) What is the maximum amount that the firm can spend on the training program per year?

   (b) What would be the effect of the training program on the hourly wage of the operator?

14. A manufacturing concern has sales offices in 5 states in addition to the main plant and offices. Each branch sales manager is paid $16,000 per year and the vice-president for sales receives $24,000 per annum. At the present time, average annual sales amount to $2,300,000. An annual sales conference is being considered which will meet at the main plant for 2 days each year.

   (a) If each branch sales manager is paid $280 for travel and other expenses, what is the total cost of the conference on the basis of a 240-day work year?

   (b) What % increase in sales attributable to the conference is necessary to justify the annual meeting if the gross profits are 12% of sales?

15. Foreman A directs the work of 12 men. He plans his work "as he goes" during the 8-hour work period. Foreman B also directs the work of 12 men, but spends 2 hours per day in planning the work to be done, thus reducing the time available for active supervision. What percentage more effective must his supervision be per unit of time so that his supervisory effectiveness equals that of Foreman A?

16. An engineer can do certain required computations in 3 hours or he can delegate the work to an engineering aid. If the work is delegated, it will take 0.75 hour to explain the computational procedure and 0.50 hour to check the results. The actual calculations will take 4 hours to do if done by the aid. If the engineer receives a salary of $18,600 per year and the aid receives $7,200 per year, what are the comparative costs for each of the methods for a working year of 2,080 hours?

17. Assume that N units of product can be made manually by one man in a year of 2,000 working hours, that the man's wage rate is W dollars per hour, and that a labor saving machine whose annual capital and operating cost is equivalent to $R \times 2,000$ hours of labor, will, if used throughout the year, reduce the amount of labor needed to produce the N units by $S \times 2,000$ hours. Write equations for the unit cost of product when manually made and when made

with the aid of the machine. Determine the ratio of R to S when the costs are equal.

18. The cost of manufacturing a certain tool is $2.75 per unit of which $0.85 arises from material cost and the balance is processing cost. A new design is under consideration which will require an initial cost of $19,500. If the new design is adopted, the cost of material will be reduced by 6% and the cost of processing will be reduced by 3.5%. If 800,000 units are to be manufactured per year, and if the benefits from the improved design are to be considered for one year only, what amount can be spent for the new design if a return of $3 is required for each $1 spent?

part two

INTEREST FORMULAS
AND EQUIVALENCE

INTEREST AND
INTEREST FORMULAS

The term *interest* is used to designate a rental amount charged by financial institutions for the use of money. Charging a rental for the use of money is a practice dating back to the time of man's earliest recorded history. The ethics and economics of interest have been a subject of discussion for philosophers, theologians, statesmen, and economists throughout the ages. The concept of interest can be extended to earning assets which "borrow" from their owner repaying through the earnings generated. This economic gain through the use of money is what gives money its time value. Because engineering projects require the investment of money it is important that the time value of the money used be properly reflected in the evaluation of these projects. In this chapter we present the interest formulas that assist in considering the time value of money in engineering economy studies.

4.1. Interest Rate, The Rate of Capital Growth

An *interest rate* or *the rate of capital growth* is the rate of gain received from an investment. Usually this rate of gain is stated on a per year basis and it represents the percentage gain realized on the money committed to the

4

undertaking. Thus, a 12% interest rate indicates that for every dollar of money used an additional $0.12 must be returned as payment for the use of those funds.

In one aspect, interest is an amount of money *received* as a result of investing funds, either by loaning it or by using it in the purchase of materials, labor, or facilities. Interest received in this connection is gain or profit. In another aspect, interest is an amount of money *paid out* as a result of borrowing funds. Interest paid in this connection is a *cost*.

Interest rate from the lender's viewpoint. A person who has a sum of money is faced with several alternatives regarding its use:

1. He may exchange the money for goods and services that will satisfy his personal wants. Such an exchange would involve the purchase of consumer goods.
2. He may exchange the money for productive goods or instruments. Such an exchange would involve the purchase of producer goods.
3. He may hoard the money, either for the satisfaction of gloating over it, or in awaiting an opportunity for its subsequent use.

4. He may lend the money asking only that the original sum be returned at some future date.
5. He may lend the money on the condition that the borrower will repay the initial sum plus interest at some future date.

If the decision is to lend the money with the expectation of its return plus interest, the lender must consider a number of factors in deciding on the interest rate. The following are perhaps the most important:

1. What is the probability that the borrower will not repay the loan? The answer to this question may be derived from the integrity of the borrower, his wealth, his potential earnings, and the value of any security granted the lender. If the chances are two in fifty that the loan will not be repaid, the lender is justified in charging 4% of the sum to compensate him for the risk of loss.
2. What expense will be incurred in investigating the borrower, drawing up the loan agreement, transferring the funds to the borrower, and collecting the loan? If the sum of the loan is $1,000 for a period of one year and the lender values his efforts at $20, then he is justified in charging 2% of the sum to compensate for the expense involved.
3. What net amount will compensate for being deprived of electing other alternatives for disposing of the money? Assume that $3 per hundred or 3% is considered as adequate return considering the investment opportunities foregone.

On the basis of the reasoning above, the interest rate arrived at will be 4% plus 2% plus 3%, or 9%. Therefore, an interest rate may be thought of, for convenience, as being made up of percentages for (1) risk of loss, (2) administrative expenses, and (3) pure gain or profit.

Interest rate from the borrower's viewpoint. In many, if not most, cases the alternatives open to the borrower for the use of borrowed funds are limited by the lender, who may grant the loan only on condition that it be used for a specific purpose. Except as limited by the conditions of a loan, the borrower has open to him essentially the same alternatives for the use of money as a person who has ownership of money, but the borrower is faced with the necessity of repaying the amount borrowed and the interest on it in accordance with the conditions of the loan agreement or suffering the consequences. The consequences may be loss of reputation, seizure of property or of other moneys, or the placing of a lien on his future earnings. Organized society provides many pressures, legal and social, to induce a borrower to repay a loan. Default may have serious and even disastrous consequences to the borrower.

The prospective borrower's viewpoint on the rate of interest will be influenced by the use he intends to make of funds he may borrow. If he borrows the funds for personal use, the interest rate he is willing to pay will be a measure of the amount he is willing to pay for the privilege of having satisfactions immediately instead of in the future.

If funds are borrowed to finance operations expected to result in a gain, the interest to be paid must be less than the expected gain. An example of this is the common practice of banks and similar enterprises of borrowing funds to lend to others. In this case it is evident that the amount paid out as interest, plus risks incurred, plus administrative expenses must be less than the interest received on the money reloaned, if the practice is to be profitable. A borrower may be expected to seek to borrow funds at the lowest interest rate possible.

4.2. The Earning Power of Money

Funds borrowed for the prospect of gain are commonly exchanged for goods, services, or instruments of production. This leads us to the consideration of the earning power of money that may make it profitable to borrow money.

Consider the example of Mr. Digg who manually digs ditches for underground cable. For this he is paid $0.25 per linear foot and averages 200 linear feet per day. Weather conditions limit this kind of work to 180 days per year. Thus, he has an income of $50 per day worked or $9,000 per year.

An advertisement brings to his attention a power ditcher that can be purchased for $4,000. He buys the ditcher after borrowing $4,000 at 12% interest. The machine will dig an average of 800 linear feet per day. By reducing the price to $0.15 per linear foot he can get sufficient work to keep the machine busy when the weather will permit.

At the end of the year the ditching machine is abandoned because it is worn out. A summary of the venture follows:

Receipts		
Amount of loan	$ 4,000	
Payment for ditches dug, 180 days × 800 lin. ft.		
× $0.15	$21,600	$25,600
Disbursements		
Purchase of ditcher	$ 4,000	
Fuel and repairs for machine	1,100	
Interest on loan, $4,000 × 0.12	480	
Repayment of loan	4,000	$ 9,580
Receipts Less Disbursements		$16,020

An increase in net earnings for the year over the previous year of $16,020 — $9,000 = $7,020 is enjoyed by Mr. Digg.

The above example is an illustration of what is commonly spoken of as the "earning power of money." It should be noted that the money borrowed was converted into an instrument of production. It was the instrument of production, the ditcher, which enabled Mr. Digg to increase his earnings. If Mr. Digg had held the money throughout the year it could have earned him nothing; also, if he had exchanged it for an instrument of production that turned out to be unprofitable, he might have lost money. Indirectly, money has earning power when exchanged for profitable instruments of production.

4.3. The Time Value of Money

Because money can earn at a certain interest rate through its investment for a period of time, it is important to recognize that a dollar received at some future date is not worth as much as a dollar in hand at present. It is this relationship between interest and time that leads to the concept of "the time value of money." For example, a dollar in hand now is worth more than a dollar received 5 years from now. Why? Because having the dollar now provides the opportunity for investing that dollar for 5 years more than the dollar to be received 5 years hence. Since money has an *earning power*, this opportunity will earn a return so that after 5 years the original dollar plus its interest will be a larger amount than the $1 received at that time. Thus, the fact that money has a time value means that equal dollar amounts at different points in time have different value as long as the interest rate that can be earned exceeds zero. This relationship between money and time is presented in Figure 4.1.

It could be argued that money also has a time value because the *purchasing power* of a dollar changes through time. During periods of inflation the amount of goods that can be bought for a particular amount of money

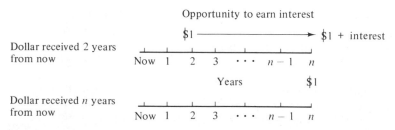

Figure 4.1. Time value of money.

decreases as the time of purchase occurs further out in the future. Although this change in the buying power of currency is important, it is more important that we limit the concept of the time value of money to the fact that money has an *earning power*. Any future reference to the time value of money will be restricted to this concept. The effects of inflation will be handled separately and explicitly.

Engineering economic analysis is concerned with the evaluation of economic alternatives. These alternatives are often described by indicating the amount and timing of estimated future receipts and disbursements that will result from each decision. Since the time value of money is concerned with the effect of time and interest on monetary amounts, it is essential that this topic be given primary attention in engineering economy. In this connection it is useful to think in terms of the types of interest used to determine the return expected from an alternative. Knowledge of the various methods of computing interest is necessary in order to accurately determine the actual effect of the time value of money in the comparison of alternative courses of action.

4.4. Types of Interest

The rental rate for a sum of money is usually expressed as the percent of the sum that is to be paid for the use of the sum for a period of one year. Interest rates are also quoted for periods other than one year, known as *interest periods*. In order to simplify the following discussion, consideration of interest rates for periods of other than one year will be deferred until later.

Simple interest. In simple interest the interest to be paid on repayment of a loan is proportional to the length of time the principal sum has been borrowed. The interest that will be earned may be found in the following manner. Let P represent the principal, n the interest period, and i the interest rate. Then

$$I = Pni.$$

Suppose that $1,000 is borrowed at simple interest at a rate of 6% per annum. At the end of one year, the interest would be

$$I = \$1,000(1)(0.06) = \$60.$$

The principal plus interest would be $1,060 and would be due at the end of the year.

A simple interest loan may be made for any period of time. Interest and principal become due only at the end of the loan period. When it is necessary to calculate the interest due for a fraction of a year, it is common to consider the year as composed of 12 months of 30 days each, or 360 days. For example, on a loan of $100 at an interest rate of 7% per annum for the period February 1 to April 20, the interest due on April 20 along with the principal sum of $100 would be 0.07($100)(80 ÷ 360) = $1.55.

Compound interest. When a loan is made for a length of time equal to several interest periods, interest is calculated at the *end* of each interest period. There are a number of loan repayment plans and these range from paying the interest when it is due to accumulating all the interest until the loan is due. For example, the payments on a 4-year loan of $1,000 at 6% interest per annum payable when due would be calculated as shown in Table 4.1.

Table 4.1. APPLICATION OF COMPOUND INTEREST WHEN INTEREST IS PAID ANNUALLY

Year	Amount Owed at Beginning of Year	Interest to Be Paid at End of Year	Amount Owed at End of Year	Amount to Be Paid by Borrower at End of Year
1	$1,000	$60	$1,060	$ 60
2	1,000	60	1,060	60
3	1,000	60	1,060	60
4	1,000	60	1,060	1,060

If the borrower does not pay the interest earned at the end of each period and if he is charged interest on the *total* amount owed (principal plus interest), the interest is said to be *compounded*. That is, the interest owed in the previous year becomes part of the total amount owed for this year. On this basis, this year's interest charge includes interest that has been earned on previous interest charges. A loan of $1,000 at 6% interest compounded annually for a 4-year period will produce the effect shown in Table 4.2.

Table 4.2. APPLICATION OF COMPOUND INTEREST WHEN INTEREST IS PERMITTED TO COMPOUND

Year	Amount Owed at Beginning of year (A)	Interest to Be Added to Loan at End of Year (B)	Amount Owed at End of Year (A + B)	Amount to Be Paid by Borrower at End of Year
1	$1,000.00	$1,000.00 × 0.06 = $60.00	$1,000(1.06) = $1,060.00	$ 00.00
2	1,060.00	1,060.00 × 0.06 = 63.60	$1,000(1.06)^2 = 1,123.60	00.00
3	1,123.60	1,123.60 × 0.06 = 67.42	$1,000(1.06)^3 = 1,191.02	00.00
4	1,191.02	1,191.02 × 0.06 = 71.46	$1,000(1.06)^4 = 1,262.48	1,262.48

Although the two financial arrangements shown in Tables 4.1 and 4.2 require that the interest be calculated on the unpaid balance, the two schemes produce different effects because of how the payments are to be paid. In the first case payment of the interest at the time it is due prevents having to pay interest on interest. The reverse is true in the second payment scheme. Thus, the effect of applying compound interest is dependent upon the size of the payments and when they are made. The following section presents interest formulas that are useful in dealing with compounding interest.

4.5. The Description of an Investment Opportunity

In many engineering economy studies only small elements of a whole enterprise are considered. For example, studies are often made to evaluate the consequences of the purchase of a single tool or machine in a complex of many facilities. In such cases it would be desirable to isolate the element from the whole by some means analogous to the "free body" diagram in mechanics. To do this, for example, with respect to a machine being considered for purchase, it would be necessary to learn all the receipts and all the disbursements that would arise from the machine. If this could be done, the disbursements could be subtracted from the receipts. This difference would represent profit or gain, from which the investment's return could be calculated.

To aid in identifying and recording the economic effects of alternative investments, a graphical description of each alternative's cash transactions may be used. This pictorial descriptor, referred to as a *cash flow diagram*, will provide all the information necessary for analyzing an investment proposal.

The cash flow diagram represents any receipts received over a period of time as an upward arrow (an increase in cash) located at the period's end. The arrow's height is proportional to the magnitude of the receipts received during that period. Similarly, the disbursements are represented by a downward arrow (a decrease in cash). These arrows are then placed on a time scale representing the duration of the proposal.

An example of a cash flow diagram is shown in Figure 4.2 representing the borrower's transactions given in Table 4.1. Another example is shown in Figure 4.3 for the transactions in Table 4.2. However, in Figure 4.3 the cash flow diagram represents the lender's view of the cash transactions. It is important to understand that it is necessary to always identify the point of view being taken when preparing cash flow diagrams.

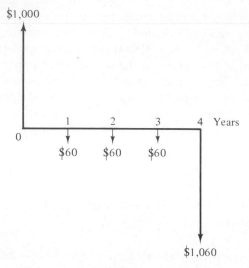

Figure 4.2. Cash flow diagram (borrower).

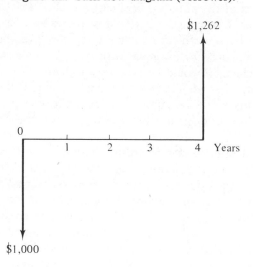

Figure 4.3. Cash flow diagram (lender).

4.6. Interest Formulas (Annual Compounding, Annual Payments)

The interest factors derived in this section apply to the common situation of annual compounding interest and annual payments. The following symbols will be used. Let

$i =$ the annual interest rate;

$n =$ the number of annual interest periods;

$P =$ a present principal sum;

$A =$ a single payment, in a series of n equal payments, made at the end of each annual interest period;

$F =$ a future sum, n annual interest periods hence, equal to the compound amount of a present principal sum P, or equal to the sum of the compound amounts of payments, A in a series.

Single-payment compound-amount factor. If an amount, P, is invested now with the amount earning at the rate i per year, how much principal and interest are accumulated after n years? The cash flow diagram for this financial arrangement is shown in Figure 4.4.

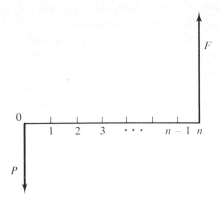

Figure 4.4. Single-present amount and single-future amount.

Since this investment does not provide any payments until the investment is terminated, interest is compounded as shown in Table 4.2. There the interest earned is added to the principal at the end of each annual interest period. By substituting general terms in place of numerical values in Table 4.2, the results shown in Table 4.3 are developed. The resulting factor, $(1 + i)^n$,

Table 4.3. DEVELOPMENT OF SINGLE-PAYMENT COMPOUND-AMOUNT FACTOR

Year	Amount at Beginning of Year	Interest Earned During Year	Compound Amount at End of Year
1	P	Pi	$P + Pi$ $\qquad = P(1 + i)$
2	$P(1 + i)$	$P(1 + i)i$	$P(1 + i) \quad + P(1 + i)i \quad = P(1 + i)^2$
3	$P(1 + i)^2$	$P(1 + i)^2 i$	$P(1 + i)^2 \quad + P(1 + i)^2 i \quad = P(1 + i)^3$
n	$P(1 + i)^{n-1}$	$P(1 + i)^{n-1} i$	$P(1 + i)^{n-1} + P(1 + i)^{n-1} i = P(1 + i)^n$ $\qquad\qquad\qquad\qquad\quad = F$

is known as the *single-payment compound-amount factor*[1] and is designated
($\overset{F/P\ i,\ n}{\quad}$). This factor may be used to find the compound-amount, F, of a
present principal amount, P. The relationship is

$$F = P(1 + i)^n$$

or

$$F = P(\overset{F/P\ i,\ n}{\quad}).$$

The factor designator used to identify the single-payment compound-amount factor is $F/P\ i,\ n$ and it appears over the space where the value of that factor is to be written. The first element in the designator, F/P, represents a ratio which identifies what the factor must be multiplied by, P, in order to find F. The i represents the interest rate per period, and the n represents the number of periods between the occurrence of P and F. For the moment, n will be restricted to years and i will be the interest rate per annum. This method of identifying interest factors is used throughout this text and all tables of factor values are designated with this functional notation system.

Referring to the example of Table 4.2, if \$1,000 is invested at 6% interest compounded annually at the beginning of year one, the compound amount at the end of the fourth year will be

$$F = \$1,000(1 + 0.06)^4 = \$1,000(1.262)$$
$$= \$1,262.$$

Or, by use of the factor designation and its associated tabular value,

$$F = \$1,000(\overset{F/P\ 6,\ 4}{1.262}) = \$1,262.$$

Single-payment present-worth factor. The single-payment compound-amount relationship may be solved for P as follows:

$$P = F\left[\frac{1}{(1 + i)^n}\right].$$

The resulting factor, $1/(1 + i)^n$, is known as the *single-payment present-worth factor* and is designated ($\overset{P/F\ i,\ n}{\quad}$). This factor may be used to find the present worth, P, of a future amount, F, for the investment described in Figure 4.4. Here the question is, "How much must be invested now at 6% compounded annually so that \$1,262 can be received 4 years hence?" This calculation is

---

[1]Values for interest factors for annual compounding interest-annual payments are given in Appendix A, Tables A.1 through A.21.

as follows:

$$P = \$1{,}262\left[\frac{1}{(1 + 0.06)^4}\right] = \$1{,}262(0.7921) = \$1{,}000.$$

Or, by using the factor designation and the interest tables

$$P = \$1{,}262\overset{P/F\,6,\,4}{(0.7921)} = \$1{,}000.$$

Note that the single-payment compound-amount factor and the single-payment present-worth factor are reciprocals.

Equal-payment-series compound-amount factor. In many engineering economy studies, it is often necessary to find the single-future value that would accumulate from a series of equal payments occurring at the end of succeeding annual interest periods. Such a series of cash flows is presented in Figure 4.5. The sum of the compound amounts of the several payments

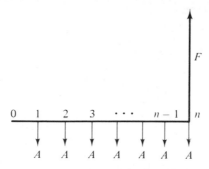

Figure 4.5. Equal-payment-series and single-future amount.

may be calculated by use of the single-payment compound-amount factor. For example, the calculation of the compound amount of a series of five $100 payments made at the end of each year at 6% interest compounded annually is illustrated in Table 4.4.

Table 4.4. THE COMPOUND AMOUNT OF A SERIES OF YEAR-END PAYMENTS

End of Year	Year-end Payment Times Compound-Amount Factor	Compound Amount at End of 5 years	Total Compound Amount
1	$100(1.06)^4	$126	
2	100(1.06)^3	119	
3	100(1.06)^2	113	
4	100(1.06)^1	106	
5	100(1.06)^0	100	$564

It is apparent that the tabular method illustrated is cumbersome for calculating the compound amount for an extensive series. Therefore, it is desirable that a compact solution for this type of problem be available. If A represents a series of n equal payments, such as the $100 series in Table 4.4,

$$F = A(1) + A(1 + i) + \ldots + A(1 + i)^{n-2} + A(1 + i)^{n-1}.$$

The total future amount F is equal to the sum of individual future amounts calculated for each payment A. Multiplying this equation by $(1 + i)$ results in

$$F(1 + i) = A(1 + i) + A(1 + i)^2 + \ldots + A(1 + i)^{n-1} + A(1 + i)^n.$$

Subtracting the first equation from the second gives

$$
\begin{array}{ll}
F(1+i) \quad\;\; = & A(1+i) + A(1+i)^2 + \ldots + A(1+i)^{n-1} + A(1+i)^n \\
-F = -A - A(1+i) - A(1+i)^2 - \ldots - A(1+i)^{n-1} \\
\hline
F(1+i) - F = -A & \hspace{4cm} + A(1+i)^n.
\end{array}
$$

Solving for F from the result above gives

$$F = A\left[\frac{(1 + i)^n - 1}{i}\right].$$

The resulting factor, $[(1 + i)^n - 1]/i$, is known as the *equal-payment-series compound-amount factor* and is designated $\left(\overset{F/A\, i,\, n}{\quad}\right)$. This factor may be used to find the compound amount, F, of an equal-payment-series, A. For example, the future amount of a $100 payment deposited at the end of each of the next 5 years and earning 6% per annum will be

$$F = \$100\left[\frac{(1 + 0.06)^5 - 1}{0.06}\right] = \$100(5.637) = \$563.70$$

which agrees with the result found in Table 4.4. Using the factor designation and the interest tables gives

$$F = \$100(\overset{F/A\, 6,\, 5}{5.637}) = \$563.70.$$

Equal-payment-series sinking-fund factor. The equal-payment-series compound-amount relationship may be solved for A as follows:

$$A = F\left[\frac{i}{(1 + i)^n - 1}\right].$$

The resulting factor, $i/[(1 + i)^n - 1]$, is known as the *equal-payment-series sinking-fund factor* and is designated ($\overset{A/F\,i,\,n}{(\quad)}$). This factor may be used to find the required year-end payments, A, to accumulate a future amount, F, as described in Figure 4.5. If, for example, it is desired to accumulate $563.70 by making a series of five equal annual payments at 6% interest compounded annually, the required amount of each payment will be

$$A = \$563.70\left[\frac{0.06}{(1 + 0.06)^5 - 1}\right]$$

$$= \$563.70(0.1774) = \$100.$$

or

$$A = \$563.70\overset{A/F\,6,\,5}{(0.1774)} = \$100.$$

The derivation of this factor and this example illustrate that the equal-payment-series compound-amount factor and the equal-payment-series sinking-fund factor are reciprocals.

Equal-payment-series capital-recovery factor. A deposit of amount P is made now at an annual interest rate i. The depositor wishes to withdraw his principal plus earned interest in a series of equal year-end amounts over the next n years. When the last withdrawal is made there should be no funds left on deposit. The cash flow diagram for this situation is presented in Figure 4.6. It has been previously shown that F is related to A by the equal-payment-series sinking-fund factor and that F and P are linked by the single-payment compound-amount factor. The substitution of $P(1 + i)^n$ for F in the equal-payment-series sinking-fund relationship results in

$$A = P(1 + i)^n\left[\frac{i}{(1 + i)^n - 1}\right]$$

$$= P\left[\frac{i(1 + i)^n}{(1 + i)^n - 1}\right].$$

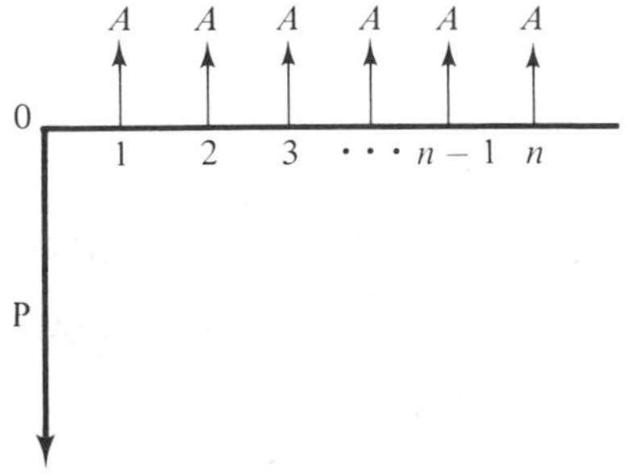

Figure 4.6. Equal-payment-series and single-present amount.

The resulting factor, $i(1 + i)^n/[(1 + i)^n - 1]$ is known as the *equal-payment-series capital-recovery factor* and is designated ($\overset{A/P\,i,\,n}{}$). This factor may be used to find the year-end payments, A, that will be provided by a present amount, P. For example, \$1,000 invested at 5% interest compounded annually will provide for eight equal year-end payments of

$$A = \$1,000\left[\frac{0.05(1 + 0.05)^8}{(1 + 0.05)^8 - 1}\right]$$

$$= \$1,000(0.1547) = \$154.72.$$

or

$$A = \$1,000(\overset{A/P\,5,\,8}{0.1547}) = \$154.72.$$

It should be realized that as each annual withdrawal is made, the amount remaining on deposit is smaller than the amount remaining after the previous withdrawal. Because the interest earned is based on the amount on deposit the interest earned each year also diminishes. The equal-payment-series capital-recovery factor accounts for these year-by-year changes and allows the straightforward calculation of what appears to be a complicated relationship between interest earned and amount of withdrawal.

Equal-payment-series present-worth factor. To find what single amount must be deposited now so that equal end-of-year payments can be made, P must be found in terms of A. The equal-payment-series capital-recovery factor may be solved for P as follows:

$$P = A\left[\frac{(1 + i)^n - 1}{i(1 + i)^n}\right].$$

The resulting factor, $[(1 + i)^n - 1]/i(1 + i)^n$, is known as the *equal-payment-series present-worth factor* and is designated ($\overset{P/A\,i,\,n}{}$). This factor may be used to find the present worth, P, of a series of equal annual payments, A, as depicted in Figure 4.6. For example, the present worth of a series of eight equal annual payments of \$154.72 at an interest rate of 5% compounded annually will be

$$P = \$154.72\left[\frac{(1 + 0.05)^8 - 1}{0.05(1 + 0.05)^8}\right]$$

$$= \$154.72(6.4632) = \$1,000.$$

or

$$P = \$154.72(\overset{P/A\,5,\,8}{6.4632}) = \$1,000.$$

This example and the derivation illustrate that the equal-payment-series capital-recovery factor and the equal-payment-series present-worth factor are reciprocals.

Uniform gradient-series factor. In many cases, annual payments do not occur in an equal-payment series. For example, a series of payments that would be uniformly increasing is $100, $125, $150 and $175 occurring at the end of the first, second, third, and fourth year. Similarly, a uniformly decreasing series would be $100, $90, $80, and $70 occurring at the end of the first, second, third, and fourth year. In general, a uniformly increasing series of payments for n interest periods may be expressed as $A_1, A_1 + G, A_1 + 2G,$ $\ldots, A_1 + (n-1)G$ as shown in Figure 4.7 where A_1 denotes the first year-end payment in the series and G the annual change in the magnitude of the payments.

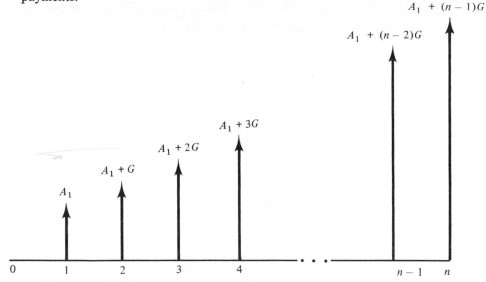

Figure 4.7. A uniform increasing gradient series.

One way of evaluating such a series is to apply the interest formulas developed previously to each payment in the series. This method will yield good results but will be very time consuming. Another approach is to reduce the uniformly increasing series of payments to an equivalent equal-payment series so that the equal-payment-series factor can be used. Let

A_1 = payment at the end of the first year;
G = annual change or gradient;
n = the number of years;
A = the equivalent equal annual payment.

A uniform gradient series may be considered to be made up of two separate series, an equal-payment series with equal annual payment A_1, and a gradient series $0, G, 2G, \ldots, (n-1)G$ at the end of successive years. Each payment in an equal-payment series equivalent to this series can be represented as

$$A = A_1 + A_2$$

where

$$A_2 = F(\overset{A/F\,i,\,n}{}) = F\left[\frac{i}{(1+i)^n - 1}\right]$$

and F is the future amount equivalent to the gradient series. The gradient series can be separated into $(n-1)$ distinct equal-payment series with annual payments of G as shown in Table 4.5. The future amount equivalent to the gradient series can be derived from the table as follows:

$$F = G(\overset{F/A\,i,\,n-1}{}) + G(\overset{F/A\,i,\,n-2}{}) + \ldots + G(\overset{F/A\,i,\,2}{}) + G(\overset{F/A\,i,\,1}{})$$

$$= G\left[\frac{(1+i)^{n-1} - 1}{i}\right] + G\left[\frac{(1+i)^{n-2} - 1}{i}\right] + \ldots + G\left[\frac{(1+i)^2 - 1}{i}\right]$$

$$+ G\left[\frac{(1+i)^1 - 1}{i}\right]$$

$$= \frac{G}{i}[(1+i)^{n-1} + (1+i)^{n-2} + \ldots + (1+i)^2 + (1+i) - (n-1)]$$

$$= \frac{G}{i}[(1+i)^{n-1} + (1+i)^{n-2} + \ldots + (1+i)^2 + (1+i) + 1] - \frac{nG}{i}.$$

Table 4.5. GRADIENT SERIES AND AN EQUIVALENT SET OF SERIES

End of Year	Equal Payment Series	Gradient Series	Set of Series Equivalent to Gradient Series
0	0	0	0
1	A_2	0	0
2	A_2	G	G
3	A_2	$2G$	$G + G$
4	A_2	$3G$	$G + G + G$
.	.	.	. . .
.	.	.	. . .
.	.	.	. . .
$n-1$	A_2	$(n-2)G$	$G + G + G + \ldots + G$
n	A_2	$(n-1)G$	$G + G + G + \ldots + G + G$

The terms in the brackets define the equal payment-series compound-amount factor for n years. Therefore,

$$F = \frac{G}{i}\left[\frac{(1+i)^n - 1}{i}\right] - \frac{nG}{i}$$

and

$$A_2 = F\left[\frac{i}{(1+i)^n - 1}\right]$$

$$= \frac{G}{i}\left[\frac{(1+i)^n - 1}{i}\right]\left[\frac{i}{(1+i)^n - 1}\right] - \frac{nG}{i}\left[\frac{i}{(1+i)^n - 1}\right]$$

$$A_2 = \frac{G}{i} - \frac{nG}{i}\left[\frac{i}{(1+i)^n - 1}\right]$$

$$A_2 = \frac{G}{i} - \frac{nG}{i}(\overset{A/F\,i,\,n}{}) = G\left[\frac{1}{i} - \frac{n}{i}(\overset{A/F\,i,\,n}{})\right].$$

The resulting factor, $\left[\frac{1}{i} - \frac{n}{i}(\overset{A/F\,i,\,n}{})\right]$, is called the *gradient factor* for annual compounding interest and will be designated $(\overset{A/G\,i,\,n}{})$.

As an example of the use of the gradient factor assume that an individual is planning to save \$1,000 from his income during this year and he feels he can increase this amount by \$200 for each of the following 9 years. Since the end-of-year convention is to be used unless otherwise stated, this series begins at the end of the first year and the last savings occurs at the end of the tenth year. If interest is 8% compounded annually, what equal-annual series beginning at the end of year 1 and ending at year 10 would produce the same accumulation at the end of year 10 as would be realized from the gradient series?

$$A = A_1 + G(\overset{A/G\,i,\,n}{})$$

$$= \$1,000 + \$200(\overset{A/G\,8,\,10}{3.8713})$$

$$= \$1,774 \text{ per year.}$$

The gradient series may also be used for a uniformly decreasing gradient. Suppose you wish to find the equal-annual series that is equivalent to the decreasing gradient series in Figure 4.8. Visualize the cash flow in Figure 4.8 as resulting from the year-by-year subtraction of an *increasing* gradient series where $G = \$600$ from an equal-annual series of \$5,000 per year. By approaching the problem in this manner no new factors are needed and the equal-

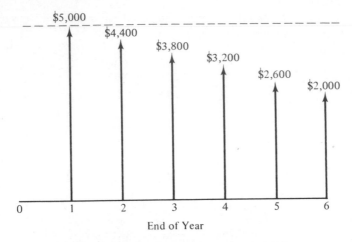

Figure 4.8. A uniform decreasing gradient series.

annual series equivalent to this decreasing gradient series at 9% per annum is

$$A = \$5{,}000 - \$600(\overset{A/G\ 9,\ 6}{2.2498})$$
$$= \$3{,}650 \text{ per year.}$$

Format for the use of interest factors. In engineering economy studies, disbursements made to initiate an alternative are considered to take place at the beginning of the period embraced by the alternative. Payments occurring during the period of the alternative are usually assumed to occur at the end of the year or interest period in which they occur. To use the several interest factors that have been developed, it is necessary that the monetary transactions conform to the format for which the factors are applicable. The schematic arrangement of the factors in Table 4.6 should be helpful in this connection.

Five important points should be noted in the use of interest factors for annual payments:

1. The end of one year is the beginning of the next year.
2. P is at the beginning of a year at a time regarded as being the present.
3. F is at the end of the nth year from a time regarded as being the present.
4. An A occurs at the *end* of each year of the period under consideration. When P and A are involved, the first A of the series occurs one year after P. When F and A are involved, the last A of the series occurs at the same time as F.
5. In the solution of problems, the quantities P, F, and A must be set up to conform with the pattern applicable to the factors used.

Table 4.6. SCHEMATIC ILLUSTRATION OF THE USE OF FACTORS

End of Year	Single Payment		Equal Payment Series				Gradient Series
	Use of Compound-Amount Factor	Use of Present-Worth Factor	Use of Compound-Amount Factor	Use of Sinking-Fund Factor	Use of Present-Worth Factor	Use of Capital-Recovery Factor	Use of Gradient-Series Factor
0	P	P	—	—	—	P	—
1			A	A	A	A	—
2			A	A	A	A	G
3			A	A	A	A	$2G$
r			A	A	A	A	$(r-1)G$
n	F	F	F	F	A	A	$(n-1)G$
	$F = P(^{F/P\,i,n})$	$P = F(^{P/F\,i,n})$	$F = A(^{F/A\,i,n})$	$A = F(^{A/F\,i,n})$	$P = A(^{P/A\,i,n})$	$A = P(^{A/P\,i,n})$	$A = G(^{A/G\,i,n})$

There are two important advantages of using the factor designations in place of the algebraic expressions for the factors. These advantages are: (1) the equations for solving problems may be set up prior to looking up any values of factors from the tables and inserting them in the parentheses, and (2) the source and the identity of values taken from the tables are maintained throughout the solution. For example, in solving a problem where it is required to find the present worth of $800 six years hence at 4% interest compounded annually, the following format may be used:

$$P = F(\overset{P/F\,i,\,n}{\ })$$

$$P = \$800(\overset{P/F\,4,\,6}{0.7903}) = \$632.24.$$

The usefulness of this factor designation system will become more apparent as more complicated cash flows are encountered.

Interest factor relationships. There are numerous relationships between the interest factors which allow the calculation of one type factor from another. Awareness of these relationships provides the means for more effective use of the interest tables along with a better understanding of how these factors reflect the time value of money. These relationships are shown below.

$$(\overset{P/F\,i,\,n}{\ }) = \frac{1}{(\overset{F/P\,i,\,n}{\ })}$$

$$(\overset{F/A\,i,\,n}{\ }) = \frac{1}{(\overset{A/F\,i,\,n}{\ })}$$

$$(\overset{P/A\,i,\,n}{\ }) = \frac{1}{(\overset{A/P\,i,\,n}{\ })}$$

$$(\overset{F/A\,i,\,n}{\ }) = 1 + (\overset{F/P\,i,\,1}{\ }) + (\overset{F/P\,i,\,2}{\ }) + \ldots + (\overset{F/P\,i,\,n-1}{\ })$$

$$(\overset{P/A\,i,\,n}{\ }) = (\overset{P/F\,i,\,1}{\ }) + (\overset{P/F\,i,\,2}{\ }) + \ldots + (\overset{P/F\,i,\,n}{\ })$$

$$(\overset{A/P\,i,\,n}{\ }) = (\overset{A/F\,i,\,n}{\ }) + i$$

$$(\overset{F/G\,i,\,n}{\ }) = (\overset{A/G\,i,\,n}{\ })(\overset{F/A\,i,\,n}{\ }).$$

Although the $(\overset{F/G\,i,\,n}{\ })$ factor is not listed in the interest tables, the above expression reveals an easy method for determining such a factor if it is needed.

Many problems arise in engineering economy where the value of n that is appropriate cannot be found in the interest tables available. Realization that

$$\left(\stackrel{F/P\,i,\,n}{}\right) = \left(\stackrel{F/P\,i,\,n_1}{}\right)\left(\stackrel{F/P\,i,\,n_2}{}\right)\ldots\left(\stackrel{F/P\,i,\,n_k}{}\right)$$

where $n = n_1 + n_2 + \ldots n_k$ provides a method for direct calculation of the factor in question. Suppose the value of $\left(\stackrel{F/P\,10,\,174}{}\right)$ is desired. From the previous relationship the value of the factor is

$$\left(\stackrel{F/P\,10,\,174}{}\right) = (13780.612)(\stackrel{F/P\,10,\,100}{789.747})(\stackrel{F/P\,10,\,70}{1.464})\stackrel{F/P\,10,\,4}{}$$

$$= 15{,}933{,}000.39.$$

Thus, $1 invested now will be worth $15,933,000.39 in 174 years if the interest is allowed to compound at 10% per year. The extraordinary power of compounding becomes evident with examples such as this.

4.7. Nominal and Effective Interest Rates

For simplicity, the discussion to this point has involved interest periods of only one year. However, agreements may specify that interest shall be paid more frequently, such as each half year, each quarter, or each month. Such agreements result in interest periods of one-half year, one-quarter year, or one-twelfth year, and the compounding of interest twice, four times, or twelve times a year, respectively.

The interest rates associated with this more frequent compounding are normally quoted on an annual basis according to the following convention. When the actual or *effective* rate of interest is 3% interest compounded each 6-month period, the annual or *nominal* interest is quoted as "6% per year compounded semiannually." For an effective rate of interest of 1.5% compounded at the end of each 3-month period the nominal interest is quoted as "6% per year compounded quarterly." Thus, the nominal rate of interest is expressed on an annual basis and it is determined by multiplying the actual or effective interest rate per interest period times the number of interest periods per year.

Discrete compounding. The effect of the more frequent compounding is that the actual interest rate per year or effective interest rate per year is higher than the nominal interest rate. For example, consider a nominal interest rate of 6% compounded semiannually. The value of $1 at the end of one year when $1 is compounded at 3% for each half year period is

$$F = \$1(1.03)(1.03)$$

$$= \$1(1.03)^2 = \$1.0609.$$

The actual interest earned on the dollar for one year is 1.0609 minus 1.0000 = 0.0609. Therefore, the effective annual interest rate is 6.09%.

An expression for the effective annual interest rate may be derived from the above reasoning. Let

r = nominal interest rate (per year);
i = the effective interest rate (per period);
c = the number of interest periods per year.

Therefore,

$$i = \text{effective annual interest rate} = \left(1 + \frac{r}{c}\right)^c - 1.$$

Continuous compounding. As a limit, interest may be considered to be compounded an infinite number of times per year, that is *continuously*. Under these conditions, the effective annual interest for continuous compounding is defined as

$$i = \lim_{c \to \infty} \left(1 + \frac{r}{c}\right)^c - 1.$$

But since

$$\left(1 + \frac{r}{c}\right)^c = \left[\left(1 + \frac{r}{c}\right)^{c/r}\right]^r$$

and

$$\lim_{c \to \infty} \left(1 + \frac{r}{c}\right)^{c/r} = e,$$

then

$$i = \lim_{c \to \infty} \left[\left(1 + \frac{r}{c}\right)^{c/r}\right]^r - 1 = e^r - 1.$$

Therefore, when interest is compounded continuously,

$$i = \text{effective annual interest rate} = e^r - 1.$$

Comparing interest rates. The effective interest rates corresponding to a nominal annual interest rate of 6% compounded annually, semiannually, quarterly, monthly, weekly, daily, and continuously are shown in Table 4.7.

Since the effective interest rate represents the actual interest earned, it is this rate that should be used to compare the benefits of various nominal rates of interest. For example, one might be confronted with the problem of determining whether it is more desirable to receive 16% compounded annually or 15% compounded monthly. The effective rate of interest per year for 16% compounded annually is of course 16%, while for 15% compounded monthly the effective annual interest rate is

$$\left(1 + \frac{0.15}{12}\right)^{12} - 1 = 16.1\%$$

Table 4.7. EFFECTIVE ANNUAL INTEREST RATES FOR VARIOUS COMPOUNDING
PERIODS AT A NOMINAL RATE OF 6%

Compounding Frequency	Number of Periods per Year	Effective Interest Rate per Period	Effective Annual Interest Rate
Annually	1	6.0000%	6.0000%*
Semiannually	2	3.0000	6.0900
Quarterly	4	1.5000	6.1364
Monthly	12	0.5000	6.1678
Weekly	52	0.1154	6.1797
Daily	365	0.0164	6.1799
Continuously	∞	0.0000	6.1837

*Note that the effective annual interest rate always equals the nominal rate when compounding occurs annually.

Thus, 15% compounded monthly yields an actual rate of interest that is higher than 16% compounded annually.

Using effective and nominal interest rates. The interest formulas for annual compounding interest-annual payments were derived on the basis of an effective interest rate for an interest period; specifically, for an annual interest rate compounded annually. However, they may be used when compounding occurs more frequently than once a year. This may be done in one of two ways: (1) find the effective annual interest rate from the relationships derived above or from the effective rates tabulated[2] and use this rate when time periods are years, or (2) match the interest rate to the interest period and use the formula directly or its corresponding tabulated value. Consider the following example in which it is desired to find the compound amount of $1,000 four years from now at a nominal annual interest rate of 6% compounded semiannually. The effective interest rate is 6.09% and may be used with the single-payment compound-amount factor as follows:

$$F = \$1,000(1 + 0.0609)^4$$
$$= \$1,000(\overset{F/P\ 6.09,\ 4}{1.267}) = \$1,267.$$

Or, since the nominal annual interest rate is 6% compounded semiannually, the interest rate is 3% for an interest period of one-half year. The required calculation is as follows:

$$F = \$1,000(1 + 0.030)^8$$
$$= \$1,000(\overset{F/P\ 3,\ 8}{1.267}) = \$1,267.$$

---

[2]Effective interest rates corresponding to nominal annual rates for various compounding frequencies are given in Appendix B.

This analysis may be used for nominal annual interest rates compounded with any frequency up to and including continuous compounding. Note, however, that compounding frequencies in excess of 52 times per year differ only slightly from the assumption of continuous compounding.

To help distinguish between effective and nominal rates of interest in this book, the letter i is used to represent effective rates of interest while the letter r is used for nominal rates of interest. The derivations of the interest formulas in Section 4.6 were based on an interest rate per period or an effective interest rate. The letter i was used in these derivations to indicate that these formulas require an effective rate of interest rather than the nominal rate of interest. Of course, when compounding is on a yearly basis the nominal interest rate can be used in those formulas since it is equal to the effective interest rate.

4.8. Interest Formulas (Continuous Compounding, Discrete Payments)

In certain economic evaluations, it is reasonable to assume that continuous compounding interest more nearly represents the true situation than does annual compounding. Also, the assumption of continuous compounding may be more convenient from a computational standpoint in some applications. Therefore, this section presents interest formulas that may be used in those cases where annual payments and continuous compounding interest seem appropriate. The following symbols will be used. Let

$r =$ the nominal annual interest rate;

$n =$ the number of annual periods;

$P =$ a present principal sum;

$A =$ a single payment, in a series of n equal payments, made at the end of each annual period;

$F =$ a future sum, n annual periods hence, equal to the compound amount of a present principal sum, P, or equal to the sum of the compound amounts of payments, A in a series.

Single-payment compound-amount factor.   When finding a future amount in years hence that will result from a present amount, it is necessary to consider the frequency of compounding. The single-payment compound-amount factor depends upon the number of compounding periods in the following way:

annual compounding $F = P(1 + r)^n$

semiannual compounding $F = P\left(1 + \dfrac{r}{2}\right)^{2n}$

$$\text{monthly compounding} \quad F = P\left(1 + \frac{r}{12}\right)^{12n}$$

In general, if there are c compounding periods per year

$$F = P\left(1 + \frac{r}{c}\right)^{cn}.$$

When interest is permitted to compound continuously, the interest earned is instantaneously added to the principal at the end of each infinitesimal interest period. For continuous compounding, the number of compounding periods per year is considered to be infinite. As a result,

$$F = P\left[\lim_{c \to \infty} \left(1 + \frac{r}{c}\right)^{cn}\right].$$

By rearranging terms

$$F = P\left\{\lim_{c \to \infty} \left[\left(1 + \frac{r}{c}\right)^{c/r}\right]^{rn}\right\}.$$

But,

$$\lim_{c \to \infty} \left(1 + \frac{r}{c}\right)^{c/r} = e = 2.7182.$$

Therefore,

$$F = Pe^{rn}.$$

The resulting factor, e^{rn}, is the *single-payment compound-amount factor*[3] for continuous compounding interest and is designated $[\overset{F/P\,r,\,n}{}]$.

Single-payment present-worth factor. The single-payment compound-amount relationship may be solved for P as follows:

$$P = F\left[\frac{1}{e^{rn}}\right].$$

The resulting factor, e^{-rn}, is the *single-payment present-worth factor* for continuous compounding interest and is designated $[\overset{P/F\,r,\,n}{}]$.

Equal-payment-series present-worth factor. By considering each payment in the series individually, the total present worth of the series is a sum

---

[3]Values for interest factors for continuous compounding interest-annual payments are given in Appendix C, Tables C.1 through C.15.

of the individual present-worth amounts as follows:

$$P = A(e^{-r}) + A(e^{-r2}) + \ldots + A(e^{-rn})$$
$$= Ae^{-r}(1 + e^{-r} + e^{-r2} + \ldots + e^{-r(n-1)})$$

which is Ae^{-r} times the geometric series $\sum_{j=0}^{n-1} \left(\dfrac{1}{e^r}\right)^j$. Therefore,

$$P = Ae^{-r}\left[\frac{1 - e^{-rn}}{1 - e^{-r}}\right]$$
$$= A\left[\frac{1 - e^{-rn}}{e^r - 1}\right].$$

The resulting factor, $(1 - e^{-rn})/(e^r - 1)$, is the *equal-payment-series present-worth factor* for continuous compounding interest and is designated $\left[\overset{P/A\,r,\,n}{}\right]$.

Equal-payment-series capital-recovery factor. The equal-payment-series present-worth relationship may be solved for A as follows:

$$A = P\left[\frac{e^r - 1}{1 - e^{-rn}}\right].$$

The resulting factor, $(e^r - 1)/(1 - e^{-rn})$, is the *equal-payment-series capital-recovery factor* for continuous compounding interest and is designated $\left[\overset{A/P\,r,\,n}{}\right]$.

Equal-payment-series sinking-fund factor. The substitution of Fe^{-rn} for P in the equal-payment-series capital-recovery relationship results in

$$A = Fe^{-rn}\left[\frac{e^r - 1}{1 - e^{-rn}}\right]$$
$$= F\left[\frac{e^r - 1}{e^{rn} - 1}\right].$$

The resulting factor, $(e^r - 1)/(e^{rn} - 1)$, is the *equal-payment-series sinking-fund factor* for continuous compounding interest and is designated $\left[\overset{A/F\,r,\,n}{}\right]$.

Equal-payment-series compound-amount factor. The equal-payment-series sinking-fund relationship may be solved for F as follows:

$$F = A\left[\frac{e^{rn} - 1}{e^r - 1}\right].$$

The resulting factor, $(e^{rn} - 1)/(e^r - 1)$, is the *equal-payment-series com-*

pound-amount factor for continuous compounding interest and is designated $F/A\ r, n$
[   ].

Gradient series, continuous compounding. The equivalent annual payment corresponding to an initial payment A_1, linear gradient, G, number of years, n, and interest rate, r, may be found in a similar manner as for annual compounding. It can be shown that

$$A = A_1 + G\left[\frac{1}{e^r - 1} - \frac{n}{e^{rn} - 1}\right].$$

The resulting factor, $\left[\dfrac{1}{e^r - 1} - \dfrac{n}{e^{rn} - 1}\right]$, is called the *gradient factor* for continuous compounding interest and is designated [$A/G\ r, n$].

4.9. Interest Formulas (Continuous Compounding, Continuous Payments)

In the previous derivations, payments were considered to be concentrated at discrete points in time. However, in many instances, it is reasonable to assume that monetary transactions occur on a relatively uniform basis throughout the year. In this case, a uniform flow of money best describes the nature of the transaction. Situations such as this involve a *funds-flow process* which may be described in terms of an annual flow rate. The following symbols will be used. Let

$r =$ the nominal annual interest rate;
$n =$ the time expressed in years;
$P =$ a present principal sum;
$\bar{A} =$ the uniform flow rate of money per year;
$F =$ a future amount equal to the compound amount of a uniform flow of money at time n.

Where there is no flow of payments, as in the case with annual payments, the compound amount and the present-worth factors are identical to those for continuous compounding interest-annual payments. Thus,

$$F = Pe^{rn}$$

as was shown in Section 4.8. Its reciprocal

$$P = Fe^{-rn}$$

was also developed previously.

Funds-flow compound-amount factor. The following symbols will be used to develop interest formulas for the funds-flow process. Let

ΔF = a future amount equal to the compound amount ΔP. This future amount occurs t years from time n as shown in Figure 4.9.

$\bar{A}$ = uniform rate of flow of money per year.

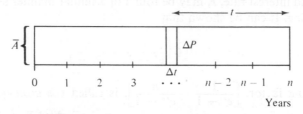

Figure 4.9. Uniform continuous funds flow.

Since it has been shown that $F = Pe^{rn}$,

$$\Delta F = \Delta P e^{rt}.$$

But,

$$\Delta P = \bar{A}\Delta t$$

so that

$$\Delta F = \bar{A}e^{rt}\Delta t.$$

By letting Δt approach zero

$$dF = \bar{A}e^{rt}dt.$$

And, for the entire interval 0 to n

$$F = \int_0^n dF = \int_0^n \bar{A}e^{rt}dt$$

$$F = \left[\frac{\bar{A}e^{rt}}{r}\right]_0^n = \bar{A}\left[\frac{e^{rn}}{r} - \frac{e^0}{r}\right]$$

$$F = \bar{A}\left[\frac{e^{rn} - 1}{r}\right].$$

The resulting factor, $(e^{rn} - 1)/r$, is called the *funds-flow compound-amount factor* and is designated $\left[\overset{F/\bar{A}r,\,n}{}\right]$.

Funds-flow sinking-fund factor. The funds-flow compound-amount relationship may be solved for $\bar{A}$ as follows:

$$\bar{A} = F\left[\frac{r}{e^{rn} - 1}\right].$$

The resulting factor, $r/(e^{rn} - 1)$, is the *funds-flow sinking-fund factor* and is designated [$\overset{\bar{A}/F\,r,\,n}{}$].

Funds-flow capital-recovery factor. By using the single-payment compound-amount relationship for continuous compounding, $F = Pe^{rn}$, and the funds-flow sinking-fund relationship just derived, it is seen that

$$\bar{A} = Pe^{rn}\left[\frac{r}{e^{rn} - 1}\right]$$

$$\bar{A} = P\left[\frac{re^{rn}}{e^{rn} - 1}\right].$$

The resulting factor $(re^{rn})/(e^{rn} - 1)$, is the *funds-flow capital-recovery factor* and is designated [$\overset{\bar{A}/P\,r,\,n}{}$].

Funds-flow present-worth factor. The funds-flow capital-recovery relationship may be solved for P as follows:

$$P = \bar{A}\left[\frac{e^{rn} - 1}{re^{rn}}\right].$$

The resulting factor, $(e^{rn} - 1)/(re^{rn})$, is the *funds-flow present-worth factor* and is designated [$\overset{P/\bar{A}\,r,\,n}{}$].

Funds-flow conversion factor. Tabulated values for the interest factors for continuous compounding interest-annual payments may be modified and used for the funds-flow factors. The required conversion factor may be derived by finding the year end equivalent of a summation of an infinite number of payments occurring during the year. The equal-payment-series compound-amount factor

$$F = A\left[\frac{e^{rn} - 1}{e^{r} - 1}\right]$$

may be modified to reflect c interest periods per year ($n = 1$) as follows:

$$F = \frac{A}{c}\left[\frac{e^{(r/c)c} - 1}{e^{r/c} - 1}\right] = \frac{A}{c}\left[\frac{e^{r} - 1}{e^{r/c} - 1}\right].$$

But,

$$\lim_{c \to \infty} \frac{A}{c}\left[\frac{e^{r} - 1}{e^{r/c} - 1}\right] = \bar{A}\left[\frac{e^{r} - 1}{r}\right]$$

$$F = \bar{A}\left[\frac{e^{r} - 1}{r}\right].$$

The above factor expresses the equivalence between a uniform continuous flow of funds for one year, $\bar{A}$, and a future amount at the end of the year, F. For a time span greater than one year the same factor also calculates the equivalence between a uniform flow of funds occurring at the rate of $\bar{A}$, a year and equal annual amounts A at the end of each year. Thus, for time spans greater than one year

$$A = \bar{A}\left[\frac{e^r - 1}{r}\right].$$

The resulting factor, $(e^r - 1)/r$, is called the *funds-flow conversion factor*[4] and is designated $[\overset{A/\bar{A}\,r}{\quad}]$. This conversion factor may be used with the interest factors for continuous compounding interest-annual payments to yield values for the funds-flow factors in the following manner:

$$[\overset{\bar{A}/P\,r,n}{\quad}] = [\overset{A/P\,r,n}{\quad}] \div [\overset{A/\bar{A}\,r}{\quad}]$$

$$[\overset{P/\bar{A}\,r,n}{\quad}] = [\overset{P/A\,r,n}{\quad}][\overset{A/\bar{A}\,r}{\quad}]$$

$$[\overset{\bar{A}/F\,r,n}{\quad}] = [\overset{A/F\,r,n}{\quad}] \div [\overset{A/\bar{A}\,r}{\quad}]$$

$$[\overset{F/\bar{A}\,r,n}{\quad}] = [\overset{F/A\,r,n}{\quad}][\overset{A/\bar{A}\,r}{\quad}]$$

As an example of the use of the funds-flow conversion factor consider the following example. Find the present amount of $800 per year flowing uniformly for a period of 6 years at an interest rate of 6% compounded continuously. The required calculations are

$$P = \bar{A}[\overset{P/\bar{A}\,r,n}{\quad}] = \bar{A}[\overset{P/A\,r,n}{\quad}][\overset{A/\bar{A}\,r}{\quad}]$$

$$= \$800[\overset{P/A\,6,6}{4.8891}][\overset{A/\bar{A}\,6}{1.030608}] = \$4,031.$$

4.10. Summary of Interest Formulas

The three groups of interest formulas derived in this chapter are summarized in Table 4.8. Each group is based on assumptions about the nature of payments and the compounding of interest. In engineering economic anal-

---

[4]Values for the funds-flow conversion factor for various interest rates are given in Appendix D.

Table 4.8. SUMMARY OF COMPOUNDING FACTORS AND FACTOR DESIGNATIONS

	Factor	Find	Given	Discrete Payments — Discrete Compounding	Discrete Payments — Continuous Compounding	Continuous Payments — Continuous Compounding
Single-Payment	Compound-Amount	F	P	$F = P(1+i)^n = P(^{F/P\,i,n})$	$F = Pe^{rn} = P[^{F/P\,r,n}]$	$F = Pe^{rn} = P[^{F/P\,r,n}]$
Single-Payment	Present-Worth	P	F	$P = F\dfrac{1}{(1+i)^n} = F(^{P/F\,i,n})$	$P = F\dfrac{1}{e^{rn}} = F[^{P/F\,r,n}]$	$P = F\dfrac{1}{e^{rn}} = F[^{P/F\,r,n}]$
Equal-Payment Series	Compound-Amount	F	A	$F = A\left[\dfrac{(1+i)^n-1}{i}\right] = A(^{F/A\,i,n})$	$F = A\left[\dfrac{e^{rn}-1}{e^r-1}\right] = A[^{F/A\,r,n}]$	$F = \bar{A}\left[\dfrac{e^{rn}-1}{r}\right] = \bar{A}[^{F/\bar{A}\,r,n}]$
Equal-Payment Series	Sinking-Fund	A	F	$A = F\left[\dfrac{i}{(1+i)^n-1}\right] = F(^{A/F\,i,n})$	$A = F\left[\dfrac{e^r-1}{e^{rn}-1}\right] = F[^{A/F\,r,n}]$	$\bar{A} = F\left[\dfrac{r}{e^{rn}-1}\right] = F[^{\bar{A}/F\,r,n}]$
Equal-Payment Series	Present-Worth	P	A	$P = A\left[\dfrac{(1+i)^n-1}{i(1+i)^n}\right] = A(^{P/A\,i,n})$	$P = A\left[\dfrac{1-e^{-rn}}{e^r-1}\right] = A[^{P/A\,r,n}]$	$P = \bar{A}\left[\dfrac{e^{rn}-1}{re^{rn}}\right] = \bar{A}[^{P/\bar{A}\,r,n}]$
Equal-Payment Series	Capital-Recovery	A	P	$A = P\left[\dfrac{i(1+i)^n}{(1+i)^n-1}\right] = P(^{A/P\,i,n})$	$A = P\left[\dfrac{e^r-1}{1-e^{-rn}}\right] = P[^{A/P\,r,n}]$	$\bar{A} = P\left[\dfrac{re^{rn}}{e^{rn}-1}\right] = P[^{\bar{A}/P\,r,n}]$
Equal-Payment Series	Uniform-Gradient-Series	A	G	$A = G\left[\dfrac{1}{i} - \dfrac{n}{(1+i)^n-1}\right] = G(^{A/G\,i,n})$	$A = G\left[\dfrac{1}{e^r-1} - \dfrac{n}{e^{rn}-1}\right] = G[^{A/G\,r,n}]$	

ysis that group which most accurately represents the situation under study should be used.

PROBLEMS

1. If $350 in interest is earned in 3 months on an investment of $15,000, what is the annual rate of simple interest?

2. For what period of time will $6,500 have to be invested to amount to $8,500 if it earns 7% simple interest per annum?

3. Compare the interest earned by $100 for 10 years at 8% simple interest with that earned by the same amount for 10 years at 8% compounded annually.

4. What is the principal amount if the principal plus interest at the end of $1\frac{3}{4}$ years is $4,049 for a simple interest rate of 7% per annum?

5. A man lends $1,500 at 8% simple interest for 3 years. At the end of this time he invests the entire amount (principal plus interest) at 7% compounded annually for 10 years. How much will he have at the end of the 13-year period?

6. What will be the amount accumulated by each of these present investments?
   (a) $8,000 in 8 years at 10% compounded annually.
   (b) $675 in 11 years at 12% compounded annually.
   (c) $2,500 in 43 years at 6% compounded annually.
   (d) $11,000 in 52 years at 8% compounded annually.

7. What is the present value of these future payments?
   (a) $5,500 6 years from now at 9% compounded annually.
   (b) $1,700 12 years from now at 6% compounded annually.
   (c) $6,200 37 years from now at 12% compounded annually.
   (d) $4,300 48 years from now at 7% compounded annually.

8. What is the present value of the following series of prospective payments?
   (a) $3,500 a year for 8 years at 7% compounded annually.
   (b) $230 a year for 37 years at 15% compounded annually.
   (c) $1,000 a month for 4 years at 12% compounded annually.
   (d) $2,500 every 6 months for 10 years at 8% compounded annually.

9. What is the accumulated value of each of the following series of payments?
   (a) $500 at the end of each year for 12 years at 6% compounded annually.
   (b) $1,400 at the end of each quarter for 10 years at 8% compounded annually.
   (c) $4,200 at the end of every year for 43 years at 9% compounded annually.
   (d) $500 at the end of each month for 67 years at 10% compounded annually.

10. What equal series of payments must be paid into a sinking fund to accumulate the following amounts?
    (a) $9,000 in 5 years at 7% compounded annually when payments are annual.
    (b) $15,000 in 8 years at 12% compounded annually when payments are annual.

(c) $4,000 in 37 years at 6% compounded annually when payments are annual.

(d) $5,200 in 58 years at 8% compounded annually when payments are annual.

11. What equal series of payments are necessary to repay the following present amounts?
(a) $5,000 in 5 years at 4% compounded annually with annual payments.
(b) $16,000 in 8 years at 10% compounded annually with semiannual payments.
(c) $37,000 in 72 years at 9% compounded annually with monthly payments.
(d) $10,000 in 110 years at 12% compounded annually with quarterly payments.

12. What annual uniform payment series are necessary to repay the following increasing or decreasing series of payments?
(a) A series of 4 end-of-year payments that begins at $1,000 and increases at the rate of $100 a year with 5% interest compounded annually.
(b) A series of 47 end-of-year payments that begins at $250 and increases at the rate of $50 a year with 9% interest compounded annually.
(c) A series of 10 end-of-year payments that begins at $5,000 and decreases at the rate of $200 a year with 12% interest compounded annually.
(d) A series of 42 end-of-year payments that begins at $10,000 and decreases at the rate of $100 a year with 8% interest compounded annually.

13. How would you compute the following amounts of payments if the interest rate is 10% compounded annually?
(a) The present value of a series of prospective payments, $300 a year for 130 years.
(b) The accumulated value of a series of prospective payments, $10 a year for 130 years.
(c) The equal series of payments to be paid into a sinking fund to accumulate $10,000 in 130 years.
(d) The equal series of payments to be made to repay a present amount of $85,000 in 130 years.

14. What rate of interest compounded annually is involved if:
(a) An investment of $10,000 made now will result in a receipt of $13,690, 8 years from now?
(b) An investment of $1,000 made 18 years ago is increased in value to $7,690?

15. How many years will it take for an investment to double itself if interest is compounded annually for an interest rate of 5%? 12%?

16. At what rate of interest compounded annually will an investment triple itself in 10 years?

17. For an interest rate of 9% compounded annually, find:
(a) How much can be loaned now if $2,000 will be repaid at the end of 4 years?
(b) How much will be required 6 years hence to repay a $1,500 loan made now?

18. How many years will be required for:
(a) An investment of $2,500 to increase to $8,200 if interest is 8% compounded annually?

(b) An investment of $1,000 to increase to $10,641 if interest is 3% compounded annually.

19. What is the equal-payment series that is equivalent to a payment series of $12,000 at the end of the first year decreasing by $300 each year over the next 14 years. Interest is 10% compounded annually.

20. What equal annual amount must be deposited for 10 years in order to provide withdrawals of $200 at the end of the second year, $400 at the end of the third year, $600 at the end of the fourth year, and so on, up to $1,800 at the end of the tenth year? The interest rate is 7% compounded annually.

21. An investor has borrowed $1,000 with an agreement to repay the loan over 4 years in equal-annual payments for an interest rate of 10%.
    (a) How much are these payments?
    (b) Calculate and present in tabular form the amount that remains on balance just after payment is made at each point in time including the time the loan is made.
    (c) Plot the unpaid balance as a function of time.

22. A widow received $10,000 from an insurance company after her husband's death. She plans to deposit this amount in a savings account that earns interest at a rate of 7% compounded annually for 5 years.
    (a) If she wants to withdraw equal annual amounts from the account for 5 years with the first withdrawal occurring one year after the deposit, how much are these disbursements?
    (b) Calculate and present in tabular form the amount that remains on balance just after withdrawal is made at each point in time.
    (c) Plot the account's balance as a function of time.

23. A man deposits a sum of money in his savings account which earns interest at a rate of 6% compounded annually. He would like to pay his life insurance premiums ($250 a year) from this bank account for 3 years.
    (a) How much should he deposit so that his third payment just depletes the balance?
    (b) Calculate and present in tabular form the amount that remains on balance just after payment is made at each point in time including the present time.
    (c) Plot the account's balance as a function of time.

24. Find the interest factor ($F/G\ i, n$) that will convert a gradient series as defined in Table 4.5 into its future equivalent at the end of the nth year.

25. Graphically illustrate the function of each of the six interest factors for annual compounding.

26. Graphically illustrate the function of the uniform gradient-series factor for annual compounding; for continuous compounding.

27. Find the interest factor ($P/G\ i, n$) that will convert a gradient series as defined in Table 4.5 to its equivalent value at the present.

28. How would you determine a desired equal-payment-series sinking-fund factor if you only had a table of:

(a) Single-payment compound-amount factors?
(b) Single-payment present-worth factors?
(c) Equal-payment-series compound-amount factors?
(d) Equal-payment-series capital-recovery factors?

29. How would you determine a desired equal-payment-series capital-recovery factor if you only had a table of:
    (a) Single-payment present-worth factors?
    (b) Equal-payment-series present-worth factors?
    (c) Equal-payment-series compound-amount factors?
    (d) Single-payment compound-amount factors?

30. Develop a formula for finding the accumulated amount F at the end of n interest periods which will result from a series of beginning-of-period payments each equal to B if the latter are placed in a sinking fund for which the interest rate per period is i, compounded each period.

31. Rewrite the formula given for the single-payment compound-amount factor to apply to the compounding of interest at the end of each period, where p represents the number of compounding periods per year, y represents the number of years, and r represents the nominal annual rate of interest. Use P as the present sum and F as the compound amount and express F in terms of P, r, p, and y.

32. What effective annual interest rate corresponds to the following?
    (a) Nominal interest rate of 8% compounded semiannually.
    (b) Nominal interest rate of 8% compounded monthly.
    (c) Nominal interest rate of 8% compounded quarterly.
    (d) Nominal interest rate of 8% compounded weekly.

33. An annual effective interest rate of 12% is desired:
    (a) What nominal rate should be asked if compounding is to be semiannually?
    (b) What nominal rate should be asked if compounding is to be quarterly?

34. The Square Deal Loan Company offers money at $\frac{1}{2}$% interest per week compounded weekly. What is the effective annual interest rate? What is the nominal interest rate?

35. What nominal interest rate is paid if:
    (a) Payments of $4,500 per year for 6 years will repay an original loan of $22,000?
    (b) Twenty-four monthly deposits of $100 will result in $2,800 at the end of 2 years?

36. What is the effective interest rate if a nominal rate of 12% is compounded continuously? If an effective interest rate of 8% is desired, what must the nominal rate be if compounding is continuous?

37. How much more desirable is 12% compounded monthly than 12% compounded yearly?

38. How many years will it take an investment to triple itself if the interest rate is 6% compounded annually; compounded continuously?

39. What is the present worth of a uniform series of year-end payments of $400 each for 10 years if the interest rate is 7% compounded continuously?

40. An interest rate of 9% compounded continuously is desired on an investment of $29,000. How many years will be required to recover the capital with the desired return if $7,000 is received each year?

41. What will be the required quarterly payment to repay a loan of $2,500 in 3 years if the interest rate is 8% compounded continuously?

42. What is the present worth of the following prospective payments?
    (a) $4,400 in 30 years at an interest rate of 6% compounded continuously.
    (b) $2,400 in 5 years at an interest rate of 12% compounded weekly.

43. What equal semiannual payment must be deposited into a sinking fund to accumulate $50,000 in 20 years at 10% interest compounded continuously?

44. What is the accumulated value of each of the following series of payments?
    (a) $700 at the end of each month for 5 years at 6% interest compounded continuously.
    (b) $500 at the end of each quarter for 39 years at 8% interest compounded continuously.

45. Find the annual uniform payment series which would be equivalent to the following increasing series of payments if the interest rate is 9% compounded annually; compounded continuously.

$200 at the end of the first year.
$300 at the end of the second year.
$400 at the end of the third year.
$500 at the end of the fourth year.
$600 at the end of the fifth year.
$700 at the end of the sixth year.
$800 at the end of the seventh year.

46. Find the equal quarterly series that would be exchanged for the following decreasing series if the interest rate is 8% compounded annually:

$5,000 at the end of the first quarter.
$4,500 at the end of the second quarter.
$4,000 at the end of the third quarter.
$3,500 at the end of the fourth quarter.
$3,000 at the end of the fifth quarter.
$2,500 at the end of the sixth quarter.
$2,000 at the end of the seventh quarter.
$1,500 at the end of the eighth quarter.

47. What is the present value of the following continuous funds flow?
    (a) $3,500 per year for 8 years at 6% compounded continuously.
    (b) $700 per month for 10 years at 9% compounded continuously.
    (c) $6,000 per quarter for 8.3 years at 8% compounded continuously.
    (d) $900 per year for 15.8 years at 7% compounded continuously.

48. What amount will be accumulated by each of these continuous funds flow?
(a) $5,000 per year in 6 years at 12% compounded continuously.
(b) $1,500 per month in 12.5 years at 6% compounded continuously.

49. For how many years must an investment of $82,870 provide a continuous flow of funds at the rate of $8,000 per year so that an annual interest rate of 15% compounded continuously is earned?

50. What annual interest rate compounded continuously will be earned on an investment of $18,300 that provides a continuous flow of funds at the rate of $275 monthly for 7.5 years?

CALCULATIONS OF EQUIVALENCE
INVOLVING INTEREST

Many calculations in engineering economy require that prospective receipts and disbursements of two or more alternative proposals be placed on an equivalent basis for comparison. This may be accomplished by the proper use of the interest formulas developed in the previous chapter. Also, it is essential that the economic meaning of equivalence be understood. This chapter illustrates the concept of equivalence and presents computational methods required when interest formulas are used in engineering economy studies.

5.1. The Meaning of Equivalence

If two or more situations are to be compared, their characteristics must be placed on an equivalent basis. Which is worth more, 4 ounces of Product A or 1,800 grains of Product A? In order to answer this question, it is necessary to place the two amounts on an equivalent basis by use of the proper conversion factor. After conversion of ounces to grains, the question becomes: Which is worth more, 1,750 grains of Product A or 1,800 grains of Product A? The answer is now obvious.

5

Two things are said to be equivalent when they have the same effect For instance, the torques produced by applying forces of 100 pounds and 200 pounds 2 feet and 1 foot, respectively, for the fulcrum of a lever are equivalent since each produces a torque of 200 foot-pounds.

Three factors are involved in the equivalence of sums of money. These are (1) the amounts of the sums, (2) the time of occurrence of the sums, and (3) the interest rate. The interest formulas developed consider time and the interest rate. Thus, they constitute a convenient way of taking the time value of money into consideration when calculating the equivalence of monetary amounts occurring at different points in time.

Equivalence between cash flows. In engineering economy the meaning of equivalence pertaining to value in exchange is of primary importance; for example, a present amount of $300 is equivalent to $478.20 if the amounts are separated by 8 years and if the interest rate is 6% per annum. This is so because a person who considers 6% to be a satisfactory rate of interest would be indifferent to receiving $300 now or $478.20 eight years from now.

This equivalence between the two cash flows presented in Figure 5.1 may be illustrated by use of the single-payment formulas for annual com-

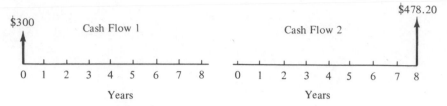

Figure 5.1. Two equivalent cash flows at 6%.

pounding interest. A sum of $300 in the present is equivalent to

$$\$300(1 + 0.06)^8 = \$300(\overset{F/P\,6,\,8}{1.594}) = \$478.20$$

8 years from now. Similarly, $478.20 to be received 8 years from now is equivalent to

$$\$478.20\left[\frac{1}{(1 + 0.06)^8}\right] = \overset{P/F\,6,\,8}{478.20(0.6274)} = \$300.00$$

at the present.

The first calculation describes the equivalent value 8 years hence of a $300 amount presently in hand. The second calculation defines the amount now that is equivalent to $478 received 8 years from now. The decision to restrict the analysis to the present or to a time 8 years from now is made only for computational convenience. Other analyses could be used since it is known *that for one cash flow to be equivalent to another their equivalent values must be equal at any point in time.* Otherwise, the person considering either of the cash flows would not be indifferent to either one of them and the cash flows could not be said to be equivalent. Using the cash flows in Figure 5.1 it is demonstrated in Table 5.1 that they are equivalent at a number of different points on the time scale. From this example it can be concluded that equivalent cash flows will be equal at any point in time.

This fact allows equivalent cash flows to be properly calculated by using a variety of expressions. For example, the solution of the following expression determines the amount F, 8 years hence, that is equivalent to $300 now when interest is 6% per annum.

$$\$300(\overset{F/P\,6,\,3}{1.191}) = F(\overset{P/F\,6,\,5}{0.7473})$$

$$F = \$478.$$

In this case it was arbitrarily decided to equate the two cash flows at the end of the third year. Usually the point in time at which the analysis is made is selected in such a way that the number of interest factors required are minimized.

Table 5.1. EQUIVALENT VALUES OF EQUIVALENT CASH FLOWS AT DIFFERENT POINTS IN TIME

Time	Equivalent Value at Time t	
t	Cash Flow 1	Cash Flow 2
0	$300	$\overset{P/F\,6,\,8}{\$478(0.6274)} = \$300$
1	$\overset{F/P\,6,\,1}{\$300(1.060)} = 318$	$\overset{P/F\,6,\,7}{478(0.6651)} = 318$
2	$\overset{F/P\,6,\,2}{300(1.124)} = 337$	$\overset{P/F\,6,\,6}{478(0.7050)} = 337$
.	.	.
.	.	.
.	.	.
7	$\overset{F/P\,6,\,7}{300(1.504)} = 451$	$\overset{P/F\,6,\,1}{478(0.9434)} = 451$
8	$\overset{F/P\,6,\,8}{300(1.594)} = 478$	$= 478$
10	$\overset{F/P\,6,\,10}{300(1.791)} = 537$	$\overset{F/P\,6,\,2}{478(1.124)} = 537$
30	$\overset{F/P\,6,\,30}{300(5.744)} = 1{,}723$	$\overset{F/P\,6,\,22}{478(3.604)} = 1{,}723$

Equivalence is not directly apparent. The relative value of several alternatives is usually not apparent from a simple statement of their future receipts and disbursements until these amounts have been placed on an equivalent basis. Consider the following example: An engineer sold his patent to a corporation and is offered a choice of $12,500 now or $1,650 per year for the next 10 years, the estimated beneficial life of the patent to the corporation. The engineer is paying 6% interest on his home mortgage and will use this rate in his evaluation. The patterns of receipts are shown in Table 5.2.

Since money has a time value it is not apparent from a cursory examination of the receipts of the two alternatives which is economically the most

Table 5.2. PATTERN OF RECEIPTS FOR TWO ALTERNATIVES

End of Year Number	Receipts Alternative A	Receipts Alternative B
0	$12,500	0
1	0	$1,650
2	0	1,650
3	0	1,650
4	0	1,650
5	0	1,650
6	0	1,650
7	0	1,650
8	0	1,650
9	0	1,650
10	0	1,650
Total Receipts	$12,500	$16,500

desirable. For instance, it is incorrect to say alternative B is more desirable than alternative A because the sum of receipts from those alternatives are $16,500 and $12,500, respectively. Such a statement would be correct only if the interest rate is considered to be zero.

The equivalence values for these two alternatives for an interest rate of 6% must be found by the use of interest formulas. One way to determine an equivalent value for alternative B is to calculate an amount at the present which is equivalent to 10 receipts of $1,650 each as

$$P = \$1,650(\overset{P/A\,6,\,10}{7.3601}) = \$12,144.$$

This amount is equivalent to 10 future payments of $1,650 each and is directly comparable with $12,500. This is because both figures represent amounts of money at the same point in time, the present. Thus, the engineer can now see that on an equivalent basis the $12,500 lump sum is most desirable.

It should be noted that the $12,144 is only an equivalent amount determined from an anticipated series of cash receipts. An actual receipt of $12,144 would not occur, even if this alternative had been chosen. The actual receipts would be $1,650 per year for 10 years.

5.2. Equivalence Calculations Requiring a Single Factor

The interest formulas derived in Chapter 4 express relationships that exist between the several elements making up the formulas. These formulas exhibit relationships between P, A, F, i, and n for annual compounding and between P, A, F, r, and n for continuous compounding. For the case where continuous funds flow is assumed the formulas exhibit the relationships between P, $\bar{A}$, F, r, and n.

The paragraphs which follow will illustrate methods for calculating equivalence where these interest formulas are involved. In the examples, the quantities P, A, and F will be set up to conform to the pattern applicable to the particular factor used, as was illustrated in Table 4.6.

Single-payment compound-amount factor calculations. The single-payment compound-amount factors yield a sum F, at a given time in the future, that is equivalent to a principal amount P for a specified interest rate i compounded annually, or r compounded continuously. For example, the solution for finding the compound amount on April 1, 1987 that is equivalent to a principal sum of $200 on April 1, 1979 for an interest rate of 5% compounded annually is

$$n = 1987 - 1979 = 8$$

$$F = P(\overset{F/P\,i,\,n}{})$$

$$= \$200(\overset{F/P\,5,\,8}{1.477}) = \$295.40.$$

If interest is compounded continuously, the solution is

$$F = P[\overset{F/P\,r,\,n}{}]$$

$$= \$200[\overset{F/P\,5,\,8}{1.492}] = \$298.40.$$

If the principal P, the compound amount F, and the number of years n are known, the interest rate i, may be determined by interpolation in the interest tables. For example, if $P = \$300$, $F = \$525$, and $n = 9$, the solution for i is

$$F = P(\overset{F/P\,i,\,n}{})$$

$$\$525 = \$300(\overset{F/P\,i,\,9}{})$$

$$(\overset{F/P\,i,\,9}{1.750}) = \frac{\$525}{\$300}.$$

A search of the interest tables for annual compounding interest reveals that 1.750 falls between the single payment compound-amount factors in the 6% and 7% tables for $n = 9$. The value from the 6% table is 1.689 and the value from the 7% table is 1.838. By linear proportion

$$i = 6 + (1)\frac{1.689 - 1.750}{1.689 - 1.838}$$

$$= 6 + \frac{0.061}{0.149} = 6.41\%.$$

The linear interpolation used for i is illustrated by Figure 5.2.

Solution for i might have been done without the use of tables as follows:

$$F = P(1 + i)^n$$

$$\$525 = \$300(1 + i)^9$$

$$(1 + i)^9 = \frac{\$525}{\$300}$$

$$i = \sqrt[9]{1.750} - 1$$

$$i = 1.0641 - 1 = 0.0641, \text{ or } 6.41\%.$$

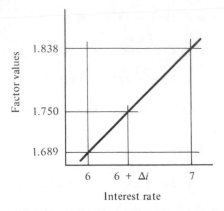

Figure 5.2. Interpolation for i.

Solutions accomplished without the use of tables are usually more time-consuming than solutions by interpolation. Since engineering economy studies usually require estimates of the future, the error introduced by interpolation will rarely be of significance. Therefore, it is recommended that the interpolation method be used unless tables of factors are not available or the calculations are being done by computer where directed calculation is faster than a table look-up operation.

If the principal sum, P, its compound amount, F, and the interest rate, i, are known, the number of years, n, may be determined by interpolation of the interest tables. For example, if $P = \$400$, $F = \$704.40$, and $i = 0.07$, the solution for n is

$$F = P(\overset{F/P\,i,\,n}{})$$

$$\$704.40 = \$400(\overset{F/P\,7,\,n}{})$$

$$(\overset{F/P\,7,\,n}{1.761}) = \frac{\$704.40}{\$400}.$$

A search of the 7% interest table reveals that 1.761 falls between the single-payment compound-amount factors for $n = 8$ and $n = 9$. For $n = 8$, the factor is 1.718 and for $n = 9$, the factor is 1.838. By linear proportion

$$n = 8 + (1)\frac{1.718 - 1.761}{1.718 - 1.838}$$

$$= 8 + \frac{0.043}{0.120} = 8.36 \text{ years.}$$

The linear interpolation used for n is illustrated by Figure 5.3.

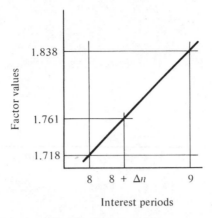

Figure 5.3. Interpolation for n.

Solution for n might have been done without the use of tables as follows:

$$F = P(1 + i)^n$$

$$\$704.40 = \$400(1 + 0.07)^n$$

$$(1.07)^n = \frac{\$704.40}{\$400} = 1.761$$

$$n = 8.37 \text{ years.}$$

The interpretation of n in this problem is that 8.37 years are required for $400 to earn enough interest so that the total amount available after 8.37 years equals $704.40. However, when compounding occurs at the end of discrete periods as it does in this example, the result (8.37 years) only approximates the time required to accumulate the amount in question. This lack of accuracy occurs because interest is paid only at the end of each period and *at least* 9 years are necessary to accumulate $704.40 or *more*. Actually, after 9 years the compound amount would be $735.20.

Single-payment present-worth factor calculations. The single-payment present-worth factors yield a principal sum P, at a time regarded as being the present, which is equivalent to a future sum F. For example, the solution for finding the present worth of a sum equal to $400 received 12 years hence, for an interest rate of 6% compounded annually, is

$$P = F\left(\overset{P/F\,i,\,n}{}\right)$$

$$= \$400(\overset{P/F\,6,\,12}{0.4970}) = \$198.80.$$

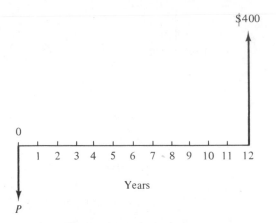

Figure 5.4. Invest P and receive $400.

Thus, if a man desires $400 at the *end* of 1992, he may deposit $198.80 at the *beginning* of 1980 in a bank account paying 6% compounded annually. This cash flow is shown in Figure 5.4 with time zero representing the beginning of 1980. For continuous compounding at 6%, the same objective could be attained with less money invested now. That is,

$$P = F[\overset{P/F\,r,\,n}{}]$$

$$= \$400[\overset{P/F\,6,\,12}{0.4868}] = \$194.72.$$

Equal-payment-series compound-amount factor calculations. The equal-payment-series compound-amount factors yield a sum F at a given time in the future, which is equivalent to a series of payments A, occurring at the end of successive years such that the last A concurs with F. The solution for finding the equivalent amount, 7 years from now, of a series of seven $40 year-end payments whose final payment occurs simultaneously with the compound amount being determined, for an interest rate of 6% is

$$F = A(\overset{F/A\,i,\,n}{})$$

$$= \$40(\overset{F/A\,6,\,7}{8.394}) = \$335.76.$$

If the compound amount F, the annual payments A, and the number of years n are known, the interest rate i may be determined by interpolation of the interest tables. For example, if $F = \$441.10$, $A = \$100$, and $n = 4$, the solution for i is

$$F = A(\overset{F/A\,i,\,n}{})$$

$$\$441.10 = \$100(\overset{F/A\,i,\,4}{})$$

$$(\overset{F/A\,i,\,4}{4.411}) = \frac{\$441.10}{\$100}.$$

This value falls between the equal-payment-series compound-amount factors in the 6% and 7% table for $n = 4$. By linear interpolation:

$$i = 6 + (1)\frac{4.375 - 4.411}{4.375 - 4.440}$$

$$= 6 + \frac{0.036}{0.065} = 6.55\%.$$

Such calculations are an integral part of retirement plans in which an individual sets aside part of his earnings on a regular basis to provide for his retirement years. If an individual places $800 per year in a retirement plan that pays 7% compounded annually, what lump sum could he withdraw providing he had participated over a 35-year time span. The cash flow is shown in Figure 5.5.

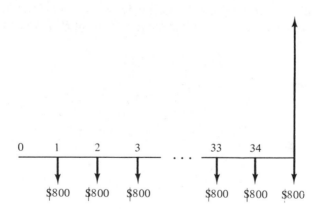

Figure 5.5 Invest $800 per year and receive F.

The single amount, F, the individual could withdraw 35 years hence is

$$F = \$800(\overset{F/A\,7,\,35}{138.237}) = \$110,590.$$

Thus, having deposited $28,000 over 35 years ($800 × 35) results in accumulated interest earnings of $82,590, yielding a total amount of $110,590. If the individual had the entire $28,000 available to invest at the beginning of the 35 years, he could accumulate an even greater amount as shown by the following calculation:

$$F = \$28,000(\overset{F/P\,7,\,35}{10.677}) = \$298,956.$$

Suppose for an equal-annual cash flow of $100 each year it is necessary to calculate how long it would take to accumulate $2,000 if the interest rate is 8% compounded continuously.

$$F = A[\overset{F/A\,r,\,n}{}]$$

$$\$2,000 = \$100[\overset{F/A\,8,\,n}{}]$$

$$[\overset{F/A\,8,\,n}{20.00}] = \frac{\$2,000}{\$100}$$

$$n = 12 + (1)\frac{19.351 - 20.000}{19.351 - 21.963}$$

$$n = 12 + \frac{0.649}{2.612} = 12.25 \text{ years.}$$

This problem can also be solved directly by using the interest formula. This direct approach requires solution for n in terms of F, A, and r as follows:

$$F = A\left[\frac{e^{rn} - 1}{e^r - 1}\right]$$

$$e^{rn} - 1 = \frac{F}{A}(e^r - 1)$$

$$rn = \log_e\left[\frac{F}{A}(e^r - 1) + 1\right]$$

$$n = \frac{\log_e\left[\frac{F}{A}(e^r - 1) + 1\right]}{r}.$$

For $F = \$2,000$, $A = \$100$, $r = 8\%$

$$n = \frac{\log_e\left[\frac{\$2,000}{100}(e^{0.08} - 1) + 1\right]}{0.08} = \frac{\log_e[20(0.0834) + 1]}{0.08}$$

$$= \frac{\log_e[2.668]}{0.08} = \frac{0.981}{0.08} = 12.26 \text{ years.}$$

When tables of interest factors are available it is usually less time consuming to calculate the value of the factor and then interpolate from the tables rather than using the approach just shown.

Equal-payment-series sinking-fund factor calculations. The equal-payment-series sinking-fund factor is used to determine the amount A of a series of equal payments, occurring at the end of successive years, which are equivalent to a future sum F. The solution for finding the amount of annual sinking-fund deposits A for the period June 1, 1979, to June 1, 1986, that are equivalent to a single amount F of $400 on June 1, 1986, at 5% interest is

$$A = F(\overset{A/Fi,\,n}{)}$$

$$= \$400(\overset{A/F\,5,\,7}{0.1228}) = \$49.13.$$

Recall that all payments are end-of-period transactions so that the first payment occurs on June 1, 1980, with the last payment coming on June 1, 1986. Solution for i and n when F, A, and n or i are known may be done by interpolation in the interest tables as was illustrated for the single-payment compound-amount factor.

Suppose a firm estimates it will require $1,000,000 six years from now for the purchase of new equipment. To accumulate this sum it is decided that an amount will be set aside each year for this purpose. If this firm is able to earn 8% compounded annually on their cash, the amount which must be deposited at the end of each of the 6 years to accumulate the $1,000,000 is

$$A = \$1,000,000(\overset{A/F\,8,\,6}{0.1363}) = \$136,300.$$

The total interest earned by the firm over the 6 years is

$$\$1,000,000 - (\$136,300)(6) = \$182,200.$$

Equal-payment-series present-worth factor calculations. The equal-payment-series present-worth factors are used to find the present worth P, of an equal-payment-series A, occurring at the end of successive years following the time taken to be the present. For example, the solution for finding the present worth P, which is equivalent to a series of five $60 year-end payments beginning at the end of the first interest period after the present for an interest rate of 10%, is

$$P = A(\overset{P/Ai,\,n}{)}$$

$$= \$60(\overset{P/A\,10,\,5}{3.7908}) = \$227.45.$$

This factor may be used to calculate the capital investment which would be justified if it would result in an annual saving each year for several years. For example, suppose that a labor-saving device is proposed which will reduce the cost of labor by $10,000 per year for 15 years. If the interest rate is 8%, the capital investment which can be justified is any amount less than

$$P = \$10,000(\overset{P/A\,8,\,15}{8.5595}) = \$85,595.$$

Equal-payment-series capital-recovery factor calculations. The equal-payment-series capital-recovery factors are used to determine the amount A of each payment of a series of payments occurring at the end of successive years which is equivalent to a present sum P. For example, the solution for finding the amount A of annual year-end payments for a 5-year period which is equivalent to an amount P of $300 at the present for an interest rate of 6% is

$$A = P(\overset{A/P\,i,\,n}{\quad})$$

$$= \$300(\overset{A/P\,6,\,5}{0.2374}) = \$71.22.$$

When purchasing items on credit it is common to require repayment by equal payments over the duration of the loan. Suppose an individual purchases a house for $50,000. If a down payment of 20% is required, the individual must borrow $40,000. Now consider that he is able to borrow that amount from a savings and loan association which charges 8% interest compounded annually (the 8% is applied to the unpaid balance at the beginning of each year). The equal end-of-year payments which must be made to repay this loan in 20 years are

$$A = \$40,000(\overset{A/P\,8,\,20}{0.1019}) = \$4,076.$$

The total amount paid over the 20-year period is $81,520 ($4,076 × 20) and thus the total interest paid on this loan will amount to $41,520 ($81,520 − $40,000). This interest amount is larger than the value of the original loan.

Suppose for the preceding problem one wishes to know how much of the loan remains to be paid after the annual payment has been made at the end of the ninth year. This unpaid amount can be easily found because it is represented by the equivalence of the *remaining* payments at the end of the ninth year. Thus, the unpaid balance (or unrecovered balance) of this loan at the end of the ninth year is

$$\$4,076(\overset{P/A\,8,\,11}{7.1390}) = \$29,099.$$

When funds are flowing continuously and the interest rate is compounded continuously it is necessary to use the funds-flow factors developed in Chapter 4. To find the funds-flow equivalent of a present sum of $600 when the time period is 12 years and the interest rate is 10%, calculate

$$\bar{A} = P[\overset{\bar{A}/P\,r,\,n}{}] = P[\overset{A/P\,r,\,n}{}] \div [\overset{\bar{A}/A\,r}{}]$$

$$\bar{A} = \$600[\overset{\bar{A}/P\,10,\,12}{}] = \$600[\overset{A/P\,10,\,12}{0.15050}] \div [\overset{A/\bar{A}\,10}{1.0517}] = \$85.86 \text{ per year.}$$

Direct calculation gives

$$\bar{A} = P\left[\frac{re^{rn}}{e^{rn} - 1}\right]$$

$$\bar{A} = \$600\left[\frac{(0.10)e^{(0.10)(12)}}{e^{(0.10)(12)} - 1}\right] = \$600\left[\frac{0.332}{2.32}\right] = \$85.86 \text{ per year.}$$

5.3. Equivalence Calculations for More Frequent Compounding

Although the interest factors derived in Section 4.4 are based on annual compounding, it is important to understand that any length of time can be selected as the compounding period. Usually, these periods are discrete time intervals that may be a day, week, month, 3 months, 6 months, or a year, depending on the financial institution or financial instrument involved. Even continuous compounding is offered by some financial institutions where the compounding periods are infinitesimal and the effective annual interest rate is the maximum that can be earned for a given nominal interest rate.

When solving interest problems there are three situations that can arise with regard to the compounding frequency and the frequency of payments received. These three situations are:

1. The compounding periods and the occurrence of payments coincide.
2. The compounding periods occur more frequently than the receipt of payments.
3. The compounding periods occur less frequently than the receipt of payments.

These three situations are discussed together with example calculations in this section.

Compounding and payment periods coincide. The interest factors of Section 4.6 require the use of the effective interest rate per period and the number of periods in the time span being considered. In Section 4.6 the effective annual interest rate is used exclusively, while the payments occurred annually. The use of compounding for other than annual periods is presented in the following paragraphs.

If payments of $100 occur semiannually at the end of each 6-month period for 3 years and the nominal interest rate is 12% compounded semi-annually, the present worth P is determined as follows:

$$i = \frac{12\%}{2 \text{ periods}} = 6\% \text{ per semiannual period}$$

$$n = (3 \text{ years})(2 \text{ periods per year}) = 6 \text{ periods}$$

$$P = A(\overset{P/A\,i,\,n}{}) = \$100(\overset{P/A\,6,\,6}{4.9173}) = \$491.73.$$

As a second example, suppose that a man borrows $2,000 and is to repay this amount in 24 equal installments of $99.80 over the next 2 years. Interest is compounded monthly on the unpaid balance of the loan. What is the effective interest rate per month and the nominal interest rate the man is paying for this loan? What is the effective annual rate of interest on the loan? The solution is formulated as follows:

$$\$99.80 = \$2,000(\overset{A/P\,i,\,24}{})$$

$$(\overset{A/P\,i,\,24}{}) = 0.0499.$$

A search of the interest tables reveals that the above factor value is found for $i = 1\frac{1}{2}\%$. Since the periods in this problem are months, the effective monthly interest rate is $1\frac{1}{2}\%$. The nominal rate of interst is

$$r = (1\frac{1}{2}\% \text{ per month})(12 \text{ months}) = 18\% \text{ per year.}$$

And, the effective annual interest rate is

$$i = \left(1 + \frac{r}{c}\right)^c - 1 = \left(1 + \frac{0.18}{12}\right)^{12} - 1 = 19.56\%.$$

Compounding more frequent than payments. There are two basic approaches for dealing with a series of receipts and disbursements with more frequent compounding than payment periods. Since these approaches are identical in a theoretical sense, the solutions will be the same. The following examples demonstrate the calculations required for each method.

Suppose a deposit of $100 is placed in a bank account at the end of each of the next 3 years. The bank pays interest at the rate of 6% compounded quarterly. How much will be accumulated in this account at the end of 3 years?

One approach to this problem is to make the calculations based on the compounding periods which are 3 months in length. These calculations are

$$i = \frac{6\%}{4 \text{ quarters}} = 1\tfrac{1}{2}\%$$

per quarter. The amount accumulated in the account is

$$F = \$100(\overset{F/P\,1\frac{1}{2},\,8}{1.127}) + \$100(\overset{F/P\,1\frac{1}{2},\,4}{1.061}) + \$100 = \$318.80.$$

The first term indicates that the first $100 deposited at the end of the first year will earn interest for the next 8 quarters. The second term indicates the second deposit will earn interest for the next 4 quarters, and the last term is the $100 deposited at the end of the third year.

The second approach is to find the effective interest rate for the payment period and then make all calculations on the basis of that period. The effective annual interest rate is

$$i = \left(1 + \frac{r}{c}\right)^c - 1.$$

In this problem $c = 4$ and $r = 6\%$. Therefore,

$$i = \left(1 + \frac{0.06}{4}\right)^4 - 1 = 6.14\%.$$

The solution is

$$F = \$100(\overset{F/A\,6.14,\,3}{3.188}) = \$318.80.$$

The interest formulas for continuous compounding interest developed in Section 4.8 can be modified to accommodate equal payments that occur more frequently than annually. When there are c payments per year, let

$$n = c \text{ (number of years)}$$

and

$$r = \frac{\text{nominal interest rate per year}}{c}.$$

Suppose it is desired to find a future amount at the end of 5 years that would result from end of month deposits of $1,000 made throughout the entire 5-year period. Assume that the interest earned on these deposits is 15% compounded continuously.

$$n = (12 \text{ periods per year})(5 \text{ years}) = 60 \text{ periods}$$

$$r = \frac{15\%}{12 \text{ periods}} = 1\tfrac{1}{4}\%.$$

Since there is no continuous compounding interest table for $1\tfrac{1}{4}\%$, it is necessary to calculate the factor in its algebraic form.

$$F = A\left[\frac{e^{rn} - 1}{e^r - 1}\right]$$

$$= \$1,000\left[\frac{e^{(0.0125)(60)} - 1}{e^{0.0125} - 1}\right]$$

$$= \$1,000\left[\frac{1.1170}{0.0126}\right] = \$88,650.$$

As another example, suppose that $10,000 is placed in a bank account where the interest rate is 8% compounded continuously. What is the size of equal annual withdrawals that can be made over the next 5 years so that the account balance will equal zero after the last withdrawal? Using the continuous-compounding, discrete-payment factors, we see that the solution is

$$A = \$10,000[\overset{A/P\ 8,\ 5}{0.2526}] = \$2,526 \text{ per year.}$$

Suppose the problem is the same as previously described with the exception that the equal withdrawals are to be made quarterly over the 5-year time span. Since the payments are quarterly, the calculations must be on that basis. The required calculations are

$$r = \frac{8\%}{4 \text{ quarters}} = 2\% \text{ per quarter compounded continuously.}$$

$n = (4 \text{ periods per year}) (5 \text{ years}) = 20 \text{ periods.}$

$$A = \$10,000[\overset{A/P\ 2,\ 20}{0.0613}] = \$613 \text{ per quarter.}$$

Compounding less frequent than payments. Most financial institutions calculate the interest to be paid for an interest period by applying the interest rate for that period to the amount of funds on deposit for the full interest period. Usually no interest is paid for funds deposited *during* an interest period. Funds deposited during an interest period begin to earn interest for the following interest period. Thus, deposits made during a period are placed at the end of that period. Similarly, funds removed during a period usually earn no interest for that period. Therefore, withdrawals during a period are placed at the beginning of that period.

Consider an individual who makes deposits and withdrawals according to the cash flow presented in Figure 5.6. If interest is compounded quarterly, the cash flows can be relocated as shown in Figure 5.7. The cash flow shown in Figure 5.7 is equivalent to the cash flow presented in Figure 5.6 for quarterly compounding. Now proceed as previously discussed for the case where the compounding periods and the payment periods coincide.

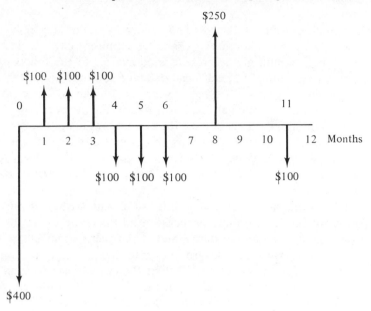

Figure 5.6. Cash flow of monthly receipts and disbursements.

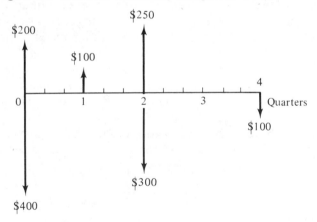

Figure 5.7. Equivalent cash flow for quarterly compounding.

5.4. Equivalence Calculations Requiring Several Factors

When a number of calculations of equivalence involving several interest factors are to be made, some difficulty may be experienced in laying out a plan of attack. Also, until considerable experience has been gained with this type of calculation it may be difficult to keep track of the lapse of time.

For complex problems, the speed and accuracy can usually be improved by a schematic representation. For example, suppose that it is desired to determine what amount at the present is equivalent to the following cash flow for an interest rate of 5%: $300 end of year 6; $60 end of years 9, 10, 11, and 12; $210 end of year 13; $80 end of years 15, 16, and 17. These payments may be represented schematically as illustrated in Figure 5.8.

The plan of attack is to determine the amount at the beginning of year 1 that is equivalent to the single payments and groups of payments comprising the cash flow described above. By converting the various payments to their equivalents at the same point in time it then is possible to determine the total equivalent amount by direct addition. Remember that when interest may be earned dollars occurring at different points in time cannot be directly added.

To use the interest formulas properly recall that P occurs at the beginning of an interest period and that F and A payments occur at the end of interest periods. For instance, the group of four $60 payments are converted to a single equivalent amount of $212.76 at the end of year 8 which is one interest period before the first $60 payment. This is in accordance with the convention of the conversion formula which requires that the amount P occur one interest period prior to the first A payment. In Figure 5.8, the $252.24 as of the end of year 17 represents the equivalent future worth of the three $80 payments. Note that the $252.24 amount concurs with the last $80. Again, this is in accordance with the convention adopted for the derivation of the factor that finds F when given a series of payments A.

The sequence of calculations in the solution of this problem is clearly indicated in the diagram. The position of the arrowhead following each multiplication represents the position of the result with respect to time. The intermediate quantities $212.76 and $252.24 need not have been found. Much time may be saved if all calculations to be made in solving a problem are indicated prior to looking up factor values from the tables and making calculations. In the above example this might have been done as follows:

$$P_1 = \overset{P/F\,5,\,6}{\$300(0.7462)} \qquad = \$223.86$$

$$P_2 = \overset{P/A\,5,\,4\ \ P/F\,5,\,8}{\$\ 60(3.5456)(0.6768)} \ = \ 144.00$$

$$P_3 = \overset{P/F\,5,\,13}{\$210(\ 0.5303\ )} \qquad = \ 111.36$$

$$P_4 = \overset{F/A\,5,\,3\ \ P/F\,5,\,17}{\$\ 80(\ 3.153\ )(\ 0.4363\ )} = \ \underline{110.05}$$

$$P = P_1 + P_2 + P_3 + P_4 \quad = \$589.27$$

If interest had been 5% compounded continuously instead of 5% compounded annually, values would have been taken from the table for contin-

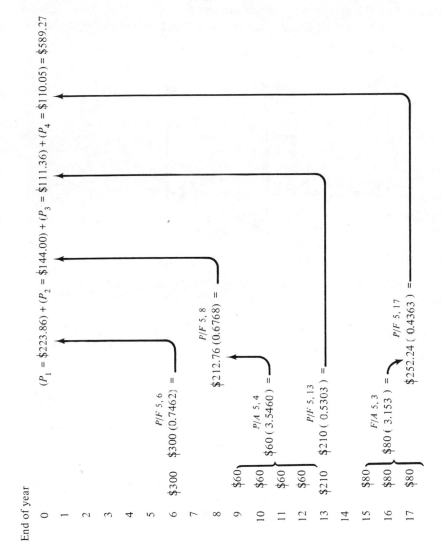

Figure 5.8. Schematic illustration of equivalence.

119

uous compounding interest. All other calculations would have remained the same.

The amount of calculation required in a given situation can be kept to a minimum by the proper selection of interest factors. Consider the example shown in Figure 5.9. By recognizing that the cash flow beginning at the end

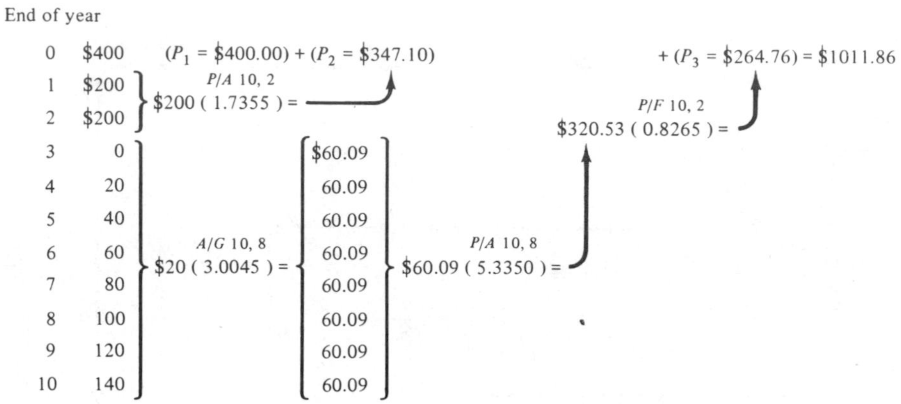

End of year

0	$400	$(P_1 = \$400.00) + (P_2 = \$347.10)$	$+ (P_3 = \$264.76) = \$1011.86$

$\left.\begin{array}{ll} 1 & \$200 \\ 2 & \$200 \end{array}\right\}$ $\overset{P/A\ 10,\ 2}{\$200\ (\ 1.7355\ ) =}$

$\overset{P/F\ 10,\ 2}{\$320.53\ (\ 0.8265\ ) =}$

$\left.\begin{array}{ll} 3 & 0 \\ 4 & 20 \\ 5 & 40 \\ 6 & 60 \\ 7 & 80 \\ 8 & 100 \\ 9 & 120 \\ 10 & 140 \end{array}\right\}$ $\overset{A/G\ 10,\ 8}{\$20\ (\ 3.0045\ ) =}$ $\left\{\begin{array}{l} \$60.09 \\ 60.09 \\ 60.09 \\ 60.09 \\ 60.09 \\ 60.09 \\ 60.09 \\ 60.09 \end{array}\right.$ $\overset{P/A\ 10,\ 8}{\$60.09\ (\ 5.3350\ ) =}$

Figure 5.9. Calculation of equivalence using the gradient factor.

of year 4 is a gradient series with a gradient of $20 per year, it is possible to convert that series into an equivalent equal annual series by the following calculation:

$$A = A_1 + G(\overset{A/G\ i,\ n}{\quad\quad})$$

$$A = \$0 + \$20(\overset{A/G\ 10,\ 8}{3.0045}) = \$60.09.$$

Note that the equal payments of $60.09 begin at the end of year 3 although the gradient series begins at the end of year 4. This situation arises because of the cash flow convention in the derivation of the gradient series factor.

The equivalence at the beginning of year 1 of the cash flow shown in Figure 5.9 is calculated as follows:

$$P_1 \quad\quad\quad\quad\quad\quad\quad\quad = \$400.00$$

$$P_2 = \$200(\overset{P/A\ 10,\ 2}{1.7355}) \quad\quad = \$347.10$$

$$P_3 = \$20(\overset{A/G\ 10,\ 8}{3.0045})(\overset{P/A\ 10,\ 8}{5.3349})(\overset{P/F\ 10,\ 2}{0.8265}) = \underline{\quad\$264.76\quad}$$

$$P = P_1 + P_2 + P_3 \quad\quad\quad\quad = \$1,011.86$$

In the previous two examples the equivalences of the cash flows were found at the beginning of year 1. In many cases the equivalence amount that is desired is a single payment at some point in time other than the present or it may be an equal payment series over some time span. The calculations required to find various cash flow equivalents are quite similar to those shown in Figures 5.8 and 5.9. For instance, to find 10 equal annual payments that are equivalent to the cash flow in Figure 5.9, it is only necessary to convert the single payment equivalent of $1,011.86 at the beginning of year 1 into the desired equivalent by

$$A = P(\overset{A/P\,i,\,n}{})$$

$$= \$1,011.86(\overset{A/P\,10,\,10}{0.1628}) = \$164.73.$$

Because there are a number of different calculations that will lead to the same solution for this type of problem, an effort should be made to minimize the amount of computation required. For example, in this problem each cash payment could have been converted to its equivalent at the beginning of year 1 by using the appropriate single-payment present-worth factor. This approach would have required nine factors while only six factors were required in the calculation shown above. The ability to reduce the amount of calculation for a particular cash flow will develop from experience.

5.5. Bonds and Interest Calculations

A bond is a financial instrument setting forth the conditions under which money is borrowed. Usually, it consists of a pledge of a borrower of funds to pay a stated amount or percent of interest on the par or face value at stated intervals and to repay the par value at a stated time. Bonds are commonly written with par values in multiples of $100 or $1,000. A typical $1,000 bond may embrace a promise to pay its holder $60, for example, one year after purchase and each succeeding year until the principal amount or par value of $1,000 is repaid on a designated date. Such a bond would be referred to as a 6% bond with interest payable annually. Bonds may also provide for interest payments to be made semiannually or quarterly. Since pledges to pay as they are embodied in bonds have value, bonds are bought and sold. The market price of a bond may range above or below its par or face value depending on prevailing market conditions.

Suppose that an individual is considering the purchase at $900 of a $1,000, 6% bond with interest payable *semiannually* and the face value due at the end of 7 years. What will be the equivalent rate of interest earned on

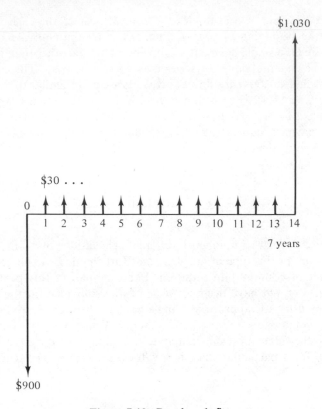

Figure 5.10. Bond cash flow.

such an arrangement if all payments stipulated on the bond are met? Figure 5.10 shows the anticipated disbursements and receipts.

The question may be answered by finding the interest rate that makes the expenditure of $900 at the present equivalent to the present worth of the receipts. This is stated as

$$\$900 = \$30(\overset{P/A\,i,\,14}{}) + \$1,000(\overset{P/F\,i,\,14}{})$$

which must be solved for i by trial and error. The present worth of the receipts at 3% is

$$\$30(\overset{P/A\,3,\,14}{11.2961}) + \$1,000(\overset{P/F\,3,\,14}{0.6611}) = \$1,010.$$

And the present worth of the receipts at 4% is

$$\$30(\overset{P/A\,4,\,14}{10.5631}) + \$1,000(\overset{P/F\,4,\,14}{0.5775}) = \$894.$$

The value of i that makes the present worth of the receipts equal to $900 lies between 3% and 4%. By interpolation,

$$i = 3\% + 1\%\left[\frac{\$1{,}010 - \$900}{\$1{,}010 - \$894}\right]$$
$$= 3\% + 0.94\% = 3.94\%$$

per semiannual period. The nominal annual interest rate is

$$r = (3.94\% \text{ semiannually})(2 \text{ six-month periods}) = 7.88\%.$$

And the effective annual interest rate earned on this bond is

$$i = \left(1 + \frac{0.0788}{2}\right)^2 - 1 = 1.0804 - 1 = 8.04\%.$$

Because the market price of this bond is less than its par value, the real return earned exceeds that of the interest stated on the bond. Therefore, the interest actually earned on bonds fluctuates as the market price of the bond varies. Thus, when the prevailing interest available to investors increases or decreases, the market value of bonds decreases or increases accordingly. This movement in bond prices is followed just as movements in common stock prices are studied.

5.6. Inflation and Interest

A look into the past performance of economies around the world reveals a general inflationary trend in the cost of goods. Of course, during particular periods this trend has been reversed, but overall there seems to be an incessant upward pressure on prices. For small rates of inflation this effect of changing prices appears to have small impact, but inflation at rates exceeding 10% can produce extremely serious consequences for both individuals and institutions.

Inflation is usually described in terms of an annual percentage that represents the rate at which this year's prices have increased over the previous year's prices. Because the rate is defined in this manner, inflation has a compounding effect. Thus, prices that are inflating at a rate of 5% per year will increase 5% the first year, and for the next year the expected increase will be 5% of these new prices. Since the new prices included the original 5% increase, the rate of increase is applied to the 5% increase already experienced. The same is true for succeeding years and therefore rates of inflation are compounded in the same manner that an interest rate is compounded.

To incorporate the effects of inflation in engineering economy studies, it is necessary to use the interest factors so that the inflationary effects on dollars occurring at different points in time can be recognized. The usual procedure for dealing with the loss in buying power that accompanies inflation is to follow these steps:

1. Estimate all the costs associated with a project in terms of today's dollars.
2. Modify the costs estimated in step 1 so that at each future date they represent the cost at that date in terms of the dollars that must be expended at that time.
3. Calculate the equivalent amount of the cash flow resulting from step 2 by considering the time value of money.

Suppose a 50-year-old man is attempting to prepare for his retirement. He plans to retire at age 65 and he estimates that he can live comfortably on $10,000 per year in terms of today's dollars. It is estimated that the future rate of inflation will be 6% per year and he is able to invest his savings at 8% compounded annually. What equal amount must this man save each year until he retires so that he can make withdrawals that will allow him to live as comfortably as he desires for 5 years beyond his retirement?

He is already aware that he needs $10,000 per year from age 66 through 70 as measured in today's dollars. Therefore, the next step is to convert these estimates into the amounts that would be required to support his lifestyle in terms of future dollars (inflated dollars). These calculations are presented in Table 5.3. If the end-of-the-year convention is used, it is observed that in this case the man requires $32,070 at age 70 to purchase the same goods that he could buy at age 50 for $10,000. This difference represents a

Table 5.3. CONVERTING PRESENT DOLLARS REQUIRED INTO INFLATED FUTURE DOLLARS

End of Year n	Age	Dollars Required at Year n to Provide $10,000 per Year in Present Dollars When Inflation at 6% per Year
16	66	$10,000($\overset{F/P\,6,16}{2.540}$) = $25,400
17	67	10,000($\overset{F/P\,6,17}{2.693}$) = 26,930
18	68	10,000($\overset{F/P\,6,18}{2.854}$) = 28,540
19	69	10,000($\overset{F/P\,6,19}{3.026}$) = 30,260
20	70	10,000($\overset{F/P\,6,20}{3.207}$) = 32,070

serious loss in purchasing power and it becomes even more serious at higher rates of inflation.

Using these future values, the cash flow that now reflects the dollars required to make the purchases desired is presented in Figure 5.11. Because money has an earning power it is now necessary to compute the amounts that must be invested so that the amounts invested plus interest provide the withdrawals needed.

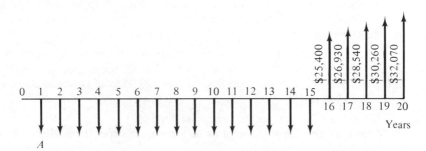

Figure 5.11. Savings and withdrawals in terms of future dollars.

To find the value A that must be saved each year, it is necessary to find the savings cash flow that is equivalent to the withdrawal cash flow. As pointed out earlier, two equivalent cash flows can be equated at any point in time. For this problem the end of year 15 is the time that is most convenient for making the calculation.

$$\overset{F/A\,8,\,15}{A(\;27.152\;)} = \$25,400 \overset{P/F\,8,\,1}{(0.9259)} + \$26,930 \overset{P/F\,8,\,2}{(0.8573)} + \$28,540 \overset{P/F\,8,\,3}{(0.7938)}$$

$$+\; \$30,260 \overset{P/F\,8,\,4}{(0.7350)} + \$32,070 \overset{P/F\,8,\,5}{(0.6806)}$$

$$A = \$113,327 \div 27.152$$

$$A = \$4,174 \text{ per year.}$$

PROBLEMS

1. From the interest tables in the text, determine the following value of the factors by interpolation:
   (a) The single-payment present-worth factor for 47 periods at $7\frac{1}{2}\%$ interest.
   (b) The equal-payment-series capital-recovery factor for 53 periods at $6\frac{1}{4}\%$ interest.

2. From the interest tables given in the text, determine the value of the following factors by interpolation:

(a) The single-payment compound-amount factor for 12 periods at $5\frac{1}{2}\%$ interest.

(b) The equal-payment-series sinking-fund factor for 42 periods at 6% interest.

(c) The equal-payment-series present-worth factor for 12 periods at $3\frac{1}{4}\%$ interest.

(d) The equal-payment-series compound-amount factor for 37 periods at 8% interest.

3. An individual is purchasing a $6,000 automobile which is to be paid for in 24 monthly installments of $300. What nominal interest rate is being paid for this financing arrangement?

4. A no-load mutual stock fund has grown at a rate of 15% compounded annually since its beginning. If it is anticipated that it will continue to grow at this rate, how much must be invested every year so that $40,000 will be accumulated at the end of 12 years?

5. For interest at 6% compounded semiannually, find:
   (a) What payment can be made now to prevent an expense of $350 every 6 months for the next 7 years.
   (b) What semiannual deposit into a fund is required to total $15,000 in 8 years.

6. An individual's salary is now $12,000 per year and he anticipates retiring in 30 more years. If his salary is increased by $840 each year and he deposits 10% of his yearly salary into a fund that earns 7% interest compounded annually, what will be the amount accumulated at the time of his retirement?

7. What single amount at the end of the fifth year is equivalent to a uniform annual series of $1,000 per year for 12 years? The interest rate is 9% compounded annually.

8. A young couple have decided to make advance plans for financing their 3-year-old son's college education. Money can be deposited at 7% compounded annually. What annual deposit on each birthday from the 4th to the 17th inclusive must be made to provide $5,000 on each birthday from the 18th to the 21st inclusive?

9. (a) An interest rate of 10% compounded annually is desired on an investment of $13,000. How many years will be required to recover the capital with the desired interest if $2,500 is received each year?
   (b) A building is priced at $45,000. If a down payment of $14,000 is made and a payment of $2,022 every 6 months thereafter is required, how many years will be necessary to pay for the building? Interest is charged at the rate of 10% compounded semiannually.

10. What will be the amount accumulated by each of these present investments?
    (a) $675 in 20 years at 4% compounded semiannually.
    (b) $11,000 in 10 years at 12% compounded quarterly.

11. What is the present value of these future payments?
    (a) $1,700 12 years from now at 6% compounded monthly.
    (b) $6,200 15 years from now at 12% compounded monthly.

12. What is the present value of the following series of prospective payments?
    (a) $1,000 a month for 4 years at 10% compounded semiannually.
    (b) $5,000 a year for 6 years at 12% compounded quarterly.

13. What is the accumulated value of each of the following series of payments?
    (a) $1,400 at the end of each quarter for 10 years at 8% compounded quarterly.
    (b) $500 at the end of each month for 2 years at 10% compounded semiannually.

14. What equal series of payments are necessary to repay the following present amounts?
    (a) $16,000 in 8 years at 7% compounded semiannually with annual payments.
    (b) $37,000 in 5 years at 9% compounded monthly with monthly payments.
    (c) $8,000 in 3 years at 12% compounded quarterly with annual payments.

15. What equal series of payments must be paid into a sinking fund to accumulate the following amounts?
    (a) $15,000 in 8 years at 12% compounded quarterly when payments are quarterly.
    (b) $4,000 in 11 years at 9% compounded semiannually when payments are annual.
    (c) $17,000 in 15 years at 8% compounded quarterly when payments are monthly.

16. A series of equal quarterly payments of $720 extends over a period of 10 years. What is the amount at the present that is equivalent to this series at 8% interest compounded annually; compounded quarterly; compounded continuously?

17. A continuous flow of funds of $3,300 per year is deposited into a sinking fund. What amount will be accumulated at the end of 5 years if the interest rate is 12% compounded annually; compounded monthly; compounded continuously?

18. A series of equal quarterly payments of $650 for 25 years is equivalent to what present amount at an interest rate of 8% compounded quarterly; compounded continuously?

19. A series of 10 annual payments of $1,500 is equivalent to three equal payments at the end of years 6, 10, and 15 at 12% interest compounded annually. What is the amount of these three payments?

20. A series of payments—$10,000, first year; $9,000, second year; $8,000, third year; $7,000, fourth year; and $6,000, fifth year—is equivalent to what present amount at 10% interest compounded annually; compounded continuously? Solve this problem using the gradient factors, and then solve it using only the single payment present-worth factors.

21. As usually quoted, the prepaid premium of insurance policies covering loss by fire and storm for a 3-year period is 2.5 times the premium for one year of coverage. What rate of interest does a purchaser receive on the additional present investment if he purchases a 3-year policy now rather than three 1-year policies at the beginning of each of the years?

22. An engineering firm is seeking a loan of $200,000 to finance production of a newly patented product line. Due to a good reception of the product at its

introductory showing, the bank has agreed to loan the firm an amount equal to 90% of the present worth of firm orders received for delivery during the next 5 years. The orders are as follows:

Year 1 25,000
Year 2 20,000
Year 3 15,000
Year 4 10,000
Year 5 5,000

If the product will sell for $4 each, will the present worth of the orders received justify the loan required? Interest is 12% compounded annually.

23. A petroleum engineer estimates that the present production of 300,000 barrels of oil during this year from a group of 10 wells will decrease at the rate of 15,000 barrels per year for the next 19 years. Oil is estimated to be worth $13 per barrel for the next 9 years and $18 per barrel thereafter. If the interest rate is 10% compounded annually, what is the equivalent present amount of the prospective future receipts from the wells?

24. A city that was planning an addition to its water supply and distribution system contracted to supply water to a large industrial user for 10 years under the following conditions: The first 5 years of service were to be paid for in advance, and the last 5 years of service were to be paid for at a rate of $25,000 a year payable at the beginning of each year.
    Two years after the system is in operation the city finds itself in need of funds and desires that the company pay off the entire contract so that the city can avoid a bond issue.
    (a) If the city uses 6% interest compounded annually in calculating a fair receipt on the contract, what amount can they expect?
    (b) If the company uses 10% interest compounded annually, how much is the difference between what the company would consider a fair value for the contract and what the city considers to be a fair value to pay for the contract?

25. A man has borrowed $5,000 which he will repay in 48 equal monthly installments. After his twenty-fifth payment he desires to pay the remainder of the loan in a single payment. At 24% interest compounded monthly what is the amount of the payment?

26. A manufacturing company purchased electrical services to be paid for $70,000 now and $15,000 per year beginning with the sixth year. After 2 years service the company, having surplus profits, requested to pay for another 5 years service in advance. If the electrical company elected to accept payment in advance, what would each company set as a fair settlement to be paid if (a) the electrical company considered 15% compounded annually as a fair return, and (b) the manufacturing company considered 12% a fair return.

27. A man has borrowed $1,500 from the ABC Loan Company to buy a used car with an agreement to repay $500 at the end of each of the first 2 years and $1,000 at the end of the third year. What rate of interest makes the receipts and disbursements equal to each other for this loan agreement?

28. A man has the following outstanding debts:
    (a) $10,000 borrowed 4 years ago with the agreement to repay the loan in 60 equal monthly payments. (There are 12 payments outstanding). Interest on the loan is 9% compounded monthly.
    (b) Twenty-four monthly payments of $400 owed on a loan on which interest is charged at the rate of 1% per month on the unpaid balance.
    (c) A bill of $2,000 due in 2 years.
    A loan company has offered to pay his debts if he will pay them $286.30 per month for the next 5 years. What monthly rate of interest is he paying if he accepts the loan company's offer? What is the nominal rate he is paying? What is the annual effective interest rate he is paying?

29. A chapter of a social fraternity is being organized and is in need of housing facilities. A local real estate man agrees to lease the fraternity a suitable house and pay all maintenance costs for $24,000 a year. The fraternity can purchase a building site for $15,000 and construct a house for $185,000. Annual maintenance, taxes, and insurance will cost the fraternity $3,560 if they own their house. With interest at 8%, how many years will it require before the new house would pay for itself, assuming that the building site will continue to be valued at $15,000 and that all other costs would be the same regardless of whether the house is leased or built? Assume the house salvage value is zero.

30. A manufacturer pays a patent royalty of $0.95 per unit of a product he manufactures, payable at the end of each year. The patent will be in force for an additional 5 years. For this year, he manufactures 8,000 units of the product, but it is estimated that output will be 11,000, 14,000, 17,000, and 20,000 in the 4 succeeding years. He is considering asking the patent holder to terminate the present royalty contract in exchange for a single payment at present or asking the patent holder to terminate the present contract in exchange for equal annual payments to be made at the beginning of each of the 5 years. If 8% interest is used, what is (a) the present single payment and (b) the beginning-of-the-year payments that are equivalent to the royalty payments in prospect under the present agreement?

31. A man is planning to retire in 30 years. He wishes to deposit a regular amount every 3 months until he retires so that beginning one year following his retirement he will receive annual payments of $10,000 for the next 20 years. How much must he deposit if the interest rate is 8% compounded quarterly?

32. A city power plant wishes to install a feed-water heater in their steam generation system. It is estimated that the increase in efficiency will pay for the heater one year after it is installed, and a contractor has promised he can install the heater in 5 months. If the venture is undertaken and it is found that the heater does pay for itself in a year by saving $1,400 per month, what amount was paid to the contractor at the last of each month of construction? Assume that the saving of

$1,400 occurs at the last of each month and that the money paid to the contractor could have been invested elsewhere at 9% compounded monthly.

33. A company is considering the purchase of an air compressor. The compressor has a first cost of $5,000 and the following end-of-year maintenance costs:

Year	1	2	3	4	5	6	7	8
Maintenance Costs	$800	$800	$900	$1,000	$1,100	$1,200	$1,300	$1,400

What is the present equivalent value of this series of costs if interest is 12%?

34. Mr. A possesses a mine property estimated to contain 120,000 tons of coal. The mine is now leased to a coal company, which pays Mr. A $1.20 royalty per ton of coal removed. Coal is removed at the rate of 20,000 tons per year. The rate is expected to continue until the mine is exhausted, at which time the mine property is estimated to be worth $25,000 Mr. A now employs a checker whose duty is to measure the coal removed and to bill the coal company for the royalty on the coal removed. The checker receives $8,000 per year.
   (a) If interest is at 8% compounded annually and taxes are neglected, for how much can Mr. A afford to sell the property?
   (b) If interest is at 6% compounded annually, how much can the coal company afford to pay for the property?
   (c) What is the most important factor causing the difference in the results obtained in (a) and (b)?

35. Oil reservoir engineers estimate the annual production of an oil well for the next 12 years to be as follows:

Year End	Annual Production in Barrels
1	45,300
2	25,700
3	10,900
4	8,500
5	6,200
6	4,500
7	3,300
8	1,400
9	800
10	430
11	250
12	60

Assuming that the oil will sell for $12 per barrel during the 12-year period, what is the present equivalent of estimated future production, if interest is compounded annually at 8%?

36. What single payment at the end of year 5 is equivalent to a uniform flow of payments of $4,500 per year beginning at the start of year 3 and ending at the end of year 15. Interest is 8% compounded continuously.

37. What uniform flow of payments for 7 years is equivalent to a series of equal end-of-year payments of $750 for 10 years at 8% compounded continuously?

38. An increasing annual uniform gradient series begins at the end of the second year and ends after the fifteenth year. What is the value of the gradient G that makes the gradient series equivalent to a uniform flow of payments of $900 per month for 7 years at 10% compounded continuously?

39. A man desires to make an investment in bonds, provided he can realize 9% on his investment. How much can he afford to pay for a $1,000 bond that pays 7% interest annually and will mature 10 years hence?

40. A $1,000, 7% bond is offered for sale for $900. If interest is payable annually and the bond will mature in 7 years, what interest rate will be received?

41. A bond is offered for sale for $1,140. Its face value is $1,000 and the interest is 9% payable annually. What rate of interest will be received if the bond matures 10 years hence?

42. How much can be paid for a $1,000, 8% bond with interest paid semiannually, if the bond matures 12 years hence? Assume the purchaser will be satisfied with 6% interest compounded semiannually, since the bonds were issued by a very stable and solvent company.

43. A $1,000 bond will mature in 10 years. The annual rate of interest is 7% payable semiannually. If the bond can be purchased for $870, what annual interest compounded semiannually will be received?

44. If the inflation rate is λ and the discount rate is i, write an expression showing the composite rate per period that would be used to compute the present equivalent of future costs that have been estimated in today's dollars. (Find the composite rate x such that the present equivalent of a future payment stated in today's dollars occurring n periods hence is $(1 + x)^n$ times the estimated payment.)

45. If for Problem 44 the rate of inflation is 7% and the discount rate is 14%, what is the single composite rate per period that would be used to compute the present equivalent of future costs estimated in terms of present-day dollars?

46. Consider a project which has the following cost series for a 5-year period:

End of Year	1	2	3	4	5
Estimated Future Cost in Present-day Dollars	$1,000	$1,000	$1,000	$1,000	$1,000

(a) If the rate of inflation is 6% and the discount rate is 10%, what is the present equivalent of this series? What is the equal-annual series of payments equivalent to this series?

(b) If the rate of inflation is 7% and the discount rate is 5%, what is the present-worth and annual equivalent of this series?

47. Suppose a young couple with a 9-year-old son attempt to save for their son's college expenses in advance. Assuming that he enters college at age 18, they estimate that an amount of $4,000 per year in terms of today's dollars will be required to support his college expenses for 4 years. It is also estimated that the future rate of inflation will be 5% per year and they can invest their savings at 7% compounded annually. Determine the equal amount this couple must save each year until they send their son to college.

48. A 20 Mw power plant now under construction is expected to be in full commercial operation 2 years later. This power plant is designed to be operated on distillate oil only. The fuel cost is a function of plant size, thermal conversion efficiency (heat rate), and plant utilization factor. However, since it is believed that the future price of oil will increase, the fuel cost in each year will be represented by the following expression:

$$F_n = (C)(H)(U)\left(\frac{8{,}760 \text{ hr./year}}{(10)^6}\right)P_r$$

where $P_n = P_{n-1}(1 + \lambda)$, if P_n = price of fuel per million BTU in year n, and

F_n = annual fuel cost in the nth year ($/year);
C = plant size in kw(1 Mw = 1,000 kw);
H = heat rate at operating conditions in BTU/kwh;
U = plant utilization factor;
λ = average annual fuel inflation rate.

If the starting price for fuel during the first year of operation is $3.5/10^6 BTU and this fuel cost is increased at the rate of 6% every year thereafter, what is the equivalent annual fuel cost for 10 years after the plant begins operation if the rate of interest is 12% per year? [Assume that $H = 9{,}300$ BTU/kwh, $u = 0.15$].

part three

ECONOMIC ANALYSIS
OF ALTERNATIVES

BASES FOR COMPARISON
OF ALTERNATIVES

All the decision criteria that are to be considered in this book incorporate some index, measure of equivalence, or *basis for comparison* that summarizes the significant differences between investment alternatives. A basis for comparison is an index containing particular information about a series of receipts and disbursements representing an investment opportunity. The reduction of alternatives to a common base is necessary so that apparent differences become real differences, with the time value of money considered. When expressed in terms of a common base, the real differences become directly comparable and may be used in decision making. The most common bases for comparison are the present-worth amount, the annual equivalent amount, the capitalized equivalent amount, the future-worth amount, and the rate of return. Other bases for comparison that are considered in this chapter are the payout period and the prospective value.

It is important to realize that the basis for comparison represents only one element of any systematic approach used to choose between economic alternatives. However, since many of the significant differences among decision criteria can be directly attributed to the bases for comparison used, it is essential that these be studied in detail.

6

6.1. Net Cash Flows for Investment Opportunities

An investment opportunity is usually described by the actual cash receipts and disbursements that are anticipated if the investment is undertaken. The representation of the amounts and timing of these cash receipts and disbursements is referred to as the investment's *cash flow*.

Consider the following investment opportunity. An individual is planning to place $1,000 in a savings account that pays him 5% interest at the end of each year. The saver anticipates that he will withdraw his initial investment at the end of 4 years. The cash flow that represents the saver's investment opportunity is shown in Table 6.1. Notice that a negative amount represents a cash cost or disbursement for the saver while a positive amount indicates a cash income or receipt.

Table 6.1 also illustrates the cash flow from the bank's point of view. The bank receives $1,000 from the saver at the beginning of the first year and it must make payments of $50 at the end of each year including the last when the bank returns the principal to the saver. Thus, it is quite important

Table 6.1. CASH FLOWS AS VIEWED BY TWO PARTIES

End of Year	Cash Flow from the Saver's Viewpoint	Cash Flow from the Bank's Viewpoint
0	$-1,000	$1,000
1	50	-50
2	50	-50
3	50	-50
4	1,050	-1,050

that the proper point of view is considered when cash flows are being developed for decision-making purposes.

When an investment opportunity has both cash receipts and disbursements occurring simultaneously, a net cash flow is usually calculated for the investment opportunity. The net cash flow is the arithmetic sum of the receipts (+) and the disbursements (−) that occur at the same point in time. The utilization of net cash flows in decision making implies that the net dollars received or disbursed have the same effect on an investment decision as does the separate consideration of an investment's total receipts and disbursements.

To facilitate describing investment cash flows the following notation will be adopted. Let F_{jt} = net cash flow for investment proposal j at time t. If $F_{jt} < 0$ then F_{jt} represents a net cash cost or disbursement. If $F_{jt} > 0$ then F_{jt} represents a net cash income or receipt.

For example, consider a cash flow with the cash receipts and disbursements shown in Table 6.2. The values of the F_{jt}'s are shown in the right-hand column.

Table 6.2. AN ILLUSTRATION OF A NET CASH FLOW

End of Year	Receipts	Disbursements	F_{jt}
0	$ 0	-$5,000	-$5,000
1	4,000	- 2,000	2,000
2	5,000	- 1,000	4,000
3	0	- 1,000	- 1,000
4	7,000	0	7,000

6.2. Present-Worth Amount

In determining a basis for comparison of investment alternatives one likely candidate would be an index that reflects the differences between alternatives by considering the time value of money. In Chapter 5 the idea of equivalence was discussed and it was seen that for a known cash flow and given interest rate an equivalent amount for the cash flow can be calculated at any point in time. Thus, it is possible to calculate a single equivalent amount at any point in time that is equivalent in *value* to a particular cash flow. Since this single value summarizes the value of a cash flow, it seems appropriate to use such a value as a basis for comparison.

The present-worth amount is an amount at the present ($t = 0$) that is equivalent to an investment's cash flow for a particular interest rate i. Thus, the present-worth of investment proposal j at interest rate i with a life of n years can be expressed as

$$PW(i)_j = F_{j0}(\overset{P/Fi,0}{\ }) + F_{j1}(\overset{P/Fi,1}{\ }) + F_{j2}(\overset{P/Fi,2}{\ }) + \ldots + F_{jn}(\overset{P/Fi,n}{\ })$$

$$PW(i)_j = \sum_{t=0}^{n} F_{jt}(\overset{P/Fi,t}{\ }).$$

But since $(\overset{P/Fi,t}{\ }) = (1+i)^{-t}$

$$PW(i)_j = \sum_{t=0}^{n} F_{jt}(1+i)^{-t}. \qquad (-1 < i < \infty)^1.$$

The present-worth amount has a number of features that makes it suitable as a basis for comparison. First, it considers the time value of money according to the value of i selected for the calculation. Second, it concentrates the equivalent value of any cash flow in a single index at a particular point in time ($t = 0$). Third, the value of the present-worth amount is always unique no matter what may be the investment's cash flow pattern. That is, any sequence of receipts and disbursements will give a unique present-worth amount for a particular value of i.

In addition, the present-worth amount is the equivalent amount by which the equivalent receipts of a cash flow exceed or fail to equal the equivalent disbursements of that cash flow. Figure 6.1 illustrates the range of present-worth values for a particular cash flow by plotting the present-worth amount *vs.* the interest rate for the cash flow described in Table 6.3.

By examining the $PW(i)$ function in Figure 6.1 considerable information useful for desicion-making purposes can be ascertained about the investment opportunity. For the range of interest rates ($0 \leq i < 22\%$) it is observed that $PW(i)$ is positive indicating that the equivalent receipts at the present exceed the equivalent disbursements. On the assumption that the cash flow estimates of Table 6.3 eventually prove to be correct, the significance of the

---

[1]For the range ($-1 < i < 0$) no interest is received and the principal invested is not completely repaid. Discussions of interest rate for most problems can be confined to the range ($0 \leq i < \infty$).

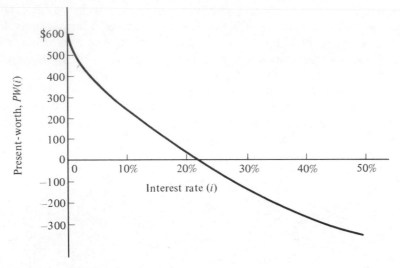

Figure 6.1. Present-worth as a function of interest rate for cash flow in Table 6.3.

Table 6.3. CALCULATION OF THE PRESENT-WORTH FOR A CASH FLOW USING DIFFERENT INTEREST RATES

End of Year	Cash Flow	i	$PW(i) = -\$1,000 + \$400\left\{\sum_{t=1}^{4}\left[\frac{1}{(1+i)^t}\right]\right\}$
0	-$1,000	0%	$600
1	400	10%	268
2	400	20%	35
3	400	22%	0
4	400	30%	-133
		40%	-260
		50%	-358
		∞	-1,000

$PW(i)$ function is that for a particular value of i, say 10%, the investment can be said to produce a 10% return on the unrecovered investment plus an equivalent receipt of $270 at the beginning of the first year.

If the plot had been extended for $i > 50\%$ it would be observed that the curve would be asymptotic to $PW(i) = -\$1,000$, the value of F_{jo}. The reason for this result becomes clear when the definition of $PW(i)$ is re-examined. As the interest rate is increased every cash flow in the future is discounted to the present by a factor of the general form $1/(1+i)^t$. As i approaches infinity it is evident that the discounting factors will approach zero for all points in time except for $t = 0$. Thus, $F_{j_o} = -\$1,000$ is the only cash flow that is not reduced to zero when $i = \infty$.

For cash flows that have initial disbursements followed by a series of positive receipts the $PW(i)$ function possesses characteristics similar to those of Figure 6.1. That is, the function will be decreasing as i increases from zero and it will intersect the horizontal axis only once. Since most of the cash flows that occur in actual decision situations have this pattern of early disbursements followed by future receipts, most of the discussions concerning the selection of investment alternatives is confined to proposals with $PW(i)$ functions similar to the one in Figure 6.1. However, it must be recognized that there are many real world investments that have cash flows which produce $PW(i)$ functions dissimilar to Figure 6.1. One such function is shown in Figure 6.2.

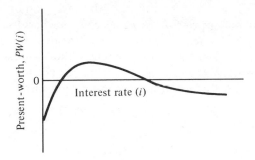

Figure 6.2

Capitalized equivalent. A special case of the present-worth basis of comparison is the so-called *capitalized equivalent, CE(i)*. In engineering economic analysis this term represents a basis of comparison that consists of finding a single amount at the present which at a given rate of interest will be equivalent to the net difference of receipts and disbursements if a given cash flow pattern is repeated in perpetuity.

This concept should not be confused with the accountant's concept of "capitalizing" an expenditure in the books of account. For the accountant an expenditure is capitalized if it is recorded as an asset or prepaid expense rather than being recorded as an expense at the time it is incurred.

To calculate the capitalized equivalent for an investment or a series of investments that are expected to produce cash flows from the present to infinity, the most common method is to first convert the actual cash flow into an equivalent cash flow of equal annual payments A that extends to infinity. Then the equal annual payments are discounted to the present by using the equal payment series present-worth factor.

$$CE(i) = PW(i) \text{ where the cash flow extends forever}$$

$$CE(i) = A(\overset{P/A\,i,\,\infty}{}) = A\left[\frac{(1+i)^{\infty} - 1}{i(1+i)^{\infty}}\right]$$

$$= A\left[\frac{1 - \dfrac{1}{(1+i)^{\infty}}}{i}\right] = \frac{A}{i}.$$

Thus,

$$CE(i) = \frac{A}{i}.$$

An intuitive understanding of this last relationship is obtained by considering what present-worth amount invested at i will enable an investor to periodically withdraw an amount A forever. If the investor withdraws more than amount A each period, he will be withdrawing a portion of the initial principal. If this initial principal is being consumed with each withdrawal, it will eventually be reduced to zero and addtional withdrawals will be impossible. However, when the amount being withdrawn each period equals the interest earned on the principal for that period, the principal remains intact. Thus, the series of withdrawals can be continued forever.

6.3. Annual Equivalent Amount

The annual equivalent amount is another basis for comparison that has characteristics similar to the present-worth amount. This similarity is evident when it is realized that any cash flow can be converted into a series of equal annual payments by first calculating the present-worth amount for the original series and then multiplying the present-worth amount by the interest factor ($\overset{A/P\,i,\,n}{}$). Thus, the annual equivalent amount for interest rate i and n years can be defined as

$$AE(i) = PW(i)(\overset{A/P\,i,\,n}{})$$

$$= \left[\sum_{t=0}^{n} F_{jt}(1+i)^{-t}\right]\left[\frac{i(1+i)^n}{(1+i)^n - 1}\right].$$

There are two important features of this relationship that need to be understood. First, if the values of i and n are finite the relationship reduces to $AE(i) = PW(i)$ times a constant. Therefore, when different cash flows are evaluated for a particular value of i and a particular value of n, the comparison of their annual equivalent amounts will yield the same relative results as those obtained from making the comparison on the basis of the present-worth amount. That is, the ratio of the annual equivalent amounts for two different cash flows will equal the ratio of the present-worth amounts for the respective cash flows.

Second, as long as i and n are finite the values of $AE(i)$ and $PW(i)$ will be zero for the same value of i. Graphically this means that the intersections of the horizontal axis $AE(i) = 0$ by the $AE(i)$ function will occur at the same value of i for which the $PW(i)$ function intersects the horizontal axis $PW(i) = 0$. Thus, the present-worth amount and the equivalent annual amount can be said to be *consistent* bases for comparison. Any particular decision criterion that uses either of these bases for direct comparison will yield the same selection of alternatives for fixed values of i and n.

The $AE(i)$ basis of comparison will sometimes be preferred to $PW(i)$ because of computational advantages that may arise when it is assumed that a particular cash flow pattern will repeat itself. The cash flow in Table 6.4 illustrates a repeating cash flow pattern that results from renewing an investment every 2 years. The cash flows enclosed by lines indicate the receipts and disbursements associated with each investment proposal. For example, the $1,000 disbursement at the end of year 2 is actually the disbursement at the *beginning* of year 3 for another proposal that will return $400 and $900 at the end of year 3 and year 4, respectively.

When converting each investment into an annual equivalent series over the 2-year period of the investment's life, the same calculations are required for each investment. It is only necessary to calculate the annual equivalent amount over the life of one investment proposal. When the investment is to be repeated the same annual equivalent amount will be repeated over the time span for which the investment's cash flow is repeated. For example, the calculation of the annual equivalent amount is $61.93 per year for an investment requiring a $1,000 disbursement followed by receipts of $400 and $900 in one and two years, respectively. If this investment is repeated seven times as shown in Table 6.4, the annual equivalent over the 14-year time span is still $61.93 per year.

The calculation of the present-worth amount for the seven investments of Table 6.4 would require the discounting of all payments in years 1 through 14 back to the present. In this instance the annual equivalent amount basis for comparison would be computationally more efficient than finding the present-worth of each receipt and disbursement.

Of course, if computational efficiency and the present-worth amount are both desired for decision-making it is a simple matter to first determine the annual equivalent as described above and then convert that quantity to a present-worth amount.

Table 6.4. THE USE OF ANNUAL EQUIVALENT FOR REPEATED CASH FLOWS

End of Year	Disbursements	Receipts	Annual Equivalent for Each Investment Proposal
0	−$1,000		
1		$400	
2	−1,000	900	$\left[-\$1,000 + \$400(0.9091)^{P/F\,10,1} + \$900(0.8265)^{P/F\,10,2}\right](0.5762)^{A/P\,10,2} = \begin{cases}\$61.93\\ \$61.93\end{cases}$
3		400	
4	−1,000	900	$\left[-\$1,000 + \$400(0.9091)^{P/F\,10,1} + \$900(0.8265)^{P/F\,10,2}\right](0.5762)^{A/P\,10,2} = \begin{cases}\$61.93\\ \$61.93\end{cases}$
.		.	.
.		.	.
.		.	.
12	−1,000	900	
13		400	$\left[-\$1,000 + \$400(0.9091)^{P/F\,10,1} + \$900(0.8265)^{P/F\,10,2}\right](0.5762)^{A/P\,10,2} = \begin{cases}\$61.93\\ \$61.93\end{cases}$
14		900	

Capital recovery with return. An asset such as a machine is a unit of capital. Such a unit of capital loses value over a period of time in which it is used in carrying on the productive activities of an enterprise. This loss of value of an asset represents actual piecemeal consumption or expenditure of capital.

Capital assets are purchased in the belief that they will earn more than they cost. One part of the prospective earnings is considered to be *capital recovery*. Capital invested in an asset is recovered in the form of income derived from the services rendered by the asset and from its sale at the end of its useful life. If the asset provided services valued at $800 per year during its 5 year life, and if $1,000 was received from its sale, a total of $5,000 would be recovered.

A second part of the prospective earnings will be considered to be *return*. Since capital invested in an asset is ordinarily recovered piecemeal, it is necessary to consider the interest on the unrecovered balance as a cost of ownership. Thus, an investment in an asset is expected to result in income sufficient not only to recover the amount of the original investment, but also to provide for a return on the diminishing investment remaining in the asset at any time during its life. This gives rise to the phrase *capital recovery with return* which will be used in subsequent chapters.

Two monetary transactions are associated with the procurement and eventual retirement of a capital asset; its first cost and salvage value. From these amounts it is possible to derive a simple formula for the equivalent annual cost of the asset for use in economy studies. Let

$P =$ first cost of the asset;
$F =$ estimated salvage value;
$n =$ estimated service life in years;
$CR(i) =$ capital recovery with return.

Then, the annual equivalent cost of the asset may be expressed as the annual equivalent first cost less the annual equivalent salvage value, or

$$CR(i) = P(\overset{A/P\,i,\,n}{}) - F(\overset{A/F\,i,\,n}{}).$$

But since

$$(\overset{A/F\,i,\,n}{}) = (\overset{A/P\,i,\,n}{}) - i$$

by substitution

$$CR(i) = P(\overset{A/P\,i,\,n}{}) - F\left[(\overset{A/P\,i,\,n}{}) - i\right]$$

and

$$CR(i) = (P - F)(\overset{A/P\,i,\,n}{}) + Fi.$$

The equivalence between an asset's loss in value and its equivalent annual cost due to that loss (capital recovery with return) is represented in Figure 6.3.

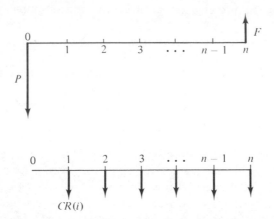

Figure 6.3. Capital recovery with return as an equivalent to an asset's loss in value.

As an example of the use of this important formula, consider the following situation. An asset with a first cost of $5,000 has an estimated service life of 5 years and an estimated salvage value of $1,000. For an interest rate of 6% the annual equivalent cost is

$$(\$5,000 - \$1,000)(\overset{A/P\,6,\,5}{0.2374}) + \$1,000(0.06) = \$1,010.$$

It should be recognized that the cost of an asset is made up of the cost resulting from the loss in value plus the cost of interest on unrecovered capital. The asset in this example suffered a loss in value of $800 per year for 5 years. In addition, the annual equivalent cost of interest on the unrecovered capital was $210 per year for 5 years.

6.4. Future-Worth Amount

The future-worth basis for comparison is an equivalent amount of a cash flow calculated at a future time for some interest rate. The future-worth amount for proposal j at some future time n years from the present is

$$FW(i)_j = F_{j0}(\overset{F/P\,i,\,n}{}) + F_{j1}(\overset{F/P\,i,\,n-1}{}) + \ldots F_{j,n-1}(\overset{F/P\,i,\,1}{}) + F_{j,n}(\overset{F/P\,i,\,0}{})$$

$$FW(i)_j = \sum_{t=0}^{n} F_{jt}(\overset{F/P\,i,\,n-t}{}).$$

But since

$$(\overset{F/P\,i,\,n-t}{}) = (1+i)^{n-t}$$

$$FW(i)_j = \sum_{t=0}^{n} F_{jt}(1+i)^{n-t}.$$

Another method of calculating the future-worth amount is to first determine the present-worth amount of the cash flow and then to convert to its future equivalent n years hence. Thus, the future-worth amount for a proposal j can be expressed as

$$FW(i)_j = PW(i)_j(\overset{F/P\,i,\,n}{}).$$

From this relationship it is clear that for given finite values of i and n the future-worth amount is merely the present-worth amount times a constant. As a consequence the relative differences between alternatives on the basis of present-worth will be the same as the relative differences between alternatives compared on the basis of future-worth as long as i and n are fixed. Therefore, an alternative which has a present-worth three times as large as another alternative's present-worth will also have a future-worth three times as large as the future-worth of that other alternative.

The future-worth amount, annual equivalent amount and the present-worth amount are consistent bases of comparison. As long as i and n are fixed and proposals A and B are being compared, the following relationships will hold.

$$\frac{PW(i)_A}{PW(i)_B} = \frac{AE(i)_A}{AE(i)_B} = \frac{FW(i)_A}{FW(i)_B}.$$

Because the present-worth amount, the annual equivalent amount, and the future-worth amount are all measures of equivalence differing only in the times at which they are stated, it is not surprising that they are consistent bases for comparison of investment alternatives. Therefore, it should be expected that any decision criterion that directly compares present-worth amounts could just as well employ future-worth amounts or annual equivalent amounts as bases for comparison without affecting the selection outcome.

6.5. Rate of Return

The internal rate of return or rate of return, as it is frequently called, is a widely accepted index of profitability. It is defined as the interest rate that reduces the present-worth amount of a series of receipts and disbursements to zero. That is, the rate of return for investment proposal j, is the interest rate i_j^* that satisfies the equation

$$0 = PW(i_j^*)_j = \sum_{t=0}^{n} F_{jt}(1 + i_j^*)^{-t}.$$

where proposal j has a life of n periods. The rate of return must lie in the interval $(-1 < i^* < \infty)$ in order to be economically relevant. For most practical problems it will be sufficient to consider the range of the rate of return in the interval $(0 < i^* < \infty)$.

From the previous discussion of the relationships between the annual equivalent amount, the future-worth amount, and the present-worth amount it follows that the rate of return (i_j^*) for a cash flow will also satisfy the expressions $0 = AE(i_j^*)_j$ and $0 = FW(i_j^*)_j$.

The meaning of rate of return. In economic terms the rate of return represents the percentage or rate of interest earned on the *unrecovered* balance of an investment. The unrecovered balance of an investment can be viewed as the portion of the initial investment that remains to be recovered after interest payments and receipts have been added and deducted, respectively, up to the point in time being considered.

By examining the calculations displayed in Table 6.5 the fundamental meaning of rate of return should become evident. Each of the two cash flows in Table 6.5 can be viewed as an arrangement in which someone has borrowed $1,000 with an agreement to pay 10% on the unpaid or unrecovered balance and to reduce the unpaid balance to zero at the time the loan is fully repaid. These cash flows could also represent the purchase of productive assets where these assets will yield a rate of profit of 10% on the amount of dollars that are unrecovered or "tied-up" in the assets during their lifetime.

If U_t = the unrecovered balance at the beginning of period t, the unrecovered balance for any time period can be found from the recursive equation

$$U_{t+1} = U_t(1 + i) + F_t$$

where

F_t = payment received at the end of period t;
$\quad i$ = interest rate earned on the unrecovered balance during period t;
U_1 = initial amount of loan or first cost of asset.

The unrecovered balances related to each cash flow appear as negative values in Table 6.5 indicating that they are amounts owed by the borrower or the amounts yet to be recovered by the lender.

Table 6.5. TWO CASH FLOWS DEMONSTRATING THE FUNDAMENTAL MEANING OF RATE OF RETURN

End of Year	Cash Flow at End of Year t	Unrecovered Balance at Beginning of Year t	Interest Earned on the Unrecovered Balance During Year t	Unrecovered Balance at the Beginning of Year $t + 1$
		Proposal $A(i_A^* = 10\%)$		
t	$F_{A,t}$	U_t	$U_t(0.10)$	$U_t(1 + 0.10) + F_{A,t} = U_{t+1}$
0	$-\$1,000$	—	—	$-\$1,000$
1	400	$-\$1,000$	$-\$100$	-700
2	370	-700	-70	-400
3	240	-400	-40	-200
4	220	-200	-20	0
		Proposal $B(i_B^* = 10\%)$		
t	$F_{B,t}$	U_t	$U_t(0.10)$	$U_t(1 + 0.10) + F_{B,t} = U_{t+1}$
0	$-\$1,000$	—	—	$-\$1,000$
1	100	$-\$1,000$	$-\$100$	$-1,000$
2	100	$-1,000$	-100	$-1,000$
3	100	$-1,000$	-100	$-1,000$
4	1100	$-1,000$	-100	0

A common misinterpretation of what the rate of return (i^*) measures is observed when the rate of return of a project is viewed as being the rate of interest earned on the initial outlay required by that project. Proposal A in Table 6.5 has a rate of return of 10% but it is clear that the cash flow of the proposal does not yield a 10% return on the $1,000 initial outlay for the full 4-year life of the proposal. In fact, Proposal A earns $100, $70, $40, and $20 in years 1, 2, 3, and 4, respectively, on an investment that decreases from an initial commitment of $1,000 to $0 at the end of the fourth year.

Proposal B represents a special type of cash flow for which the rate of return (i^*) does indicate the return earned on the initial investment. That is, the rate of return is 10% and the $100 amounts earned each year are 10% of the initial investment. The reason for this result is the fact that for this unique cash flow the unrecovered balance throughout the proposal's life always equals the amount of the proposal's initial outlay.

Thus, the fundamental concept of rate of return emerges; it is the rate of interest earned on the unrecovered balance of an investment so that the remaining balance is zero at the end of the investment's life.

Computing the rate of return. The computation of rate of return generally requires a trial-and-error solution. For example, to calculate the rate of return for the cash flow shown below it is necessary to find the value i^* that sets the present-worth amount to zero.

End of Year t	Cash Flow F_t
0	−$1,000
1	−800
2	500
3	500
4	500
5	1200

That is, find the value of i that satisfies

$$0 = PW(i)$$
$$= -\$1,000 - \$800(\overset{P/F\,i,\,1}{}) + \$500(\overset{P/A\,i,\,4}{})(\overset{P/F\,i,\,1}{}) + \$700(\overset{P/F\,i,\,5}{}).$$

Instead of trying to solve for i^* directly from this equation, a trial-and-error solution should be attempted. Try $i = 0\%$

$$PW(0) = -\$1,000 - \$800(1) + \$500(4)(1) + \$700(1)$$
$$PW(0) = \$900.$$

With the present-worth amount greater than zero at $i = 0$ the next step is to examine the cash flow in order to see how the next rate selected will affect the present-worth amount. Since all the positive cash flows are further in the future than the negative cash flows, an increase in the interest rate will reduce the present-worth of the receipts more than the present-worth of the outlays. Thus, the total present-worth amount will be decreased toward zero. Try $i = 12\%$

$$PW(12) = -\$1,000 - \$800(\overset{P/F\,12,\,1}{0.8929}) + 500(\overset{P/A\,12,\,4}{3.0374})(\overset{P/F\,12,\,1}{0.8929})$$
$$+ 700(\overset{P/F\,12,\,5}{0.5674})$$
$$PW(12) = \$39.$$

Since $PW(12)$ is still greater than zero, try a larger interest rate. With $i = 15\%$

$$PW(15) = -\$1,000 - \$800(\overset{P/F\,15,\,1}{0.8696}) + \$500(\overset{P/A\,15,\,4}{0.8550})(\overset{P/F\,15,\,1}{0.8696})$$
$$+ \$700(\overset{P/F\,15,\,5}{0.4972})$$
$$PW(15) = -\$106.$$

Thus, it is determined that the rate of return will lie between 12% and 15%. By interpolation

$$i^* = 12\% + 3\%\left[\frac{39 - 0}{39 - (-106)}\right] = 12\% + 3\%\left[\frac{39}{145}\right] = 12.8\%.$$

Since the solution for the rate of return of a cash flow with a life of n periods is the solution of an nth degree polynomial, there exist various mathematical methods that systematically converge on the roots or values of i that satisfy such a polynomial. Thus, the trial-and-error approach is not the only method of solving for rate of return.

To reduce the computational effort required to determine the rate of return computer solutions of the present-worth equations are often used. As a result the computational effort associated with the rate of return is a significant consideration only where manual calculations are required.

One of the obvious differences between rate of return and the other bases previously discussed is that no interest rate need be known in order to determine the rate of return. The present-worth amount, annual equivalent amount, and the future-worth amount are all functions of an interest rate and in order to calculate a particular value of these bases a particular value of i must be known. For investment situations where knowledge about the future and future interest rates is highly uncertain, rate of return can be a workable way to compare the economic desirability of alternative investments.

By definition, the rate of return is related to the present-worth amount as shown in Figure 6.4. The value of i where the $PW(i)$ function intersects the horizontal axis is i^* the value of i that sets the present-worth amount equal

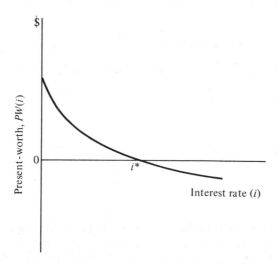

Figure 6.4. Rate of return and its relationship to present-worth.

to zero. Note that for the $PW(i)$ function in Figure 6.4 the present-worth is positive for all values of i less than i^*. This relationship also exists for the annual equivalent and future-worth amounts.

Cash flows with no rate of return. It should be recognized that there are certain cash flows for which *no* rate of return exists in the interval $(-1 < i < \infty)$. The most common example of this situation is when the cash flow consists of either all receipts or all disbursements with the initial receipt or disbursement occurring at the beginning of year 1.

In practice, investment proposals are frequently described by cost cash flows when the alternatives are assumed to provide the same service, benefit, or revenue. Since it is impossible to calculate a meaningful rate of return for such a cash flow pattern, means other than direct calculation of rate of return must be utilized for decision making with cost cash flows. These methods are discussed in Chapter 7.

Cash flows with a single rate of return. Because of the desirability of having a present-worth function with a single rate of return, and the form of the function shown in Figure 6.4, it is important to have a method for predicting whether a particular cash flow can produce such a present-worth function. One useful generalization that can be made is that any proposal cash flow with an initial disbursement or a series of disbursements starting at the present *followed* by a series of positive receipts will always have a present-worth function similar to Figure 6.4 if the absolute sum of the receipts is *greater* than the absolute sum of the disbursements. Table 6.6 presents the cash flows of two proposals (A and B) that satisfy these requirements and two cash flows (C and D) that do not.

Table 6.6. FOUR CASH FLOW PATTERNS

End of Year	Proposal A	Proposal B	Proposal C	Proposal D
0	−$1,000	−$1,000	−$2,000	−$1,000
1	500	− 500	0	4,700
2	400	− 500	10,000	− 7,200
3	300	− 500	0	3,600
4	200	1,500	0	0
5	100	2,000	−10,000	0

For Proposal A the sum of the receipts ($1,500) is greater than the sum of the disbursements ($1,000) and for Proposal B the sum of the receipts ($3,500) exceeds the sum of the disbursements ($2,500). Thus, these two cash flows have a single rate of return with a present-worth plot of the form shown in Figure 6.4. Most practical proposals have estimated cash flows that have patterns similar to Proposal A and Proposal B since most investments require an initial commitment of funds which is followed by the income series resulting from the productivity of the project.

Proposals C and D represent a more unusual class of cash flow patterns. These cash flows consist of disbursements, receipts, more disbursements, receipts, etc. Such cash flows do not follow the pattern of the class of cash flows that include Proposals A and B so there is no assurance that their present-worth plots will resemble the present-worth function of Figure 6.4.

Cash flows with multiple rates of return. Figure 6.2 exhibited a situation where the cash flow has more than one value of i for which the present-worth amount will equal zero. This result should be expected since the present-worth equation is an nth degree polynomial of the form

$$PW(i) = 0 = F_0 + F_1 x + F_2 x^2 + F_3 x^3 + \ldots + F_n x^n$$

where $x = \dfrac{1}{(1 + i)}$.

For this polynomial there may be n different roots or values of x which satisfy this equation. In order to have a rate of return that has economic relevance the value of i must lie in the interval $(-1 < i < \infty)$. Thus, x must be in the interval $(0 < x < \infty)$. The number of positive real roots of x that satisfy the above equation equals the number of meaningful rates of return that will be produced by a cash flow.

For decision-making purposes cash flows which have a unique rate of return and behave similarly to the example shown in Figure 6.4 are much simpler to handle than cash flows with multiple rates of return. When multiple rates of return occur, questions arise such as, "Which rate of return is the correct one?" and "Are the decision rules most frequently used for investment selection applicable when multiple rates of return occur?" The answers to these and other questions that relate to decision criteria are deferred until the next chapter.

The question that should be answered at this time is, "What effect does the cash flow pattern have on the number of positive real roots of the above equation?" By understanding something about these relationships an intelligent determination can be made concerning the conditions under which the rate of return is an appropriate basis for comparison.

A rule that can be helpful in identifying the possibility of multiple rates of return is Descartes' rule of signs for an nth degree polynomial. This rule states that the number of real positive roots of an nth degree polynomial with real coefficients is never greater than the number of changes of sign in the sequence of its coefficients

$$F_0, F_1, F_2, F_3, \ldots, F_{n-1}, F_n$$

and if less, always by an even number. For example, the sequence of signs of the cash flows for Proposal A and Proposal B shown in Table 6.6 changes only once while the sequence of signs for Proposal C and D changes two and three times, respectively. The sequence of signs for Proposal C has one change from the initial negative value to positive at the end of year 2. (A zero cash flow can be considered signless for the purpose of applying the rule of signs.) The sequence of signs remains unchanged until the end of year 5 when the sequence changes from positive back to negative.

For both Proposal A and Proposal B the rule of signs indicates that there is no more than one rate of return in the interval $(-1 < i < \infty)$. This result is verified since both cash flows are known to have a unique rate of return as their present-worth functions are of the form shown in Figure 6.4. However, it should be understood that although the rule of signs may indicate a maximum of one rate of return there is no assurance that the present-worth function will be of the form shown in Figure 6.4.

The rule of signs indicates for Proposal C that the maximum possible number of positive, real roots is two. Figure 6.5 depicts the present-worth

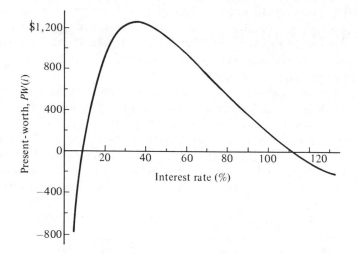

Figure 6.5. Present-worth versus i for cash flow C in Table 6.6.

function for Proposal C and it is seen that in fact this cash flow does have two distinct interest rates for which the present-worth is zero. The two rates of return for Proposal C are 9.8% and 111.5%.

It must be understood that the rule of signs can give only an indication as to the possibility of multiple rates of return because the rule only predicts the *maximum* number of *possible* rates of return that may occur. Thus, numerous cash flows exist that have multiple changes in their sequence of signs but still possess a present-worth function similar to Figure 6.4.

Therefore, the decision maker, who develops a decision procedure based on the assumption that the alternatives being considered will have a present-worth function similar to Figure 6.4, must assure himself that the cash flows do indeed produce such a function. The first test to make is to observe if the cash flow has a series of disbursements that is followed by a series of receipts and the arithmetic sum of the series is positive. If the cash flow has such a pattern then the present-worth function is confirmed to be of the normal form. If the cash flow does not meet this requirement or if the rule of signs indicates that there is the possibility of multiple rates of return, the decision maker should plot $PW(i)$ vs. i for those investment proposals. Using the plot of the present-worth function one can properly interpret the desirability of the proposed investment by observing the values of i for which the present-worth amount is greater than zero. For Proposal D in Table 6.6 the present-worth plot shown in Figure 6.6 reveals that this proposal is desirable for

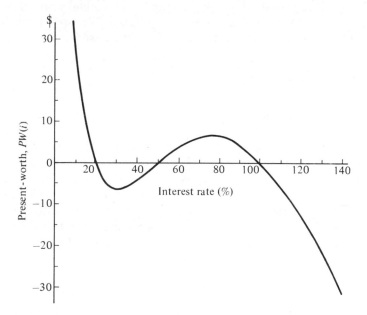

Figure 6.6. Present-worth function for cash flow D in Table 6.6.

values of i less than 20% and for values of i between 50% and 100%. To attempt to interpret multiple rates of return using normal procedures will in most cases be meaningless without the supplementary use of the present-worth function.

It is simple matter to construct a cash flow pattern that produces selected multiple rates of return. Suppose one desires to find a cash flow pattern that has rates of return of 20%, 50%, and 100%. All that is required is to multiply the factors that yield these rates of return. The result of such a multiplication is an equation that represents a future-worth calculation. This equation must be equal to zero since rates of return are defined as those interest rates that set the present-worth or future-worth of a cash flow equal to zero. An example of such a calculation is shown below.

$$FW(i^*) = 0 = [(1 + i) - 1.2][(1 + i) - 1.5][(1 + i) - 2.0]$$
$$0 = (1 + i)^3 - 4.7(1 + i)^2 + 7.2(1 + i) - 3.6$$

Multiplying by $- \$1,000$ gives

$$0 = - \$1,000(1 + i)^3 + 4,700(1 + i)^2 - \$7,200(1 + i) + \$3,600.$$

The coefficients in this last equation are the individual cash receipts and disbursements of Proposal *D* described in Table 6.6. The present-worth plot in Figure 6.6 confirms that the rates of return of this cash flow are 20%, 50%, and 100%.

6.6. Payout Period

Expressions like "This investment will pay itself in less than three years" are common in industry and they are indicative of the tendency to evaluate assets in terms of a *payback* or *payout* period. The payout period is most commonly defined as the length of time required to recover the first cost of an investment from the net cash flow produced by that investment for an interest rate equal to zero. That is, if P = first cost of the investment and if F_t = the net cash flow in period t then the payout period is defined as the value of n that satisfies the equation

$$P = \sum_{t=1}^{n} F_t.$$

It should be noted that in actual practice there are other methods that are variations of the above approach that are also referred to as payout period. However, the pertinent limitations and merits of most payout period methods as a basis for comparison can be made clear by examining the payout period as defined.

For example, Table 6.7 presents the cash flows for three investment

Table 6.7. THREE PROPOSALS WITH A PAYOUT PERIOD OF THREE YEARS

End of Year	Proposal *A*	Proposal *B*	Proposal *C*
0	− $1,000	− $1,000	− $700
1	500	200	− 300
2	300	300	500
3	200	500	500
4	200	1000	0
5	200	2000	0
6	200	4000	0
Sum of Cash Flow	600	7000	0

proposals for each of which the payout period is 3 years. A look at the proposals presented in Table 6.7 reveals the fact that the payout period as a measure of investment desirability does have some serious shortcomings. Certainly no one would feel that under normal circumstances these three proposals have equal economic desirability although they have equal payback periods.

In general, the most serious deficiencies of the payback period are that it fails to consider:

1. The time value of money.
2. The consequences of the investment following the payback period including the magnitude and timing of the cash flows and the expected life of the investment.

Because of the limitations just mentioned the payout period tends to favor shorter lived investments. Experience has generally indicated that this bias is unjustifiable and in many cases economically unsound.

Nevertheless, it must be said that the payout period does give some measure of the rate at which an investment will recover its initial outlay. For those situations where there is a high degree of uncertainty concerning the future and a firm is interested in its cash position and borrowing commitments, the payout period can supply useful information about investments that are under consideration. As a result, this measure of investment desirability is frequently used to supplement the bases for comparison that have been discussed earlier.

6.7. Prospective Value

A recently developed basis for comparison that differs from those just presented requires the knowledge of two interest rates before it can be calculated.[2] Called the prospective value (PV) this measure of investment desirability considers the fact that in practice the rate representing a minimum desirable return from investments is nearly always greater than the rate at which money can be invested so that it is available on demand. For a proposal j with life n the prospective value is expressed as

$$PV(\bar{m}, i_\delta)_j = F_{j_o}(1 + i_\delta)(1 + \bar{m})^{n-1} + \sum_{t=1}^{n} F_{jt}(1 + \bar{m})^{n-t}$$

where $\bar{m}$ is an average marginal rate at which it is anticipated the net receipts will be invested in the future and i_δ is a rate at which highly liquid funds can be invested i.e., a bank savings account.

The basic assumption that is unique to this approach is the supposition that if proposal j is not accepted the money available for investment, F_{j_o}, will be invested at an interest rate i_δ for one period and then withdrawn and invested at the rate $\bar{m}$ for the remainder of the $n - 1$ periods. Thus, the future-worth of this opportunity foregone by the acceptance of proposal j is represented by the term $F_{j_o}(1 + i_\delta)(1 + \bar{m})^{n-1}$. The other terms $\sum_{t=1}^{n} F_{jt}(1 + \bar{m})^{n-t}$ represent the future-worth of the money to be realized from the investment of F_{j_o} dollars if proposal j is undertaken and the net receipts from that proposal are invested at rate $\bar{m}$ until the end of its life.

The sum of these two parts represent the net future-worth of accepting proposal j. That is, if proposal j is undertaken the future-worth to be derived is $\sum_{t=1}^{n} F_{jt}(1 + \bar{m})^{n-1}$. This amount must be reduced by the future-worth of the opportunity foregone when proposal j is accepted. The future-worth of the foregone opportunity is $F_{j_o}(1 + i_\delta)(1 + \bar{m})^{n-1}$. Since F_{j_o} is normally a disbursement and therefore negative, this reduction is accomplished simply by summing the two amounts.

By following this line of reasoning it is seen that if $PV(\bar{m}, i_\delta)_j > 0$ the opportunity associated with the investment of money in proposal j is worth more than the opportunity that is to be foregone. Similarly, if $PV(\bar{m}, i_\delta)_j < 0$ it is more profitable to invest in highly liquid funds for one period and then invest those proceeds in the future at an anticipated rate of interest $\bar{m}$.

In order to simplify this basis for comparison it is generally easier to compute this measure discounted to the present rather than examining it as a future-worth amount. Thus, we can write

$$P(\bar{m}, i_\delta)_j = PV(\bar{m}, i_\delta)_j(\overset{P/F\,\bar{m},\,n}{}) = PV(\bar{m}, i_\delta)_j[1/(1 + \bar{m})^n]$$

---

[2]Oakford, R. V., and Thuesen, G. J., "The Maximum Prospective Value Criterion," *The Engineering Economist*, Vol. 13, No. 3, 1968.

$$P(\bar{m}, i_\delta)_j = [F_{jo}(1 + i_\delta)(1 + \bar{m})^{n-1} + \sum_{t=1}^{n} F_{jt}(1 + \bar{m})^{n-t}][1/(1 + \bar{m})^n]$$

$$P(\bar{m}, i_\delta)_j = F_{jo}\frac{(1 + i_\delta)}{(1 + \bar{m})} + \sum_{t=1}^{n} F_{jt}(1 + \bar{m})^{-t}.$$

It is seen from this expression that the present-worth amount which is defined as

$$PW(i)_j = \sum_{t=0}^{n} F_{jt}(1 + i)^{-t}$$

is a special case of $P(\bar{m}, i_\delta)_j$ when $\bar{m}$ and $i_\delta = i$. When $\bar{m} = i$ and $\bar{m} > i_\delta$ the only difference between the present-worth amount and $P(\bar{m}, i_\delta)_j$ is in the first term. For $P(\bar{m}, i_\delta)_j$ this term is $F_{jo}[(1 + i_\delta)/(1 + \bar{m})]$ and for the present-worth amount it is F_{jo}.

Nevertheless, this difference is significant since $P(\bar{m}, i_\delta)_j$ can be greater than zero even if the rate of return i_j^* is less than the rate $\bar{m}$. When $P(\bar{m}, i_\delta)_j > 0$, the acceptance of proposal j produces more revenue than the opportunity that would be undertaken if proposal j was not accepted. For the present-worth amount, $PW(i)_j$, can never be greater than zero as long as the proposals' rates of return are less than i, the interest rate used in the present-worth calculation. (This statement is based on the assumption that the $PW_j(i)$ function for the proposals being considered is of the form described in Figure 6.4.) Of course, for proposals with $i^* > \bar{m}$ or i, $P(\bar{m}, i_\delta)_j$ and $PW(i)_j$ will always be greater than zero.

Under what conditions is a proposal with a rate of return in the interval $(i_\delta < i^* < \bar{m})$ deemed economically desirable? That is, what causes $P(\bar{m}, i_\delta)_j$ to be greater than zero for a proposal with $i_\delta < i^* < \bar{m}$? It is the speed with which the proposed investment promises to generate its returns; i.e., the faster a proposal liquidates itself the greater the chance of $P(\bar{m}, i_\delta)_j$ being greater than zero.

A simple example of this would be a proposal with $i^* = 8\%$ where the investment proposal is a cash flow with a single disbursement now and a single receipt one period from now. If $i_\delta = 5\%$ and $\bar{m} = 15\%$ it would be a sound decision to accept the proposal earning 8% rather than investing the money in a bank account at 5%. The 5% investment or the 8% investment are equally liquid since the money is available for investment in the very next period.

When the lives of the proposals are longer than one period the decision is not so obvious. For example Proposals *A*, *B*, and *C* shown in Table 6.8 have the same initial cost, the same rate of return (10%), but different lives. The values of $P(\bar{m}, i_\delta)_j$ for these proposals are also shown in Table 6.8. It is assumed that $i_\delta = 5\%$ and $\bar{m} = 15\%$.

Table 6.8. THREE PROPOSALS WITH VARYING LIQUIDITY

End of Year	Proposal *A*	Proposal *B*	Proposal *C*
0	−$1,000	−$1,000	−$1,000
1	576	402	315
2	576	402	315
3		402	315
4			315
$P_j(0.15, 0.05)$	$23	$5	−$14

PROBLEMS

1. An investor is considering a business opportunity which requires the receipts and disbursements shown below. Calculate the net cash flow representing this investment and then find the present worth of the net cash flow for an interest rate of 12%.

End of Year	Disbursements (*thousands*)	Receipts (*thousands*)
0	$100	$ 0
1	15	5
2	20	20
3	5	45
4	0	60
5	0	30

2. Al, Bill, and Charlie started a business, and the following financial transactions occurred:

End of Year	Transactions
0	A, B, and C each invested $10,000
1	B paid A $2,000; business paid C $3,000
2	C paid A $1,000; B paid A $1,000; business paid $2,000 each to B and C
3	Business paid $2,000 each to A, B, and C; B paid $8,000 each to A and C for their interest in business
4	B received $36,000 from the business for his interest in the business

Show the *net cash flow* each period for each individual and the business.

3. Given the cash flow shown below find the present worth amount as a function of interest rate and graph the results.

End of Year	0	1	2	3
Cash Flow	−$1,000	$500	$700	$400

(a) Assume the interest rate is compounded annually.
(b) Assume the interest rate is compounded continuously.
(c) Repeat Problem 3 for the annual equivalent amount.
(d) Repeat Problem 3 for the future-worth amount.

4. Calculate and graph the present-worth amount as a function of the interest rate for the following cash flow:

End of Year	0	1	2	3	4	5
Cash Flow	−$10,000	$2,500	$2,500	$2,500	$2,500	$2,500

(a) Assume the interest rate is compounded annually.
(b) Assume the interest rate is compounded continuously.
(c) Repeat Problem 4 for the annual equivalent amount.
(d) Repeat Problem 4 for the future-worth amount.

5. For the following cash flow calculate and graph the present worth as a function of the interest rate:

End of Year	0	1	2	3	4	5	6	7
Cash Flow	−$50,000	$5,000	$8,000	$11,000	$14,000	$17,000	$20,000	$23,000

(a) Assume the interest rate is compounded annually.
(b) Assume the interest rate is compounded continuously.
(c) Repeat Problem 5 for the annual equivalent amount.
(d) Repeat Problem 5 for the future-worth amount.

6. Graph the present-worth amount as a function of interest rate for the following cash flows:

(a) End of Year	0	1	2	3	4
Cash Flow	−$8,000	$2,000	$2,000	$2,000	$2,000
(b) End of Year	0	1	2	3	4
Cash Flow	$1,000	$1,500	$2,000	−$2,500	−$3,000

7. Investment proposals A and B have the net cash flows shown below.

End of Year	0	1	2	3	4
A	−$100	$30	$30	$70	$60
B	−$100	$60	$60	$30	$30

(a) Compare the present worth of A with the present worth of B for $i = .05$. Which has the higher value?

(b) Now let $i = .15$ and compare the two. Which has the higher value?

(c) On the same axis, graph the present-worth amounts for each of these proposals as a function of the interest rate.

8. A company can invest in one of two mutually exclusive alternatives. The life of both alternatives is estimated to be 5 years with the following initial investments and salvage values:

Alternative	A	B
Investment	−$10,000	−$12,000
Salvage Value	1,500	3,500

(a) What is the capital recovery with return for each alternative?

(b) Determine the salvage value at the end of the project life for Alternative B which will result in the same capital recovery with return for both alternatives. Assume that the rate of interest is 15%.

9. A new college freshman wants to buy a scientific calculator for her engineering class. With a hope to resell the calculator at the time of graduation, she can purchase the calculator at $85 with an estimated salvage value of $25 at the end of 4 years. For an interest rate of 6%, what is the equivalent annual cost of this investment?

10. A firm requires power shovels for its open-pit mining operation. The mining equipment with a first cost of $250,000 has an estimated salvage value of $35,000 at the end of 10 years service. If the firm uses a rate of interest of 12% for the project evaluation, how much must be earned on an equivalent annual basis so that the firm recovers its invested capital plus earns a return on the capital committed to the equipment during its lifetime?

11. A man is considering giving an endowment to a university in order to provide payments of $5,000, $4,000, $3,000, and $2,000, respectively, at the end of the first, second, third, and fourth quarters during a year. If the interest rate is 12% compounded quarterly, what is the capitalized equivalent that must be deposited now so that the quarterly payments can be repeated forever?

12. A tunnel to transport water through a mountain range requires periodic maintenance. If the maintenance costs are as shown below for each 6-year mainte-

nance cycle, what is the capitalized equivalent of these expenses? The interest rate is 8%.

End of Year	1	2	3	4	5	6
Maintenance Costs	$35,000	$35,000	$35,000	$40,000	$40,000	$40,000

13. Consider the cash flows given below and assume that $i = 10\%$.

	End of Year				
	0	1	2	3	4
Cash Flow					
A	-100	$50	$50	$50	$50
B	-100	40	40	60	60

Calculate the present-worth amounts, annual equivalents, and future-worth amounts of these two cash flows. Next calculate PW_A/PW_B, AE_A/AE_B, and FW_A/FW_B. Then compare these ratios. What important observation can be made from this comparison?

14. An automobile owner is concerned about the increasing cost of gasoline. He feels that the cost of gasoline will be increasing at the rate of 5% per year over the present price of 16¢ per liter. His experience with his car indicates that he averages 9 kilometers per liter of gasoline. Since he expects to drive an average of 20,000 kilometers each year, what is the present worth of the cost of fuel for this individual for the next 4 years? What is the annual equivalent cost of fuel over this period of time?

15. Find the rate of return for the following cash flows by using the interest factors:

Year	0	1	2	3	4
(a)	$-$10,000	$5,000	$5,000	$5,000	$5,000
(b)	$-$ 200	100	200	300	400
(c)	$-$ 500	500	1,000	1,000	1,000
(d)	$-$ 100	25	25	25	25
(e)	$-$ 1,000	800	400	700	600

16. Graph the present-worth of the following cash flow as a function of interest rate. What is the rate of return for this cash flow?

(a) *End of Period*	*Net Cash Flow*
0	$-$1,000
1	2,230
2	$-$ 1,242

(b) *End of Period*	*Net Cash Flow*
0	−$ 70,000
1	170,000
2	− 102,300

17. Graph the present-worth amount as a function of the interest rate for each cash flow shown below. The range of interest-rate values for which these graphs should be drawn extends from 0% to 50%. From the graphs determine the rates of return for each cash flow.

			End of Year		
	0	1	2	3	4
Cash Flow					
(a)	− $4,000	$ 0	$ 0	$40,000	− $40,000
(b)	− 1,000	2,500	− 1,540		
(c)	−10,000	50,000	−93,500	77,500	− 24,024
(d)	− 1,000	3,900	− 5,030	2,145	
(e)	− 1,000	3,600	− 4,320	1,728	
(f)	− 1,000	1,000	− 100	1,000	

18. If a cash flow is composed of a series of disbursements in its early life followed by a series of receipts that total to an amount greater than the disbursements, what can be said about the relationship between rate of return and the present-worth amount?

19. Apply Descartes' rule of signs to the cash flows in Problem 17 and find the maximum number of real, positive roots that can be found for each cash flow.

20. Design a cash flow having a present-worth function that intersects the horizontal axis only once although Descartes' rule of signs indicates that the cash flow will have a maximum number of positive, real roots that exceeds one.

21. Calculate for the following cash flows the present worth amount and the prospective value, $PV(\bar{m}, i_\delta)$ for $i = \bar{m} = 15\%$ and $i_\delta = 6\%$. Use the interest factors to reduce the amount of computation required.

			End of Year		
	0	1	2	3	4
Cash Flow					
(a)	− $ 1,000	$ 350	$ 350	$ 350	$ 350
(b)	− 8,000	1,300	2,300	3,300	4,300
(c)	− 11,000	2,000	2,000	2,000	5,000
(d)	− 5,000	3,000	2,500	2,000	1,500
(e)	− 2,000	1,000	200	1,000	200

22. A proposal with an initial cost of $5,000 is expected to earn a rate of return of 15%. If a company is calculating the economic worth of this proposal using the prospective value for $\bar{m} = 20\%$ and $i_\delta = 5\%$, what is the longest time over which the proposal can receive equal annual payments and still appear economically desirable? What is the answer to the preceding question if the rate of return on the proposal is expected to be 8%?

23. If the prospective value, $PV(\bar{m}, i_\delta)_j$, for investment proposal j is (a) positive, (b) negative, what is indicated about the economic desirability of proposal j? The desirability of proposal j is measured with respect to what investment strategy?

24. As the difference between $\bar{m}$ and i_δ increases, what is the effect on the prospective value, $PV(\bar{m}, i_\delta)$? Does such an increase make a proposal appear more or less desirable? For the cash flow $-\$100, \$40, \$40, \$40, \$40$ as i_δ decreases, what happens to the prospective value?

DECISION MAKING
AMONG ALTERNATIVES

Once a set of investment alternatives has been submitted for consideration the question that arises is: What investment opportunity should be selected? To answer this question it is necessary to study various decision criteria that describe how investment decisions should be made.

A decision criterion is a rule or procedure that describes how to select investment opportunities so that particular objectives can be achieved. The degree to which these objectives are realized depends on the efficacy of the decision criterion. Therefore, it is important that the strengths and weaknesses of the most commonly used decision criteria be fully understood. To develop this understanding, this chapter examines decision criteria with respect to the fundamentals of economic comparison.

7.1. Types of Investment Proposals

An *investment proposal* in this book is considered to be a single project or undertaking which is being considered as an investment possibility. It is important to distinguish an investment proposal from an *investment alternative* which is defined to be a decision option. According to the above de-

7

finitions every investment proposal can be considered to be an investment alternative. However, an investment alternative can consist of a group or set of investment proposals. It can also represent the option of "doing nothing." Thus, if a decision maker is considering two proposals, A and B, it is possible for him to have four decision options or alternatives, as shown below:

Alternatives	X_A[1]	X_B
Do nothing (Reject A, Reject B)	0	0
Accept only A	1	0
Accept only B	0	1
Accept A and B	1	1

Independent proposals. When the acceptance of a proposal from a set of proposals has no effect on the acceptance of any of the other proposals contained in the set, the proposal is said to be *independent*. Even though few

---

[1]For each investment proposal there is a binary variable X_j that will have the value 0 or 1 indicating that proposal j is rejected (0) or accepted (1). Each row of binary numbers represents an investment alternative. This convention is used throughout this chapter.

proposals in a firm are truly independent, for practical purposes it is quite reasonable to assume that certain proposals are independent. For instance, the decision to air-condition a company's production facility would normally be considered to be independent of the decision to undertake an advertising campaign since each investment has a different function.

It is almost always possible to recognize some relationship between proposals that are functionally different. For example, it might be expected that the proposed investment to air-condition the plant will lead to a lower product cost through increased productivity. This decrease in cost may have an effect on the selling price of the product which in turn may affect consumer demand. The advertising campaign is also intended to influence consumer demand. Even though such relationships may exist between investment proposals, they are usually very difficult to trace and in many cases their effect is negligible. Therefore, unless such relationships are rather direct they are usually ignored. In addition, if the amounts of money available for investment in the production or marketing operations are not immediately transferable it is reasonable to assume that any financial dependency between the proposals can also be disregarded.

Usually, if proposals are functionally different and there are no other obvious dependencies between them it is reasonable to consider the proposals as independent. For example, proposals concerning the purchase of a numerically controlled milling machine, a security system, office furniture, and fork lift trucks would be considered independent under most circumstances.

Dependent proposals. For many decision problems a group of investment proposals will be related to one another in such a way that the acceptance of one of the proposals will influence the acceptance of the others. Such interdependencies between proposals occur for a variety of reasons.

First, if the proposals contained in the set of proposals being considered are related so that the acceptance of one proposal from the set precludes the acceptance of any of the other proposals in the set, the proposals are said to be *mutually exclusive*. Mutually exclusive proposals usually occur when a decision maker is attempting to fulfill a need and there are a variety of proposals each of which will satisfy that need.

For example a road building contractor may require additional earth moving capability. As shown in Table 7.1 there may be a number of types of equipment, each of which could perform the function desired. Although these proposals may have different first costs and different operating characteristics, they are still considered to be mutually exclusive for decision-making purposes since the selection of one precludes the selection of the others.

Another type of relationship between proposals arises from the fact that once some initial project is undertaken there are a number of other auxiliary

Table 7.1. A SET OF MUTUALLY EXCLUSIVE PROPOSALS

Proposal	Type of Equipment
A1	Roadgrader—Model B, Manufacturer 1
A2	Roadgrader—Model C, Manufacturer 1
A3	Roadgrader—Model A, Manufacturer 2
A4	Caterpillar Tractor—Model D, Manufacturer 3
A5	Caterpillar Tractor—Model D, Manufacturer 4

investments that become feasible as a result of the initial investment. Such auxiliary proposals are called *contingent* proposals because their acceptance is conditional on the acceptance of another proposal. Thus, the purchase of a computer magnetic tape drive unit is contingent on the purchase of the computer central processing unit. The construction of the third floor of a building is contingent on the construction of the first and second floors. A contingent relationship is a one-way dependency between proposals. That is, the acceptance of a contingent proposal is dependent on the acceptance of some prerequisite proposal but the acceptance of the prerequisite proposal is independent of the contingent proposals.

When there are limitations on the amount of money available for investment and the initial cost of all the proposals exceeds the money available for investment, financial interdependencies are introduced between proposals. These interdependencies are usually complex and they will occur whether the proposals are independent, mutually exclusive, or contingent. Thus, whenever a budget constraint is imposed on some decision problem it is important to realize that interdependencies which are not obvious are being introduced. For instance assume the three proposals shown in Table 7.2 are

Table 7.2. THREE INDEPENDENT PROPOSALS

End of Year	Cash Flow		
	Proposal *A*	Proposal *B*	Proposal *C*
0	−$1,000	−$3,000	−$5,000
1	600	1,500	2,000
2	600	1,500	2,000
3	600	1,500	2,000

independent. If the decision maker has only $5,500 to spend it is seen that the acceptance of Proposal *C* eliminates Proposals *A* and *B* from possible acceptance. In addition, the acceptance of Proposal *A* or Proposal *B* now precludes the acceptance of Proposal *C*. Depending on the budget size, the number of proposals, and their first costs these interdependencies can be rather complex. For the example just cited the possible arrangements of proposals that are feasible with regard to the budget constraint are shown in Table 7.3.

Table 7.3. ACCEPTABLE ARRANGEMENTS OF INDEPENDENT PROPOSALS FOR A
BUDGET CONSTRAINT OF $5,500

Possible Arrangements of Proposals	Proposals			Total Money Required for each Arrangement	Budget Remaining
	X_A	X_B	X_C		
1	0	0	0	0	$5500
2	1	0	0	$1,000	4500
3	0	1	0	3,000	2500
4	0	0	1	5,000	500
5	1	1	0	4,000	1500
6	1	0	1	6,000	− 500*
7	0	1	1	8,000	−2500*
8	1	1	1	9,000	−3500*

*Infeasible arrangements

7.2. Mutually Exclusive Alternatives and Decision Making

To facilitate the discussion of decision criteria in the next section the
selection of investment proposals will be viewed as a problem of selecting
a single economic alternative from a set of alternatives. Because the acceptance of one of these alternatives will preclude the acceptance of any of the
other alternatives being considered, the alternatives are considered to be
mutually exclusive.

Comparing mutually exclusive alternatives. When comparing mutually
exclusive alternatives it is the future *difference* between the alternatives that
is relevant for determining the economic desirability of one compared to the
other. It is this fundamental concept that is the basis for the discussion concerning decision criteria in this chapter.

The reason why the difference between alternatives is so fundamental
in the comparison of alternatives is demonstrated by the comparison of
alternatives $A1$ and $A2$ shown in Table 7.4. To compare the two alternatives

Table 7.4. DIFFERENCES BETWEEN MUTUALLY EXCLUSIVE ALTERNATIVES

End of Year	Alternative $A1$	Alternative $A2$	Cash Flow Difference $(A2 − A1)$
0	− $1,000	− $1,500	− $500
1	800	700	− 100
2	800	1,300	500
3	800	1,300	500

described in Table 7.4 it is sufficient to examine the cash flow that represents the difference between $A1$ and $A2$ because the advantage or disadvantage of Alternative $A2$ over Alternative $A1$ is completely described by the cash flow representing $A2$ subtracted from $A1$. The cash flow representing Alternative $A2$ can be viewed as being the sum of two separate and distinct cash flows as shown in Figure 7.1. One of these cash flows is identical to the cash flow of Alternative $A1$. The other is the cash flow representing the difference

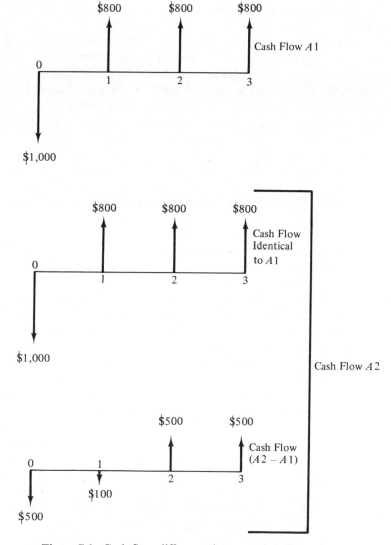

Figure 7.1. Cash flow difference between two alternatives.

between Alternative $A1$ and Alternative $A2$. To decide which of the two alternatives is economically superior it is sufficient to utilize the following simple decision rule:

> If cash flow $(A2 - A1)$ is economically *desirable*,
> Alternative $A2$ is preferred to Alternative $A1$.
> If cash flow $(A2 - A1)$ is economically *undesirable*,
> Alternative $A1$ is preferred to Alternative $A2$.

If the cash flow representing the differences between the alternatives is economically desirable, then Alternative $A2$ consisting of a cash flow that is the sum of a cash flow like Alternative $A1$ and a desirable cash flow is clearly economically superior to Alternative $A1$. On the other hand, if the difference between alternatives is considered to be undesirable, Alternative $A2$ is inferior to Alternative $A1$. The reader should confirm the reasonableness of this argument since it is fundamental to most decision processes.

For the example cited in Table 7.4, the decision to undertake Alternative $A2$ rather than Alternative $A1$ requires an additional or incremental investment of $500 now and $100 one year hence. The extra receipts expected from the extra investment are $500 at the end of years 2 and 3. Do the extra receipts justify the extra investment? This is the question that must be answered to determine which of the two alternatives is economically more desirable.

Forming mutually exclusive alternatives. In Section 7.1 it was pointed out that investment proposals can be independent, mutually exclusive, or contingent and that additional interdependencies between proposals can result if there is a limited amount of money to invest. To devise special rules to incorporate each of these different relationships in a decision criterion would produce a complicated, difficult to apply criterion.

In order to provide a simple method of handling the various types of proposals and to provide some insight into mathematical programming formulations of the decision problem a general approach is undertaken. This approach requires that all investment *proposals* be arranged so that the selection decision involves the consideration of the cash flows of *mutually exclusive alternatives* only.

All that is required is the enumeration of all the feasible combinations of the proposals under consideration. Each combination of proposals represents a mutually exclusive alternative since each combination is unique and the acceptance of one combination of proposals precludes the acceptance of any of the other combinations. The cash flow of each alternative is determined simply by adding, period by period, the cash flows of each proposal contained in the alternative being considered.

For example consider the three independent investment proposals described in Table 7.2. Table 7.5 shows the cash flow that would be realized for each mutually exclusive alternative produced from those proposals. As previously explained, each row of binary numbers represents a mutually exclusive alternative where the value of X_j indicates whether proposal j is rejected (0) or accepted (1).

Table 7.5. MUTUALLY EXCLUSIVE ALTERNATIVES FOR THE THREE INDEPENDENT PROPOSALS IN TABLE 7.2

Mutually Exclusive Alternatives	Proposals			Cash Flow			
	X_A	X_B	X_C	End of Year			
				0	1	2	3
1	0	0	0	$ 0	$ 0	$ 0	$ 0
2	1	0	0	−1,000	600	600	600
3	0	1	0	−3,000	1,500	1,500	1,500
4	0	0	1	−5,000	2,000	2,000	2,000
5	1	1	0	−4,000	2,100	2,100	2,100
6	1	0	1	−6,000	2,600	2,600	2,600
7	0	1	1	−8,000	3,500	3,500	3,500
8	1	1	1	−9,000	4,100	4,100	4,100

Now suppose that the decision maker is considering two independent sets of mutually exclusive proposals. That is, proposals $A1$ and $A2$ are mutually exclusive while proposals $B1$ and $B2$ are mutually exclusive. However, the selection of any proposal from the set of proposals $A1$ and $A2$ is independent of the selection of any proposal from the set of proposals $B1$ and $B2$. For example, the decision problem may be to select one lathe where

Table 7.6. MUTUALLY EXCLUSIVE ALTERNATIVES FOR TWO INDE-PENDENT SETS OF MUTUALLY EXCLUSIVE PROPOSALS

Mutually Exclusive Alternatives	Proposals			
	X_{A1}	X_{A2}	X_{B1}	X_{B2}
1	0	0	0	0
2	1	0	0	0
3	0	1	0	0
4	0	0	1	0
5	0	0	0	1
6	1	0	1	0
7	1	0	0	1
8	0	1	1	0
9	0	1	0	1

two lathes are under consideration and to select one fork lift truck where two fork lift trucks are being considered. The decision options available to the decision maker can be represented as the set of mutually exclusive alternatives shown in Table 7.6.

If proposals are contingent it is again possible to arrange the proposals into a single set of mutually exclusive alternatives by enumerating the combination of proposals that are feasible considering the relationships that exist between the proposals. Suppose that three proposals are being considered where Proposal C is contingent on the acceptance of both proposals A and B, and Proposal B is contingent on the acceptance of Proposal A. Table 7.7 indicates the mutually exclusive alternatives that can be developed from a group of proposals with the described contingency relationships.

Table 7.7. MUTUALLY EXCLUSIVE ALTERNATIVES FOR CONTINGENT PROPOSALS

Mutually Exclusive Alternatives	Proposals		
	X_A	X_B	X_C
1	0	0	0
2	1	0	0
3	1	1	0
4	1	1	1

Now suppose there is a decision problem where there are independent, mutually exclusive and contingent proposals and in addition there is a limitation on the amount of money available for investment. As before, all the possible combinations of proposals will be listed and the cash flows for these mutually exclusive alternatives will be determined. The only additional procedure required to account for the constraint on the budget amount is the elimination of each mutually exclusive alternative that requires more money that the total amount available. An example of this procedure is demonstrated in Table 7.3 where a total budget of $5,500 was assumed for the three independent proposals in Table 7.2. By examining the first cost of each mutually exclusive alternative it is seen that Alternatives 6, 7, and 8 are infeasible and therefore should be eliminated from consideration.

For problems that concern a small number of proposals the general technique just presented for arranging various types of proposals into mutually exclusive alternatives can be computationally practical. However for larger numbers of proposals the number of mutually exclusive alternatives becomes quite large and therefore this approach becomes computationally cumbersome. For instance, the maximum number of mutually exclusive alternatives, N, that can be obtained where

$S =$ the maximum number of sets of proposals that are independent;
$M_j =$ the maximum number of proposals within each set j where each proposal in that set is mutually exclusive is

$$N = \prod_{j=1}^{s} (M_j + 1) = (M_1 + 1)(M_2 + 1)(M_3 + 1) \ldots (M_S + 1).$$

Suppose a decision maker had the following proposals to consider:

$$
\begin{array}{llllll}
A1 & A2 & A3 & A4 & A5 & A6 \\
B1 & B2 & B3 & & & \\
C1 & C2 & C3 & C4 & & \\
D1 & D2 & & & &
\end{array}
$$

where the proposals in each row are mutually exclusive and the set of proposals on each row is independent from any other set or row of proposals. That is, proposals $A1$ and $A2$ are mutually exclusive but proposals $A1$ and $B1$ are independent. The maximum number of mutually exclusive alternatives that can be obtained from the group of proposals in this example is found to be 420.

$$S = 4 \text{ (the number of rows)}$$
$$M_1 = 6, \ M_2 = 3, \ M_3 = 4, \ M_4 = 2$$

so that

$$N = (M_1 + 1)(M_2 + 1)(M_3 + 1)(M_4 + 1)$$
$$= (7)(4)(5)(3) = 420.$$

Thus, for this relatively small problem the number of mutually exclusive alternatives is sizable.

The important point that must be realized is that the approach just discussed makes possible the consideration of a variety of proposal relationships in a single form: the mutually exclusive alternative. Therefore, any decision criteria that are designed to make decisions about mutually exclusive alternatives can also handle proposals that are independent, mutually exclusive, or contingent merely by arranging the proposals into mutually exclusive alternatives. In addition, the imposition of a budget constraint can also be quite easily incorporated into such decision criteria. Thus, a conceptually simple and consistent approach has been presented encompassing a wide variety of investment situations.

7.3. Decision Criteria for Mutually Exclusive Alternatives

All the decision criteria discussed in this section have as their objective the maximization of equivalent profit given that all investment alternatives must yield a return that exceeds some *minimum attractive rate of return* (MARR). This cut-off rate is usually the result of a policy decision made by the management of the firm.

The minimum attractive rate of return can be viewed as a rate at which the firm can always invest since it has a large number of opportunities that yield such a return. Thus, whenever any money is committed to an investment proposal an opportunity to invest that money at the MARR has been foregone. For this reason the minimum attractive rate of return is sometimes considered to be an "opportunity" cost.

If this view of unlimited investment opportunities yielding a return at the MARR is extended into the future, it can be assumed that the proceeds produced by current investments can be invested at the minimum attractive rate of return. Under these circumstances the minimum attractive rate of return has been called a "reinvestment" rate since the future income received from current investments is thought to be invested or "reinvested" at this rate.

Selecting an interest rate. Over the years there has been much discussion of how to select the MARR. Unfortunately, there is yet to be offered a completely satisfactory method for precisely determining this rate. Because the rate that is selected represents the firm's profit objectives, it is usually based on the judgment of the firm's top management. This judgment is in turn based on the management's view of the firm's future opportunities along with the firm's financial situation.

If the MARR selected is too high, many investments that have good returns may be rejected. On the other hand, a rate that is too low may allow the acceptance of a large number of proposals, some of which are marginally productive or result in economic loss. Thus, when choosing a MARR there is a trade-off between being too selective or not being selective enough.

One method for selecting a MARR is to examine the proposals available for investment and to identify the maximum rate that can be earned if the funds are *not* invested in the proposals under consideration. For example, an individual should avoid selecting a MARR that is less than the interest rate banks are paying on savings accounts. This is because the individual always has the opportunity to invest at the bank rate regardless of his other investment opportunities. For this reason the Do Nothing alternative (which represents the return earned if all proposals under consideration are rejected) assumes that all available funds are invested at the MARR.

Another consideration in choosing a MARR relates to the rationing of the scarce resource, investment capital. For example, a large firm may want to assure that funds allocated to various divisions within the firm are used effectively. If there is considerable variance in the quality of proposals produced by one division compared to another, the appropriate MARR will prevent investing in unproductive proposals. This allows for the redistribution of the uninvested funds to the divisions that do have high return proposals.

This concept of capital rationing can also be applied to investment decisions to be made over some time span. The fact that there are business cycles produces fluctuations in the quality of investment proposals available at various points in time. The proper selection of the MARR can prevent investing in marginally productive proposals during the "down" years. These unspent funds can then be made available for financing the higher quality proposals that are available in the "up" years.

The minimum attractive rate of return should not be confused with the cost of capital: a composite rate that represents the cost of providing money from external sources through the sale of stock, the sale of bonds, and by direct borrowing. Normally the minimum attractive rate of return is substantially higher than the cost of capital. Where a firm's cost of capital may be 9% its minimum attractive rate of return may be 15%. This difference occurs because few firms are willing to invest in projects that are expected to earn slightly more than the cost of capital due to the risk elements in most projects and because of uncertainty about the future.

The Do Nothing alternative. In many engineering economy studies it is assumed that if the funds available are *not* invested in the projects being considered they will be invested in the Do Nothing alternative. The Do Nothing alternative does not mean that the funds would be "hidden under a mattress" thereby yielding no return. What it does mean is that the investor will "do nothing" about the projects being considered and that the available funds will be placed in investments that yield a rate of return equal to the MARR.

It was shown previously that the present-worth, annual equivalent, and future-worth amounts are zero for any cash flow yielding a rate of return equal to the interest rate at which these indexes are evaluated. Therefore, when the rate of return of any alternative equals the MARR (as assumed for the Do Nothing alternative) the equivalent profit will be zero. This fact simplifies the comparison of alternatives because the cash flow pattern of the Do Nothing alternative does not have to be known and it can be assumed for computational purposes that no cash flow is associated with the Do

Nothing alternative. This procedure is shown in Table 7.8 and it will lead to the proper conclusion.

Table 7.8. CASH FLOWS REPRESENTING FOUR MUTUALLY EXCLUSIVE ALTERNATIVES

End of Year	Alternatives			
	Do Nothing	A1	A2	A3
0	$0	− $5,000	− $8,000	− $10,000
1–10	0	1,400	1,900	2,500

Present-worth on total investment. This criterion is one of the most frequently used criteria for selecting an investment alternative from a set of mutually exclusive alternatives. Since the stated objective of the selection of alternatives is to choose the alternative with the maximum present-worth amount, the rules for this criterion are rather simple. All that is required is to calculate the present-worth amount for the cash flow representing each alternative. Then select the alternative that has the maximum present-worth amount provided this amount is positive. The present-worth amount must be positive to assure that the alternative yields a return that is greater than the minimum attractive rate of return.

To see the computational simplicity of this criterion, it is applied to the mutually exclusive alternatives described in Table 7.8. Using a MARR = 15% the calculations of the present-worth amounts gives

$$PW(15)_0 = \$ \quad 0.00$$

$$PW(15)_{A1} = -\$ \ 5,000 + \$1,400(\overset{P/A\ 15,\ 10}{5.0188}) = \$2,026.32$$

$$PW(15)_{A2} = -\$ \ 8,000 + \$1,900(\overset{P/A\ 15,\ 10}{5.0188}) = \$1,535.72$$

$$PW(15)_{A3} = -\$10,000 + \$2,500(\overset{P/A\ 15,\ 10}{5.0188}) = \$2,547.00.$$

It is seen that the maximum value of the present-worth amounts for these four alternatives is $2,547.00, the present-worth of Alternative A3. Although for this example the alternative selected happened to have the largest first cost, it is certainly possible for alternatives with the smaller first costs to have present-worths greater than those alternatives with the larger first costs. For example, if Alternative A3 is excluded from consideration it is seen that Alternative A1 has a larger present-worth than Alternative A2 even though it requires less initial outlay.

When the receipts from a number of alternatives are assumed to be equal, it is common to describe the cash flows of the alternatives by showing only their costs. If the costs are shown as positive numbers, then the decision rule for this criterion is to select the alternative that minimizes the present-worth amount of the costs. The Do Nothing alternative is not meaningful when cost only cash flows are being considered.

In Chapter 6 it is shown that the present-worth amount, the annual equivalent amount, and the future-worth amount are consistent bases for comparing alternatives. That is, if the present-worth amount of Alternative A is greater than the present-worth amount of Alternative B, then

$$AE(i)_A > AE(i)_B$$

and

$$FW(i)_A > FW(i)_B.$$

Therefore, if either the annual equivalent amount or the future-worth amount is substituted for the present-worth amount as the basis for comparison in this criterion the same conclusion will result. By applying the annual equivalent on total investment criterion or the future-worth on total investment criterion to the alternatives in Table 7.8 the selection of Alternative $A3$ is again selected as expected.

$$AE(15)_0 = \$\ 0$$

$$AE(15)_{A1} = -\$5,000(\overset{A/P\ 15,\ 10}{0.1993}) + \$1,400 = \$403.50$$

$$AE(15)_{A2} = -\$8,000(\overset{A/P\ 15,\ 10}{0.1993}) + \$1,900 = \$305.60$$

$$AE(15)_{A3} = -\$10,000(\overset{A/P\ 15,\ 10}{0.1993}) + \$2,500 = \$507.00.$$

or

$$FW(15)_0 = \$\ 0.00$$

$$FW(15)_{A1} = -\$5,000(\overset{F/P\ 15,\ 10}{4.046}) + \$1,400(\overset{F/A\ 15,\ 10}{20.304}) = \$8,195.60$$

$$FW(15)_{A2} = -\$8,000(\overset{F/P\ 15,\ 10}{4.046}) + \$1,900(\overset{F/A\ 15,\ 10}{20.304}) = \$6,209.60$$

$$FW(15)_{A3} = -\$10,000(\overset{F/P\ 15,\ 10}{4.046}) + \$2,500(\overset{F/A\ 15,\ 10}{20.304}) = \$10,300.00.$$

An examination of the calculations of the future-worth amounts indicates that the receipts from the investment are actually invested at the minimum attractive rate of return from the time they are received to the end of the life of the alternative. Thus it is said that future-worth calculations *explicitly* consider the investment or "reinvestment" of the future receipts generated by investment alternatives. Because the three decision criteria just discussed are consistent and lead to the same selection of alternatives, it follows that use of the present-worth amount and the annual equivalent amount *implicitly* assumes the investment or "reinvestment" of alternatives" receipts at the minimum attractive rate of return.

At first glance it appears that the criterion, present-worth on total investment, violates the basic decision rule that requires the consideration of the differences between alternatives. (See Section 7.2.) The fact that the differences between the alternatives are reflected in the comparison of the present-worths on total investment becomes clear in the following example. Suppose there are two mutually exclusive alternatives $A1$ and $A2$ as shown below.

End of Year	Alternative A1	Alternative A2	A2 − A1
0	−$1,000	−$1,500	−$500
1–5	500	900	400

Alternative $A2$ can be visualized as consisting of a cash flow identical to Alternative $A1$ plus the cash flow representing the difference between alternatives $A2$ and $A1$. The cash flow portion of $A2$ that is identical to $A1$ will have the same present-worth as Alternative $A1$. Therefore, the only difference between the present-worths for the total investment $A1$ and the total investment $A2$ is represented by the present-worth for the incremental cash flow $(A2 − A1)$. The realization that the present-worth on total investment criterion does in fact examine the differences in alternatives will be further substantiated in the discussion of the decision criterion, present-worth on incremental investment.

Present-worth on incremental investment. When making decisions between mutually exclusive alternatives it is the differences between alternatives that are relevant for decision-making purposes. The present-worth on incremental investment criterion provides an example of this rule since it requires that the incremental differences between alternative cash flows actually be calculated.

When comparing one alternative to another the first task is to determine the cash flow representing the difference between the two cash flows. Then the decision whether to select a particular alternative rests on the determination of the economic desirability of the additional increment of investment required by one alternative over the other. The incremental investment is considered to be desirable if it yields a return that exceeds the minimum attractive rate of return. In other words, if the present-worth amount for the incremental investment is greater than zero, the increment is considered desirable and the alternative requiring this additional investment is deemed best. As long as the present-worth function for the cash flow representing the difference between alternatives is of the form shown in Figure 6.4, any positive present-worth assures that such a cash flow will yield a return greater than the MARR.

If the cash flows representing the differences between alternatives do not have present-worth plots similar to the one shown in Figure 6.4, the rules of this criterion most probably will not be applicable. This deficiency is not generally serious since for most decision problems the cash flows are such that their differences consist of a disbursement followed by a series of receipts. It has been pointed out previously that such a cash flow pattern will produce a present-worth function of the form shown in Figure 6.4.

To apply this decision criterion to a set of mutually exclusive alternatives such as shown in Table 7.8 the following steps must be utilized:

1. List the alternatives in ascending order of their equivalent first cost or initial disbursements.

	Alternatives			
End of Year	*Do Nothing*	*A1*	*A2*	*A3*
0	$0	−$5,000	−$8,000	−$10,000
1–10	0	1,400	1,900	2,500

2. Select as the initial "current best" alternative the one which requires the smallest first cost. In most cases the initial "current best" alternative will be the alternative, Do Nothing, as it is in this example. All too frequently investment alternatives are compared without including the possibility of not undertaking the project at all. The exclusion of the Do Nothing alternative can lead to the investment of a scarce resource, money, in unproductive activities, that is, activities that yield a return that is less than the MARR.

3. Compare the initial "current best" alternative and the first "challenging" alternative. The challenger is always the next highest alternative in order of first cost that has not been previously involved in a comparison. The comparison is accomplished by examining the differences between the two cash flows. If the present-worth of the incremental cash flow evaluated at the MARR is greater than zero the challenger becomes the new "current best" alternative. If the present-worth is less than or equal to zero the "current best" alternative remains unchanged and the challenger in the comparison is eliminated from consideration. The new challenger is the next alter-

native in order of first cost that has not been a challenger previously. Then the next comparison is made between the alternative that is the "current best" and the alternative that is currently the challenger.

4. Repeat the comparisons of the challengers to the "current best" alternative as described in Step 3. These comparisons are continued until every alternative other than the initial "current best" alternative has been a challenger. The alternative that maximizes present-worth and provides a rate of return that exceeds the MARR is the last "current best" alternative.

Step 3 and 4 lead to the following calculations for the alternatives being considered in Table 7.8. Assume that the MARR is equal to 15%.

The first comparison to be made in this example is between Alternative $A1$ (the first challenger) and the Do Nothing alternative (the initial "current best" alternative). The subscript notation in $PW(15)_{A1-0}$ indicates the present-worth amount is for the cash flow representing the difference between Alternative $A1$ and Do Nothing. The Do Nothing alternative is signified by zero.

$$PW(15)_{A1-0} = -\$5,000 + \$1,400(\overset{P/A15,10}{5.0188}) = \$2,026.32.$$

Note that when comparing an alternative to the Do Nothing alternative the cash flow representing the incremental investment is the same as the cash flow on the total investment. Thus, for such a situation the calculations of the present-worth on total investment criterion are exactly the same as for this criterion.

Because the present-worth amount of the differences between the cash flows is greater than zero ($2,026.32), Alternative $A1$ becomes the new "current best" alternative as dictated by Step 3. The second challenger becomes Alternative $A2$. Alternative $A2$ is then compared to Alternative $A1$ on an incrementel basis as follows:

$$PW(15)_{A2-A1} = -\$3,000 + \$500(\overset{P/A\ 15,\ 10}{5.0188}) = -\$490.60.$$

Since this value is negative Alternative $A2$ is dropped from further consideration and Alternative $A1$ remains the "current best" alternative. The third challenger is Alternative $A3$. Comparing the "current best" with the next challenger yields

$$PW(15)_{A3-A1} = -\$5,000 + \$1,100(\overset{P/A\ 15,\ 10}{5.0188}) = \$520.68.$$

The present-worth on the additional investment required by Alternative $A3$ over Alternative $A1$ is positive and therefore that increment is economically desirable. Thus, Alternative $A3$ becomes the "current best" alternative and the list of alternatives has been exhausted so that there is no new challenger possible. According to Step 4 when all challengers have been considered, the "current best" alternative is the alternative that maximizes present-worth and provides a return greater than the MARR. Therefore, Alternative $A3$ is the optimum selection from the set of alternatives shown in Table 7.8.

The present-worth on incremental investment criterion is also appropriate for comparing alternatives that are described by cash flows that exclude the alternative's earnings. Table 7.9 presents the cash flows for a set of

Table 7.9. NET CASH FLOWS FOR FOUR ALTERNATIVES PROVIDING THE SAME SERVICE

End of Year	Alternatives			
	B1	B2	B3	B4
0	−$10,000	−$12,000	−$12,000	−$15,000
1	−2,500	−1,500	−1,200	−400
2	−2,500	−1,500	−1,200	−400
3	1,000*	1,500	1,500	3,000

*Positive values can arise as the result of the salvage value received when the asset is sold at the end of its life.

alternatives where it is assumed that the services provided by the alternatives are identical. Since these cash flows do not reflect the income produced by these alternatives, the Do Nothing alternative is not a meaningful alternative, i.e., no one would ever accept a cash flow that resulted in only disbursements if the opportunity to not invest was available. Thus, it is assumed that it is mandatory to select one of the alternatives listed in Table 7.9.

Applying the present-worth on incremental investment criterion to these alternatives produces the following calculations for the MARR $= 10\%$.

$$PW(10)_{B2-B1} = -\$2,000 + \$1,000(\overset{P/A\,10,\,2}{1.7355}) + \$500(\overset{P/F\,10,\,3}{0.7513}) = \$111.15.$$

Since this value is positive Alternative $B2$ becomes the "current best" alternative and Alternative $B1$ is eliminated from further consideration. The next comparison pits Alternative $B3$ against Alternative $B2$ as

$$PW(10)_{B3-B2} = \$0 + \$300(\overset{P/A\,10,\,2}{1.7355}) + \$0 = \$520.65.$$

Thus, Alternative $B3$ is accepted and Alternative $B2$ is dropped from consideration. Now compare Alternative $B4$ to Alternative $B3$ yielding

$$PW(10)_{B4-B3} = -\$3,000 + \$800(\overset{P/A\,10,\,2}{1.7355}) + \$1,500(\overset{P/F\,10,\,3}{0.7513}) = -\$484.65.$$

The present-worth of the increment $(B4 - B3)$ is negative and therefore Alternative $B4$ is unacceptable and it is eliminated from further consideration. Since there are no more alternatives to become challengers the decision process is completed and Alternative $B3$, the last "current best" alternative, is the optimum selection.

The above calculations do consider each increment of investment but compared to the present-worth on total investment criterion the computations are more time consuming. It has been shown by example that both criteria lead to the same solution. In fact, it is easy to prove generally that both methods will yield the same selection of alternatives. All that is required is to show that the present-worth of any Alternative B minus the present-worth of any Alternative A is equal to the present-worth of the difference between Alternative A and B. That is, demonstrate that

$$PW(i)_B - PW(i)_A = PW(i)_{B-A}$$

By definition

$$PW(i)_j = \sum_{t=0}^{n} F_{jt}(1 + i)^{-t}$$

so that

$$
\begin{aligned}
PW(i)_B - PW(i)_A &= \sum_{t=0}^{n} F_{Bt}(1 + i)^{-t} - \sum_{t=0}^{n} F_{At}(1 + i)^{-t} \\
&= F_{B0} - F_{A0} + F_{B1}(1 + i)^{-1} - F_{A1}(1 + i)^{-1} + \dots \\
&\quad + F_{Bn}(1 + i)^{-n} - F_{An}(1 + i)^{-n} \\
&= F_{B-A,0} + F_{B-A,1}(1 + i)^{-1} + \dots + F_{B-A,n}(1 + i)^{-n} \\
&= \sum_{t=0}^{n} F_{B-A,t}(1 + i)^{-t} \\
&= PW(i)_{B-A}.
\end{aligned}
$$

Thus, the relationship between the decision rules of the two present-worth criteria should be evident. If the objective is to maximize present-worth and $PW(i)_B > PW(i)_A$, the present-worth on total investment criterion says to select Alternative B. If $PW(i)_B > PW(i)_A$ then $PW(i)_{B-A}$ must be positive and the decision rule for the present-worth on incremental investment criterion is to accept B rather than A if $PW(i)_{B-A} > 0$.

The calculation of the present-worths on total investment for the alternatives described in Table 7.9 demonstrates that the incremental approach does select the alternative that minimizes total cost.

$$PW(10)_{B1} = -\$10,000 - \$2,500(\overset{P/A\ 10,\ 2}{1.7355}) + \$1,000(\overset{P/F\ 10,\ 3}{0.7513}) = -\$13,587.45$$

$$PW(10)_{B2} = -\$12,000 - \$1,500(\overset{P/A\ 10,\ 2}{1.7355}) + \$1,500(\overset{P/F\ 10,\ 3}{0.7513}) = -\$13,476.30$$

$$PW(10)_{B3} = -\$12,000 - \$1,200(\overset{P/A\,10,\,2}{1.7355}) + \$1,500(\overset{P/F\,10,\,3}{0.7513}) = -\$12,955.65$$

$$PW(10)_{B4} = -\$15,000 - \$400(\overset{P/A\,10,\,2}{1.7355}) + \$3,000(\overset{P/F\,10,\,3}{0.7513}) = -\$13,440.30$$

As has been pointed out for the present-worth on total investment criterion, the substitution of the annual equivalent amount or the future-worth amount for the present-worth amount as the basis for comparison for incremental decision making will lead to consistent solutions. The relationships shown below confirm this fact and they can be proved in a manner similar to that used for the present-worth amount.

$$AE(i)_B - AE(i)_A = AE(i)_{B-A}$$

and

$$FW(i)_B - FW(i)_A = FW(i)_{B-A}$$

Using the incremental approach and the annual equivalent amount requires the following calculations for the set of alternatives presented in Table 7.8.

$$AE(15)_{A1-0} = -\$5,000(\overset{A/P\,15,\,10}{0.1993}) + \$1,400 = \$403.75$$

$$AE(15)_{A2-A1} = -\$3,000(\overset{A/P\,15,\,10}{0.1993}) + \$500 = -\$97.75$$

$$AE(15)_{A3-A1} = -\$5,000(\overset{A/P\,15,\,10}{0.1993}) + \$1,100 = \$103.75$$

By following the decision rules for an incremental analysis the optimum solution is to select Alternative *A*3. This is the same decision given by the total investment criteria in the preceeding section.

Rate of return on incremental investment. This particular decision criterion is based on the same type of incremental analysis applied to the previously discussed criterion, present-worth on incremental investment. The only difference in the decision rules between these two criteria is the decision rule in Step 3 that determines whether an increment of investment is economically desirable. For the rate of return on incremental investment criterion the increment of investment is considered desirable if the rate of return resulting from the increment is greater than the minimum attractive rate of return ($i_{B-A}^* >$ MARR). Figure 7.2 shows the present-worth function of the increment $B - A$ and as long as it has the general form shown in Figure 7.2 the decision rules for both criteria are consistent.

To apply rate of return on an incremental basis it is first necessary to rank the alternatives in order by increasing equivalent first cost and then to select the initial "current best" alternative. Using the set of alternatives in

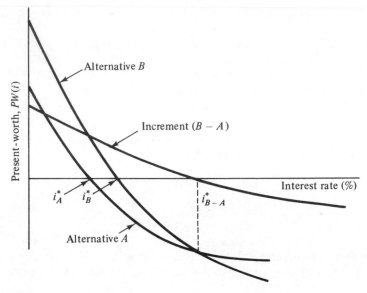

Figure 7.2. The present-worth function of an increment of investment.

Table 7.8, Steps 3 and 4 of the incremental analysis procedure require the following calculations. Find the value i^* of i so that the equation representing the present-worth of the incremental cash flow is set equal to zero. Again the MARR is assumed to be 15%. For increment $A1 - 0$

$$0 = -\$5,000 + \$1,400(\overset{P/A\,i,\,10}{})$$
$$i^*_{A1-0} = 25.0\%.$$

Because the rate of return on the increment is greater than the MARR, Alternative $A1$ becomes the initial "current best" alternative and the Do Nothing alternative is dropped from further consideration. Next compare Alternative $A2$ to Alternative $A1$. For the increment $(A2 - A1)$

$$0 = -\$3,000 + \$500(\overset{P/A\,i,\,10}{})$$
$$i^*_{A2-A1} = 10.5\%.$$

Because the rate of return of this increment is less than the MARR, Alternative $A1$ remains the "current best" and Alternative $A2$ is rejected. Then compare Alternative $A3$ to Alternative $A1$, the "current best" alternative. For increment $A3 - A1$

$$0 = -\$5,000 + \$1,100(\overset{P/A\,i,\,10}{\quad})$$

$$i^*_{A3-A1} = 17.6\%.$$

Alternative $A3$ becomes the "current best" alternative and Alternative $A1$ is removed from consideration. Since all the alternatives have been compared, Alternative $A3$, the last "current best" alternative, is the optimum solution. This is the same solution given by the present-worth on total investment criterion and the present-worth on incremental investment criterion.

In general, the three criteria discussed to this point will yield identical solutions for most types of investment decision problems. If present-worth plots other than the general form shown in Figure 7.2 occur for the cash flows being considered these three criteria could lead to inconsistent results unless a more detailed analysis is undertaken. For most problems encountered in practice the three criteria that have been discussed will select the alternative that maximizes total present-worth and which provides a return greater than the MARR.

One very important point to recognize from this example is that selecting the alternative with the highest rate of return on its total cash flow may *not* lead to the alternative that will maximize the total present-worth at the MARR. The rates of return for the *total* cash flows of the alternatives in Table 7.8 are

$$i^*_0 = 15\%, \quad i^*_{A1} = 25\%, \quad i^*_{A2} = 19.9\%, \quad i^*_{A3} = 21.9\%.$$

If Alternative $A1$ is selected because it has the maximum rate of return, the total present-worth will *not* be maximized for a MARR $= 15\%$. It has already been shown that Alternative $A3$ will maximize the present-worth for that minimum attractive rate of return.

The reason for the inconsistency introduced by examining the rates of return on total investment can be demonstrated in a number of ways. First, the relationship between rate of return on total cash flows and rate of return on incremental cash flows is not the same as the relationship between present-worth amounts. That is, for two alternatives A and B

$$i^*_B - i^*_A \text{ does not necessarily equal } i^*_{B-A}.$$

Therefore, even though using rate of return on an incremental basis does produce the optimum solution there is no relationship between the incremental approach and the total cash flow approach to assure an optimum solution from the latter approach.

Second, a review of the decision making procedures for the incremental approach shows that the acceptance of the increment of investment ($A3$ —

A1) yields a rate of return of 17.6%. This rate is higher than the MARR of 15% but it is lower than the rate of return of 25% that would be earned if Alternative A1 is accepted. Since it is always desirable to continue investing additional funds as long as they earn more than the MARR, the increment (A3 − A1) is acceptable. Therefore, Alternative A3 is more desirable than Alternative A1 since a return of 25% is earned on the portion of its cash flow that is identical to Alternative A1 and a 17.6% return is earned on the portion of its cash flow represented by the increment (A3 − A1). It is expected that the combination of these two cash flows would yield a return less than 25% and greater than 17.6%. In fact, Alternative A3 has a rate of return of 21.9% for its total cash flow.

Third, Figure 7.3 indicates a situation in which the rate of return on total investment produces a solution contrary to the objective of maximizing present-worth for a particular MARR. For any MARR that is less than the interest rate at which the present-worths of Alternative A and B are

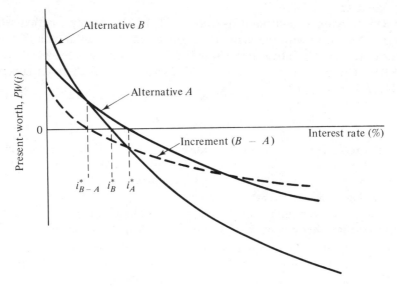

Figure 7.3. Rates of return on total investment and incremental investment.

equal, (i^*_{B-A}), Alternative B maximizes the present-worth amount. However, the rates of return for alternatives A and B indicate that i^*_A is greater than i^*_B, a contradiction to our objective to maximize present-worth for the MARR.

The application of rate of return on incremental investment when the income cash flows are assumed to be equal requires no changes in the criterion's decision rules. For example, the set of alternatives described in Table 7.9 are analyzed in the following way. After placing the alternatives in order of ascending first cost, each increment of investment is compared to the "current best" alternative. In this case the initial "current best" alternative is Alternative $B1$ and the MARR is 10%

$$0 = PW(i)_{B2-B1} = -\$2,000 + \$1,000(\overset{P/A\,i,\,2}{}) + \$500(\overset{P/F\,i,\,3}{})$$

$i^*_{B2-B1} = 13.5\%$ (B_2 becomes "current best" alternative)

$$0 = PW(i)_{B3-B2} = \$0 + \$300(\overset{P/A\,i,\,2}{}) + \$0$$

$i^*_{B3-B2} = \infty$ (B_3 becomes "current best" alternative)

$$0 = PW(i)_{B4-B3} = -\$3,000 + \$800(\overset{P/\,Ai,\,2}{}) + 1,500(\overset{P/F\,i,\,3}{})$$

$i^*_{B4-B3} = 1.5\%$ (B_3 remains "current best" solution).

Alternative $B3$ is the optimum selection as has been previously shown.

7.4. Applying Decision Criteria When Money Is Limited

Up to now the decision criteria that have been discussed have been applied to sets of mutually exclusive alternatives and it has been assumed that sufficient money is available to undertake all of the proposals. Now it will be assumed that the total money available for investment is a fixed amount thus restricting the number of feasible alternatives.

To illustrate the technique required to incorporate a budget constraint in the decision process, the present-worth on total investment criterion is applied to the proposed investments described in Table 7.10. Each set of proposals with the same letter designation ($A1$, $A2$) is considered to be independent of the other sets having different letter designations ($B1$, $B2$).

Table 7.10. CASH FLOWS FOR FIVE INVESTMENT PROPOSALS

Proposal	First Cost	Net Income (Years 1–10)	Salvage Value (Year 10)	$PW(8)_j$
$A1$	$-\$10,000$	\$2,000	\$1,000	\$3,883
$A2$	$-12,000$	2,100	2,000	3,018
$B1$	$-20,000$	3,100	5,000	3,117
$B2$	$-30,000$	5,000	8,000	7,255
$C1$	$-35,000$	4,500	10,000	-173

Therefore, no more than one proposal with the same letter can be accepted, but proposals having different letters can be accepted together. The problem is to select the proposal or proposals that maximize total present-worth if the amount of money available for investment is $35,000 and the minimum attractive rate of return equals 8%.

No changes at all are required in order to include the budget constraint in the decision process. All that is required is that the proposals be rearranged into mutually exclusive alternatives as described in Section 7.2. To reduce the number of mutually exclusive combinations that must be considered it is usually worthwhile to first calculate the present-worth for the cash flows of each proposal. If any proposal has a present-worth that is not positive it can be eliminated from consideration immediately. Clearly, any proposal with a negative present-worth when combined with other proposals to make a mutually exclusive alternative will reduce the total present-worth of that alternative.

The present-worth amounts at 8% for each proposal are shown in Table 7.10 in the far right-hand column. All the present-worths are positive except for Proposal C1 which can be immediately dismissed from further consideration. The remaining proposals are arranged into mutually exclusive alternatives and these alternatives along with their cash flows and present-worth amounts are shown in Table 7.11.

Next, those alternatives that have a first cost that exceeds the budget amount of $35,000 must be dropped from consideration. Thus, the alternative to accept A1 and B2 and the alternative to accept A2 and B2 are eliminated since they require more funds than are available.

The last step is to select the remaining alternative that maximizes the present-worth. It is seen that Alternative 5 representing the acceptance of Proposal B2 meets this objective with a present-worth equal to $7,255. Thus, the optimum solution to this problem is to select Proposal B2 and to reject all the other proposals. It is assumed that the $5,000 that is still available for investment after Alternative B2 is undertaken will be invested at the minimum attractive rate of return.

Usually, the present-worth on total investment criterion is the most efficient criterion to use when solving this type of problem by enumerating all the mutually exclusive alternatives. It is important to realize that it would be just as easy and just as correct to use annual equivalent or future-worth on the basis of total investment.

In addition, the incremental approaches described for present-worth, annual equivalent, future-worth, and rate of return would also produce the same solution. The only difference from the approach used in these examples would be that the mutually exclusive alternative formed from the proposals under consideration must be placed in order of ascending first cost. In Table 7.11 this would mean switching Alternatives 7 and 8. Then additional cal-

Table 7.11. CASH FLOWS FOR THE MUTUALLY EXCLUSIVE ALTERNATIVES FROM PROPOSAL IN TABLE 7.10

Alterna-tives	Proposals				Proposals Accepted	First Cost	Net Income Years 1-10	Salvage Value Year 10	$PW(8)_j$
	X_{A1}	X_{A2}	X_{B1}	X_{B2}					
1	0	0	0	0	None	$ 0	$ 0	$ 0	$ 0
2	1	0	0	0	A1	-10,000	2,000	1,000	3,883
3	0	1	0	0	A2	-12,000	2,100	2,000	3,018
4	0	0	1	0	B1	-20,000	3,100	5,000	3,117
5	0	0	0	1	B2	-30,000	5,000	8,000	7,255**
6	1	0	1	0	A1, B1	-30,000	5,100	6,000	7,000
7	1	0	0	1	A1, B2	-40,000	7,000	9,000	11,138*
8	0	1	1	0	A2, B1	-32,000	5,200	7,000	6,135
9	0	1	0	1	A2, B2	-42,000	7,100	10,000	10,273*

*Infeasible because the alternative requires more money than is available for investment ($35,000).

**Alternative that maximizes present-worth for a limited budget.

culations would be required to find the incremental cash flows representing the differences between the mutually exclusive alternatives. The decision rules would be exactly the same as those used in the incremental analysis of the examples in Section 7.4.

Linear integer programming formulation of the budget constraint prob-lem. It is obvious that when there are large numbers of proposals under consideration the number of mutually exclusive alternatives that can be formed is quite large. (See Section 7.2.) For instance, the number of mutually exclusive alternatives that can be formed from 100 independent proposals is approximately 1.268×10^{30}. Therefore, to solve many problems of practical significance it is necessary to be able to utilize mathematical techniques that can consider all possible alternatives without having to make calculations for each alternative.

One such technique is linear integer programming. This technique requires that the problem under consideration be formulated according to a particular format. Once the problem has been properly formulated there are a number of solution procedures or algorithms that are available to solve problems of such a structure. These algorithms usually converge to the optimum solution in a highly efficient manner. Many of these algorithms are available as computer programs and the problem solutions generally require only a few minutes of computer time. Because an analysis of the mathematics of these algorithms is beyond the scope of this book, the emphasis of this discussion is on the formulation of the investment decision problem as a linear integer programming problem.

The general format of the linear integer programming problem is written as follows:

$$\text{Maximize } Z = c_1 x_1 + c_2 x_2 + \ldots + c_n x_n$$

subject to

$$a_{11} x_1 + a_{12} x_2 + \ldots + a_{1n} x_n \leq b_1$$
$$a_{21} x_1 + a_{22} x_2 + \ldots + a_{2n} x_n \leq b_2$$
$$\vdots \qquad \vdots \qquad \vdots$$
$$a_{m1} x_1 + a_{m2} x_2 + \ldots + a_{mn} x_n \leq b_m$$
$$x_i \text{ for } i = 1, 2, \ldots n \text{ must be an integer} \geq 0$$

The c's, a's, and b's are constants and the x's are the decision variables representing the values to be determined. The solution to this type of problem is given by the values of the x's such that Z will be maximized and all the constraints (the equations under "subject to") are satisfied.

The problem of making decisions between alternatives is easily converted to the linear integer programming format. The decision variables X_j rather than being any integer greater than or equal to zero are confined to the integers 0, (reject Proposal j) or 1, (accept Proposal j). The value of c_j is the present-worth of Proposal j at the minimum attractive rate of return, $PW(MARR)_j$. The constraints for this type of problem are the result of two different types of relationships. One type of constraint reflects the limitation of the amount of money available for investment. That is, the total first cost of the proposals undertaken must not exceed the amount of the budget. The second type of constraint reflects the relationships between proposals such as whether the proposals are mutually exclusive, independent, or contingent. Thus, for the decision problem related to the proposals in Table 7.10 the integer programming formulation is as follows:

Maximize $Z = \$3,883 X_{A1} + \$3,018 X_{A2} + \$3,117 X_{B1} + \$7,255 X_{B2} - \$173 X_{C1}$

subject to

(budget constraint)

$\$10,000 X_{A1} + \$12,000 X_{A2} + \$20,000 X_{B1} + \$30,000 X_{B2}$

$$+ \$35,000 X_{C1} \leq \$35,000$$

(mutually exclusive) constraint

$$X_{A1} + X_{A2} \leq 1$$

(mutually exclusive) constraint

$$X_{B1} + X_{B2} \leq 1$$

(zero—one constraint) $X_j = 0$ or 1 for all proposals.

Solving this problem with existing techniques yields the solution[2]

$$X_{A1} = 0, \; X_{A2} = 0, \; X_{B1} = 0, \; X_{B2} = 1, \; X_{C1} = 0.$$

Thus, Proposal $B2$ is accepted and all the other proposals are rejected as previously shown in Table 7.11. The value of Z for this solution is $\$7,255$.

7.5. Other Decision Criteria

There exist decision criteria other than those previously discussed for selecting alternatives from a set of alternatives. Two of these criteria are discussed in this section to illustrate some significant differences that exist

---

[2]Discussion of other linear programming models appears in Section 19.2, and techniques for solving integer linear programming problems are presented in *Introduction to Operations Research*, 2nd ed., by Hillier and Lieberman, Holden-Day, Inc., 1974, Chapter 17.

between the various criteria that are used in economic analyses. The two criteria that are examined are the rank on rate of return criterion and the maximum prospective value criterion.

Rank on rate of return. The decision rules of this criterion are accurately described by its name: rank on rate of return. The rate of return is calculated for each proposal and then the proposals are ranked in descending order of rate of return. The proposal with the highest rate of return is ranked first, the proposal with the second highest second, etc. The decision rule is to move down the ranked proposals accepting each proposal until there are no more proposals with a rate of return greater than the MARR. Although this criterion is widely used it has some important deficiencies.

For example, ranking on the rate of return will guarantee to select the set of proposals that maximize the total present-worth amount only if all the proposals are independent and there is no limitation on the money available for investments. The fact that rank on rate of return is a reliable criterion only for the conditions just described is easily demonstrated by a comparison with the decision criterion, rate of return on incremental investment.

First, it must be recognized that since there are no interdependencies between the proposals the decision about each proposal is independent of the decisions regarding any of the other proposals. Associated with each independent decision are two mutually exclusive alternatives: accept the proposal or reject the proposal (Do Nothing). Thus, if rate of return is applied to such a set of proposals on an incremental basis, each proposal can be independently accepted or rejected on the basis of the rate of return on its increment of investment. This increment of investment represents the difference between accepting the proposal and doing nothing. As a result, the rate of return on the increment of investment is the *same* as the rate of return on total investment.

Recall that the decision rule for rate of return on incremental investment is to accept each increment of investment as long as its rate of return exceeds the MARR. This decision rule is effectively the same as the rank on rate of return decision rule that says to accept every proposal with a total investment rate of return greater than the MARR. Therefore, under the stated conditions it is seen that rank on rate of return is the same as rate of return on incremental investment. It has been previously shown that the latter criterion does guarantee an optimum solution as long as the cash flows have a normal present-worth function.

However, if mutually exclusive relationships are introduced or a budget constraint limits the money available for investment this ranking scheme faulters. As soon as interdependencies exist between the proposals it becomes necessary to compare one alternative to another rather than comparing each proposal to the MARR as is done when all proposals are independent. It has already been demonstrated in the discussion of the rate of return on incremental investment criterion in Section 7.3 that comparing one alternative to another on the basis of rate of return on total investment may lead to non-optimum solutions. (See Figure 7.3)

In addition, if the proposals are considered to be indivisible (no fractional portion of the project can be accepted) the ranking procedure of this criterion can lead to nonoptimum solutions. This is easily seen by considering the following three independent proposals with a budget constraint of $50,000 and a MARR = 10%.

Proposal	First Cost	Rate of Return
A	$ 1,000	25%
B	20,000	24%
C	30,000	23%

If these proposals were selected in strict order of ranking then proposals *A* and *B* would be selected. However, it should be clear that it is economically more desirable to have $30,000 earning a return of 23% than to have $1,000 yielding a 25% return while 10% is earned on the remaining $29,000. Therefore, the optimum selection from this set of proposals consists of Proposal *B* and Proposal *C*. It is this sort of difficulty that results from any decision criterion that uses a ranking procedure. When there are numerous independent proposals being considered and each of their first costs is a small proportion of the total budget this effect is mitigated.

Maximum prospective value criterion. The objective of this criterion is to select from a set of proposed investments those proposals that maximize the total prospective value. In Section 6.8 the basis for comparison use by the maximum prospective value criterion (MPV) is defined as

$$P(\bar{m}, i_{\delta})_j = F_{jo}\frac{(1 + i_{\delta})}{(1 + \bar{m})} + \sum_{t=1}^{n} F_{jt}(1 + \bar{m})^{-t}.$$

The selection of the two rates of interest ($\bar{m}$, i_{δ}) required for the calculation of the prospective value is important to the successful implementation of this decision criterion. The rate of interest i_{δ} represents the return that can be received on the money that is invested so that it is available for investment at the time the next decision is made. Usually, this rate reflects the return that can be earned on bank accounts, short-term government securities, or other similar investments that are easy to convert to cash.

The rate of interest $\bar{m}$ represents the rate at which the cash flow *differences* between the two best alternative set of proposals are expected to be invested during the future. A detailed discussion of the rate selection for this criterion is not presented here but it is available elsewhere.[3] For most practical investment decisions $\bar{m}$ can be approximated by the minimum attractive rate of return. Following are three conditions for which the MARR is a good approximation to $\bar{m}$:

1. When many proposals are being considered simultaneously.
2. When each proposal requires only a small proportion of the money available for investment.
3. When the proposals return their investment on a regular periodic basis.

The effectiveness of the MPV criterion has been tested and compared against the present-worth on total investment criterion, the rank on rate of return criterion, and other criteria.[4] In general, this criterion does a better job of maximizing the capital growth rate because the criterion considers the advantage of investing now or not investing in light of future opportunities that are anticipated. Also, the two-rate formulation of this criterion incorporates relationships that are frequently found in actual practice.

7.6. Comparison of Alternatives With Unequal Service Lives

Up to now all the examples that have been utilized to demonstrate the application of the various decision criteria have consisted of alternatives that have equal service lives. Often it is necessary to compare alternatives for which the time span of service will not be equal. In such situations it is necessary to make certain assumptions about the service interval so that the techniques of decision making just discussed are applicable.

When comparing alternatives with unequal lives the principle that all alternatives under consideration must be compared over the same time span is basic to sound decision making. The time span over which alternatives are considered must be equal so that the effect of undertaking one alternative can be considered to be identical to the effect of undertaking any of the other alternatives. Clearly, the direct comparison of Alternative *A* with a 5-year life and Alternative *B* with an 11-year life fails to consider the possible invest-

---

[3]Oakford, R. V., and Thuesen, G. J., "The Maximum Prospective Value Criterion," *The Engineering Economist*, Vol. 13, No. 3, 1968.

[4]Oakford, R. V., and Thuesen, G. J., "The Effectiveness of the Maximum Prospective Value Criterion," *The Proceedings of the 19th Annual Institute Conference and Convention American Institute of Industrial Engineers*, May 1968.

ments that could be undertaken during the 6 years following Alternative *A*'s termination. There are two basic approaches that can be used so that alternatives with different lives can be compared over an equal time span.

Study period approach. This approach confines the consideration of the effects of the alternatives being evaluated to some study period that is usually the life of the shortest-lived alternative. To illustrate this approach suppose a decision must be made as to which alternative should be selected from the two alternatives described in Table 7.12. It is assumed that these two alternatives provide the same service for each year that they are in existence.

Table 7.12. TWO ALTERNATIVES WITH UNEQUAL LIVES

End of Year	Alternative *A*1	Alternative *A*2
0	− $15,000	− $20,000
1	−7,000	−2,000
2	−7,000	−2,000
3	−7,000	−2,000
4	−7,000	—
5	−7,000	—

The study period for this example is chosen to be 3 years, the life of Alternative *A*2. Using the annual equivalent on total investment for an interest rate of 7% yields

$$\overset{A/P\,7,\,5}{AE(7)_{A1}} = -\$15,000(0.2439) - \$7,000 = -\$10,659 \text{ per year.}$$

The $15,000 first cost of Alternative *A*1 is distributed over its entire life to find its equivalent cost per year.

For Alternative *A*2

$$\overset{A/P\,7,\,3}{AE(7)_{A2}} = -\$20,000(0.3811) - \$2,000 = -\$9,622 \text{ per year.}$$

The cost advantage of Alternative *A*2 over Alternative *A*1 is $1,037 per year for the first 3 years. For years 4 and 5 Alternative *A*1 costs $10,659 more than Alternative *A*2 which provides no service for those last 2 years. Since the study period has been selected as 3 years the cost advantage of Alternative *A*2 over Alternative *A*1 is stated as $1,037 per year for 3 years. The costs occurring after the study period are disregarded since the equivalent costs are being compared only for the study period indicated.

The costs occurring after the study period would be considered when

Alternative $A2$'s successor is to be compared to continuing with Alternative $A1$. The decision about $A2$'s successor is assumed to be separable from the original decision when the study period approach is used. An implication of this approach is that for any alternatives with a life longer than the study period the unrecovered balance of their first cost at the end of the study period is the assumed salvage value for these alternatives. For the example just discussed the assumed salvage for Alternative $A1$ after 3 years would have to be

$$\overset{A/P\,7,\,5}{\$15,000(0.2439)}\overset{P/A\,7,\,2}{(1.8080)} = \$6,615.$$

Estimating future alternatives. The second approach to the problem of unequal lives is to estimate the future sequence of events that are anticipated for each alternative being considered so that the time span is the same for each alternative. Two methods that are frequently used to accomplish this end are

1. The explicit consideration of future alternatives over the same time span.
2. The assumption that an investment opportunity will be replaced by an identical alternative until a common multiple of lives is reached.

To illustrate the first method suppose that it is anticipated that after Alternative $A2$ in Table 7.12 is terminated the service it was providing is continued by incurring costs of \$15,000 at the end of years 4 and 5. Now the service is provided over equal time spans of 5 years and the annual equivalent costs for Alternative $A2$ and the additional expenditures required in years 4 and 5 are

$$AE(7) = \left[-\$20,000 - \overset{P/A\,7,\,3}{\$2,000(2.6243)} \right]\overset{A/P\,7,\,5}{(0.2439)} - \$15,000(\overset{F/A\,7,\,2}{2.070})\overset{A/F\,7,\,5}{(0.1739)}$$

$$= -\$11,558.$$

The annual equivalent cost for Alternative $A1$ has been computed for a life of 5 years to be \$10,659 per year. Now Alternative $A1$ has an annual cost advantage of \$11,558 less \$10,659 over Alternative $A2$ and its replacement. This advantage is stated as \$899 per year for 5 years.

The second method that can be used to equate alternatives with unequal lives is to assume that each opportunity will be replaced by itself until a common multiple of lives is reached. For the alternatives described in Table 7.12 this assumption produces the cash flows presented in Table 7.13.

The annual equivalent comparison should be applied when such an assumption is made since it is computationally the most efficient approach. Because the cash flows for each alternative consist of identical repeated

Table 7.13. TWO ALTERNATIVES WITH IDENTICAL REPLACEMENTS FOR A COMMON MULTIPLE OF LIVES

End of Year	Alternative $A1$		Alternative $A2$	
0	$-\$15,000$		$-\$20,000$	
1	$-7,000$		$-2,000$	
2	$-7,000$		$-2,000$	
3	$-7,000$		$-2,000$	$-20,000$
4	$-7,000$		$-2,000$	
5	$-7,000$	$-15,000$	$-2,000$	
6	$-7,000$		$-2,000$	$-20,000$
7	$-7,000$		$-2,000$	
8	$-7,000$		$-2,000$	
9	$-7,000$		$-2,000$	$-20,000$
10	$-7,000$	$-15,000$	$-2,000$	
11	$-7,000$		$-2,000$	
12	$-7,000$		$-2,000$	$-20,000$
13	$-7,000$		$-2,000$	
14	$-7,000$		$-2,000$	
15	$-7,000$		$-2,000$	

cash flows, it is only necessary to calculate the annual equivalent for the original alternative. That is, the 5-year equivalent annual cost for Alternative $A1$ described in Table 7.12 equals the 15-year equivalent annual cost for Alternative $A1$ presented in Table 7.13. Thus, under the assumption of repeated replacements the annual equivalents for the two alternatives in Table 7.13 are

$$AE(7)_{A1} = -\$15,000(\overset{A/P\,7,\,5}{0.2439}) - \$7,000 = -\$10,659 \text{ per year}$$

and

$$AE(7)_{A2} = -\$20,000(\overset{A/P\,7,\,3}{0.3811}) - \$2,000 = -\$9,622 \text{ per year.}$$

The lowest common multiple of years for these two alternatives is 15 years. Therefore, when using this method of examining alternatives over equal time spans the cost advantage of Alternative $A2$ over Alternative $A1$ is stated as $1,037 per year for 15 years. If, in fact, the alternatives are replaced with similar alternatives as assumed, this approach is sound. However, it is infrequent that a sequence of alternatives will repeat themselves since technological progress can lead to improved alternatives in the future. This method of comparing alternatives tends to overstate the differences between the alternatives when it assumes that the differences will occur over a time span that exceeds the service lives of the current alternatives.

To use present-worth calculations for the method just discussed requires additional computation. The annual equivalent can be calculated for the life of each alternative and it is then converted to a present-worth amount over the same time period.

$$PW(7)_{A1} = -\$10,659(\overset{P/A\,7,\,15}{9.1079}) = -\$97,081$$

$$PW(7)_{A2} = -\$9,622(\overset{P/A\,7,\,15}{9.1079}) = -\$87,636.$$

An even more laborious way of making these present-worth calculations is to describe the repeated cash flows so that the receipts and disbursements of the alternative and its successor are known year by year over the number of years that is the common multiple. Such a cash flow is shown in Table 7.13. Direct calculation of the present-worth amount for each of these alternatives will produce the present-worth amounts just computed.

It should be clear that *to calculate the present-worth for cash flows of unequal duration is incorrect.* That is, for the example just presented the following calculations are incorrect for comparing alternatives A1 and A2 of Table 7.12.

$$PW(7)_{A1} = -\$15,000 - \$7,000(\overset{P/A\,7,\,5}{4.1002}) = -\$43,701$$

$$PW(7)_{A2} = -\$20,000 - \$2,000(\overset{P/A\,7,\,3}{2.6243}) = -\$25,249.$$

Such a calculation and comparison imply that for years 4 and 5 Alternative A2 will provide at no cost a service or income equal to that of Alternative A1. Thus, when present-worth comparisons are made, it is essential that the alternatives be compared over the same time span. This same principle holds when making rate of return comparisons on an incremental basis.

PROBLEMS

1. The estimated annual incomes and costs of a prospective venture are as follows:

Year End	Income	Cost
0	$ 0	$1,175
1	600	100
2	700	200
3	800	250

Determine if this is a desirable venture by the (a) present-worth comparison, (b) the annual equivalent comparison, and (c) the rate-of-return comparison for a minimum attractive rate of return of 10%; 20%.

2. A prospective venture is described by the following receipts and disbursements:

Year End	Receipts	Disbursements
0	$ 0	$3,565
1	1,000	600
2	1,200	500
3	1,800	400
4	2,500	200

For an interest rate of 12% determine the desirability of the venture on the basis of:
(a) The present-worth cost comparison.
(b) The annual equivalent comparison.
(c) The rate-of-return comparison.

3. A silver mine can be purchased for $400,000. On the basis of estimated production, an annual income of $55,000 is foreseen for a period of 15 years. After 15 years, the mine is estimated to be worthless. What annual rate of return is in prospect? If the minimum attractive rate of return is 15%, should the mine be purchased?

4. A temporary warehouse with a zero salvage value at any point in time can be built for $15,000. The annual value of the storage space less annual maintenance and operating costs is estimated to be $2,500. If the interest rate is 12% and the warehouse is used 8 years, will this be a desirable investment? For what life will this warehouse be a desirable investment?

5. A special lathe was designed and built for $80,000. It was estimated that the lathe would result in a saving in production cost of $10,500 per year for 15 years. With a zero salvage value at the end of 15 years, what was the expected rate of return? Actually, the lathe became inadequate after 6 years of use and was sold for $20,000. What was the actual rate of return?

6. An engineering graduate estimated that his education had cost the equivalent of $21,000, as of the date of graduation, considering his increased expenses and loss of earnings while in college. He estimated that his earnings during the first decade after leaving college would be no greater than if he had not gone to college. If, by virtue of his added preparation, $3,000, $5,000, and $7,000 additional per year is earned in succeeding decades, what is the rate of return realized on his $21,000 investment in education?

7. As usually quoted for a popular magazine, the prepaid subscription rate for a 3-year period is 2.2 times the one-year subscription rate. What rate of interest

does a subscriber receive on the additional present investment if he purchases a 3-year subscription now rather than three 1-year subscriptions at the beginning of each year? If the interest rate is 10%, which alternative would be most attractive?

8. A manufacturer pays a patent royalty of $0.95 per unit of a product he manufactures, payable at the end of each year. The patent will be in force for an additional 4 years. Previously he manufactured 8,000 units of the product per year but it is estimated that output will be 10,000 12,000, 14,000, and 16,000 in the 4 succeeding years. He is considering (a) asking the patent holder to terminate the present royalty contract in exchange for a single payment at present or (b) asking the patent holder to terminate the present contract in exchange for equal annual payments to be made at the beginning of each of the 4 years. If 8% interest is used, what is (a) the present single payment and (b) the beginning-of-the-year payments that are equivalent to the royalty payments in prospect under the present agreement?

9. The heat loss through the exterior walls of a building costs $215 per year. Insulation that will reduce the heat loss cost by 93% can be installed for $127 and, insulation that will reduce the heat loss cost by 89% can be installed for $90. Determine which insulation is most desirable if the building is to be used for 8 years and if the interest rate is 12%.

10. An industrial firm can purchase a special machine for $22,000. A down payment of $2,500 is required and the balance can be paid in 5 equal year-end installments plus 8% interest on the unpaid balance. As an alternative the machine can be purchased for $19,000 in cash. If the firm's minimum attractive rate of return is 10%, determine which alternative should be accepted. Use the present-worth on incremental investment approach.

11. A needed service can be purchased for $102 per unit. The same service can be provided by equipment which costs $100,000 and which will have a salvage value of $25,000 at the end of 10 years. Annual operating expense will be $5,500 per year plus $31 per unit.
    (a) If these estimates are correct, what will be the incremental rate of return on the investment if 400 units are produced per year?
    (b) What will be the incremental rate of return on the investment if 250 units are produced per year?
    (c) If the firm providing this service has an interest rate of 12% what would be the alternative to select for the production levels in part (a) and part (b)?

12. Every year the stationery department of a large concern uses 1,200,000 sheets of paper with three holes drilled for binding and 250,000 sheets that have the corners rounded. At present the drilling and corner cutting are done by a commercial printing establishment at a cost of $0.35 and $0.30 per thousand sheets, respectively.

    Two alternatives are being considered. Alternative A consists of the purchase of a paper drill for $600, and Alternative B consists of the purchase of a

combination paper drill and corner cutter for $800. Obviously the two alterna-
tives do not provide equal service. The following data apply to the two machines:

	Drill	Combined Drill and Cutter
Life	15 Years	15 Years
Salvage value	$50.00	$65.00
Annual maintenance	5.00	6.00
Annual space charge	11.00	11.00
Annual labor to drill	35.00	40.00
Annual labor to cut corners	—	24.00
Interest rate	8%	8%

(a) Alter one or the other of the alternates given above so that they may be
compared on an equitable basis. Calculate the equivalent annual cost of
each of the revised alternatives.
(b) What other alternative or alternatives should be considered?

13. It is estimated that the annual heat loss cost in a small power plant is $520. Two
competing proposals have been formulated which will reduce the loss. Proposal
A will reduce heat loss cost by 60% and will cost $300. Proposal B will reduce
heat loss cost by 55% and will cost $250. If the interest rate is 8%, and if the
plant will benefit from the reduction in heat loss for 10 years, which proposal
should be accepted?
(a) Use present-worth on total investment.
(b) Use present-worth on incremental investment.
(c) Use annual equivalent on total investment.
(d) Use annual equivalent on incremental investment.
(e) Use rate of return on incremental investment.

14. A manufacturing plant and its equipment are insured for $700,000. The present
annual insurance premium is $0.86 per $100 of coverage. A sprinkler system
with an estimated life of 20 years and no salvage value at the end of that time
can be installed for $18,000. Annual operation and maintenance cost is estimat-
ed at $360. Taxes are 0.8% of the initial cost of the plant and equipment. If the
sprinkler is installed and maintained, the premium rate will be reduced to $0.38
per $100 of coverage.
(a) How much of an incremental rate of return is in prospect if the sprinkler
system is installed?
(b) If the minimum attractive rate of return is 15%, which alternative should be
selected?

15. An engineering student who will soon receive his B.S. degree is contemplating
continuing his formal education by working toward an M.S. degree. The student estimates that his average earnings for the next 6 years with a B.S. degree

will be $13,000 per year. If he can get an M.S. degree in one year his earnings should average $14,500 per year for the subsequent 5 years. His earnings while working on the M.S. degree will be negligible and his additional expenses will be $4,000.

The engineering student estimates that his average per year earnings in the three decades following the initial 6-year period will be $13,500, $16,000, and $18,500 if he does not stay for an M.S. degree. If he receives an M.S. degree his earnings in the three decades can be stated as $13,500 + x, $16,000 + x, and $18,500 + x. For an interest rate of 10% find the value of x for which the extra investment in formal education will pay for itself.

16. A 100-horsepower motor is required to power a large capacity blower. Two motors have been proposed with the following engineering and cost data:

	Motor A	Motor B
Cost	$4,500	$4,000
Life	12 Years	12 Years
Salvage value	0	0
Efficiency $\frac{1}{2}$ load	85%	83%
Efficiency $\frac{3}{4}$ load	92%	89%
Efficiency full load	89%	88%
Hours use per year at $\frac{1}{2}$ load.................	800	800
Hours use per year at $\frac{3}{4}$ load.................	1,000	1,000
Hours use per year at full load	600	600

Power cost per kilowatt-hour is $0.06. Annual maintenance, taxes, and insurance will amount to 1.6% of the original cost. Interest is 10%.
(a) What is the equivalent annual cost for each motor?
(b) What will be the return on the additional amount invested in Motor A?

17. In a hydroelectric development under consideration, the question to be decided is the height of the dam to be built. The function of the dam is to create a head of water. Because of the width of the proposed dam site at different elevations, heights of the dam under consideration are 173, 194, and 211 feet; costs for these heights are estimated at $1,860,000, $2,320,000, and $3,020,000, respectively. The capacity of the power plant is based on the minimum flow of the stream of 1,760 cubic feet per second. This flow will develop $[(h \times 1,760 \times 62.4) \div 550]0.75$ horsepower where h equals the height of the dam in feet. A horsepower-year is valued at $31. The cost of the power plant, including building and equipment, is estimated at $180,000 for the building and $34 per hp. of capacity for the equipment.

To be conservative, the useful life of the dam and buildings is estimated at 40 years with no salvage value. Life of the power equipment is also estimated at 40 years with no salvage value. Annual maintenance, insurance, and taxes on the dam and buildings are estimated at 2.8% of first cost. Annual maintenance,

insurance, and taxes on the equipment are estimated at 4.7% of first cost. Operation costs are estimated at $38,000 per year for each of the alternatives. Deter-
mine the rate of return for each height and the rate of return on the added
investment for each added height. To which height should the dam be built if
10% is required on all investments?

18. A firm has identified three viable but mutually exclusive investment proposals.
The life of all three alternatives is estimated to be 5 years with negligible salvage
value. The minimum attractive rate of return is 7%.

Proposal	A1	A2	A3
Investment proposal	$5,000	$7,000	$8,500
Annual net income	1,319	1,942	2,300
Return on total investment	10%	12%	11%

Find the alternative that should be selected by using the following:
(a) Rate of return on incremental investment.
(b) Present-worth on incremental investment.

19. Three mutually exclusive proposals requiring different investments are being
considered. The life of all three alternatives is estimated to be 20 years with no
salvage value. The minimum rate of return that is considered acceptable is 4%.
The cash flows representing these three proposals are shown below.

Proposal	A1	A2	A3
Investment proposal	−$70,000	−$40,000	−$100,000
Net income per year	5,620	4,075	9,490
Return on total investment	5%	8%	7%

Find the investment that should be selected using (a) rate of return on incremental investment, (b) present-worth on incremental investment, and (c)
present-worth on total investment.

20. A firm is considering the purchase of a new machine to increase the output of an
existing production process. Of all the machines considered the management
has narrowed the field to the machines represented by the cash flows shown as
follows:

Machine	Initial Investment	Annual Operating Cost
1	$ 50,000	$22,500
2	60,000	20,540
3	75,000	17,082
4	80,000	15,425
5	100,000	11,374

If each of these machines provides the same service for 8 years and the minimum attractive rate of return is 12%, which machine should be selected? Solve by using the rate of return on incremental investment. Compare this result to the result obtained by applying the annual equivalent on total investment.

21. The state highway department is considering six locations for a new interstate highway. Listed below are the estimated construction costs, maintenance costs, and the user costs associated with each location.

Location	Construction Costs per Kilometer	Annual Maintenance Costs per Kilometer	Annual Users Cost per Kilometer
A1	$500,000	$3,263	$150,000
A2	562,500	3,075	145,754
A3	625,000	2,894	142,368
A4	700,000	2,659	133,442
A5	750,000	2,213	123,508
A6	812,500	2,133	118,698

The life of the highway is expected to be 25 years with no salvage value. If the initial interest rate is 8%, which highway location is most desirable?
(a) Solve using incremental analysis.
(b) Solve using total investment analysis.

22. A wholesale distributor is considering the construction of a new warehouse to serve a geographic region that he has been unable to serve until now. There are six cities where the warehouse could be built. After extensive study the expected income and costs associated with locating the warehouse in a particular city have been determined.

City	Initial Cost	Net Annual Income
A	$1,000,000	$407,180
B	1,120,000	444,794
C	1,260,000	482,377
D	1,420,000	518,419
E	1,620,000	547,771
F	1,900,000	562,476

The life of the warehouse is estimated to be 15 years. If the minimum attractive rate of return is 12%, where should the wholesaler locate his warehouse?
(a) Solve this problem using an incremental approach.
(b) Solve this problem using a total investment approach.
(c) What city would be selected if the alternative that maximized rate of return

on total investment had been used? Does this conform to the results in part (a) or part (b)?

23. A shipping firm is considering the purchase of a materials handling system for unloading ships at the dock. The firm has reduced their choice to five different systems, all of which are expected to provide the same unloading speed. The initial costs and the operating costs estimated for each system are described below.

System	Initial Cost	Annual Operating Expenses
A7	$650,000	$ 91,810
B3	780,000	52,569
D8	600,000	105,000
K2	750,000	68,417
E5	720,000	74,945

The life of each system is estimated to be 5 years and the firm's minimum attractive rate of return is 15%. If the firm must select one of the materials handling systems, which one is the most desirable?
(a) Solve using the total investment approach.
(b) Solve using an incremental approach.

24. Assume that the proposals in Problem 19 are independent proposals rather than mutually exclusive proposals. If the minimum attractive rate of return is 4% and the amount of money available to invest is $170,000, which proposal or proposals should be selected? Which proposal or proposals should be selected if the amount of money available is $100,000?

25. The production manager of a plant has received the sets of proposals listed below from the supervisors of three independent production activities. The proposals related to a particular production activity are identified by the same letter and they are mutually exclusive. If the proposals are expected to have a life of 8 years with no salvage value and the minimum attractive rate of return is 15%, what proposals should be selected if the amount of money available for investment is (a) unlimited, (b) $40,000, and (c) $20,000.

Proposal	Initital Investment	Net Annual Income
Activity A		
A1	$10,000	$3,004
A2	20,000	6,530
A3	30,000	7,970
Activity B		
B1	$ 5,000	$1,006
B2	10,000	5,312
B3	15,000	6,209
B4	20,000	7,077

Proposal	Initital Investment	Net Annual Income
Activity C		
C1	$15,000	$4,506
C2	30,000	7,829

26. A company is considering a group of research proposals that are related to either Product A, Product B, or Product C. It has been decided that one proposal will be selected from each set of proposals related to a particular product. The research proposals that are concerned with Product A are identified by the letter A, and those concerned with Product B are identified by the letter B, etc. The company expects the research to extend over a 5-year period. In the past the company has considered a return on investment of at least 10% to be satisfactory. Since it is believed that all projects will be equally beneficial to the company only the costs related to each project are shown below. If the money available is (a) unlimited, (b) $100,000, and (c) $80,000, what proposals should be selected?

Proposal	Initial Cost	Annual Expenses
A1	$40,000	$ 8,100
A2	62,000	2,134
B1	20,000	8,200
B2	25,000	6,528
B3	30,000	5,115
C1	15,000	16,200
C2	20,000	15,013
C3	35,000	7,840
C4	50,000	5,530

27. A regional sales manager has received 11 proposals for future expenditures from the 4 sales districts in his region. The proposals listed below are expected to span 10 years and the sales manager uses a minimum attractive rate of return of 12% to determine the acceptability of investment proposals. The proposals from each district are designated by a different letter. The acceptance of a proposal from one district does not affect the acceptance of proposals from the other districts unless money is limited. The proposals related to a particular district are mutually exclusive so it is impossible to select more than one proposal from a particular district. What proposal or proposals should the sales manager select if the money available for investment is (a) unlimited, (b) $700,000, (c) $450,000, and (d) $350,000?

Proposals	Initial Costs	Net Annual Revenues
Q1	$100,000	$19,925
Q2	120,000	24,695
Q3	130,000	26,688
Q4	140,000	29,488
R1	150,000	35,778
R2	180,000	41,755
S1	200,000	32,550
S2	240,000	57,245
S3	300,000	48,825
T1	400,000	95,408
T2	500,000	123,415

28. Write the linear integer programming formulation of the decision problem described in (a) Problem 24, (b) Problem 25, (c) Problem 26, and (d) Problem 27.

29. The Plasco Corporation is considering six investment alternatives for investment. The six proposals under consideration by management:

Proposal	Required Initial Investment	Net Annual Revenue	Expected Life (years)
A1	$1,200,000	$ 240,000	40
B1	1,500,000	450,000	35
C1	2,400,000	820,000	45
C2	2,600,000	840,000	38
D1	3,800,000	1,200,000	30
D2	5,000,000	1,500,000	35

It is expected the net salvage value at the end of the life of each proposal will be zero and the pretax minimum attractive rate of return used by Plasco is 20%. Which proposal or proposals should be accepted if there is no limitation on the amount of money available? Which if the budget is limited to $8,000,000? List any assumptions that you make. Proposals with the same letters are mutually exclusive, e.g., (C1, C2).

30. A refinery can provide for water storage with a tank on a tower or a tank of equal capacity placed on a hill some distance from the refinery. The cost of installing the tank and tower is estimated at $102,000. The cost of installing the tank on the hill, including the extra length of service lines, is estimated at $83,000. The life of the two installations is estimated at 40 years, with negligible salvage value for either. The hill installation will require an additional investment of $9,500 in pumping equipment, whose life is estimated at 20 years with a salvage value of $500 at the end of that time. Annual cost of labor, electricity, repairs, and insurance incident to the pumping equipment is estimated at $1,000. The interest rate is 7%.

(a) Compare the present-worth cost of the two plans.

(b) Compare the two plans on the basis of equivalent annual cost.

(c) Compare the two plans on the basis of their capitalized costs.

31. It is estimated that a manufacturing concern's needs for storage space can be met by providing 240,000 square feet of space at a cost of $8.30 per square foot now and providing an additional 60,000 square feet of space at a cost of $55,000 plus $8.30 per square foot of space 6 years hence. A second plan is to provide 300,000 square feet of space now at a cost of $8.00 per square foot. Either installation will have zero salvage value when retired some time after 6 years, and if taxes, maintenance, and insurance cost $0.15 per square foot and the interest rate is 12%, which plan should be adopted?

32. A logging concern has two proposals under consideration which will provide identical service. Plan A is to build a water slide from the logging site to the saw mill at a cost of $380,000. Plan B consists of building a $150,000 slide to a nearby river and allowing the logs to float to the mill. Associated machinery at a cost of $100,000 and salvage value of $25,000 after 10 years will have to be installed to get the logs from the river to the mill. Annual cost of labor, maintenance, electricity, and insurance of the machinery will be $9,800. The life of the slides is estimated to be 30 years with no salvage value. The interest rate is 8%.

(a) Compare the two plans on the basis of equivalent annual cost.

(b) Compare the two plans on the basis of 30 years of service.

33. A firm has two alternatives for improvement of its current production system. The data are as follows:

	Machine A	Machine B
Initial installment cost	$1,500	$2,500
Annual operating cost	800	650
Service life	5 years	8 years
Salvage value	$0	$0

Determine the rate of return on the extra investment in Machine B and select the best alternative for an interest rate of 15%.

34. A firm is considering the purchase of one of two new machines. The data on each are described below.

	Machine A	Machine B
Initial cost	$3,400	$6,500
Service life	3 years	6 years
Salvage value	$100	$500
Net operating cost after taxes	$2,000/ year	$1,800/ year

If the firm's MARR is 12%, which alternative should be selected when using the following methods?
(a) Equivalent annual cost.
(b) Present-worth comparison.
(c) Incremental rate-of-return comparison.

EVALUATING REPLACEMENT
ALTERNATIVES

Mass production has been found to be the most economical method of satisfying human wants. However, mass production necessitates the employment of large quantities of capital assets which become consumed, inadequate, obsolete, or in some way become candidates for replacement. The failure to continuously upgrade these assets can result in serious loss of operating efficiency. A sound program of replacement analysis can ultimately affect the financial success of an enterprise.

When replacement decisions are being considered there are two courses of action available. The first possibility is to retain the asset presently owned for an additional period of time. The other alternative requires the immediate removal of the existing asset with its subsequent replacement by another asset. As with other economic alternatives, the economic future of the present asset can be represented by a cash flow of estimated receipts and disbursements. Since the economic future of a possible replacement can also be represented by cash flows, the methods of analysis described in Chapter 7 are appropriate for comparing the cash flows of the present asset and its challenger. However, there are certain concepts and techniques in replacement analysis such as sunk cost, economic life, and unused value that require special attention. This chapter presents these concepts and provides examples of their application in evaluating replacement alternatives.

8

8.1. The General Nature of Replacement Analysis

To facilitate the discussion of the principles involved in replacement analysis, it is necessary to introduce some important terms commonly used in replacement studies. The two terms given below represent interpretations that are widely accepted by practitioners involved in replacement analyses.

Defender: The existing old asset being considered as the asset to be replaced

Challenger: The asset proposed to be the replacement

Because the economic characteristics of the defender and the challenger are usually so dissimilar special attention is required when these two options are compared. One obvious feature of replacement alternatives is that the duration and the magnitude of cash flows for old existing assets and new assets are quite different. New assets characteristically have high capital costs and low operating costs. The reverse is usually true for assets which are being considered for retirement. Thus, capital costs for an asset to be replaced may be expected to be low and decreasing while operating costs are usually high and increasing.

In addition, the remaining life of an asset being considered for replacement is usually short, and the future of the asset can be estimated with relative certainty. There is also the advantage that a decision not to replace it now may be reversed at any time in the future. Thus, a decision may be made on the basis of next year's cost of the old asset, and if it is not replaced, a new decision can be made on the basis of next year's cost a year later and so forth.

Basic reasons for replacement. There are two basic reasons for considering the replacement of a physical asset: physical impairment and obsolescence. Physical impairment refers only to changes in the physical condition of the asset itself. Obsolescence is used here to describe the effects of changes in the environment external to an asset. Physical impairment and obsolescence may occur independently or they may occur jointly in regard to a particular asset.

Physical impairment may lead to a decline in the value of service rendered, increased operating cost, increased maintenance cost, or a combination. For example, physical impairment may reduce the capacity of a bulldozer to move earth and consequently reduce the value of the service it can render. Fuel consumption may rise, thus increasing its operating cost, or the physical impairment may necessitate increased expenditure for repairs.

Little useful data are available relative to how such costs occur in relationship to length of service of assets. A storage battery may render perfect service and require no maintenance up to the moment it fails. Water pipes, on the other hand, may begin to acquire deposits on installation, which reduce their capacity in some proportion to the time they have been in service. Many assets are composites of a number of elements of different service lives. Roofs of buildings usually must be replaced before side walls. The basic structure of bridges ordinarily outlasts several deck surfacings.

Obsolescence occurs as a result of the continuous improvement of the tools of production. Often, the rate of improvement is so great that it is an economy to replace a physical asset in good operating condition with an improved unit. In some cases, the activity for which a piece of equipment has been used declines to the point that it becomes advantageous to replace it with a smaller unit. In either case, replacement is due to obsolescence and necessitates disposing the remaining utility of the present asset in order to allow for the employment of the more efficient unit. Therefore, obsolescence is characterized by changes external to the asset and is used as a distinct reason in itself for replacement where warranted.

Replacement should be based on economic factors. When the success of an economic venture is dependent upon profit, replacement should be based upon the economy of future operation. Although production facilities are, and should be, considered as a means to an end, that is, production at lowest cost, there is ample evidence that motives other than economy often enter into analysis concerned with the replacement of assets.

The idea that replacement should occur when it is most economical rather than when the asset is worn out is contrary to the fundamental concept of thrift possessed by many people. In addition, existing assets are often venerated as old friends. People tend to derive a measure of security from familiar old equipment and to be skeptical of change, even though they may profess a progressive outlook. Replacement of equipment requires a shift of enthusiasm. When a person initiates a proposal for new equipment, he must ordinarily generate considerable enthusiasm to overcome inertia standing in the way of its acceptance. Later, enthusiasm may have to be transferred to a replacement. This is difficult to do, particularly if one must confess to having been overenthusiastic about the equipment originally proposed.

Part of the reluctance to replace physically satisfactory but economically inferior units of equipment has roots in the fact that the import of a decision to replace is much greater than that of a decision to continue with the old. A decision to replace is a commitment for the life of the replacing equipment. But a decision to continue with the old is usually only a deferment of a decision to replace that may be reviewed at any time when the situation seems clearer. Also, a decision to continue with old equipment that results in a loss will usually result in less censure than a decision to replace it with new equipment that results in an equal loss.

The economy of scrapping a functionally efficient unit of productive equipment lies in the conservation of effort, energy, material, and time resulting from its replacement. The unused remaining utility of an old unit is sacrificed in favor of savings in prospect with a replacement. Consider, by way of illustration, a shingle roof. Even a roof that has many leaks will have some utility as a protection against the weather and may have many sound shingles in it. The remaining utility could be made use of by continual repair. But the excess of labor and materials required to make a series of small repairs over the labor and materials required for a complete replacement may exceed the utility remaining in the roof. If so, labor and materials can be conserved by a decision to replace the roof.

When a new unit of equipment is purchased, a number of additional expenses beyond its purchase price may be incurred to put the unit into operation. Such expense items may embrace freight, cartage, construction of foundations, special connection of wiring and piping, guard rails, and personal services required during a period of test or adjustment. Expenses for such items as those mentioned are first-cost items and for all practical purposes represent an investment in a unit of equipment under consideration. For this reason all first-cost items necessary to put a unit of equipment into operation should be considered as part of the total original investment in the unit.

When a unit of equipment is replaced, its removal may entail considerable expense. Some of the more frequently encountered items of removal expense are dismantling, removal of foundations, haulage, closing off water and electrical connections, and replacing floors or other structural elements. The sum of such costs should be deducted from the amount received for the old unit to arrive at its net salvage value. It is clear that this may make the net salvage value a negative quantity. When the net salvage value of an asset is less than zero, it is mathematically correct to treat it as a negative quantity in depreciation calculations.

8.2. Evaluation of Replacements Involving Sunk Costs

The method of treating data relative to an existing asset should be the same as that used in treating data relative to a possible replacement. In both cases only the future of the assets should be considered and *sunk costs should be disregarded*. Thus, the value of the defender that should be used in a study of replacement is not what it cost when originally purchased, but what it is worth at the present time.

The following example will be used to illustrate correct and incorrect methods of evaluating replacements where sunk costs are involved. Suppose that Machine *A* was purchased 4 years ago for $2,200. It was estimated to have a life of 10 years and a salvage value of $200 at the end of its life. Its operating expense had been found to be $700 per year, and it appeared that the machine would serve satisfactorily for the balance of its estimated life. Presently a salesman is offering Machine *B* for $2,400. Its life is estimated at 10 years and its salvage value at the end of its life is estimated to be $300. Operating costs are estimated at $400 per year.

The operation for which these machines are used will be carried on for many years in the future. Equipment investments are expected to justify a 15% minimum attractive rate of return in accordance with the policy of the company concerned. The salesman offers to take the old machine in on trade for $600. This appears low to the company, but the best offer received elsewhere is $450. All estimates relative to both machines above have been carefully reviewed and are considered sound.

In order to make a proper comparison of alternatives the analysis may be undertaken from the standpoint of a person who has a need for the service that Machine *A* or Machine *B* will provide but owns neither. In attempts to purchase a machine he finds that he can purchase Machine *A* for $600 and Machine *B* for $2,400. This analysis of which to buy will not be biased by the past since he was not part of the original transaction for Machine *A* and, therefore, will not be forced to admit a sunk cost. With this *outsider viewpoint*, the appropriate cash flows are presented in Figure 8.1. The important effect of using the outsider viewpoint is that the old machine's present market

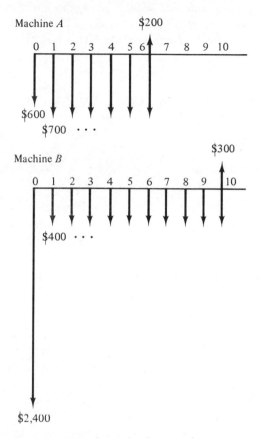

Figure 8.1. Outsider viewpoint for a replacement problem.

value is identified as the investment required to continue its use. This method of analysis is correct even though the retention of the old machine requires no actual disbursement at the present.

    Comparison based on outsider viewpoint. If Machine *A* is retired, its original investment of $2,000 should be ignored. If an outsider were to consider this problem, he would have to anticipate paying $600 for Machine *A* because this figure represents its worth at the present. That is, the $600 amount that would be received if Machine *B* is purchased and Machine *A* is "sold" represents the present best estimate of its worth. If the outsider were to purchase Machine *B*, he would pay its present market price of $2,400 since he has no asset to trade in. Thus, the logical alternatives are (1) to consider Machine *A* to have a value of $600 and to continue with it for 6 years and (2) to purchase Machine *B* for $2,400 and use it for 10 years. Because these alternatives have different service lives, the study period approach discussed in Section 7.6 is applicable. A study period of 6 years is assumed.

The equivalent annual cost to continue with Machine *A* for 6 years is calculated as follows:

$$A/P\,15,6$$
Annual capital recovery with return, ($600 − $200)(0.2642)
+ $200(0.15) $136
Annual operating cost 700
$836

The equivalent annual cost to dispose of Machine *A*, purchase Machine *B*, and use it for 10 years is calculated as follows:

$$A/P\,15,10$$
Annual capital recovery with return, ($2,400 − $300)(0.1993)
+ $300(0.15) $464
Annual operating cost 400
$864

If the alternative to continue with Machine *A* is adopted, the annual saving prospect for the next 6 years is $864 − $836 = $28. For the next 4 years after that time the amount of savings will be dependent upon the characteristics of the machine that might have been purchased 6 years from the present to replace Machine *A*. If it is assumed that Machine *A* will be replaced after 6 years by a machine identical to Machine *B*, the equivalent annual costs of the two alternatives will be the same after the first 6 years.

Calculation of comparative use value. A second method of comparison, which is particularly good for demonstrating the correctness of the comparison above to skeptical people, is to calculate the value of the machine to be replaced which will result in an annual cost equal to the annual cost of operation with the replacement. In this calculation, let *X* equal the present value of Machine *A* for which annual cost with Machine *A* equals annual cost with Machine *B*. Then

$$(X - \$200)(\overset{A/P\,15,6}{0.2642}) + \$200(0.15) + \$700$$

$$= (\$2,400 - \$300)(\overset{A/P\,15,10}{0.1993}) + \$300(0.15) + \$400.$$

Solving for X results in

$$X = \$707.$$

Machine A has a comparative use worth in comparison with Machine B of $707. Thus, it is obvious that Machine A should be retained if it can be disposed of for only $600. Compare this result with that obtained in the previous section. Note that $707 — $600 = $107 is equivalent to

$$\overset{P/A\,15,\,6}{\$28(\,3.7845\,)} = \$106.$$

Difficulties when using actual cash flow as a basis for comparison. In Chapter 4 it was emphasized that the actual cash flows associated with an alternative are all that are necessary to describe the economic effects of that alternative. So why take the outsider viewpoint when comparing replacement alternatives? Why not just define the actual cash flows and make the comparison as previously described. The answer is that there are pitfalls that accompany this approach because of the special nature of replacement alternatives. Unless the analyst is extremely careful these pitfalls can lead to erroneous conclusions or extra calculations.

To illustrate the most common error that can occur, two replacement situations are analyzed. First, suppose that Machine C was purchased for $3,400 a year ago and it had an estimated life of 6 years at that time. Its salvage value is estimated to be $400 with operating expenses of $3,200 per year. At the end of the first year a salesman offers Machine D for $4,600. This machine has an estimated life of 5 years, a salvage value of $600, and, owing to improvements it embodies, an operating cost of only $2,200. The salesman offers to allow $1,400 for Machine C on the purchase price of Machine D and the interest rate is 8%. The *actual* cash flows for the defender and challenger are presented in Figure 8.2.

Since these two alternatives have *equal* lives the direct comparison of their annual costs gives the correct result. The equivalent annual cost of Machine C for its 5 years of service is

$$AE(8)_C = \$3,200 - \overset{A/F\,8,\,5}{\$400(0.1705)} = \$3,131.$$

And the equivalent annual cost of Machine D is expressed as

$$AE(8)_D = (\$4,600 - \overset{A/P\,8,\,5}{\$1,400)(0.2505)} + \$2,200 - \overset{A/F\,8,\,5}{\$600(0.1705)}$$
$$= \$2,899.$$

The advantage of the challenger over the defender in this case is $232 per year for 5 years. Using the outsider viewpoint for this problem will lead to

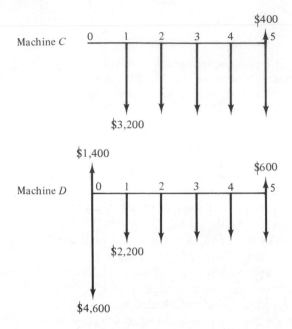

Figure 8.2. Actual cash flows for replacement alternatives.

exactly the same result. That is, the annual equivalent difference between the two options is $232 per year. This consistency between these methods holds only as long as the two alternatives have equal lives.

To see what difficulties arise when the lives of the defender and challenger are unequal let's compare Machine A and Machine B shown in Figure 8.1 on the basis of their actual cash flows. If Machine A is retained there is no transfer of cash related to the $600 trade-in, and, therefore, the equivalent annual cost for the defender is

$$AE(15)_A = \$700 - \$200(\overset{A/F\,15,6}{0.1142}) = \$677.$$

Using the $600 salvage to be realized from Machine A to reduce the cost of Machine B will lead to an *incorrect* conclusion about the economic differences between the defender and challenger. This error occurs because the annual equivalent cost of the challenger for its actual cash flow is found from the expression

$$AE(15)_B = (\$2,400 - \$600)(\overset{A/P\,15,\,10}{0.1993}) + \$400 - \$300(\overset{A/F\,15,\,10}{0.0493})$$
$$= \$744.$$

According to this comparison, the advantage of retaining Machine A is valued at $67 per year for 6 years. However, this result does not compare with that obtained by using the outsider viewpoint. Recall that when the outsider's view was applied, the advantage of Machine *A* over Machine *B* was $28 per year for 6 years.

In this case, the $67 figure is incorrect because an improper method of analysis has been used. It is erroneous to assume that both the $2,400 first cost of Machine *B* and the $600 trade-in to be received from Machine *A* should be annualized over the 10-year life of the challenger. In fact, the $600 represents the worth of Machine *A* at the present and this amount should be annualized for the life of that machine, namely, 6 years. Thus, for this example, the mistake of not associating an asset's worth with its particular life leads to an overstatement of the advantages of the defender compared to the challenger.

The use of the outsider viewpoint prevents the type of error just discussed. In addition, it facilitates analysis when a number of mutually exclusive alternatives are being compared to an existing asset. This savings in effort occurs because there is no need to deduct the present salvage of the defender from each of the other alternatives.

Fallacy of including sunk cost in a replacement analysis. In spite of the fact that sunk cost cannot be recovered, a face saving practice of charging the sunk cost of a machine to the cost of its contemplated replacement is often employed. This practice, human but unrealistic, will be illustrated by the following situation.

Three years ago *A*, who authorizes machine purchases in a manufacturing concern, was approached by *B* for authorization to purchase a machine. *B* pleaded his cause in glowing terms and with enthusiasm. He had many figures and arguments to prove that an investment in the machine he proposed would easily pay out. *A* was at first skeptical, but he also became enthusiastic about the purchase as the profits in prospect were calculated, and authorized the purchase. After 3 years *B* realized that the machine was not coming up to expectations and would have to be replaced, at a loss of $1,200.

B was well aware of the necessity of admitting this sunk cost when he went to *A* to get authorization for a replacement. He realized the difficulty of trying to establish confidence in his arguments for the replacement and at the same time admit an error in judgment that had resulted in a loss of $1,200. But he hit on the expedient of focusing attention on the proposed machine by emphasizing that the $1,200 loss could be added to the cost of the new machine and that the new machine had such possibilities for profit that it would pay out shortly, even though burdened with the loss of the previous machine. Such improper handling of sunk cost is merely deception designed to make it appear that an error in judgment has been corrected.

As a numerical example of the fallacy of adding sunk cost of an old machine to the cost of a replacement, consider the replacement situation previously described for Machine *C* and Machine *D*. (See Figure 8.2.) Remember that Machine *C* was purchased a year ago at a cost of $3,400. Since the present salvage of Machine *C* is $1,400 there has been a decrease in the asset's value of $2,000 over the last year.

Annual cost with Machine *C*, on the basis of its present trade-in value and estimated salvage value 5 years hence, is

$$A/P\,8,\,5$$

Annual capital recovery with return ($1,400 − $400)(0.2505)
+ $400(0.08) .... $ 283
Annual operating cost 3,200
                                                            ‾‾‾‾‾‾
                                                            $3,483

Annual cost with Machine *D*, as incorrectly calculated when the loss in value of $2,000 of Machine *C* is added to the cost of Machine *D*, is

$$A/P\,8,\,5$$

Annual capital recovery with return ($6,600 − $600)(0.2505)
+ $600(0.08) $1,551
Annual operating cost 2,200
                                                            ‾‾‾‾‾‾
                                                            $3,751

On the basis of this *incorrect* result, Machine *C* is continued for the next year on the erroneous belief that $3,751 less $3,483, or $268 is being saved annually. Annual cost with Machine *D* as correctly calculated is

$$A/P\,8,\,5$$

Annual capital recovery with return ($4,600 − $600)(0.2505)
+ $600(0.08) $1,050
Annual operating cost 2,200
                                                            ‾‾‾‾‾‾
                                                            $3,250

On this correct basis, purchase of Machine *D* should result in an annual saving of $3,483 less $3,250 or $233, as previously shown.

8.3. Replacement Analysis for Unequal Lives

Because most replacement decisions are concerned with the replacement of old assets by new ones, the economic alternatives being examined are seldom of equal duration. The use of the study period approach described in Section 7.6 is appropriate when the service lives of the assets are known.

It is usually presumed that each event in history is dependent upon previous events. Thus, in theory it is necessary, for accurate comparison of a pair of alternatives, to consider the entire future or a period from the present to a point in the future when the effect of both alternatives will be identical. It is rarely feasible to consider all links in the chain of events in the future. It also is frequently impossible to be able to discern a point in the future at which the selection of one of a pair of alternatives in the present will have the same effect as the selection of the other.

In the paragraphs that follow, a general method for placing alternatives on a comparative basis involving the selection of more or less arbitrary study periods will be illustrated. With this method, comparison of alternatives is made on the basis of costs and income that occur during a selected period in the future. The effect of values occurring after the selected study period is eliminated by suitable calculations. Study periods for all alternatives in a given comparison should be equal.

As an illustration of the use of a selected study period, consider the following example. A certain operation is now being carried on with Machine *E* whose present salvage value is estimated to be \$2,000. The future life of Machine *E* is estimated at 5 years, at the end of which its salvage value is estimated to be zero. Operating costs with Machine *E* are estimated at \$1,200 per year. It is expected that Machine *E* will be replaced after 5 years by Machine *F* whose initial cost, life, final salvage value, and annual operating costs are estimated to be, respectively, \$10,000, 15 years, zero, and \$600. It should be realized that estimates relating to Machine *F* may turn out to be grossly in error.

The desirability of replacing Machine *E* with Machine *G* is being considered. Machine *G*'s estimated initial cost, life, final salvage value, and annual operating costs are estimated to be respectively, \$8,000, 15 years, zero, and \$900. The interest rate is taken to be 10%. Detailed investment and cost data for Machines *E*, *F*, and *G* are given in Table 8.1.

Analysis based on a fifteen-year study period, recognizing unused value. Because of the difficulty of making further estimates into the future, a study period of 15 years coinciding with the life of Machine *G* is selected. This will necessitate calculations that will bring both plans to equal status at the end of 15 years.

Under Plan I, the study period embraces 5 years of service with Machine *E* and 10 years of service with Machine *F*, whose useful life extends 5 years beyond the study period. Thus, an equitable allocation of the costs associated with Machine *F* must be made for the period of its life coming within and after the study period. In assuming that annual costs associated with this unit of equipment are constant during its life, the present-worth cost of service during the study period may be calculated as follows.

Table 8.1. ANALYSIS BASED ON A SELECTED STUDY PERIOD

Year End Number	Plan I		Plan II	
	Machine Investment	Operating Costs	Machine Investment	Operating Costs
0	Machine E, $2,000		Machine G, $8,000	
1		$1,200		$900
2		1,200		900
3		1,200		900
4		1,200		900
5	Machine F, $10,000	1,200		900
6		600		900
7		600		900
8		600		900
9		600		900
10		600		900
11		600		900
12		600		900
13		600		900
14		600		900
15		600		900
16		600		
17		600		
18		600		
19		600		
20		600		

(Fifteen year study period — years 0 through 15)

The equivalent annual cost for Machine F during its life is equal to

$$AE(10)_F = \$10,000(\overset{A/P\,10,\,15}{0.1315}) + \$600 = \$1,915.$$

The present-worth cost of 15 years of service in the study period is equal to

$$PW(10)_I = \$2,000 + \$1,200(\overset{P/A\,10,\,5}{3.791}) + \$1,915(\overset{P/A\,10,\,10}{6.1446})(\overset{P/F\,10,\,5}{0.6209})$$

$$= \$13,856.$$

By distributing the first cost of Machine F over the entire 15 years of its estimated service life, the calculations reflect that 5 years (years 16 through 20) of Machine F's value is unused for the 15-year study period assumed. On this basis the value remaining in Machine F at the end of the study period may be calculated as a matter of interest as follows:

$$\$10,000(\overset{A/P\,10,\,15}{0.1315})(\overset{P/A\,10,\,5}{3.791}) = \$4,985.$$

Under Plan II, the life of Machine *G* coincides with the study period. The present-worth cost of 15 years service in the study period is equal to

$$PW(10)_{\text{II}} = \$8,000 + \$900(\overset{P/A\,10,\,15}{7.606}) = \$14,845.$$

On the basis of present-worth costs of $13,856 and $14,845 for a study period of 15 years, Plan I should be chosen.

Analysis based on a fifteen-year study period, not recognizing unused value. Values remaining in assets at the end of a selected study period are sometimes disregarded in order to simplify the calculations necessary for making a comparison. The effect of disregarding values remaining in an asset at the end of a study period is to assume that the asset will be retired at the end of the study period.

On the basis of this assumption, the present-worth cost of 15 years of service for Plan I may be calculated as follows. The equivalent annual cost for Machine *F* during its life is equal to

$$AE(10)_F = \$10,000(\overset{A/P\,10,\,10}{0.1628}) + \$600 = \$2,228$$

and the present-worth cost of 15 years of service in the study period is equal to

$$PW(10)_{\text{I}} = \$2,000 + \$1,200(\overset{P/A\,10,\,5}{3.791}) + \$2,228(\overset{P/A\,10,\,10}{6.1446})(\overset{P/F\,10,\,5}{0.6209})$$
$$= \$15,050.$$

This result is significantly different from that calculated for 15 years of service for Plan I, where the initial investment required for Machine *F* was distributed over its estimated useful life. In this case the entire first cost of Machine *F* is allocated to its first 10 years even though the asset is expected to provide service for 15 years. This approach is clearly *unfair* to Machine *F* since it is being penalized by having to recover its initial cost over the first two-thirds of its life.

The practice of disregarding values remaining in an asset at the end of a study period introduces error equivalent to the actual value of the asset at that time. This practice is difficult to defend for it does not greatly reduce the burden of making comparisons, and it does not necessarily produce results in the direction of conservatism.

Analysis on the basis of a five-year study period. Lack of information often makes it necessary to use rather short study periods. For example the characteristics of the successor to Machine *E* in Table 8.1 might be vague. In that case a study period of 5 years might be selected to coincide with the estimated retirement date of Machine *E*.

The equivalent annual cost of continuing with Machine *E* during the next 5 years is

$$AE(10)_E = \$2,000(\overset{A/P\,10,\,5}{0.2638}) + \$1,200 = \$1,728.$$

The equivalent annual cost of Machine *G* based on a life of 15 years is

$$AE(10)_G = \$8,000(\overset{A/P\,10,\,15}{0.1315}) + \$900 = \$1,952.$$

The $224 per year cost advantage of Machine *E* over Machine *G* can be interpreted in two ways. It can be said that the retention of Machine *E* will produce such savings for a 5-year period if the future events after 5 years are ignored. That is, the commitment to Machine *G* for its remaining 10 years is not reflected in the savings of $224 per year for 5 years.

Another interpretation of this annual savings figure is to assume that each machine is replaced by an identical successor for the shortest period for which machine lives are common multiples. Thus, a saving of $224 per year would be realized for 15 years if Machine *E* is purchased and replaced by identical successors every 5 years as opposed to purchasing Machine *G* for 15 years. Such an assumption concerning replacement by identical successors is implicit in making direct economic comparisons of annual equivalent amounts.

In general, the longer the study period, the more significant the results. But the longer the study period, the more likely that estimates are in error. Thus, the selection of a study period must be based on estimate and judgment.

8.4. The Economic Life of an Asset

The preceding section has discussed the types of analyses that may be applied when the service life is known. However, there are many instances when the length of time a particular asset will be retained is only conjecture. Since replacement analyses are usually sensitive to the lives assumed, it is prudent to consider each alternative in its most favorable circumstances. Thus, *when comparing an existing asset and its possible replacement the lives that should be assumed are the lives that are most favorable to each asset.* In other words, the comparison should be made on the basis of each alternative's *economic life.*

The economic life of an asset is the time interval that minimizes the asset's total equivalent annual costs or maximizes its equivalent annual net income.

The economic life is also referred to as the minimum cost life or the optimum replacement interval.

One of the important determinants of an asset's economic life is the pattern of costs incurred by operating and maintenance (O&M) activities. This relationship can be observed in the following discussion of sporadic, constant, and increasing O&M costs.

Sporadic maintenance costs. Assume that a machine is purchased for $400 and that its salvage value is zero at any age at which it may be retired. Assume that the interest rate is zero. Then the pertinent facts related to Machine A may be set down as in Table 8.2. This table brings out the fact

Table 8.2. ECONOMIC HISTORY OF A MACHINE WITH SPORADIC MAINTENANCE COSTS

End of Year Number A	Maintenance Cost for End of Year Given B	Summation of Maintenance Costs, $\sum B$ C	Average Cost of Maintenance Through Year Given, $C \div A$ D	Average Capital Cost If Retired at Year End Given, $\$400 \div A$ E	Average Total Cost Through Year Given, $D + E$ F
1	$100	$ 100	$100	$400	$500
2	100	200	100	200	300
3	300	500	167	133	300
4	100	600	150	100	250
5	100	700	140	80	220
6	100	800	133	67	200
7	100	900	129	57	186
8	300	1,200	150	50	200
9	100	1,300	144	44	188
10	100	1,400	140	40	180

that capital costs decrease in some inverse proportion to the length of life. This is also true for interest rates other than zero and for any pattern of salvage value normally encountered.

The fact that maintenance costs are averaged in Column D tends to smooth out the effect of sporadic large maintenance costs. In the example, the ratio of the cost of the asset and its maintenance cost is relatively high. In spite of this, the average total cost in Column F is generally downward. Unless there is a rising trend in sporadic maintenance cost, there will be no "minimum" cost in a given year that will not be bettered in a future year. But it is clear that if replacement is to be made, it is desirable to do so immediately prior to a large expenditure for maintenance.

Constant maintenance costs. When maintenance costs are constant in succeeding years they will never justify replacement. When neither interest nor salvage value is involved an equation for the average cost of a year of service can be written as follows:

$$C = \frac{P}{n} + M$$

where

C = average annual cost of capital recovery and maintenance;
P = initial cost of asset;
F = salvage value of asset;
M = constant yearly cost of maintenance;
n = life of asset in years.

It is apparent that the value of C will never reach a minimum value.

When interest and salvage value are involved, an expression for equivalent annual cost, C, may be written as follows:

$$C = (P - F)(\overset{A/P\,i,\,n}{}) + Fi + M.$$

A glance at a table of value for $(\overset{A/P\,i,\,n}{})$ shows that C will decrease with an increase in n if the salvage value remains constant through time. However if, from period to period there are large decreases in F, this trend may be reversed.

Constantly increasing maintenance costs. An understanding of the replacement problem may also be gained from considering situations in which maintenance costs increase constantly with the age of an asset. Assume that a machine has been purchased for $800, that its salvage value is zero at any age, that its maintenance cost is zero the first year and rises at a constant rate of $100 per year thereafter. If it is assumed that the interest rate is zero the facts concerning the machine may be represented by Table 8.3.

Because there is a rising trend in maintenance cost, there will be a minimum average total cost at some point in the life of the asset. This point occurred in the fourth year in the example presented.

When a rising trend in O&M cost exists, it is possible to formulate an idealized model that will express the economic life for an asset. Neglecting interest, the average annual cost for an asset with increasing O&M cost may be expressed as follows:

$$C = \frac{P}{n} + Q + (n - 1)\frac{m}{2}$$

where

Table 8.3. ECONOMIC HISTORY OF A MACHINE WITH CONSTANTLY INCREASING
MAINTENANCE COSTS

End of Year Number A	Maintenance Cost for End of Year Given B	Summation of Maintenance Costs, $\sum B$ C	Average Cost of Maintenance Through Year Given, $C \div A$ D	Average Capital Cost If Retired at Year End Given, $\$800 \div A$ E	Average Total Cost Through Year Given, $D + E$ F
1	$ 0	$ 0	$ 0	$800	$800
2	100	100	50	400	450
3	200	300	100	267	367
4	300	600	150	200	350
5	400	1,000	200	160	360
6	500	1,500	250	133	383

C = average annual cost;
P = initial cost of asset;
Q = annual constant portion of operating cost of asset (is equal to first
year operation cost, of which maintenance is a part);
m = the amount by which maintenance costs increase each year;
n = life of asset in years.

This expression, if differentiated with respect to n, set equal to zero, and
solved for n, results in the following:

$$\frac{dC}{dn} = -\frac{P}{n^2} + \frac{m}{2} = 0$$

$$n = \sqrt{\frac{2P}{m}}.$$

For the example presented in Table 8.3, $P = \$800$, $Q = 0$, and $m = \$100$. Therefore, the minimum cost life is

$$n = \sqrt{\frac{2(\$800)}{\$100}} = 4 \text{ years}$$

as is shown in Figure 8.3. The minimum cost shown in Table 8.3 may be
verified as follows:

$$C = \frac{\$800}{4} + 3\left(\frac{100}{2}\right) = \$350.$$

Figure 8.3 graphically depicts the trade-offs between the increasing maintenance costs and the decreasing costs of capital recovery that produce a minimum cost life for the asset.

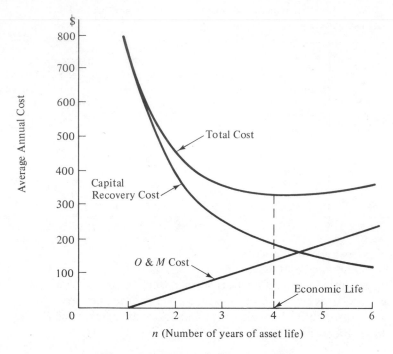

Figure 8.3. Economic life of an asset.

Finding the economic life of an asset. Because in replacement analysis the defender and challengers should be compared on the basis of the lives most favorable to each, some time needs to be devoted to describing how to calculate the economic life of an asset. If the future could be predicted with certainty, it would be possible to accurately predict the economic life for an asset at the time of its purchase. The analysis would simply involve the calculation of the total equivalent annual cost at the end of each year in the life of the asset. Selection of the total equivalent annual cost that is a minimum would specify a minimum cost life for the asset. The application of this approach is demonstrated by the following example.

The economic history of an asset whose first cost is $5,000, whose salvage value at any time is zero, and whose cost of maintenance is zero the first year and increases at a constant rate of $100 for an interest rate of 6%, is shown in Table 8.4. The minimum equivalent annual cost occurs when the cost of extending an asset's life one more year exceeds the equivalent annual cost to date. Since the salvage value is zero at any time the incremental cost of one additional year's operation is reflected in Column *B*. The incremental cost of providing service for year 12 ($1,100) exceeds the equivalent annual cost for 11 years ($1,076) and therefore the economic life of the asset is 11 years.

Table 8.4. EQUIVALENT ANNUAL COST OF MAINTENANCE PLUS CAPITAL RECOVERY WITH A RETURN OF AN ASSET FOR CONSTANTLY INCREASING MAINTENANCE

End of Year Number	Maintenance Cost at End of Year Designated	Present-Worth Factor for Year Designated, $(P/F_{i,n})$	Present-Worth as of Beginning of Year No. 1, of Maintenance for Year Designated, $B \times C$	Summation of Present Worths of Maintenance Through Year Designated, $\sum D$	Capital-Recovery Factor for Year Designated, $(A/P_{i,n})$	Equivalent Annual Cost of Maintenance Through Year Designated, $E \times F$	Equivalent Annual Cost of Capital Recovery and Return Through Year Designated, $F \times \$5,000$	Total Equivalent Annual Cost Through Year Designated, $G + H$
A	B	C	D	E	F	G	H	I
1	\$ 0	0.9434	\$ 0	\$ 0	1.06000	\$ 0	\$5,300	\$5,300
2	100	0.8900	89	89	0.54544	48	2,727	2,775
3	200	0.8396	167	256	0.37411	96	1,870	1,966
4	300	0.7921	237	493	0.28859	142	1,442	1,585
5	400	0.7473	298	791	0.23740	188	1,187	1,375
6	500	0.7050	352	1,143	0.20336	233	1,016	1,249
7	600	0.6651	399	1,542	0.17914	276	895	1,172
8	700	0.6274	439	1,984	0.16104	319	805	1,124
9	800	0.5919	473	2,457	0.14702	361	735	1,096
10	900	0.5584	502	2,960	0.13587	402	679	1,081
11	1,000	0.5268	526	3,487	0.12679	442	633	1,076
12	1,100	0.4970	546	4,033	0.11928	481	596	1,077
13	1,200	0.4688	560	4,588	0.11296	518	564	1,083
14	1,300	0.4423	574	5,162	0.10758	556	537	1,094

Table 8.4 illustrates a method for determining the equivalent annual cost of maintenance and the equivalent annual cost of capital recovery with return for lives ranging from 1 to 14 years. The sum of these costs is a minimum for a life of 11 years. The quantities in Columns G, H, and I have been plotted to reveal trends in Figure 8.4.

Study of the total equivalent annual cost curve reveals that it is rather flat in the region of the minimum. It may, therefore, be concluded that a deviation of one or two years from the minimum cost life will result in relatively small increases in total equivalent annual cost.

The mathematical model for determining the economic life of an asset assumes an interest rate of zero. For the example of Table 8.4, with an interest rate of zero, the minimum cost life is

$$n = \sqrt{\frac{2P}{m}} = \sqrt{\frac{2(\$5,000)}{\$100}} = 10 \text{ years.}$$

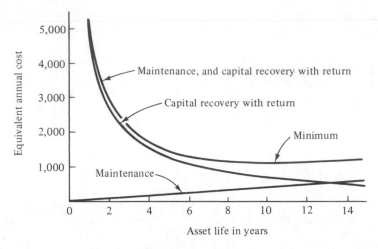

Figure 8.4. Minimum cost life of an asset.

Therefore, this equation may be used to approximate the economic life of an asset for cases involving interest.

In addition to the general procedure just discussed, there are two special situations for which the economic life can be discovered without lengthy calculations.

1. One of these situations occurs whenever the annual operating costs of an asset remain constant over its life while the asset's future salvage values remain the same over the service life. As previously shown for the case where maintenance costs remain constant, the minimum cost action is to retain the asset as long as possible. In many cases an existing old asset has a zero salvage value with no expectation of a change in its future salvage. Thus, there are many real world situations that meet these requirements and as a result, the economic life for such assets is equal to the service life. That is, the longer the asset is in service, the lower its equivalent annual costs.

2. Another special situation develops if the present and future salvage values *always equal* each other and the annual operating and maintenance costs are *always increasing*. For this set of circumstances, the economic life is the shortest possible life, namely, one year (or one period depending upon the frequency of the replacement studies). This fact is evident from the relationship describing the total costs of the asset for any year.

Total equivalent annual costs = capital recovery with return
+ equivalent annual operating costs.

The capital recovery portion of total costs will be constant for any asset for which $P = F$ no matter how long the asset is in service.

$$\text{Capital recovery with return} = (P - F)(\overset{A/P\,i,\,n}{}) + Fi.$$

The equivalent annual operating costs for an asset will be ever increasing as long as each year's operating expense is greater than the proceeding year's expense. Thus, for these two conditions the total equivalent annual costs will be minimized for the shortest time the asset might be reasonably retained. In this book this time will be considered to be the time between the present and the next time replacement would be reconsidered: one year.

8.5. Replacement Analysis Based on Economic Life

Two assets must be evaluated at the time replacement is being considered: the defender and its challenger. However, because of the cost patterns normally associated with these assets, there may be a number of alternatives that must be considered. For each asset, it may be necessary to evaluate the equivalent annual cost of keeping each asset for 1, 2, 3, etc., years. Having computed the equivalent costs associated with each of these alternatives, the economic life for the defender and the challenger are easily identified. Then the comparison is reduced to just two alternatives, the defender and the challenger retained for their respective economic lives.

To summarize the basic elements that should be considered when undertaking a replacement study, consider the following items:

1. Sunk costs should be ignored (use the outsider viewpoint).
2. Find the economic lives of the assets under consideration (use the lives most favorable to each asset).
   a. If annual costs are constant and the future salvage values are equal, select the longest possible life.
   b. If annual costs are always increasing and the present and future salvage values are equal, select the shortest possible life.
3. Compare the replacement alternatives (use the principles of comparison for assets with unequal lives presented in Sections 7.6 and 8.3).

To illustrate the comparison of alternatives on the basis of their economic life, consider the following example. Three years ago a chemical processing plant installed a system at a cost of $20,000 to remove pollutants from the waste water that is discharged into a nearby river. The present system has no present salvage value and will cost $14,500 to operate next year, with operating costs expected to increase at the rate of $500 per year thereafter. A new system has been designed to replace the existing system, and it is expected its installed cost will be $10,000. The new system is expected to have first-year operating costs of $9,000, with these costs increasing at a rate of $1,000 per year. The new system is estimated to have a useful life of 12 years. Because the original system and the new system are specially designed for this particular chemical process, their salvage values at any future time are expected to be equal to zero. Should the company replace the existing pollution control system if their minimum attractive rate of return is 12%?

First, the $20,000 initial cost of the existing plant is ignored as a sunk cost. Second, the economic life of the old system is found. However, since the present and future salvage value for the old system are zero, the shortcut method may be applied.

The total equivalent annual costs that will be incurred if the old system is retained for n more years is as follows:

Total equivalent annual costs = capital recovery with return
+ equivalent annual operating costs.

$$\text{Total equivalent annual costs for } n \text{ years} = [(P - F)(\overset{A/P\ 12,\,n}{})$$
$$+ F(0.12)] + [\$14,500 + \$500(\overset{A/G\ 12,\,n}{})].$$

Since $P = F = \$0$, the capital recovery with return for any n is equal to zero. Thus, for the pollution control system presently in service, the annual operating costs are the total costs to be incurred if the system is retained. Since the annual operating costs are increasing each year, the equivalent annual operating costs are also increasing for each additional year the present system is retained. With this pattern of increasing costs, the life for which the total equivalent annual costs will be minimized is the shortest possible life, one year. The total equivalent annual costs for the old system retained one more year is $14,500.

Next, it is necessary to find the economic life for the new system. In this case, the total equivalent annual costs are calculated for the first 6 years as shown in Table 8.5. These costs would continue to increase if the new system is operated for more than 6 years. The economic life for the new system is 5 years and the total annual equivalent cost is $13,544.

These alternatives can now be compared on a basis that is most favorable to each alternative. The methods presented in Sections 7.6 and 8.3 for comparing assets with unequal lives should now be applied. The conclusion

Table 8.5. EQUIVALENT ANNUAL COST FOR THE NEW SYSTEM

n	Capital Recovery with Return	Equivalent Annual Operating Costs	Total Equivalent Annual Cost
1	$11,200	$ 9,000	$20,200
2	5,917	9,470	15,387
3	4,164	9,920	14,084
4	3,292	10,360	13,652
5	2,774	10,770	13,544*
6	2,432	11,170	13,602

*Economic life.

is that the new system, if kept for its economic life, is preferred to operating the old system one more year. If the new system is adopted, then its cost pattern should be reviewed with regard to any future systems that might be an improvement over the existing system. This review procedure should occur periodically.

Considering sequences of future challengers. One approach to replacement analysis is to explicitly consider the consequences of retaining an existing asset and replacing it by a sequence of successors as compared to the consequences of accepting the present challenger and its sequence of successors. As a result the consequences of retaining the existing asset or accepting the challenger can be considered over a relatively long time span.

In order to quantify the cash flows that are to be expected from future challengers various assumptions have been made to describe the effects of technological change and physical impairment on these cash flows. As technological innovation and improvement continue it is expected that there will be a decrease in the cost of providing a future service similar to the service presently provided. Thus, because of obsolescence the longer an asset is retained, the greater the disparity between it and possible future replacements. These are not the only costs that increase relative to an asset's service life. Usually, those costs associated with physical impairment also increase as the life of an asset is extended.

When considering the sequences of challengers that are assumed to be the successors of an existing asset or the present challenger it is necessary to determine how frequently future challengers should be replaced in light of the effects of obsolescence and physical impairment on their future cash flows. The replacement decision can then be made considering those alternatives that assure the most favorable replacement policy for the future successors.

8.6. Examples of Replacement Analyses

The main considerations leading to replacement may be classified as inadequacy, excessive maintenance, declining efficiency, and obsolescence. Any of the above may lead to replacement, but usually two or more are involved when replacement is considered. In the sections that follow, examples illustrating an approach to replacement analysis for each of these considerations will be presented.

Replacement because of inadequacy. A physical asset that is inadequate in capacity to perform its required services is a logical candidate for replacement. For example, a boring mill used almost exclusively to face and bore pulleys has a maximum capacity of machining pulleys 54 inches in diameter. At the time the mill was purchased, the largest pulley ordered was well below the capacity of the mill, but at the present time orders are being received for pulleys up to 72 inches in diameter and these orders seem to be on the increase.

Orders for pulleys between 54 and 72 inches are subcontracted to another concern. Not only is this costly but it occasions delays that are detrimental to the reputation of the company. The factor entering into consideration of replacement in this example is inadequacy. Although the present boring mill is up to date, efficient, and in excellent condition, consideration of its replacement is being forced by the need for a boring mill of greater capacity.

When there is inadequacy, a usable piece of equipment, often in excellent condition, is on hand. Often, as in the case of the boring mill, the desired increased capacity can be met only by purchasing a new unit of equipment of the desired capacity.

In many cases, such as with pumps, motors, generators, and fans, the increased capacity desired can be met by purchasing a unit to supplement the present machine, should this alternative prove more desirable than purchasing a new unit of the desired capacity.

The method of comparing alternatives when inadequacy is the principal factor will be illustrated by the following example. One year after a 10-h.p. motor has been purchased to drive a belt coal conveyor, it is decided to double the length of the belt. The new belt requires 20 h.p. The needed power can be supplied either by adding a second 10-h.p. motor or by replacing the present motor with a 20-h.p. motor.

The present motor cost $420 installed and has a full load efficiency of 88%. An identical motor can now be purchased and installed for $440. A 20-h.p. motor having an efficiency of 90% can be purchased and installed for $780. The present 10-h.p. motor will be accepted as $270 on the purchase price of the 20-h.p. motor. Current costs $0.03 per kw-hr., and the conveyor system is expected to be in operation 2,000 hours per year.

Maintenance and operating costs other than for current of each 10-h.p. motor are estimated at $35 per year and for the 20-h.p. motor at $50 per year. Taxes and insurance are taken as 1% of the purchase price. Interest will be at the rate of 6%. The service lives of the new motors in the present application are taken as 10 years, with a salvage value of 20% of their original cost at that time. The present motor will be considered to have a total life of 11 years, an approximation that will introduce little practical error in the analysis. Most likely all motors will outlast the period of service they will have in the application under consideration.

Alternative *A* will involve the purchase of the 20-h.p. motor for $780 and the disposal of the present motor for $270. The annual cost for this alternative is computed as follows:

$$
\begin{array}{ll}
\text{Capital recovery and return, } (\$780 - \$156)(\overset{A/P\,6,\,10}{0.1359}) + \$156(0.06) & \$\quad 94.16 \\[4pt]
\text{Current cost, } \dfrac{20\ \text{h.p}}{0.90\ \text{eff.}} \times \dfrac{0.746\ \text{kw}}{\text{h.p.}} \times \dfrac{\$0.03}{\text{kw-hr.}} \times 2{,}000\ \text{hr.} \dots\dots & 994.67 \\[4pt]
\text{Maintenance and operating cost } \dots\dots\dots\dots\dots\dots\dots\dots & 50.00 \\[4pt]
\text{Taxes and insurance, } \$780 \times 0.01 \dots\dots\dots\dots\dots\dots\dots & \underline{7.80} \\[4pt]
\qquad \text{Total equivalent annual cost } \dots\dots\dots\dots\dots\dots\dots\dots & \$1{,}146.63 \\
\end{array}
$$

Alternative *B* will involve the purchase of an additional 10-h.p. motor for $440. The annual cost for this alternative is computed as follows:

Present 10-h.p. Motor:

$$
\begin{array}{ll}
\text{Capital recovery and return, } (\$270 - \$84)(\overset{A/P\,6,\,10}{0.1359}) + \$84(0.06) & \$\quad 30.32 \\[4pt]
\text{Current cost } \dfrac{10}{0.88} \times 0.746 \times \$0.03 \times 2{,}000 \dots\dots\dots\dots & 508.64 \\[4pt]
\text{Maintenance and operating cost } \dots\dots\dots\dots\dots\dots\dots & 35.00 \\[4pt]
\text{Taxes and insurance, } \$420 \times 0.01 \dots\dots\dots\dots\dots\dots\dots & 4.20 \\
\end{array}
$$

New 10-h.p. Motor:

$$
\begin{array}{ll}
\text{Capital recovery and return, } (\$440 - \$88)(\overset{A/P\,6,\,10}{0.1359}) + \$88(0.06) & 53.12 \\[4pt]
\text{Current cost, } \dfrac{10}{0.88} \times 0.746 \times \$0.03 \times 2{,}000 \dots\dots\dots & 508.64 \\[4pt]
\text{Maintenance and operating cost } \dots\dots\dots\dots\dots\dots\dots & 35.00 \\[4pt]
\text{Taxes and insurance, } \$440 \times 0.01 \dots\dots\dots\dots\dots\dots\dots & \underline{4.40} \\[4pt]
\qquad \text{Total equivalent annual cost } \dots\dots\dots\dots\dots\dots\dots & \$1{,}179.32 \\
\end{array}
$$

On the basis of the analysis above, the advantage of replacing the 10-h.p. motor rather than supplementing it is equivalent to $1,179 less $1,147, or

$32 per year. Because of the incorrect decision to purchase a 10-h.p. motor a year ago, a loss equal to $420 less $270 is incurred. This loss has been revealed, rather than caused, by the present analysis. Since engineering economy analyses are concerned with the future, this sunk cost must not enter into the analysis.

The trade-in value of $270 was taken as the present value of the original 10-h.p. motor because if it is replaced $270 will be received for it. Thus, its value is a necessary element in the comparative analysis. The annual charge for taxes and insurance was based on the original cost, because taxes and insurance charges are usually based upon book value; but for simplicity no reduction was made in these items to correspond to the expected decline in book values.

The alternative to supplement the present motor will require an investment of $270 in the present motor plus $440 in a new 10-h.p. motor or a total investment of $710. The second alternative can be implemented by an investment of $780.

The fact that $270 can be realized from the sale of a capital asset and applied upon the purchase price of the 20-h.p. motor does not reduce the expenditure necessary to acquire the motor or the amount invested in it. Thus, the analysis above reveals that an additional investment of $780 − $710 = $70 will result in a return of 6% on the additional investment plus $32 per annum.

Replacement because of excessive maintenance.   A machine rarely has all of its elements wear out at one time. Experience has proven that it is economical to repair many types of assets in order to maintain and extend their usefulness. Some repairs are of a current nature and minor in extent. Others are periodic and extensive.

An extensive periodic repair is not usually contemplated until it becomes necessary to extend the life of the unit of equipment in question. Usually, for example, an engine is not overhauled until its failure to provide acceptable service has occurred or is believed to be imminent. Thus, the cost of an extensive periodic repair may be considered to be an expenditure to purchase additional service by extending the life of a unit of equipment. This view holds even when a program of preventative maintenance is in effect.

Before an expenditure for major repairs is made to extend the service life of a machine or structure, analysis should be made to determine if the needed service might be more economically provided by other alternatives.

In this connection consider the following situation. The main roadway through an oil refinery, six-tenths of a mile long and 20 feet wide and made of concrete, is badly in need of repair to continue in service. The maintenance department of the refinery estimates that repairs which will extend the life of the roadway for 3 years can be made for $15,000. A contractor has offered to replace the present roadway with a type of pavement estimated to have a life of 20 years for $65,000.

Current maintenance cost on the repaired pavement is estimated to average $1,200 per year and that on the replacing pavement is estimated to average $200 per year. Other items are considered to be equal or negligible. The company's minimum attractive rate of return is considered to be 12%. The salvage value of the present pavement is considered to be nil if it is replaced. The annual cost comparison for the two alternatives follows:

Repair Pavement to Extend Its Life 3 Years

Capital recovery and return $15,000 (0.4164) *A/P* 12, 3	$6,246
Average annual repair cost	1,200
Total...	$7,446

Replace with Pavement with Estimated Life of 20 Years

Capital recovery and return $65,000 (0.1339) *A/P* 12, 20	$8,704
Average annual repair cost	200
Total...	$8,904
Annual advantage of repairing over replacing	$1,458

In some classes of equipment current repairs increase with age. Maintenance may be slight at first but increases at a progressive rate. Thus, a point in time is ultimately reached when it is more economical to replace than to continue maintenance. To illustrate the economy of this situation, consider the following example.

A piping system in a chemical plant was installed at a cost of $32,000. This system deteriorated by corrosion until it was replaced at the end of 6 years. The salvage value of the system was nil. Maintenance records show that maintenance costs in the past have been as given in Column *B* of Table 8.6. It may be noted that annual cost of maintenance increases with lapse of

Table 8.6. ANALYSIS OF MAINTENANCE COSTS

Year A	Cost of Maintenance for Year B	Sum of Maintenance Cost to End of Year, $\sum B$ C	Cost of *n* Years of Service, $32,000 + C$ D	Average Annual Cost of Service to End of Year, $D \div A$ E
1	$ 1,260	$ 1,260	$33,260	$33,260
2	3,570	4,830	36,830	18,415
3	6,480	11,310	43,310	14,437
4	9,840	21,150	53,150	13,287
5	14,230	35,380	67,380	13,476
6	19,820	55,200	87,200	14,533

time. This is typical of many classes of equipment and may be the primary reason for replacement. Total expenditures for repairs are given to the end of any year in Column *C*. The sum of the maintenance costs given in Column *C* and the original cost of the equipment is equal to the cost of providing the number of years' service designated in Column *A*.

The piping system could have been scrapped at the end of any year. Column *E* gives the average annual cost of service that would have resulted from scrapping the system at the end of any year. Thus, if the system had been scrapped at the end of the first year, the cost for a year of service would have been $33,260. If it had been scrapped at the end of the second year, the 2 years of service would have cost $36,830, as given in Column *D*, and the average annual cost would have been, $18,415.

The least average annual cost, $13,287, occurs for a 4-year life. If interest had been considered and equivalent annual costs had been used, the quantities in Column *E* would have been somewhat larger than those given. But the general pattern would have been much the same. Although the lowest annual cost in the example above occurs for a 4-year life, it does not necessarily follow that greatest economy would have resulted from scrapping the system after 4 years of service. The economy of replacement depends upon a number of additional factors such as the need for services of a piping system in the future, changes in levels of maintenance cost, and the characteristics and cost of a replacement. A decision to replace the present equipment should be based upon an analysis of costs in prospect with the present equipment and with a possible replacement.

Replacement because of declining efficiency. Equipment usually operates at peak efficiency initially and suffers a loss of efficiency with usage and age. A gasoline engine usually reaches its maximum efficiency after a short run-in period after which its efficiency declines as cylinder walls, pistons, piston rings, and carburetors wear and the ignition system deteriorates.

When loss of efficiency is due to the malfunctioning of only a few parts of a whole machine, it is often economical to replace them periodically and in this way maintain a high level of efficiency over a long life.

There are a number of facilities that decline in efficiency with use and age but which are not feasible to repair. Pipes that carry hot water, for example, often fill with scale. As their internal diameter decreases, the amount of energy required to force a given quantity of water through them increases. Pipe lines often decline in efficiency as carriers of fluid or gas because of increasing loss by leakage due to external or internal corrosion with age. When it is not economical to restore efficiency by maintenance, the entire system should be replaced at intervals on the basis of economy. Consider the following example. The buckets on a conveyor are subject to wear that reduces

Table 8.7. ANALYSIS OF DECLINING EFFICIENCY

Year Number A	Efficiency at Beginning of Year B	Average Efficiency During Year C	Annual Hour of Operation 1200 ÷ C D	Annual Cost of Operation Exclusive of Replacement of Buckets, D × $6.40 E	Sum of Operation Costs to End of Year, ΣE F	Average Annual Cost of Service to End of Year, ($960 + F) ÷ A G	Equivalent Annual Cost of Service to End of Year for 7% Interest H
1	1.00	0.97	1,237	$7,917	$ 7,917	$8,877	$8,973
2	0.94	0.91	1,319	8,442	16,359	8,659	8,701
3	0.88	0.86	1,395	8,928	25,287	8,749	8,773
4	0.84	0.82	1,463	9,363	34,650	8,903	8,904
5	0.80						

the capacity of the conveyor in accordance with the data given in Table 8.7.

As the capacity of the buckets becomes smaller, it is necessary to run the conveyor for longer periods of time, thus increasing operating costs. When the buckets are in new condition, the desired annual quantity of material can be handled in 1,200 hours of operation. The hours of operation required for various efficiency levels are shown in Column D. At $6.40 per hour of operation the annual cost of operation is given in Column E. The average annual cost in Column G is based on a bucket replacement cost of $960.

The example above is typical of many kinds of equipment whose efficiency declines progressively when it is not feasible to arrest the decline with maintenance. In the example the least cost of operation occurs when efficiency is permitted to decline to 88%, corresponding to a life of 2 years before replacement takes place.

Although least cost of operation occurs for a life of 2 years in the example above, this is not conclusive evidence that least cost of operation will result from a policy to replace buckets at 2-year intervals unless the replacing buckets will duplicate the buckets being replaced in first and subsequent costs. But determination of the economic life for a unit of equipment and casual consideration of subsequent replacement is often sufficient and as far as it is practical to go in many situations.

Suppose an existing conveyor system is being considered for replacement by a new conveyor with the operation characteristics shown in Table 8.7. If the salvage value of the old conveyor is nil and the estimated cost of operation for the next year is $8,900 with anticipated increases in future operating expenses, the old conveyor's economic life is one year. For an interest rate of 7% it is seen in Column *H* that the economic life of the new conveyor is 2 years. Thus, the new conveyor should be selected since an equivalent saving of at least $199 ($8,900 — $8,701) per year will be realized if the new conveyor is kept for the optimum time. Even if the new conveyor is retained for a period longer than its economic life, say for 4 years, there is still an advantage to replacing the old conveyor since its future annual operating expenses are expected to increase.

Replacement because of obsolescence. As an illustration of the analysis involving replacement because of obsolescence, consider the following example. A manufacturer produces a hose coupling consisting of two parts. Each part is machined on a turret lathe purchased 13 years ago for $6,300 including installation. A new turret lathe is proposed as a replacement for the old. Its installed cost will be $15,000.

The production times per 100 sets of parts with the new and old machine are as follows:

Part	Present Machine	New Machine
Connector............	2.92 hours	2.39 hours
Swivel	1.84 hours	1.45 hours
Total	4.76 hours	3.84 hours

The company's sales of the hose couplings average 40,000 units per year and are expected to continue at approximately this level. Machine operators are paid $8.50 per hour. The old and the proposed machine require equal floor space. The proposed machine will use power at a greater rate than the present one, but since it will be used fewer hours, the difference in cost is not considered worth figuring. This is also considered true of general overhead items. Interest is to be taken at 12%. The salesman for the new machine has found a small shop that will purchase the old machine for $1,200. The prospective buyer estimates the life of the new machine at 10 years and its salvage value at 10% of its installed cost of $15,000. The old turret lathe is estimated to be physically adequate for 2 more years and to have a salvage value of $250 at the end of that time.

The equivalent annual cost of operation if the present turret lathe is retained will be as follows:

$$A/P\,12,2$$
Capital recovery with return, ($1,200 — $250)(0.5917)
+ $250(0.12) .... $ 592
Direct labor, (4.76 ÷ 100)(40,000)($8.50)....................... 16,184
$16,776

The equivalent annual cost of operation if the new turret lathe is purchased will be as follows:

$$\overset{A/P\,12,\,10}{\text{Capital recovery with return, (\$15,000} - \text{\$1,500)(0.1770)}}$$

Capital recovery with return, ($15,000 − $1,500)($\overset{A/P\,12,\,10}{0.1770}$)
    + $1,500(0.12) .... $ 2,570
Direct labor, (3.84 ÷ 100)(40,000)($8.50)....................... 13,056
                                                               ─────────
                                                                $15,626

The annual amount in favor of the new machine is $1,150. It should be noted that the new machine will be used (3.84 ÷ 100) × 40,000 = 1,536 hours per year. No cognizance is taken of the fact that it is available for use many more hours per year; the unused capacity is of no value until used. Since, however, the additional capacity is potentially of value and may prove a safeguard against inadequacy, it should be considered an irreducible in favor of the new machine.

Replacement because of a combination of causes. In most situations, a combination of causes rather than a single cause leads to replacement consideration. As an item of equipment ages, its efficiency may be expected to decline and its need for maintenance to increase. More efficient units of equipment may become available. Moreover, it frequently happens that changes in activities result in a unit being either too large or too small for maximum economy.

Regardless of the cause or combination of causes that lead to consideration of replacement, analysis and decision must be based upon estimates of what will occur in the future. The past is irrelevant in the contemplated analysis.

PROBLEMS

1. A soft drink bottler purchased a bottling machine 2 years ago for $16,800. At that time it was estimated to have a service life of 7 years with no salvage value. Annual operating cost of the machine amounted to $4,400. A new bottling machine is being considered which would cost $20,000 but would match the output of the old machine for an annual operating cost of $1,800. The new machine's service life is 5 years with no salvage value. An allowance of $5,000 would be made for the old machine on the purchase of the new machine. The interest rate is 10%.
   (a) List the receipts and disbursements for the next 5 years if the old machine is retained; if the new machine is purchased. Compare the present worth of receipts and disbursements.

(b) Take the "outsider" viewpoint and calculate the equivalent annual cost for each of the two alternatives.

(c) What is the use value of the old machine in comparison with the new machine?

(d) Should the new machine be purchased? Why?

2. A set of magnetic tape units for a computer cost $150,000 when the computer was purchased 2 years ago. New technology has made available an improved set of tape units that can decrease the processing time of the computer system by 15%. The manufacturer of the new tape units offers to allow 25% of the old unit's first cost as a trade-in value. The new tape units cost $350,000. It is anticipated that the present computer system will be completely replaced in 4 years by a new generation computer. The salvage values of the old and new tape units at that time are estimated to be $25,000 and $40,000, respectively. The computer will operate 8 hours a day for 20 days per month. If computer time saved is valued at $300 per hour and the interest rate is 12% compounded monthly, should the existing tape units be replaced? Maintenance costs are considered to be the same for both old and new units. If it is anticipated that the operating time is to be 10 hours per day, would this change the decision?

3. A municipality 3 years ago purchased a pump for its sewage treatment plant at $1,800. This pump had annual operating costs of $1,000 and these are expected to continue. This pump is expected to continue to operate satisfactorily for 5 additional years, at which time it may be expected to have negligible salvage value. The municipality has an opportunity to purchase a new pump for $2,700. The new pump is estimated to have a life of 5 years, negligible salvage value at the end of its life, and an annual operating cost of $400. If the new pump is purchased the old pump will be sold for $200. The interest rate is 6%.

(a) What error in equivalent annual costs will result if the municipality errone-ously adds the sunk cost it has suffered to the cost of the new pump in making a comparison of the financial desirability of the two pumps?

(b) Calculate the comparative-use value of the old pump.

4. A machine was purchased 3 years ago for $12,000. Its present value is $5,000 and its operating expenses are expected to continue at $1,000 a year. A second-hand machine costing $2,000 is available but its operating expenses are expected to be $1,600 per year. It is anticipated that the machines will be in service for 6 more years with $1,000 salvage value for the present machine and zero for the second-hand machine. Using the rate-of-return approach and a minimum attractive rate of return of 15%, find the best course of action.

5. A hydroelectric plant utilizing a continuous flow of 11 cubic feet of water per second with an absolute head of 860 feet was built 4 years ago. The 18-inch pipeline in the system cost $92,000 for pipe, installation, and right of way and has a loss of head due to friction of 81 feet. Additional water rights have been acquired which will result in a total of 22 cubic feet per second of water flow. The following plans are under consideration for utilizing the total flow. Plan A:

Use the present pipeline. This will entail no additional expense but will result in a total loss of head due to friction of 346 feet resulting from the increased velocity of the water. Plan B: Add a second 18-inch pipeline at a cost of $68,000. The loss of head for this line will be 81 feet. Plan C: Install a 26-inch pipeline at a cost of $91,000 and remove the existing line, for which $3,800 can be realized. The loss of head due to friction for the 26-inch pipeline will be 63 feet.

The energy of the water delivered to the turbine is valued at $64 per horsepower year, where horsepower $= h \times F \times 62.4 \div 550$. In this equation, h is the head in feet and F is the flow in cubic feet per second. Insurance and taxes amount to 2% of first cost. Operating and maintenance costs are essentially equal for all three plans. The interest rate is 10%. If all lines, including the one now in use, will be retired in 30 years with no salvage value, what is the comparable equivalent annual costs of the three alternatives?

6. Plot the data in Columns D, E, and F of Table 8.3. If an interest rate had been used the data would be slightly different. Indicate with a superimposed dashed line the change that interest would make.

7. At the end of the seventh year, replacement of the asset described in Table 8.2 is being considered. The salvage value of the asset has been determined to be zero at the end of the seventh year. What will be the average yearly cost to operate the asset 1, 2, or 3 years more?

8. At the end of the ninth year, replacement of the asset described in Table 8.4 is being considered. At that time, the salvage value of the asset was estimated to be zero. What will be the equivalent annual cost of operating the asset 1, 2, or 3 years more?

9. A special milling machine is being installed at a first cost of $10,000. Maintenance cost is estimated to be $5,500 for the first year and will increase by 6% each year. If interest is neglected and the salvage value is $2,000 at any time, for what service life will the average annual cost be a minimum?

10. Use the same data given in Problem 9, but assume that the interest rate is 12% compounded continuously. Find the economic life of the machine.

11. The maintenance cost of a certain machine is zero the first year and increases by $200 per year for each year thereafter. The machine costs $2,000 and has no salvage value at any time. Its annual operating cost is $1,000 per year. If the interest rate is zero what life will result in minimum average annual cost? Solve by trial and error showing yearly costs in tabular form.

12. The data are the same as in Problem 11 except that the interest rate is 10%. Solve using the tabular method.

13. A special-purpose machine is to be purchased at a cost of $20,000. The following table shows the expected annual operating costs, maintenance costs, and salvage values for each year of service. If the rate of interest is 10%, what is the economic life for this machine?

Year of Service	Operation Cost for Year	Maintenance Cost for Year	Salvage Value at End of Year
1	$ 2,000	200	$10,000
2	3,000	300	9,000
3	4,000	400	8,000
4	5,000	500	7,000
5	6,000	600	6,000
6	7,000	700	5,000
7	8,000	800	4,000
8	9,000	900	3,000
9	10,000	1000	2,000
10	11,000	1100	1,000

14. Five years ago a conveyor system was installed in a manufacturing plant at a cost of $27,000. It was estimated that the system, which is still in good condition, would have a useful life of 20 years. Annual operating costs are $1,350. The number of parts to be transported have doubled and will continue at the higher rate for the rest of the life of the system. An identical system can be installed for $22,000, or a system with a 20-year life and double the capacity can be installed for $31,000. Annual operating cost is expected to be $2,500. The present system can be sold for $6,500. Either of the three systems will have a salvage value at retirement of 10% of original cost. The minimum attractive rate of return is 12%. Compare the two alternatives for obtaining the required services on the basis of equivalent annual cost over a 15-year study period, recognizing any unused value remaining in the systems at the end of that time.

15. Four years ago an ore-crushing unit was installed at a mine at a cost of $86,000. Annual operating costs for this unit are $3,540, exclusive of charges for interest and depreciation. This unit was estimated to have a useful life of 10 years and this estimate still appears to be substantially correct. The amount of ore to be handled is to be doubled and is expected to continue at this higher rate for at least 20 years. A unit that will handle the same amount of ore and have the same annual operating cost as the one now in service can be installed for $80,000. A unit with double the capacity of the one now in use can be installed for $125,000. Its life is estimated at 10 years and its annual operating costs are estimated at $4,936. The present realizable value of the unit now in use is $32,000. All units under consideration will have an estimated salvage value at retirement age of 12% of the original cost. The interest rate is 15%. Compare the two possibilities of providing the required service on the basis of equivalent annual cost over a study period of 6 years, recognizing unused value remaining in the unit at the end of that time.

16. Two bridge designs proposed for the crossing of a small stream are to be compared by the present-worth method for a period of 40 years. The wooden design has a first cost of $6,000 and an estimated life of 8 years. The steel design has a first cost of $11,000 and an estimated life of 20 years. Each structure has zero salvage value at the end of the given period and it is estimated that the annual expenditures will be the same regardless of which design is selected. Using an

interest rate of 8%, does the increased life of the steel design justify the extra investment? Make a present-worth comparison.

17. A small manufacturing company leases a building for machining of metal parts used in their final product. The annual rental of $10,000 is paid, in advance on January 1. The present lease runs until December 31, 1983 unless terminated by mutual agreement of both parties. The owner wishes to terminate the lease on December 31, 1979 and offers the company $2,000 if it will comply with the request. If the company does not agree, the lease will remain in effect at the same rate until the 1983 termination date. The company owns a suitable building lot and has a firm contract for construction of a building for a total cost of $170,000 to be completed in one year. These figures are firm whether the building is constructed in 1979.

If the company elects to stay in the leased building, it will spend $8,000, $6,500, and $7,000 in 1980, 1981, and 1982, respectively, on the facility with no salvage value resulting. It is estimated that operating expenses will be $3,500 less per year in the new building for a comparable level of output. Taxes, insurance, and maintenance will cost 3.5% of the first cost of the building per year. The life of the building is estimated to be 25 years with no salvage value and the interest rate is 6%. The decision is to be made on January 1, 1979 on the basis of the present worth of the two plans as of December 31, 1979.

(a) What is the present worth of the two plans as of December 31, 1979 if the study period is 25 years?

(b) What is the present worth of the two plans as of December 31, 1979 if the unused value is not recognized?

(c) The company considers the privilege of waiting 4 years to build to have a present worth of $6,000. On the basis of this fact, and the results of part (a), which plan should be adopted?

18. A manufacturer is considering the purchase of an automatic lathe to replace one of two turret lathes. The turret lathes were purchased 10 years ago at a cost of $3,400. The automatic lathe can be purchased for $15,800 and the turret lathe can be sold for $700. Other pertinent data are as follows:

	Turret Lathes	Automatic Lathe
Annual output, Part A	40,000 each	80,000
Annual use other than on Part A	400 hours each	0
Production, units of Part A per hour	34 each	82
Labor (one man per machine)	$5.10 per hour	$4.20 per hour
Estimated annual maintenance	$850 each	$1,600
Power cost per hour of operation	$0.15	$0.30
Taxes and insurance, 1.6% of	Present value	Original cost
Space charges per year	$300	$450

The turret lathes are in good mechanical condition and may be expected to serve an additional 10 years before maintenance rises appreciably. Their salvage value will probably never drop below $300 each. If one of the turret lathes is replaced by the automatic lathe, it is assumed that the automatic lathe will be used to produce the entire 80,000 units of Part A and that the remaining turret lathe will be used 800 hours per year. If the interest rate is 8%, how many years will be required for the automatic lathe to pay for itself?

19. Two years ago a centrifugal pump driven by a direct-connected induction motor was purchased to meet a need for a flow of 2,000 gallons of water per minute for an industrial process. The unit cost $2,300 and had an estimated salvage value of $400 at the end of 6 years of use. It consumed electric power at the rate of 47 kw and was used 12 hours per day for 300 days a year. The process for which the water is required has been changed and in the future a flow of only 800 gallons per minute is needed 3 hours per day for 300 days a year. At the decreased flow both pump and motor are relatively inefficient and the current consumption is 36 kw.

A new unit of a capacity conforming to future needs will cost $1,350. This new unit will have an estimated life of 8 years with an estimated salvage value of $200 at the end of that time. Its current consumption will be 19 kw. The present unit can be sold for $700. The original estimates of useful life and salvage value are still believed to be reliable. Insurance, taxes, and maintenance are estimated at 6% of the original cost for both units. The cost of power is $0.022 per kilowatt-hour and the minimum attractive rate of return is 10%.

(a) Should the old unit be retained or should the new unit be purchased?

(b) At what number of hours per day for 300 days a year with a requirement of 800 gallons of water per minute will the future equivalent annual cost of the two units be equal?

20. A chemical processing plant secures its water supply from a well which is equipped with a 6-inch, single-stage centrifugal pump that is currently in good condition. The pump was purchased 3 years ago for $3,000 and has an expected life of 10 years. Because of design improvements, the demand for a pump of this type is such that its present value is only $1,200. It is anticipated that 7 years from now the pump will have a trade-in value of $400. An improved pump of the same type can now be purchased for $3,700 and will have an estimated life of 10 years with a trade-in value of $280 at the end of that time.

The pumping demand is 225 cubic feet per minute against an average head of 200 feet. The old pump has an efficiency of 75% when furnishing the demand above. The new pump has an efficiency of 81% when furnishing the same demand. Power costs $0.026 per horsepower-hour and either pump must operate 2,400 hours per year. Do the improvements made in design justify the purchase of a new pump if interest of 15% is required?

21. A private hospital is considering the replacement of one of its artificial kidney machines. The machine being considered for replacement cost $35,000 4 years ago. If the present machine is kept one more year its operating and maintenance costs are expected to be $25,000. Operating and maintenance expenses in the second and third years are expected to be $27,000 and $29,000, respectively. The new machine, if purchased now, will cost $42,000 and will have an annual

operating expense of $19,000. Its economic life is anticipated to be 5 years, and its salvage value at that time is estimated to be $10,000. The company selling the new machine will allow $9,000 on the old machine for trade-in. If the hospital delays its purchase for 1, 2, or 3 years the trade-in value on the existing machine is expected to decrease to $7,000, $5,000, and $3,000, respectively. If the interest rate is considered to be 12%, what decision would be most economical? Use an annual cost comparison.

22. The replacement of a machine is being considered by the ABC Co. The new improved machine will cost $30,000 installed and will have an estimated economic life of 12 years and $2,000 salvage. It is estimated that annual operating and maintenance costs will average $1,000 per year. The present machine had an original cost of $20,000 4 years ago and at the time it was purchased its service life was estimated to be 10 years with an estimated salvage of $5,000. An offer of $7,000 has just been received for the present machine. Its estimated costs for the next 3 years are shown below:

Year	Salvage Value at End of Year	Book Value at End of Year	Operating Maintenance Costs During Year
1	$4,000	$4,500	$3,000
2	3,000	3,000	3,429
3	2,000	1,500	6,000

Using interest at 15%, make an annual cost comparison to determine whether or not it is economical to make the replacement.

23. Two years ago a soft-drink distributor purchased a materials handling system for his warehouse. Originally, the system cost $80,000. It was anticipated that in 7 years from the time of this purchase the warehouse would be inadequate and therefore it would be sold. The materials handling equipment that is now owned is expected to have no market value in the future. Annual operating expenses for the existing system are projected to increase through time. These annual expenses in order are $2,000, $10,000, $18,000, $26,000, and $34,000 for the remaining 5 years of its service life. A firm selling materials handling equipment is presently offering a new system costing $70,000 and they are offering $40,000 for trade-in of the old equipment. This new system is expected to cost $8,000 per year to operate over its service life of 5 years. What is the economic life for the new system if its prospective salvage value is zero at any future time? For a MARR $= 8\%$, prepare a table showing the equivalent annual costs for both alternatives for each of the 5 years these systems might be used. What are your conclusions?

24. A textile firm is considering the replacement of its 3-year-old knitting machine, which has a current market value of $8,000. Because of a rapid change in fashion styles, the need for the existing machine is expected to last only 5 more years.

The estimated operating costs and its salvage values for the old machine are given as follows:

Year End	1	2	3	4	5
Operating cost	$1,000	$1,500	$2,000	$2,500	$3,000
Salvage value	6,000	4,000	3,000	2,000	0

As an alternative, a new improved machine is available on the market at a price of $10,000 and has an estimated useful life of 6 years. The pertinent cost information can be summarized as follows:

Year End	1	2	3	4	5	6
Operating cost	$ 700	$ 900	$1,100	$1,300	$1,500	$3,000
Salvage value	8,000	6,000	4,000	2,000	0	0

If the rate of interest is 15%, determine which alternative should be selected and how long the selected machine should be kept in service.

25. A shipping company which is engaged mainly in the transportation of coal from nearby mines to steel mills is concerned about the replacement of one of its old steamships with a new diesel-powered ship. The old steamship was purchased 15 years ago at a cost of $250,000. If the steamship is sold now, the fair market value is estimated to be $15,000. Once a steamship reaches this age, no major changes in its future salvage value are expected. Currently, the annual operating and maintenance costs for the steamship amount to approximately $200,000 next year and these costs are expected to increase at a rate of $15,000 a year thereafter. As an alternative to retaining the steamship, a new diesel-powered ship can be purchased at the quoted price of $470,000. In addition to this initial investment, the company would need to invest another $30,000 in a basic spare-parts inventory. The annual operating and maintenance costs and its salvage values are estimated in thousands of dollars as follows:

End of Year	1	2	3	4	5	6	7	8	9	10	11	12	13	14	15
O&M	100	100	100	130	140	150	160	170	180	190	200	210	220	230	240
Engine Overhaul					50							50			
Salvage Value	430	370	320	280	250	220	190	160	130	100	70	50	40	30	20

The useful life of the diesel ship is estimated to be 15 years. The expenditure for a general engine overhaul is treated as an expense under the firm's current accounting practice. If the firm's MARR is 12%, what decision should be made?

26. A manufacturer of cans and packaging for the food industry is considering the replacement of some of its current production equipment. A new plan is to install equipment in an existing plant facility to produce a new two-piece, thin-steel container that consumes less energy and metal than the conventional three-piece soldered cans. The equipment presently in operation was installed 5 years ago at a cost of $100,000 and can presently be sold for $35,000. Because of the rapid obsolescence of production equipment as customers switch to lighter, more economical containers, the future salvage value of the present equipment is expected to decline by $4,000 a year. If the present equipment is retained one more year, its operating and maintenance costs are expected to be $65,000, with increases of $3,000 a year thereafter. The new equipment will cost $130,000 installed. Its economic life is predicted to be 8 years with a salvage value of $10,000. Annual operating disbursements will be $49,000. If the firm's MARR is 15%, make a recommendation as to the desirability of installing the new equipment.

BREAK-EVEN AND
MINIMUM COST ANALYSIS

In many situations encountered in engineering economic analysis, the cost of an alternative may be a function of a single variable. When two or more alternatives are a function of the same variable, it may be desirable to find the value of the variable that will result in equal cost for the alte natives considered. The value of such a variable is known as the *break-even* point.

If the cost of a single alternative is a function of a variable that may take on a range of values it may be useful to determine the value of the variable for which the cost of the alternative is a minimum. The value of such a variable is known as the *minimum cost* point. Multiple alternatives which depend upon the same variable can be compared on the basis of their minimum cost points. Several facets of the break-even and minimum cost aspect of engineering economic analysis are presented in this chapter.

9.1. Break-Even Analysis, Two Alternatives

When the cost of two alternatives is affected by a common variable there may exist a value of the variable for which the two alternatives will

9

incur equal cost. The costs of each alternative can be expressed as functions of the common independent variable and will be of the form

$$TC_1 = f_1(x) \quad \text{and} \quad TC_2 = f_2(x)$$

where

$TC_1 = $ a specified total cost per time period, per project, or per piece applicable to Alternative 1;

$TC_2 = $ a specified total cost per time period, per project, or per piece applicable to Alternative 2;

$x = $ a common independent variable affecting Alternative 1 and Alternative 2.

Solution for the value of x resulting in equal cost for Alternative 1 and Alternative 2 is accomplished by setting the cost functions equal, $TC_1 = TC_2$. Therefore,

$$f_1(x) = f_2(x)$$

which may be solved for x. The resulting value for x yields equal cost for the alternatives considered and is, therefore, designated the break-even point.

Break-even point, mathematical solution. When the cost of each alternative can be mathematically expressed as a function of a common variable the break-even point may be found mathematically. For example, assume that a 20-h.p. motor is needed to drive a pump to remove water from a tunnel. The number of hours that the pump will operate per year is dependent upon the rainfall and is, therefore, uncertain. The pump unit will be needed for a period of 4 years.

Two alternatives are under consideration. Proposal A calls for the construction of a power line and the purchase of an electric motor, at a total cost of $1,400. The salvage value of this equipment at the end of the 4 year period is estimated at $200. The cost of current per hour of operation is estimated at $0.84, maintenance is estimated at $120 per year, and the interest rate is 10%. No attendant will be needed since the equipment is automatic. Let

$$TC_A = \text{total equivalent annual cost of Proposal } A;$$
$$CR(i)_A = \text{equivalent annual cost of capital recovery with return}$$
$$= (\$1,400 - \$200)(\overset{A/P\ 10,\ 4}{0.3155}) + \$200(0.10) = \$399;$$
$$M = \text{annual maintenance cost} = \$120;$$
$$C = \text{current cost per hour of operation} = \$0.84;$$
$$t = \text{number of hours of operation per year.}$$

Then

$$TC_A = CR(i)_A + M + Ct.$$

Proposal B calls for the purchase of a gasoline motor at a cost of $550. The motor will have no salvage value at the end of the 4-year period. The cost of fuel and oil per hour of operation is estimated at $0.42, maintenance is estimated at $0.15 per hour of operation, and the cost of wages chargeable to the engine when it runs is $0.80 per hour. Let

$$TC_B = \text{total equivalent annual cost of Proposal } B;$$
$$CR(i)_B = \text{equivalent annual cost of capital recovery with return}$$
$$= \$550(\overset{A/P\ 10,\ 4}{0.3155}) = \$174;$$
$$H = \text{hourly cost of fuel and oil, operator, and maintenance}$$
$$= \$0.42 + \$0.80 + \$0.15 = \$1.37;$$
$$t = \text{number of hours of operation per year.}$$

Then

$$TC_B = CR(i)_B + Ht.$$

There is a value of N for which the two alternatives will incur equal cost. This value may be found by setting $TC_A = TC_B$ and solving for N

as follows:

$$CR(i)_A + M + Ct = CR(i)_B + Ht$$

$$t = \frac{CR(i)_B - [CR(i)_A + M]}{C - H}$$

Substituting

$$t = \frac{\$174 - (\$399 + \$120)}{\$0.84 - \$1.37} = 651 \text{ hours.}$$

That the total equivalent annual cost is equal for the two alternatives is shown as follows:

$$TC_A = TC_B$$

$$CR(i)_A + M + Ct = CR(i)_B + Ht$$

$$\$399 + \$120 + 651(\$0.84) = \$174 + 651(\$1.37)$$

$$\$1,066 = \$1,066.$$

For the cost data given, the annual cost of the two alternatives is calculated to be equal for 651 hours of operation per year. If the equipment is used less than 651 hours per year, selection of the gasoline motor is most economical; for more than 651 hours of operation per year, the electric motor is most economical. The total annual cost for each alternative, as a function of the number of hours of operation per year, is shown graphically in Figure 9.1.

The difference in equivalent annual cost between the two alternatives may be calculated for any number of hours of operation. For example,

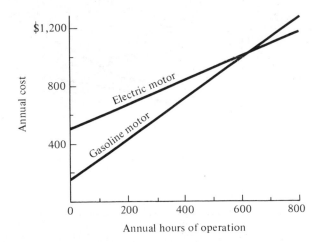

Figure 9.1. Total annual cost as a function of the number of hours of operation per year.

suppose the equipment is to be operated 100 hours per year. Then

$$TC_A - TC_B = \Delta TC$$
$$CR(i)_A + M + Ct - [CR(i)_B + Ht] = \Delta TC.$$

Substituting

$$\Delta TC = \$399 + \$120 + 100(\$0.84) - \$174 - 100(\$1.37)$$
$$= \$292.$$

Break-even point, graphical solution. There are many situations in which setting up equations to represent the cost patterns of two alternatives is either too difficult or too time-consuming to be a feasible approach for the determination of a break-even point. In situations for which the cost patterns of two alternatives can be established by determining a number of points of the pattern by calculation or experiment, the break-even point may be determined graphically.

Consider the following example. Two methods, *A* and *B*, are under consideration for packing different lengths of display material in paper cartons of 24 inches in girth and 26, 36, 48, and 62 inches in length, respectively. The average times required to pack a number of cartons of the given lengths by Method *A* and Method *B* were found by time studies. The results are shown graphically in Figure 9.2. The length of carton resulting in an equal operation time for the methods under consideration may be found by inspection from the graph.

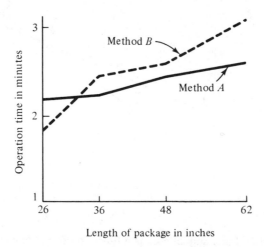

Figure 9.2. Operation time in minutes as a function of carton length.

9.2. Break-Even Analysis, Multiple Alternatives

In the examples considered thus far, break-even analysis has been applied in the case where only two alternatives confront the decision maker. This section illustrates the application of break-even analysis in cases where multiple alternatives are being considered. As an example consider an architectural engineering firm which has been asked to prepare preliminary plans for the construction of a one-story building. After careful analysis, three types of construction seem feasible.

In attempting to arrive at some quantitative basis for recommending a type of construction, the engineering firm developed fixed and variable cost data. These data are given below and are assumed to represent the cost of construction and operation for a building containing between 2,000 and 6,000 basic square feet.

Concrete and Brick

First cost per square foot	$ 24
Annual maintenance	$5,600
Annual climate control	$2,400
Estimated life in years	20
Estimate salvage value is zero.	

Steel and Brick

First cost per square foot	$ 29
Annual maintenance	$5,000
Annual climate control	$1,500
Estimated life in years	20
Estimated salvage value is 3.2% of first cost.	

Frame and Brick

First cost per square foot	$ 35
Annual maintenance	$3,000
Annual climate control	$1,250
Estimated life in years	20
Estimated salvage value is 1.0% of first cost.	

The total cost for each type of construction will be a function of the number of basic square feet enclosed by the building, A. For an interest rate of 8%, the total cost for each alternative is as follows:
Concrete and brick:

$$TC = \$24(A)(\overset{A/P\,8,\,20}{0.1019}) + \$8,000$$
$$= \$2.44(A) + \$8,000.$$

Steel and brick:

$$TC = \$29(A)(0.968)(\overset{A/P\,8,\,20}{0.1019}) + \$29(A)(0.032)(0.08) + \$6,500$$
$$= \$2.934(A) + \$6,500.$$

Frame and brick:

$$TC = \$35(A)(0.99)(\overset{A/P\,8,\,20}{0.1019}) + \$35(A)(0.01)(0.08) + \$4,250$$
$$= \$3.558(A) + \$4,250.$$

Solution for the respective break-even points may be done mathematically by considering the alternatives in pairs. Or, by graphing the total cost of each alternative as a function of the area in square feet, it is possible to determine the break-even points by inspection. Each total cost function developed is graphed in Figure 9.3.

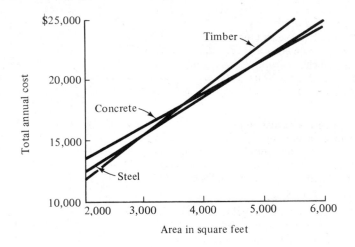

Figure 9.3. Total annual cost as a function of area in square feet.

Suppose that the client was considering a building measuring 40 feet by 100 feet. In this case, the engineering firm would recommend construction from steel and brick. However, if the client required a building of more than 5,000 square feet it would be most economical to use concrete and brick for the basic structure. For a building of less than 3,000 square feet, construction from frame and brick would be most economical. The extension of break-even analysis to cases where there are more than three alternatives follows the same reasoning.

9.3. Minimum Cost Analysis

An alternative may possess two or more cost components that are modified differently by a common variable. Certain cost components may vary directly with an increase in the value of the variable while others may vary inversely. When the total cost of an alternative is a function of increasing and decreasing cost components, most likely a value exists for the common variable that will result in a minimum cost for the alternative.

The general solution of the situation outlined above may be demonstrated for an increasing cost component and a decreasing cost component as follows:

$$TC = Ax + \frac{B}{x} + C$$

where

TC = a specified total cost per time period, per project, etc.;
x = a common variable;
A, B, and C = constants.

Taking the first derivative, equating the result to zero, and solving for x results in

$$\frac{dTC}{dx} = A - \frac{B}{x^2} = 0$$

$$x = \sqrt{\frac{B}{A}}.$$

The value for x found in this manner will be a minimum and is, therefore, designated the minimum cost point.

Minimum cost point, mathematical solution. A classical example of minimum cost analysis is given by the increasing and decreasing cost components involved in the choice of the cross sectional area of an electrical conductor. Since resistance is inversely proportional to the size of the conductor, it is evident that the cost of power loss will decrease with increased conductor size. However, as the size of the conductor increases, an increased investment charge will be incurred. For some given conductor cross section, the sum of the two cost components will be a minimum.

As an example, suppose a copper conductor is being considered to transmit the daily electrical load at a sub-station, and estimates call for transmission of 1,920 amperes for 24 hours per day, respectively, for 365 days per year. The following engineering and cost data apply to the conductor installation:

length of conductor, 140 feet; installed cost, $160 + $0.60 per pound of copper; estimated life, 20 years; salvage value, $0.50 per pound of copper. Electrical resistance of a copper conductor 140 feet long and of 1 square inch cross section is 0.0011435 ohm, and the electrical resistance is inversely proportional to the area of the cross section. The energy loss in kilowatt-hours in a conductor due to resistance is equal to $I^2R \times$ number of hours $\div$ 1,000, where I is the current flow in amperes and R is the resistance of the conductor in ohms. Copper weighs 555 pounds per cubic foot. The energy lost is valued at $0.007 per kilowatt-hour; taxes, insurance, and maintenance are negligible; the interest rate is 6%.

The I^2R loss in dollars per year:

$$(1,920)^2(24)\left(\frac{365}{1,000}\right)\left(\frac{0.0011435}{A}\right)(\$0.007) = \frac{\$258.49}{A}.$$

Weight of conductor in pounds:

$$\frac{[(140)(12)(A)(555)]}{1,728} = 539.6A.$$

Capital recovery plus return in dollars per year:

$$[\$160 + (\$0.60 - \$0.50)(539.6)(A)](\overset{A/P\,6,\,20}{0.0872})$$
$$+ \$0.50(539.6)(A)(0.06) = \$20.89A + \$13.94.$$

The total cost per year:

$$TC = \$20.89A + \frac{\$258.49}{A} + \$13.94$$

$$\frac{dTC}{dA} = \$20.89 - \frac{\$258.49}{A^2} = 0$$

$$A = \sqrt{\frac{258.49}{20.89}} = 3.52 \text{ square inches.}$$

Therefore, the selection of a conductor with a cross-sectional area of 3.52 square inches will result in a minimum total cost. Note that this example exhibits a simple case involving an increasing and a decreasing cost component. The nature of these components may be tabulated from the expressions for I^2R loss cost and investment cost. These costs and the resulting total cost are given in Table 9.1. and graphed in Figure 9.4.

Table 9.1. TOTAL ANNUAL COST AS A FUNCTION OF CROSS-SECTIONAL AREA

Cost	Cross Sectional Area (in.²)				
	2	3	4	5	6
Investment cost	$ 55.72	$ 76.61	$ 97.50	$118.39	$139.28
I^2R loss cost	$129.37	$ 86.15	$ 64.68	$ 51.75	$ 43.12
Total annual cost..	$185.09	$162.76	$162.16	$170.14	$182.40

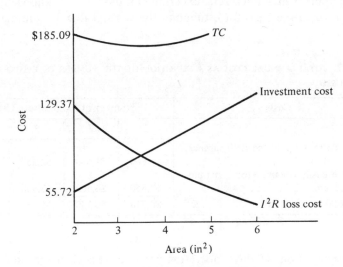

Figure 9.4. Total annual cost curve.

It was first indicated by Lord Kelvin that the most economical cross-sectional area for a conductor was one in which investment cost just equals the annual cost of lost energy. This is known as Kelvin's law.

Minimum cost point, tabular solution. There are many situations for which minimum cost points are sought but for which it is very difficult to set up equations that truly characterize the existing cost relationships. Since finding minimum cost points from cost equations involves differentiation, equations that only approximate true situations may result in grossly misleading results.

The graphical method for finding minimum cost points will be illustrated in the following example. The manufacturer of a pharmaceutical product plans to use an evaporative process. In this process one or several evaporators in multiple may be used. It is known that variable operating costs, of which the chief item is the cost of steam, will be approximately inversely proportional to the number of evaporators used in the installation. Fixed costs will be approximately in proportion to the number of evaporators used. Tests had demonstrated that a rather complex equation would be required to express accurately the cost relationships that existed, so a graphical solution was sought.

Estimates were made of the variable and the fixed operation costs that would be obtained if one, two, three, or four evaporators were used. Results from equipment manufacturers, experimental data, and a knowledge of the process formed the basis for estimating the costs. Table 9.2 summarizes the

Table 9.2. TOTAL ANNUAL COST AS A FUNCTION OF THE NUMBER OF EVAPORATORS USED

Costs	Number of Evaporators Used			
	1	2	3	4
Fixed costs (capital cost, insurance, taxes, etc.)	$ 860	$1,680	$2,350	$3,030
Variable costs (steam, labor, maintenance, etc.)	$7,850	$4,560	$3,610	$3,190
Total annual cost	$8,710	$6,240	$5,960	$6,220

estimates based on 200 days' operation per year. It is clear that, on the basis of 200 days' operation per year, the lowest annual cost of $5,960 results when three evaporators are used.

9.4. Minimum Cost Analysis, Multiple Alternatives

As an illustration of the application of minimum cost analysis where two alternatives are proposed, consider the following example. A double track railroad bridge is to be constructed for a 1,200 foot crossing. Two girder designs have been proposed. The first will result in the weight of the superstructure per foot to be $W_1 = 22(S) + 800$, where S is the span between piers. The second will result in a superstructure weight per foot of $W_2 = 20(S) + 1,000$. Regardless of the girder design chosen, the piers will cost $220,000 each. The superstructure will be erected at a cost of $0.22 per pound. All other costs for the competing designs are the same.

In order to choose a girder design on the basis of minimum total cost for the supersturcture and piers, it will be necessary to find the minimum cost pier spacing for each design. The general relationship that exists may be described as follows. As the number of piers increases, the amount of superstructure required decreases; and conversely, as the number of piers decreases, the amount of superstructure required increases. Therefore, this situation involves increasing and decreasing cost components, the sum of which will be a minimum for a certain number of piers.

The total cost for the superstructure and piers for girder Design 1 is as follows:

$$TC_1 = [22(S) + 800](\$0.22)(1,200) + \left(\frac{1,200}{S} + 1\right)\$220,000$$

$$= 5,810(S) + \frac{264,000,000}{S} + 431,000.$$

The minimum cost span between piers is

$$\frac{dTC}{dS} = 5,810 - \frac{264,000,000}{S^2} = 0$$

$$S = \sqrt{\frac{264,000,000}{5,810}} = 213 \text{ feet.}$$

And the minimum total cost for the superstructure and piers is

$$TC_1 = 5,810(213) + \frac{264,000,000}{213} + 431,000$$

$$= \$2,908,000.$$

The total cost for the superstructure and piers for girder Design 2 is as follows:

$$TC_2 = [20(S) + 1,000](\$0.22)(1,200) + \left(\frac{1,200}{S} + 1\right)\$220,000$$

$$= 5,280(S) + \frac{264,000,000}{S} + 484,000.$$

The minimum cost span between piers is

$$\frac{dTC}{dS} = 5,280 - \frac{264,000,000}{S^2} = 0$$

$$S = \sqrt{\frac{264,000,000}{5,280}} = 224 \text{ feet.}$$

And the minimum total cost for the superstructure and piers is

$$TC_2 = 5,280(224) + \frac{264,000,000}{224} + 484,000$$

$$= \$2,846,000.$$

On the basis of this analysis, the designer would choose girder Design 2 as the design that will result in a minimum total cost for the bridge. The pier spacing chosen for the design would be as approximate to 224 feet as possible. In this case, $1200 \div 224 = 5.36$ spans so 6 piers would be used for a span between piers of 240 feet. The nature of the increasing and decreasing cost components is given for various numbers of piers in Table 9.3.

Table 9.3. TOTAL COST AS A FUNCTION OF THE NUMBER OF PIERS FOR TWO COMPETING BRIDGE DESIGNS

Piers	Cost of Piers	Superstructure Cost Design 1	Superstructure Cost Design 2	Total Cost Design 1	Total Cost Design 2
4	$880,000	$2,534,000	$2,376,000	$3,414,000	$3,336,000
5	1,100,000	1,953,600	1,848,000	3,053,600	2,948,000
6	1,320,000	1,605,120	1,531,000	2,925,120	2,851,000
7	1,540,000	1,372,800	1,320,000	2,912,800	2,860,000
8	1,760,000	1,210,176	1,172,160	2,970,176	2,932,160
9	1,980,000	1,082,400	1,056,000	3,062,400	3,036,000

The extension of minimum cost analysis to cases where there are more than two alternatives follows the same procedure. Suppose three plans for constructing a large earth-filled dam are under consideration by a prime contractor. Plan I will require a greater amount of labor costs than either of the other two plans, while Plan III requires the largest expenditure for construction equipment. The total equivalent annual costs for each plan is described in Figure 9.5 as a function of the time required to complete the project.

From the annual cost curves in Figure 9.5 it is seen that the overall minimum cost plan ($12 million per year) would be Plan III if the job were completed in 3.8 years. If it was desired to complete the project in less than 3.3 years the least expensive plan would be Plan I. The least cost plan if the project were completed after 4.1 years would be Plan II. Any expected completion date between 3.3 years and 4.1 years would favor the selection of Plan III. This type of minimum cost analysis for multiple alternatives pro-

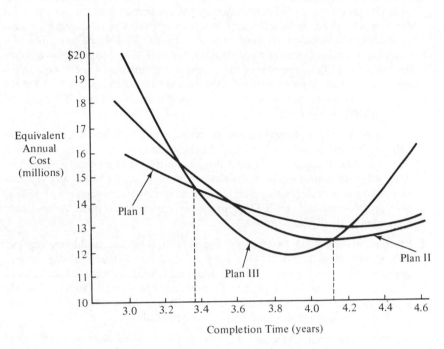

Figure 9.5. Minimum costs for three alternatives.

vides information that can be extremely helpful in the process of selecting the preferred alternative.

PROBLEMS

1. In considering the purchase of an automobile with air-conditioning, it is estimated that when the air-conditioning is off the mileage will be 14.0 miles per gallon and with air-condition on it will be 12.3 miles per gallon. The air-conditioning will add $500 to the first cost of the automobile and will increase its trade-in value by $70 after a service life of 5 years. The air-conditioning will be used 20% of the time and it will add $60 per year to the maintenance of the automobile. The benefit of having air-conditioning when needed is considered to be worth $0.04 per mile. If gasoline costs $0.60 per gallon and if the interest rate is 8%, how many miles per year must be driven for the air-conditioner to pay for itself?

2. A company may furnish a car for use by a service man for transportation between its properties, or the company may pay the service man for the use of his car at a rate of $0.16 per kilometer for this purpose. The following estimated data apply to company furnished cars: A car costs $4,000 and has a life of 4 years and a trade-in value of $800 at the end of that time. Monthly storage cost for the car is $20 and the cost of fuel, tires, and maintenance is $0.06 per kilometer. How many kilometers must a service man travel annually by car for the cost of the two methods of providing transportation to be equal if the interest rate is 15%?

3. An engineering consulting firm can purchase a small electronic computer for $30,000. It is estimated that the life and salvage value of the computer will be 6 years and $4,000, respectively. Operating expenses are estimated to be $50 per day, and maintenance will be performed under contract for $3,000 per year. As an alternative, sufficient computer time can be rented at an average cost of $130 per day. If the interest rate is 12%, how many days per year must the computer be needed to justify its purchase?

4. Machine A costs $200, has zero salvage value at any time, and labor cost per unit of product on it is $1.14. Machine B costs $3,600, has zero salvage value at any time, and labor cost per unit of product on it is $0.83. Neither machine can be used except to make the product in question. The interest rate is 5%. If the annual rate of production is 3,000 units, in how many years will the two machines break even in equivalent annual costs? Solve by trial and error.

5. A coffee concern has a weigher for filling cans with coffee. Because of its lack of sensitivity the weigher must be set so that the average filled can contains 1.008 kilograms in order to insure that no cans contain less than 1 kilogram of coffee. The present weigher was purchased 10 years ago at a price of $1,500 and is expected to last an additional 10 years. A new weigher is being considered whose sensitivity is such that the average "overage" per kilogram can is 3.5 grams. This weigher will cost $6,400. $300 will be allowed for the present weigher on the purchase price of the new weigher. The salvage value of either weigher 10 years from now is considered to be negligible and their operating costs are estimated to be equal. The value of the coffee is $2.50 per kilogram. What is the minimum number of 1 kilogram packages packed annually for which the purchase of the new weigher is justified, if the interest rate is 9%?

6. The cost of constructing a motel, including materials, labor, taxes, and carrying charges during construction and other miscellaneous items, is $1,100,000. The lots on which the motel is located cost $300,000. Furniture and furnishings cost $200,000. Working capital of 30 days' gross income at 100% capacity is required. The investment in the furniture and furnishings should be recovered in 7 years and the investment in the motel should be recovered in 25 years. The land on which the motel is built is considered not to depreciate in value.

   If the motel operates at 100% capacity, the gross annual income will be $2,000 per day for 365 days. The motel has fixed operating expenses (exclusive of capital recovery and interest) amounting to $115,000 a year and a variable operating cost of $85,000 for 100% capacity. Assume that the variable cost varies directly with the level of operation. If interest is taken at 10% compounded annually, at what percent of capacity must the motel operate to break even?

7. A certain assembly requires rods 0.10 square inches varying in length from 0.25 to 4 inches. The rods may be made from either brass or steel. The machining cost of brass rods is $0.0059 + $0.006L per piece, where L is the length of the rod in inches. For steel the machining cost may be expressed as $0.0085 + $0.010L per piece. Brass costs $0.70 per pound and steel costs $0.25 per pound. The weight of brass and steel is 0.309 and 0.283 pounds per cubic inch, respectively. What length will equalize the cost of producing the brass and steel rods? What type of rod should be used?

8. Two brands of a protective coating are being considered. Brand A costs $7.90 per gallon. Past experience with Brand A has revealed that it will cover 350 square feet of surface per gallon, will give satisfactory service for 3 years, and can be applied by a workman at the rate of 70 square feet per hour. Brand B, which costs $10.20 per gallon, is estimated to cover 400 square feet of surface per gallon and can be applied at the rate of 80 square feet per hour. The wage rate of the workman is $8.50 per hour. If an interest rate of 10% is used, how long should Brand B last to provide service at equal cost with that provided by Brand A?

9. A certain area can be irrigated by piping water from a nearby river. Two competing installations are being considered for which the following engineering and cost data apply:

	6-inch System	8-inch System
Size motor required	25 hp	10 hp
Energy cost per hour of operation	$ 0.40	$ 0.15
Cost of motor installed	$ 360	$ 160
Cost of pipe and fittings	$2,050	$2,640
Salvage value at end of 10 years.............	$ 80	$ 100

On the basis of a 10-year life with an interest rate of 12%, determine the number of hours of operation per year for which the two systems will break even.

10. An electronic manufacturer is considering two methods for producing a required circuit board. The board can be hand wired at an estimated cost of $1.20 per unit and with an annual equipment cost of $300. A printed equivalent of the required circuit can be produced with an investment of $4,500 in printed circuit processing equipment which will have an expected life of 9 years and a salvage value of $150. It is estimated that labor cost will be $0.52 per unit and that the processing equipment will cost $150 per year to maintain. If all other costs are assumed equal, and if the interest rate is 10%, how many circuit boards must be

produced each year for the two methods to break even in equivalent annual cost?

11. A contractor is offered his choice of either a gasoline, diesel, or butane engine to power a bulldozer he is to purchase. The gasoline engine will cost $2,000, will have an estimated maintenance cost of $200 per year, and will consume $3.60 worth of fuel per hour of operation. The diesel engine will cost $2,800, will cost an estimated $240 per year to maintain, and will consume $3.30 worth of fuel per hour. The butane engine will cost $3,300, will cost $315 per year to maintain, and will consume $2.90 worth of fuel per hour of operation. Since the salvage value of each engine will be identical, it may be neglected. All other costs associated with the three engines are equal and the interest rate is 15%. The service life of each engine is 5 years.
    (a) Plot the total annual cost of each engine as a function of the number of hours of operation per year.
    (b) Find the range of number of hours of operation for which it would be most economical to specify the gasoline engine; the diesel engine; the butane engine.

12. A 3-phase, 220-volt, squirrel cage induction motor is to be used to drive a pump which will fill a series of storage tanks. It is estimated that 300 horsepower-hours will be required each day for 365 days per year. Motors having the following characteristics may be leased:

Size (h.p.)	Lease Rate (per Year)	Operating Cost (per h.p.-hr.)
20	$ 928	$0.044
40	1,000	0.037
75	1,142	0.031
100	1,250	0.027
150	1,510	0.025

Plot the total yearly cost as a function of the size of the motor and select the size that will result in a minimum cost per year.

13. An hourly electric load of 1,600 amperes is to be transmitted from a generator to a transformer in a certain power plant. A copper conductor 150 feet long can be installed for $160 + $0.64 per pound, will have an estimated life of 20 years, and can be salvaged for $0.50 per pound. Power loss from the conductor will be a function of the cross-sectional area and may be expressed as $25,875 \div A$ kilowatt-hours per year. Energy lost is valued at $0.008 per kilowatt-hour; taxes, insurance, and maintenance are negligible; the interest rate is 10%. Copper weighs 555 pounds per cubic foot.

(a) Plot the total annual cost of capital recovery with a return and power loss cost for conductors for cross sections of 1, 2, 3, 4, and 5 square inches.

(b) Find the minimum cost cross section mathematically and check the result against the minimum point found in (a).

14. An overpass is being considered for a certain railroad crossing. The superstructure design under consideration will be made of steel and will have a weight per foot depending upon the span between piers in accordance with $W = 32(S) + 1,850$. Piers will be made of concrete and will cost \$215,000 each. The superstructure will be erected at a cost of \$0.45 per pound. If the number of piers required is to be one less than the number of spans, find the number of piers that will result in a minimum total cost for piers and superstructure if $L = 1,275$ feet.

15. It has been found that the heat loss through the ceiling of a building is 0.13 Btu per hour per square foot of area per degree Fahrenheit. If the 2,500-square-foot ceiling is insulated, the heat loss in Btu per hour per degree temperature difference per square foot of area is taken as being equal to

$$\frac{1}{\frac{1}{0.13} + \frac{t}{0.27}}$$

where t is the thickness in inches. The in-place cost of insulation 1, 2, and 3 inches thick is \$0.08, \$0.11, and \$0.15 per square foot, respectively. The building is heated to 75 degrees 3,000 hours per year by a gas furnace with an efficiency of 50%. The mean outside temperature is 45 degrees and the natural gas used in the furnace costs \$1.80 per 1,000 cubic feet and has a heating value of 2,000 Btu per cubic foot. What thickness of insulation, if any, should be used if the interest rate is 6% and the resale value of the building 6 years hence is enhanced \$180 if insulation is added, regardless of the thickness?

16. The government is considering charging a polluter's fee on the amount of pollutants discharged into our rivers in order to significantly reduce the pollution of our waterways. The polluter's fee is to be based on the amount of pollutant remaining in the effluent being discharged. The indicator to be used to measure the level of pollution of industrial wastes is biochemical oxygen demand, BOD. A firm is presently discharging 4,000,000 pounds of BOD per year into a nearby river because it does not treat its waste products. The cost of installing and operating a waste treatment system that reduces the BOD output to a percent of its present level is shown on page 268.

It is expected that the treatment system will last 10 years and have zero salvage value at that time. The firm's minimum attractive rate of return is 15%.

(a) If the polluter's fee is \$0.10 per pound of BOD remaining, what percent of BOD remaining will minimize the firm's annual costs?

Percent BOD Remaining	Initial Investment	Operating Expenses
5%	$250,000	$40,000
10%	100,000	25,000
15%	85,000	15,000
20%	75,000	10,000
25%	75,000	5,000

(b) If the polluter's fee is $0.05 per pound of BOD remaining, what percent of BOD remaining will minimize the firm's costs?

(c) Plot a graph of the optimum level of BOD remaining as a function of the polluter's fee. Let the polluter's fee range from $0.02 per pound of BOD to $0.20. If you were charged with deciding the amount of the polluter's fee for this firm, what amount would you choose?

17. The daily electrical load to be transmitted by a conductor in a power plant is 1,900 amperes per day for 365 days per year. Two conductor materials are under consideration, copper and aluminum. The following information is available for the competing materials:

	Copper	Aluminum
Length	120 ft.	120 ft.
Installed cost	200 + $0.64/lb	200 + $0.40/lb
Estimated life	10 yrs.	10 yrs.
Salvage value	$0.55/lb	$0.10/lb.
Electrical resistance of conductor 120 ft. by 1 sq. in. cross section	0.000982 ohms	0.001498 ohms
Density	555 lb./ft.^{3}	168 lb./ft.^{3}

The energy loss in kilowatt-hours in a conductor due to resistance is equal to I^2R times the number of hours divided by 1,000, where I is the current flow in amperes and R is the resistance in the conductor in ohms. The electrical resistance is inversely proportional to the area of the cross section. Lost energy is valued at $0.007 per kilowatt-hour.

(a) Plot the total annual cost of capital recovery and return plus power loss cost for each material for cross sections of 2, 3, 4, 5, 6, and 7 square inches if the interest rate is 15%.

(b) Solve mathematically for the optimum cross section of each material.

(c) Recommend the minimum cost conductor material and specify the cross-sectional area.

18. Ethyl acetate is made from acetic acid and ethyl alcohol. Let x = pounds of acetic acid input, y = pounds of ethyl alcohol input, and z = pounds of ethyl

acetate output. The relationship of output to input is

$$\frac{z^2}{(1.47x - z)(1.91y - z)} = 3.91.$$

(a) Determine the output of ethyl acetate per pound of acetic acid, where the ratio of acetic acid to ethyl alcohol is 2, 1.5, 1, 0.67, and 0.50, and graph the result.
(b) Graph the cost of material per pound of ethyl acetate for each of the ratios given and determine the ratio for which the material cost per pound of ethyl acetate is a minimum if acetic acid costs $0.10 per pound and ethyl alcohol costs $0.06 per pound.

THE EVALUATION OF
PUBLIC ACTIVITIES

The standards by which private enterprise evaluates its activities are markedly different from those that apply in the evaluation of public activities. In general, private activities are evaluated in terms of profit whereas public activities are evaluated in terms of the general welfare. The general welfare, as collectively and effectively expressed, is the primary basis for evaluating public activities. A basis for evaluating public activities is necessary for an understanding of the characteristics of the governmental agencies that sponsor them.

The government of the United States and its several subdivisions engage in innumerable activities—all predicated upon the thesis of promotion of the general welfare. So numerous are the services available to individual citizens, associations, and private enterprises that books are required to catalog them. This chapter presents concepts and methods of analysis applicable to the evaluation of such activities.

10.1. General Welfare Aim of Government

A national government is a super-organization to which all agencies of the government and all organizations in a nation, including lesser political

10

subdivisions such as states, counties, cities, townships, and school districts as well as private organizations and individuals, are subordinate. In some of its aspects the government of the United States may be likened to a huge corporation. Its citizens play a role similar to that of stockholders. Each, if he chooses, may have a voice in the election of the policymaking group, the Congress of the United States, which may be likened to a board of directors.

In the United States the lesser political subdivisions, such as states, counties, cities, and school districts, carry on their functions in much the same way as does the United States, for in them each citizen may have a voice in determining their policy. Each of these lesser political subdivisions has certain freedom of action, although each is in turn subordinate to its superior organization. The subdivisions of the government are delineated for the most part as continuous geographical areas that are easily reconized.

It is a basic tenet that the purpose of government is to serve its citizens. The chief aim of the United States as stated in its Constitution is the *national defense* and the *general welfare* of its citizens. For convenience in discussion, these aims may be considered to be embraced by the single term—general welfare. This simply stated aim is, however, very complex. To discharge it perfectly requires that the desires of each citizen be fulfilled to the greatest extent and in equal degree with those of every other citizen.

Since the general welfare is the aim of the United States, the super-organization to which the lesser political subdivisions are subordinate, it follows that the latter's aims must conform to the same general objective regardless of what other specific aims they may have.

The general welfare aim as seen by the citizenry. Since each citizen may have a voice, if he will exercise it, in a government, the objectives of the government stem from the people. For this reason the objectives taken by the government must be presumed to express the objectives necessary for attainment of the general welfare of the citizenry as perfectly as they can be expressed. This must be so, for there is no superior authority to decide the issue.

Thus, when the United States declares war, it must be presumed that this act is taken in the interest of the general welfare. Similarly, when a state votes highway bonds, it also must be presumed to be in the interest of the general welfare of its citizens. The same reasoning applies to all activities undertaken by any political subdivision; for, if an opposite view is taken, it is necessary to assume that people collectively act contrary to their wishes.

Broadly speaking, the final measure of the desirability of an activity of any governmental unit is the judgment of the people in that unit. The exception to this is when a subordinate unit attempts an activity whose objective is contrary to that of a superior unit, in which case the final measure of desirability will rest in part in the people of the superior unit. Also, it must be clear that governmental activities are evaluated by a summation of judgments of individual citizens whose basis for judgment has been the general welfare as each sees it. The objectives of most governmental activities appear to be primarily social in nature, although economic considerations are often a factor. Public activities are proposed, implemented, and judged by the same group, namely, the people of the governmental unit concerned.

The situation of the private enterprise is quite different. Those in control of private enterprise propose and implement services to be offered to the public, which judges whether the services are worth their cost. To survive, a private business organization must, at least, balance its income and costs; thus profit is of necessity a primary objective. For the same reason a private enterprise is rarely able to consider social objectives except to the extent that they improve its competitive position.

The general welfare aim as seen by the individual. Public activities are evaluated by a summation of judgments of individual citizens, each of whose basis for judgment has been the general welfare as he sees it. Each citizen is the product of his unique heredity and environment; his home, cultural patterns, education, and aspirations differ from those of his neighbor. Because of this and the additional fact that human viewpoints are rarely logically determined, it is rare for large groups of citizens to see eye to eye on the desirability of proposed public activities.

The father of a family of several active children may be expected to see more point to expenditures for school and receational facilities than to expenditures for a street-widening program planned to enhance the value of downtown property. It is not difficult for a person to extol the value of aviation to his community if a proposed airport will increase the value of his property or if he expects to receive the contract to build it. Many public activities have no doubt been strongly supported by a few persons primarily because they would profit handsomely thereby.

But it is incorrect to conclude that activites are supported only by those who see in them opportunity for economic gain. For example, schools and recreational facilities for youth are often strongly supported by people who have no children. Many public activities are directed to the conservation of national resources for the benefit of future generations.

It is clear that the benefits of public activities are very complex. Some that are of great general benefit may spell ruin for some persons and vice versa. Lack of knowledge of the long-run effect of proposed activites is probably the most serious obstacle in the way of the selection of those activities that can contribute most to the general welfare.

10.2. The Nature of Public Activities

Governmental activities may be classified under the general headings of protection, enlightenment and cultural development, and economic benefits. Included under protection are such activities as the military establishments, police forces, the system of jurisprudence, flood control, and health services. Under enlightenment and cultural development are such services as the public school system, the Library of Congress and other publicly supported libraries, publicly supported research, the postal service, and recreation facilities. Economic benefits include harbors and canals, power development, flood control, research and information service, and regulatory bodies.

The list above, although incomplete, shows that there is much overlapping in classification. For example, the educational system is considered by many to contribute to the protection, the enlightenment, and the economic benefit of people. Consideration of the purposes of governmental activities as suggested by the classification above is necessary in considering the pertinency of economic analysis to public activities.

Impediments to efficiency in public activities. There are two major impediments to efficiency in public activities. First, the person who pays taxes and receives services has little or no knowledge about the value of the transactions occurring between him and his government. Therefore, he has no practical way of evaluating what he receives in return for his tax payment. His tax payments go into a common pool and lose their identity. The taxpayer, with few exceptions, receives nothing in exchange at the time or place at which he pays his taxes on which to base a comparison of the worth of what he pays in and what benefits he will receive as the result of his payment.

Since governmental units are exclusive franchises, the taxpayer has no choice as to which unit he must pay taxes. Thus, he does not have an opportunity to evaluate the effectiveness of tax units on the basis of comparative performance nor an opportunity to patronize what he believes to be the most efficient unit.

In addition, the recipients of the products of tax supported activities cannot readily evaluate the products in reference to what they cost. Where no direct payment is exchanged for products, a person may be expected to accept them on the basis of their value to the recipient only. Thus, the products of governmental activities will tend to be accepted even though their value to the recipient is less than the cost to produce them.

The second deterrent to efficiency in governmental activities is the lack of competitive forces required to instill sufficient concern about the efficient use of resources. This situation is the natural consequence of certain characteristics of government and the working environment within government. Probably the most important of these characteristics is that government avoids the pressures of a market mechanism that would induce greater efficiencies in its activities. Thus, government will not "go out of business" if resources are inefficiently applied. In fact, the federal government has the unique ability to spend more than it receives.

In addition, the costs resulting from poor decisions are not recovered from the pocket of those responsible. Few direct economic pressures are felt and seldom are promotions and salary increases related to efficiency. In many instances insufficient alternatives are considered because the particular government agency is overly concerned with its own continuance.

Multiple-purpose projects. Many projects undertaken by government have more than one purpose. A good example of this arises in connection with the public lands, such as a forest reserve. For example, suppose that a new road is being planned for a certain section of a national forest. Since public land is managed under the multiple use concept, several benefits will result if the road is put into service. Among these are scenic driving opportunities, camping opportunities, improved fire protection, ease of timber removal, etc.

Justifying a public works project is normally easier if the project is to serve several purposes. This is especially true if the project is very costly and must rely upon the support from several groups. The forest road will probably appeal to persons who like to drive and camp, the U.S. Forest Service who is responsible for fire control, and the timber industry who will be granted contracts to harvest timber resources from time to time.

There are a number of problems which arise in connection with multiple-use projects. Foremost among these is the problem of evaluating the aggregate benefit to be derived from the project. What is the benefit of a scenic drive or a camping trip? How can the benefit of improved accessibility for fire protection be measured? What is the benefit of easier timber removal? Each of these questions must be answered in quantitative monetary terms if an economic analysis of the project is to be performed. They must also be answered so that each group which benefits can share in the cost of the project.

A second problem arising from multiple-purpose projects is the possibility for conflict of interest between the purposes. These conflicts frequently become political issues. A primary motive of every public servant is to get elected or re-elected. By demonstrating that direct benefits have been obtained for the parties concerned, a candidate obtains votes. Because the desire is to show that the direct benefits are not very costly, there is a tendency to allocate project costs to those benefit categories which are deemed essential by all. For example, a major portion of the cost of the road may be allocated to the U.S. Forest Service under the categories of improved fire protection. By so doing it is easy to show that the project is desirable to the general public due to its low cost in connection with scenic driving and camping opportunities.

10.3. Financing Public Activities

Two basic philosophies in the United States greatly influence the collection of funds and their expenditure by governmental subdivisions. These are collection of taxes on the premise of *ability to pay* and the expenditure of funds on the basis of *equalizing opportunity* of citizens. Application of the ability-to-pay viewpoint is clearly demonstrated in our income and property tax schedules. The equalization-of-opportunity philosophy is apparent in federal assistance to lesser subdivisions to help them provide improved educational and health programs, highway systems, old-age assistance, and the like.

Because of the two basic tenets of taxation on the basis of ability to pay and expenditure of tax funds on the basis of equalization of opportunity, there often is little relationship between the benefits that an individual receives and the amount he pays for public activities. This is in large measure true of such major activities as government itself, military and police protection, the highway system, and most educational activities.

Methods of financing. Funds to finance public activities are obtained through (1) the assessment of various taxes, (2) borrowing, and (3) charges for services. Federal receipts are derived chiefly from corporation, individual, excise, and estate taxes and from duties on imports. State income includes corporate, individual, gasoline, sales, and property taxes and vehicle licensing fees. Cities rely on income, sales, property taxes, and license fees.

Selling bonds is a common method of raising funds for a wide variety of governmental activities. Borrowing at the state and local level is usually confined to financing capital improvement projects or self-supporting activities. These "municipal" bonds are usually exempt from federal income taxes and, therefore, the interest paid is generally lower than federal and corporate bonds. This tax advantage encourages those in high-income tax brackets to invest and provides the "municipality" a low-cost source of funds.

There are numerous types of bonds but the two most common types are (1) general obligation bonds and (2) revenue bonds. The *general obligation bond* is secured by the issuer's credit and taxing power. Thus, the bond holder's risk is lessened because he has the taxing power of the government pledged to meeting the interest payments of the bond. These bonds generally offer the greatest security and the lowest interest rates. General obligation bonds usually require a vote of the citizens within the taxing authority before they may be issued and the support required ranges from two-thirds majority to a simple majority. School bonds are normally general obligation bonds.

Revenue bonds are backed by the anticipated revenues to be generated by the project being financed. This type of bond is limited to revenue-producing projects such as toll roads and bridges, housing authorities, and water and sewer systems. Because of the increased risk (the project may fail to produce sufficient revenues) revenue bonds normally have higher interest rates than those found on general obligation bonds.

Most activities financed by the federal government receive their money from the general fund. This fund is supported from various taxes and borrowings (treasury bonds, notes, and bills). Thus, at the federal level it is more difficult to identify the source of the funds that are available for investment.

Considerable income on some governmental levels is derived from fees collected for services. Examples of such incomes on the national level are incomes from the postal services and sale of electricity from hydroelectric projects. On the city level, incomes are derived from supplying water and sewer services and from levies on property owners for sidewalks and pavements adjacent to their property.

Relating benefits to the cost of financing. Many user taxes are structured so that there is a relationship between the benefits derived from the project and the project cost. The most obvious of these user-related taxes are the family of taxes which provide the revenue for state highway projects. Highway-user taxation is designed to recover from the highway users those costs that can be appropriately identified with them.

One concept of valuing the benefits received by the user considers that operating expenses provide an accurate assessment of services received. That is, the more one drives, the more use is made of the highway system. The gasoline tax which is based on this concept certainly provides revenues in relationship to the amount of use. Of course, those vehicles with lower fuel consumption pay a lower amount in comparison to the less efficient vehicles.

A second approach to measuring benefits requires that the differential costs of providing for different classes of vehicles be considered. That is, if heavier vehicles are to use the roadway, it may have to be built thicker (at additional cost) and the rate of wear and tear will be increased. An example of how this effect can be considered is illustrated below.

Suppose a state has 1,000 miles of paved highways which have been designed for heavy vehicles .The characteristics of the highway surface necessary to carry the various type vehicles in the state are shown below.

Class of Vehicle	Surface Thickness (Inches)	Cost per Mile	Incremental Cost
Passenger cars	5.5	$600,000	$600,000,000
Light trucks	6.0	650,000	50,000,000
Medium trucks	6.5	700,000	50,000,000
Heavy trucks	7.0	740,000	40,000,000

With the vehicle registration in the state shown below

Passenger cars	2,000,000
Light trucks	200,000
Medium trucks	50,000
Heavy trucks	20,000

the following scheme will allocate the incremental cost of construction to the class of vehicles responsible for those costs:

Allocation of Increment per Vehicle	Passenger Cars	Light Trucks	Medium Trucks	Heavy Trucks
$600,000,000/2,000,000	$300	$300	$ 300	$ 300
50,000,000/200,000		250	250	250
50,000,000/50,000			1,000	1,000
40,000,000/20,000				2,000
Total	$300	$550	$1,550	$3,550

If it is desired to collect taxes on the basis of the cost of service, a suitable tax plan must be devised. This may be accomplished by assessing a fuel tax and a vehicle license tax of proper amounts.

10.4. Public Activities and Engineering Economy

Engineering is a major factor in nearly all public activities because of the high levels of technological input required in most public ventures. It is evident that the complexity of systems and projects being undertaken by the federal and local governments is increasing instead of decreasing.

The contribution of engineering to the nation's space program, national defense, pollution control, urban renewal, and highway construction is well established. Engineering input to these activities involves engineers as employees of governmental agencies and as consultants to these agencies. Thus, public activities are a concern of all engineers as citizens and of many engineers as outlets for their talents.

The engineering process described in Chapter 1 involved the determination of objectives, identification of strategic factors, determination of means, evaluation of engineering proposals, and assistance in decision making. Each of these phases of the engineering process are applicable in public activities. The main modification required is the substitution of benefit or general welfare for profit.

For example, suppose that a municipality has under consideration two projects, one a swimming pool and the other a library. The municipality has resources for one or the other, but not for both. The selection cannot be made on the basis of profit, since no profit is in prospect for either venture. The selection must be made on the basis of which will contribute most to the general welfare as expressed by the citizens of the community, perhaps by a vote. There is no superior basis for evaluating the contribution of each alternative to the general welfare.

As a second example, consider the development of a weapons system to aid in national defense. Often several technically feasible weapons systems are under consideration. When this is the case it is desirable to evaluate the effectiveness of each with the thought that weapons system effectiveness and the general welfare are related. By considering the cost of each system in relation to its effectiveness, a basis for choice is established.

It should be noted that evaluations of public activities in terms of the general welfare encompass both the benefits to be received from and the cost of the proposed activity. No matter how subjective an evaluation of the contribution of an activity to the general welfare may be, its cost may often be determined quite objectively. It may be fairly simple to determine the immediate and subsequent costs for the swimming pool, the library, or the weapons systems. A knowledge of the costs in prospect for benefits to be gained may be expected to result in sounder selection of public activities in either the civil or the defense sector.

10.5. Benefit-Cost Analysis

Because of the spectacular growth in the size of government and the absence of competitive pressures for the more efficient use of government resources, there is an increased need to understand the economic desirability of using these resources. The general decision problem is to use the available resources in such a manner that the general welfare of the citizenry is maximized. To help accomplish this goal many agencies in federal, state, and local governments have relied on methods that in some manner quantitatively measure the desirability of particular programs and projects. Of these methods, the most widely utilized is a method referred to as *benefit-cost analysis.*

When applying benefit-cost analysis, the measure of a project's contribution toward the general welfare is normally stated in terms of the benefits "to whomsoever they may accrue" and the cost to be incurred. In order for a project to be considered desirable the benefits must exceed the costs or the ratio of benefits to costs must be larger than one. Otherwise, the government unit would be derelict in its responsibility by applying public resources in a manner that would produce a net decrease in the general welfare of its citizenry.

Considering alternative public projects. As in all economy studies, it is crucial that any alternative being considered be analyzed from the *proper point of view.* Otherwise, the description of the alternative will fail to represent all of the significant effects associated with that alternative. Thus, the general rule is to assume a point of view that includes all the important consequences of the project being considered. This point of view can be geographical or it may be restricted to classes of people, organizations, or other identifiable groups.

Usually, the easiest method for determining the appropriate point of view is to identify who is to receive the benefits and who is to pay for them. The point of view that encompasses these two groups is the one that should be selected. Listed below are some examples of particular projects and the point of view that seems most reasonable.

Point of View	Project
National	Interstate highway system, major water resource projects, mass transit systems
Regional	Regionally funded air quality control projects
State	State funded educational programs, state highway programs
County	County funded medical services
Municipality	City funded water supply system, parks, fire protection
Governmental agency	Agency purchase of communications and computing equipment

For practical reasons there is usually a tendency to reduce the scope of the problem under consideration. To analyze on a national basis the effects of building a new library financed from city funds represents an extreme attempt to consider the most far-reaching effects of this project. On the other hand, to analyze an urban mass transit system that is primarily funded by the federal government on the basis of the direct benefits and costs to the municipality is to erroneously understate the true costs of the system. Unfortunately, many state and local governments have the view that money supplied by outside sources are "free" funds. The result is that actions are taken that provide benefits to some at the expense of others with no *net* improvement in the general welfare.

When alternatives are evaluated in the private sector, the costs and benefits of the alternatives are based on the viewpoint of the firm or organization making the analysis. Such a point of view can lead to a misleading description of alternatives when applied by a governmental unit. That is, if a state highway department analyzes highway improvements from its point of view instead of from the state's point of view, there are many effects that will fail to be included in the description of alternatives.

Another important consideration when defining an alternative is to develop a basis for reference for identifying the impact of the project on the nation or any other sub-unit involved. Thus, for any project it is important to observe what the state of the nation or sub-unit would be *with* or *without* the project. This base of reference provides the framework for identifying all the important benefits and costs associated with the project.

It should be recognized that this approach is not the same as examining the state of things before and after the project is installed. For example, the improvements in navigation of an inland waterway may increase the growth of barge traffic. However, if some of this growth would have been expected without the improvements, it is unfair to credit the total change in traffic realized from before the project to that occurring after the project. Thus, it is the change that is attributable to the project itself that is of primary importance when describing the benefit and costs of an alternative.

Because benefit-cost analyses are intended to assist in the allocation of resources it must be realized that promoting the general welfare must reflect the numerous objectives of society. While the economic betterment of the people is one important objective, others include the desire for clean air and water, pleasant surroundings, and personal security.

Some of the benefits and disbenefits associated with these multiple objectives can be stated in economic terms while others cannot. It is important that those benefits that have a market value be represented in monetary terms. It is equally important that those benefits for which there is no market value also be included in the analysis. However, it is improper to force the statement of noneconomic objectives in terms of monetary value. For example, it would be misleading to value a grove of hardwood trees in a park on the basis of board-feet of lumber contained in these trees and the market price for that lumber.

Selecting an interest rate. Expenditures for capital goods are made on the promise that they will ultimately result in more consumer goods than can be had for a present equal expenditure. Interest represents the expected difference. Not to consider interest in the evaluation of public activities is equivalent to considering a future benefit equal to a present similar benefit. This appears to be contrary to human nature. Therefore, when considering future economic benefits and costs it is appropriate that an interest rate properly reflecting the time value of money be utilized. This rate should reflect at least the government's cost of borrowed money.

Since activities financed through taxation require payments of funds from citizens, the funds expended for public activities should result in benefits comparable with those which the same funds would bring if expended in private ventures. It is almost universal for individuals to demand interest or its equivalent as an inducement to invest their private funds. To maintain public and private expenditures on a comparative basis, it seems logical that the interest rate selected should represent the opportunity foregone when taxes are paid. That is, the interest rate should reflect the rate that could have been earned if the funds had not been removed from the private sector.

Some public activities are financed in whole or in part through the sale of services or products. Examples of such activities are power developments, irrigation and housing projects, and toll bridges. Many such services could be carried on by private companies and are in general in competition with private enterprise. Again, since private enterprise must of necessity consider interest, it seems logical to consider the opportunity foregone in the benefit-cost analysis of public activities that compete in any way with private enterprise.

The interest rate to use in an economy study of a public activity is a matter of judgment. The rate used should not be less than that paid for funds borrowed for the activity. In many cases, particularly where the activity is comparable or competitive with private activities, the rate used should be comparable with that used in private evaluations.

The benefit-cost ratio. A popular method for deciding upon the economic justification of a public project is to compute the benefit-cost ratio. This ratio may be expressed as

$$BC(i) = \frac{\text{benefits to the public}}{\text{cost to the government}}$$

where the benefits and the costs are present or equivalent annual amounts computed using the cost of money. Thus, the BC ratio reflects the users' equivalent dollar benefits and the sponsors' equivalent dollar cost. If the ratio is 1, the equivalent benefits and the equivalent costs are equal. This represents the minimum justification for an expenditure by a public agency.

Considerable care must be exercised in accounting for the benefits and the costs in connection with benefit-cost analysis. Benefits are defined to mean all the advantages, less any disadvantages, to the users. Many proposals which embrace valuable benefits also result in inescapable disadvantages. It is the net benefits to the users which are sought. Similarly, costs are defined to mean all costs, less any savings, that will be incurred by the sponsor. Such savings are not benefits to the users but are reductions in cost to the government. It is important to realize that adding a number to the numerator does not have the same effect as subtracting the same number from the denominator of the BC ratio. Thus, incorrect accounting for the benefits and costs can lead to a ratio that may be misinterpreted. Therefore, the benefit-cost ratio is normally defined as follows:

$$BC(i) = \frac{\text{equivalent benefits}}{\text{equivalent costs}}$$

where

Benefits: All the advantages, less disadvantages to the user

Costs: All the disbursements, less any savings to the sponsor

To better understand the implications of this definition, let's split the equivalent costs into two components. One component is the equivalent capital initially invested by the sponsor. The other component is the equivalent annual operating and maintenance costs less any annual revenue produced by the project. This redefinition gives

$I =$ equivalent capital invested by the sponsor
$C =$ net equivalent annual costs to the sponsor
$B =$ net equivalent benefits to the user

The benefit-cost ratio can then be expressed as

$$BC(i) = \frac{B}{I + C}.$$

For any project to remain under consideration its benefit-cost ratio must exceed 1. Therefore, the first test of a project is to determine if it is minimally acceptable by observing whether or not the equivalent benefits exceed the equivalent costs. It is seen below that using such a criterion will eliminate all those projects whose net equivalent amount is less than zero.

If

$$BC(i) > 1$$

then

$$\frac{B}{I + C} > 1$$

giving

$$B - (I + C) > 0.$$

There is an alternative method of expressing the benefit-cost ratio that will appear in some benefit-cost analyses. Although the most widely accepted definition is the one previously discussed, it is important to understand the relationship between these two ratios. The only difference between the two ratios is that the alternative ratio reflects the net benefits less the annual costs of operation of the project divided by the investment cost. This is expressed

$$BC'(i) = \frac{B - C}{I}.$$

The advantage of having the benefit-cost ratio defined in this manner is that it provides an index that indicates the net gain expected per dollar invested.

Again, it is required that for a project to remain under consideration the alternative benefit-cost raio must be larger than 1.

If

$$BC'(i) > 1$$

then

$$\frac{B - C}{I} > 1$$

giving

$$B - (I + C) > 0.$$

Therefore, either ratio will lead to the same conclusion on the initial acceptability of the project (as long as I and $I + C$ are greater than zero).

As an example of the application of benefit-cost analysis and the BC ratio consider the following situation which is of interest to a state highway department. The accidents involving motor vehicles on a certain highway have been studied for a number of years. The calculable costs of such accidents embrace lost wages, medical expenses, and property damage. On the average there are 35 nonfatal accidents and 240 property damage accidents for each fatal accident. The average equivalent present cost of these three classes of accidents is calculated to be as follows:

Fatality per person	$400,000
Nonfatal injury accidents	14,000
Property damage accidents	3,000

From the data above the aggregate cost of motor-vehicle accidents per death may be calculated as follows:

Fatality per person	$400,000
Nonfatal injury accident $14,000 × 35	490,000
Property damage accident $3,000 × 240	720,000
Total	$1,610,000

The death rate on the highway in question has been 8 per 100,000,000 vehicle miles. A proposal to add a third lane is under consideration. It is estimated that the cost per mile will be $900,000, the service life of the improvement will be 30 years, and the annual maintenance will be 3 % of the first cost. The traffic density on the highway is 10,000 vehicles per day, and the cost of money is 7%. It is estimated that the death rate will decrease to 4 per 100,000,000 vehicle miles. Although there are other benefits which will result from widening the highway, it is argued that the reduction in accidents is sufficient to justify the expenditure.

To verify the economic desirability of the widening project, the highway department performs the following calculations. The equivalent annual benefit per mile to the public is

$$\frac{(8 - 4)(10,000)(365)(\$1,610,000)}{100,000,000} = \$235,060.$$

And the equivalent annual cost per mile to the state is

$$\$900,000(\overset{A/P\,7,\,30}{0.0806}) + \$900,000(0.03\%) = \$99,540$$

This results in a benefit-cost ratio of

$$BC(7) = \frac{\$235,060}{\$99,540} = 2.36$$

Thus, widening the highway is justified based on the benefits to be derived from a reduction in accidents alone. Other benefits,such as the reduction in trip time, have not been included in the analysis and would increase the ratio.

Calculating the alternative benefit-cost ratio for this problem requires that the annual equivalent operating costs be included in the numerator rather than in the denominator. This approach yields

$$BC'(7) = \frac{\$235,060 - \$27,000}{\$72,540} = \frac{\$208,060}{\$72,540} = 2.87.$$

The result indicates that a net savings of $2.87 for each dollar invested will be realized from the widening project. Both ratios indicate that substantial benefits will occur from such an expenditure of funds.

Benefit-cost analysis for multiple alternatives. The example of the previous paragraphs illustrated a situation in which the sponsoring agency had the simple choice of widening the highway or leaving it as is. Usually, however, a sponsoring agency finds that different benefit levels and different costs result in meeting a specific objective. When this is the case, the problem of interpreting the corresponding benefit-cost ratios presents itself. A hypothetical situation and the correct method of analysis is presented in the following paragraphs.

Suppose that four mutually exclusive alternatives have been identified for providing recreational facilities in a certain urban area. The equivalent annual benefits, equivalent annual costs, and benefit-cost ratios are given in Table 10.1. Inspection of the BC ratios might lead one to select Alternative

Table 10.1. BENEFIT-COST RATIOS FOR FOUR ALTERNATIVES

Alternative	Equivalent Annual Benefits	Equivalent Annual Costs	BC Ratio
A	$182,000	$91,500	1.99
B	167,000	79,500	2.10
C	115,000	78,500	1.46
D	95,000	50,000	1.90

B because the ratio is a maximum. Actually, this choice is *not* correct. The correct alternative can be selected by applying the principle of incremental analysis as described in Sections 7.2 and 7.3. In this instance, the additional increment of outlay is economically desirable if the incremental benefit realized exceeds the incremental outlay. Thus, when comparing mutually exclusive alternatives $A1$ and $A2$ the decision rule is as follows:

$$\text{BC}(i)_{A2-A1} > 1 \quad \text{accept Alternative } A2.$$
$$\text{BC}(i)_{A2-A1} \leq 1 \quad \text{reject Alternative } A2 \text{ and}$$
$$\text{retain Alternative } A1.$$

Just as described in Section 7.3 the alternatives should be arranged in order of increasing outlay. Thus, the alternative with the lowest initial cost should be first, the alternative with the next lowest initial cost second, and so forth.

Applying these decision rules to the alternatives described in Table 10.1 indicates that Alternative A and not Alternative B is the most desirable alternative. The calculations of the incremental benefit-cost ratios are summarized in Table 10.2.

Table 10.2. INCREMENTAL BENEFIT-COST RATIOS

Alternative	Incremental Annual Benefits	Incremental Annual Costs	Incremental BC Ratio	Decision
D	95,000	50,000	1.90	Accept D
$C - D$	20,000	28,500	0.70	Reject C
$B - D$	72,000	29,500	2.44	Accept B
$A - B$	15,000	12,000	1.25	Accept A

If the Do Nothing alternative is to be considered, assume that the cash flow associated with that alternative is zero. When comparing an alternative to the Do Nothing option, the incremental benefit-cost is computed using this assumption and the decision rules just described are applied.

The sequence of calculations required to produce the results presented in Table 10.2 are as follows: For example, the Do Nothing alternative is considered to be a feasible alternative. To compare the alternative with the smallest initial investment to Do Nothing, compute the benefit-cost ratio using the total benefits and total costs for Alternative D as follows:

$$\text{BC}(i)_{D-0} = \frac{\$95,000}{\$50,000} = 1.90.$$

This procedure is identical to an incremental comparison where the cash flows for the Do Nothing alternative are considered to be zero. Since this benefit-cost ratio is greater than 1, Alternative D is seen to be preferred to the Do Nothing alternative. Therefore, the Do Nothing alternative (the initial "current best" alternative) is rejected and Alternative D becomes the new "current best" alternative. It is the alternative with the lowest investment of the four alternatives being considered (exclusive of the Do Nothing alternative) that is acceptable.

Next, it is necessary to determine whether the incremental benefits that would be realized if Alternative C were undertaken would justify the additional expenditure. Therefore, compare Alternative C to Alternative D as follows:

$$BC(i)_{C-D} = \frac{\$115,000 - \$95,000}{\$78,000 - \$50,000} = \frac{\$20,000}{\$28,500} = 0.52.$$

The incremental benefit-cost ratio is less than 1, and therefore Alternative C is rejected and Alternative D remains as the "current best" alternative.

Next, compare Alternative B to Alternative D as follows:

$$BC(i)_{B-D} = \frac{\$167,000 - \$95,000}{\$79,500 - \$50,000} = \frac{\$72,000}{\$29,000} = 2.44.$$

The incremental benefit-cost ratio in this instance exceeds 1, and therefore Alternative B is preferred to Alternative D. Alternative B becomes the new "current best" alternative.

Alternative A is now compared to Alternative B as follows:

$$BC(i)_{A-B} = \frac{\$182,000 - \$167,000}{\$91,500 - \$79,500} = \frac{\$15,000}{\$12,000} = 1.25.$$

Since the incremental benefit-cost ratio for this comparison is greater than 1, Alternative A is preferred to Alternative B. Alternative A becomes the "current best" alternative and there are no more comparisons to be made. The alternative that should be selected is the current best alternative that remains after the final comparison. Therefore, Alternative A is the preferred of the four alternatives. Selection of this alternative will assure that the equivalent annual benefits less the equivalent annual costs are maximized and that its BC ratio is greater than 1. To demonstrate that the alternative chosen is the one that will maximize the equivalent benefits less the equivalent costs, Table 10.3 presents this net figure for each alternative.

Table 10.3. BENEFITS LESS COSTS FOR FOUR ALTERNATIVES

Alternative	Equivalent Annual Benefits	Equivalent Annual Costs	Net Improvement of General Welfare
A	$182,000	$91,500	$90,500*
B	167,000	79,500	87,500
C	115,000	78,500	36,500
D	95,000	50,000	45,000

*Alternative A provides the maximum net improvement.

Remember that it has been shown in Chapter 7 that, when comparing mutually exclusive alternatives, present worth (annual equivalent) on total investment, present worth (annual equivalent) on incremental investment, and rate of return on incremental investment are all consistent decision criteria. That incremental benefit-cost analysis will lead to the same conclusions as these decision criteria can be confirmed analytically. To show this, let's demonstrate that incremental analysis using the benefit-cost ratio will lead to the same selection of projects that would be indicated by one of these criteria. Let's choose present worth on total investment because it is probably the most familiar.

Suppose that two mutually exclusive alternatives ($A1$ and $A2$) are being considered for investment. Assume that the present worth on total investment is known for each alternative and that the following relationship exists:

$$PW(i)_{A2} > PW(i)_{A1}$$

That is, the net present worth for $A2$ exceeds $A1$ and, therefore, $A2$ is economically more desirable than $A1$. Now define

$B(i)_j$ = present worth of benefits for alternative j
$I(i)_j$ = present worth of investment for alternative j
$C(i)_j$ = present worth of operating costs for alternative j

It follows that the incremental benefit-cost ratio is

$$BC(i)_{A2-A1} = \frac{B(i)_{A2} - B(i)_{A1}}{I(i)_{A2} - I(i)_{A1} + [C(i)_{A2} - C(i)_{A1}]}.$$

This ratio should be greater than 1 if $A2$ is economically preferred to $A1$ as was initially assumed. Examine the result of assuming that the incremental benefit-cost ratio is greater than 1. Does it lead to the same conclusion as the present worth on total investment criterion?

If

$$BC(i)_{A2-A1} > 1$$

then

$$\frac{B(i)_{A2} - B(i)_{A1}}{I(i)_{A2} - I(i)_{A1} + [C(i)_{A2} - C(i)_{A1}]} > 1$$

which gives

$$B(i)_{A2} - B(i)_{A1} > I(i)_{A2} - I(i)_{A1} + C(i)_{A2} - C(i)_{A1}$$

and by transposing

$$B(i)_{A2} - I(i)_{A2} - C(i)_{A2} > B(i)_{A1} - I(i)_{A1} - C(i)_{A1}$$

which by definition is

$$PW(i)_{A2} > PW(i)_{A1}.$$

This same argument can be used for the alternative benefit-cost ratio. Thus, it can be demonstrated that incremental analysis will again provide results consistent with the decision criteria presented in Section 7.3.

10.6. Identifying Benefits, Disbenefits, and Costs

As indicated in Section 10.5, it is extremely important how the accounting for benefits and costs is accomplished. First, the traditional definition of the benefit-cost ratio requires that the net benefits to the user be placed in the numerator and the net costs to the sponsor be placed in the denominator. To find the net benefits it is necessary to identify those consequences which are favorable and unfavorable to the user. These unfavorable benefits are usually referred to as *disbenefits*. When deducted from the positive effects to be realized by the user, the resulting figure represents the net "good" to be engendered by the project.

To determine the net cost to the sponsor it is necessary to identify and classify the outlays required and the revenues to be realized. These revenues or savings usually represent income generated from the sale of products or services that are developed from the project. These "costs" include both disbursements and receipts related to the project's initial investment and to its annual operation.

Presented below is an example of the classification of benefits and costs that would be related to the completion a new toll road through a rural area in order to substantially shorten the distance between two large communities.

Benefits to Public
  Reduced vehicle operating costs (excluding fuel tax)
  Reduced commercial and noncommercial travel time
  Increased safety
  Increased accessibility between communities
  Ease of driving
  Appreciation of land values

Disbenefits to the Public
  Land removed from agricultural production
  Damages resulting from changes of water flow
  Decreased movement of livestock across highway
  Increased air pollution and litter

Cost to State
  Construction costs
  Maintenance costs
  Administrative costs
Savings to State
  Toll revenues
  Increased taxes due to land appreciated and increased business activity

Types of benefits. One of the important questions in the classification of benefits is, "To what length should one go to trace all the consequences of a project?" Not only is the answer to this question critical when attempting to quantify a project's contribution to the general welfare, but it also can substantially affect the cost of undertaking the benefit-cost analysis. To distinguish between those benefits that can be directly attributable to the project and those which are less directly connected, benefits are generally classified as follows:

Primary benefits represent the value of the direct products or services realized from the activities for which the project was undertaken.
Secondary benefits represent the value of those additional products and services realized from the activities of or stimulated by the project.

Most public projects provide both primary and secondary benefits. Irrigation projects increase the crop yield (primary benefits) along with increasing the economic strength of the farming community (secondary benefit). A benefit-cost analysis should always consider the primary benefits and whenever appropriate should consider the secondary benefits. When to include secondary benefits should be a function of their effect compared to that of the primary benefits and to the cost of determining them.

Valuation of benefits. Public activities provide such a variety of benefits that it is impossible to always value benefits in monetary terms. What is important is that both benefits and costs be represented by measures that are most meaningful to those who are involved in project assessment. A solid benefit-cost analysis not only compares the quantifiable consequences but it also describes the irreducibles and nonquantifiable characteristics in whatever terms are feasible.

It should be recognized that there are certain benefits and costs where the market price accurately reflects their true value. There are other benefits for which there is a market price, but this price fails to realistically represent their actual value (e.g., products or services that are subsidized, price supported, or artificially restrained from trade).

In addition, there are benefits for which there is no market value available but an economic value may be inputed. One approach for determining this type benefit is to consider the least expensive means of achieving the same service. Another method is to infer what a user is willing to pay for a service by observing the amount he spends to take advantage of this service. This latter approach is frequently applied to ascertain the economic worth associated with recreation. Thus, to determine the value of recreation for a water resources project, an analysis is made of what the user spends to avail himself of the project's recreation opportunities.

Last, there are benefits and costs for which it would be impossible to assign economic values. If a benefit can be quantified in realistic and meaningful terms, then it should be quantified, (e.g., number of trees, classified by height, as a measure of the aesthetics of a hardwood grove). For benefits for which suitable measures cannot be discovered, qualitative descriptions will suffice. However, it is important that all significant benefits and costs be included regardless of the degree to which they can be quantified.

Consideration of taxes. Many public activities result in loss of taxes through the removal of property from tax rolls or by other means, as, for example, the exemption from sales taxes. In a nation in which free enterprise is a fundamental philosophy, the basis for comparison of the cost of carrying on activities is the cost for which they can be carried on by well-managed private enterprises. Therefore, it seems logical to take taxes into consideration in a benefit-cost analysis, particularly when the activities are competitive with private enterprise.

The federal government agreed to pay $300,000 annually for 50 years to each of the states of Arizona and Nevada in connection with the Hoover Dam project. These payments are to partially compensate for tax revenue which would accrue to these states if the project had been privately constructed and operated. These payments are a cost and were considered in the economic justification of the project.

Because governments do not pay taxes, it is possible to omit them from consideration in some cases. For example, when the government is comparing its proposals to each other the net change to the general welfare is unchanged. (Many tax payments are a transfer of economic value from one group to another.) In highway studies it is common to exclude the fuel tax from the vehicular operating costs.[1] This exclusion then reduces this cost to the user by the amount of the fuel taxes.

Benefits and costs for multiple-purpose projects: As described in Section 10.2 the nature of public activities is such that many public ventures are multiple-purpose projects. In particular, water resource projects that provide a variety of functions including electric power, flood control, irrigation, navigation, and recreation fit this category.

One of the difficulties that may arise when analyzing multipurpose pro-

---

[1]Winfrey, Robley, *Economic Analysis for Highways* (Scranton, Pa.: International Textbook Co., 1969).

jects is the assessment of the desirability of each of its functions. For example, the electric power generated by the stored water may be in competition with private power companies. Should turbines be included in the dam or should electric power be supplied from private sources? To answer such a question it is necessary to isolate the costs directly identifiable with power generation. The dam which also provides flood control, water for irrigation, and other benefits is also an integral part of power generation. Thus, the cost of the dam must be *jointly* distributed among all the project functions. Unfortunately, the inability to allocate joint costs accurately is a fact of life. As a consequence, many procedures have been developed to assist in this allocation and none can be considered to be perfect.

The same problem exists when accounting for the benefits and disbenefits of multipurpose projects. Many of these benefits are inseparable from the project as a whole. For example, the multiple-purpose project just discussed introduces a disbenefit representing the land removed from agricultural production when it is flooded by the reservoir. How should this disbenefit be apportioned among the various project functions? It becomes evident that the separate justification of integral units within a multiple-purpose project does not provide completely satisfactory conclusions. The following table shows some of the costs (joint and otherwise) and the benefits (separable and otherwise) that might be included in a benefit-cost analysis of the construction of a large multipurpose dam.

Function	Benefits	Disbenefits	Costs	Saving
Hydroelectric power	Increased availability	Land flooded	Investment and operating	Sales of power
Flood control	Reduced flood damage	Land flooded	Investment and operating	Flood relief costs avoided
Irrigation	Increased crop yield	Land flooded	Investment and operating	Water revenue
Navigation	Savings on shipping costs	Loss of railroad traffic	Investment and operating	Vessel berth charges
Recreation	Increased accessibility	Destruction of scenic river	Investment and operating	Use charges

10.7. Cost-Effectiveness Analysis

Cost-effectiveness analysis had its origin in the economic evaluation of complex defense and space systems. Its predecessor, benefit-cost analysis had its origin in the civilian sector of the economy and may be traced back to the

Flood Control Act of 1936. Much of the philosophy and methodology of the cost-effectiveness approach was derived from benefit-cost analysis and, as a result there are many similarities in the techniques. The basic concepts inherent in cost-effectiveness analysis are now being applied to a broad range of problems in both the defense and the civil sectors of public activities.

In applying cost-effectiveness analysis to complex systems three requirements must be satisfied. First, the systems being evaluated must have common goals or purposes. The comparison of cargo aircraft with fighter aircraft would not be valid, but comparison with cargo ships would if both the aircraft and ships were to be utilized in military logistics. Second, alternate means for meeting the goal must exist. This is the case with cargo ships being compared with the cargo aircraft. Finally, the capability of bounding the problem must exist. The engineering details of the systems being evaluated must be available or estimated so that the cost and effectiveness of each system can be estimated.

The cost-effectiveness approach. There are certain steps which constitute a standardized approach to cost-effectiveness evaluations.[2] These steps are useful in that they define a systematic methodology for the evaluation of complex systems in economic terms. The following paragraphs summarize these steps.

First, it is essential that the desired goal or goals of the system be defined. In the case of military logistics mentioned earlier, the goal may be to move a certain number of tons of men and supporting equipment from one point to another in a specified interval of time. This may be accomplished by a few relatively slow cargo ships or a number of fast cargo aircraft. Care must be exercised in this step to be sure that the goals will satisfy mission requirements. Each delivery system must have the capability of delivering a mix of men and equipment that will meet the requirements of the mission. Comparison of aircraft that can deliver only men against ships that can deliver both men and equipment would not be valid in a cost-effectiveness study.

Once mission requirements have been identified, alternate system concepts and designs must be developed. If only one system can be conceived, a cost-effectiveness evaluation cannot be used as a basis for selection. Also, selection must be made on the basis of an optimum configuration for each system. In the previous chapter the minimum cost pier spacing was found for two competing bridge designs before a choice was made. A similar comparable basis must be established for a ship cargo system and an aircraft cargo system.

---

[2]Kazanowski, A. D., "A Standardized Approach to Cost-Effectiveness Evaluations," Chapter 7 in J. Morley English, Editor, *Cost Effectiveness* (New York: John Wiley and Sons, Inc., 1968).

System evaluation criteria must be established next for both the cost and the effectiveness aspects of the system under study. Ordinarily, less difficulty exists in establishing cost criteria than in establishing criteria for effectiveness. This does not mean that cost estimation is easy. It simply means that the classifications and basis for summarizing cost are more commonly understood. Among the categories of cost are those arising throughout the system life cycle which include costs associated with research and development, engineering, test, production, operation, and maintenance. The phrase *life cycle costing* is often used in connection with cost determination for complex systems. Life cycle cost determination for a specific system is normally on a present or equivalent annual basis. The Department of Defense has adopted the policy of procuring systems on the basis of life cycle cost as opposed to first cost, as had been the practice in the past.

System evaluation criteria on the effectivenss side of a cost-effectiveness study are quite difficult to establish. Also, many systems have multiple purposes which complicate the problem further. Some general effectiveness categories are utility, merit, worth, benefit, and gain. These are difficult to quantify and, therefore, such criteria as mobility, availability, maintainability, reliability, and others are normally used. Although precise quantitative measures are not available for all of these evaluation criteria they are useful as a basis for describing system effectiveness.

The next step in a cost-effectiveness study is to select the fixed cost or the fixed effectiveness approach. In the fixed cost approach, the basis for selection is the amount of effectiveness obtained at a given cost The selection criterion in the fixed effectiveness approach is the cost incurred to obtain a given level of effectiveness. When multiple alternatives which provide the same service are compared on the basis of cost, the fixed effectiveness approach is being used. This was the case with the two competing bridge designs presented in the previous chapter.

Candidate systems in a cost-effectiveness study must be analyzed on the basis of their merits. This may be accomplished by ranking the systems in order of their capability to satisfy the most important criterion. For example, if the criterion in military logistics is the number of tons of men and equipment moved from one point to another in a specified interval of time this criterion becomes the one of most importance. Other criteria, such as maintainability, would be ranked in a secondary position. Often this procedure will eliminate the least promising candidates. The remaining candidates can then be subjected to a detailed cost and effectiveness analysis. If the cost and the effectiveness for the top contender are both superior to the respective values for other candidates, the coice is obvious. If criteria values for the top two contenders are identical, or nearly identical, and no significant cost difference exists, either may be selected based on irreducibles. Finally, if system costs differ significantly, and effectiveness differs significantly, the selection must be made on the basis of intuition and judgment. This latter outcome is the most common in cost-effectiveness analysis directed to complex systems.

The final step in a cost-effectiveness study involves documentation of the purpose, assumptions, methodology, and conclusions. This is the communication step and it should not be treated lightly. No wise decision maker would base a major expenditure of capital on a blind trust of the analyst.

A cost-effectiveness example. As an example of some aspects of cost-effectiveness analysis consider the goal of moving men and equipment from one point to another as discussed in the previous paragraphs. Suppose that only the cargo aircraft mode and the ship mode are feasible. Also suppose that some design flexibility exists within each mode, so that the effectiveness in tons per day may be established through design effort.

Assume that the Department of Defense has convinced Congress that such a military logistic system should be developed and that Congress has authorized a research and development program for a system whose present life cycle cost is not to exceed 1.2 billion dollars. The Department of Defense, in conjunction with a nonprofit research and engineering firm decided that three candidate systems should be conceived and costed. Table 10.4, shows

Table 10.4. COST AND EFFECTIVENESS FOR THREE SYSTEMS

System	Present cost in billions of $	Effectiveness in tons per day
Aircraft I	1.2	1,620
Ship	1.2	1,410
Aircraft II	1.0	1,410

the resulting present life cycle cost and corresponding effectiveness in tons per day for each system at maximum utilization. These data are also exhibited in Figure 10.1.

The three candidate systems were studied because of the following logic. First, using the fixed cost approach the study team projected the configuration of Aircraft System I and the Ship System and found that these systems would have effectivenesses of 1,620 and 1,410 tons per day, respectively. If the study had stopped here the conclusion would be to spend the 1.2 billion dollars for Aircraft System I. Realizing, however, that the Congress and the Department of Defense would welcome estimates of effectiveness for other levels of expenditure, the study team chose to determine the cost for Aircraft System II which would be designed to have an effectivenss equal to that of the Ship System costing 1.2 billion dollars. The cost of this system was estimated to be

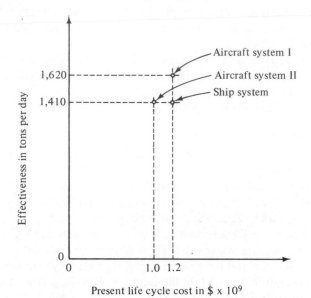

Figure 10.1. Cost and effectiveness for three systems.

1.0 billion dollars. This is an example of the fixed effectiveness approach.

At this point a decision must be made between Aircraft System I and Aircraft System II. The Ship System has a lower effectiveness of the same cost as Aircraft System I and a higher cost for the same effectiveness as Aircraft System II and so will not be a candidate in the decision process. In choosing between Aircraft System I and Aircraft System II, the Congress in conjunction with the Department of Defense must decide if the increase in effectiveness of 210 tons per day is worth a present life cycle cost of 0.2 billion dollars.

This example is quite simplified in its assumptions and analysis. Nothing has been said about how the study team was able to determine that a certain effectiveness would result from a system with a present life cycle cost of 1.2 billion dollars. Also, the example did not point out any secondary measures of effectiveness which would complicate the choice.

QUESTIONS AND PROBLEMS

1. Outline the function of government and the nature of public activities.
2. Contrast the criteria for evaluation of private and public activities.
3. Describe the impediments to efficiency in governmental activities.

4. Describe the meaning of the general welfare objective as it relates to engineering economy analysis applied to public activities.

5. Name the elements that should be considered in deciding on an interest rate to be used in the evaluation of public activities.

6. Show that project selection based upon the alternative incremental benefit-cost ratio $\left(\dfrac{B - C}{I}\right)$ leads to present worth maximization.

7. Briefly describe the factors to be considered in identifying benefits, disbenefits, and costs in the evaluation of public activities.

8. List some noneconomic indicators that could be useful in evaluating whether or not public activities are in the interest of the general welfare.

9. List the factors that would have a significant effect on a city's decision to undertake a mass transportation system. Indicate which factors could be quantified and which factors would be considered nonquantifiable.

10. A state whose population is 4,500,000 has 12 state-supported colleges and universities with a total enrollment of 60,000 students and an annual budget for instructional programs approximating $120,000,000 per year.
    (a) Most students enrolled in the colleges state that prospective increased earnings is an important reason for attending. What percentage of the state's population do you believe desire to support the program of higher education through taxation for this reason?
    (b) What prospective annual increase in earnings must you achieve to recover the cost of your education? Assume the state provides $2,000 per student per year and that the interest rate is 8%.
    (c) Indicate any benefits that may accrue to the state from its investment in education.

11. Analysis of accidents in one state indicates that increasing the width of highways from 6 meters to 7.5 meters may decrease the accident rate from 125 per 100,000,000 to 72 per 100,000,000 vehicle kilometers. Calculate the average daily number of vehicles that should use a highway to justify widening on the basis of the following estimates: Average loss per accident, $500; per kilometer cost of widening pavement 1.5 meters, $15,000; useful life of improvement, 25 years; annual maintenance, 3% of first cost; interest rate, 5%.

12. A suburban area has been annexed by a city which henceforth is to supply water service to the area. The prospective growth of the suburb has been estimated and on the basis of this the requirements for pipelines needed to meet the demand for water are as follows:

Years from Now	Pumping Cost Per Year	Pipe Diameter (Inches)	First Cost
0	$25,000	10	$300,000
14	28,000	14	390,000
25	31,000	18	450,000

From the standpoint of water-carrying capacity, a 14-inch pipe is equivalent to two 10-inch pipes and an 18-inch pipe is equivalent to three 10-inch pipes. On the basis of a minimum equivalent expenditure over the next 40 years, compare the following plans if the interest rate is 6% and if used pipeline can be sold for 20% of its original cost.

Years from Now	Plan A	Plan B	Plan C	Plan D	Plan E
0	Install 10″ Sell 10″	Install 10″	Install 14″	Install 14″	Install 18″
14	Install 14″ Sell 14″	Install 10″	—	— Sell 14″	—
25	Install 18″	Install 10″	Install 10″	Install 18″	—

13. A toll road has been opened between two cities. The distance between the entrances of the two cities is 93 miles via the highway and 104 miles via the shortest alternate free highway. From the following data determine the economic advantage, if any, of using the toll road for the following conditions applicable to the operation of a light truck: toll cost, $2.50; driver cost, $6.50 per hour; average driving rate between entrances via toll road and free road respectively, 55 m.p.h. and 50 m.p.h.; estimated average cost of operating truck per mile via toll road, $0.16, via free road, $0.17.

14. Let it be assumed that a certain state is contemplating a highway development embracing the construction of 8,000,000 square yards of pavement. Vehicle registration in the state is as follows:

Passenger cars 1,000,000
Light trucks   200,000
Medium trucks   40,000
Heavy trucks   10,000

The characteristics and pavements necessary to carry the vehicles are taken as follows:

Class of Vehicle	Pavement Thickness (Inches)	Cost Per Sq. Yard
Passenger cars	5.5	$10.50
Light trucks	6.0	11.80
Medium trucks	6.5	12.50
Heavy trucks	7.0	13.00

On the assumption that paving costs should be distributed on the basis of the number of vehicles in each class and the incremental costs of paving required for each class of vehicle, what should be the taxes per vehicle for each vehicle class?

15. Two sections of a city are separated by a marsh area. It is proposed to connect the sections by a four-lane highway. Plan A consists of a 2.4 mile highway directly over the marsh by the use of earth fill. The initial cost will be $11,500,000 and the required annual maintenance will be $14,000. Plan B consists of improving a 4.2 mile road skirting the swamp. The initial cost will be $5,500,000 with an annual maintenance cost of $7,500. A traffic survey estimates the traffic density to be as follows:

Years after Construction	Traffic Density in Vehicles per Hour
1– 5	150
6–10	800
11–20	2,200

The estimated average speed under these densities would be 55 m.p.h., 45 m.p.h., and 30 m.p.h., respectively. The traffic consists of 80% noncommercial vehicles with an operating cost of $0.12 per mile and 20% commercial vehicles with an operating cost of $0.16 per mile and $6.50 per hour. If Plan B is accepted the development of the property adjacent to the highway will result in an increase in tax revenue of $400,000 per year.

(a) Compare the alternatives on the basis of a 20-year period and 6% interest.

(b) At what traffic density are the plans equivalent for an interest rate of 6%?

16. Two mutually exclusive projects are being considered for investment. Project A1 requires an initial outlay of $100,000 with net receipts estimated to be $30,000 per year for the next 5 years. The initial outlay for Project A2 is $200,000, and net receipts have been estimated at $50,000 per year for the next 7 years. The minimum attractive rate of return is 10% and there is no salvage value associated with either project.

Using the benefit cost ratio $\left(\dfrac{\text{net benefits}}{\text{cost}} \right)$, which project would you select?

17. Suppose that there are 100 boulders in a path to a spring used regularly by 50 cave men. Each cave man considers the inconvenience and concludes that the present worth of the benefit to him would be $20 if the boulders were removed. Since the cost of removing each boulder is $2, each abandons the thought of clearing the path. An entrepreneur, perceiving the potential benefit of organization, convinces the group that they should coordinate their efforts. If each cave man removes his share of the 100 boulders, how much can the entrepreneur charge for his service if the benefit-cost ratio must be at least 2?

18. An inland state is presently connected to a seaport by means of a railroad system. The annual goods transported is 500 million ton miles. The average transport charge is 6 mills per ton mile. Within the next 20 years, the transport is likely to increase by 25 million ton miles per year.

It is proposed that a river flowing from the state to the seaport be improved at a cost of $250,000,000. This will make the river navigable to barges and will reduce the transport cost to 2 mills per ton mile. The project will be financed by

80% federal funds at no interest, and 20% raised by 7% bonds at par. There would be some side effects of the change-over as follows: (1) The railroad would be bankrupt and be sold for no salvage value. The right of way, worth about 30 million, will revert to the state; (2) 3,600 employees will be out of employment. The state will have to pay them welfare checks of $200 per month; (3) The reduction in the income from the taxes on the railroad will be compensated by the taxes on the barges.

(a) What is the benefit-cost ratio based on the next 20 years of operation?

(b) At what average rate of transport per year will the two alternatives be equal?

19. In a certain city the fire insurance premium rate is $0.57 per $100 of the insured amount. There are approximately 8,000 dwellings of an average valuation of $20,000 in the city. It is estimated that the insured amount represents 70% of the value of the dwellings. The city commissioners have been advised that the fire insurance premium rate will be reduced to $0.50 per $100 if the following improvements are made to reduce fire losses.

Improvement	Cost	Life
Increase capacity of trunk water lines from pumping station	$80,000	40 years
Increase capacity of pumps	$ 8,000	20 years
Add supply tank to increase pressure in remote sections of town	$60,000	40 years
Purchase additional fire truck and related equipment	$50,000	20 years
Add two firemen	$40,000 per year	

Operation costs of added improvements are expected to be offset by decreased pumping cost brought about by the enlargement of trunk water lines. The city is in a position to increase its bonded indebtedness and can sell bonds bearing 7% interest at par. Should the above improvements be made on the basis of prospective savings in insurance premiums paid by the home owners? What is the benefit-cost ratio?

20. The federal government is planning a hydroelectric project for a river basin. In addition to the production of electric power, this project will provide flood control, irrigation, and recreation benefits. The estimated benefits and costs that are expected to be derived from the three alternatives under consideration are listed below.

Alternatives	A	B	C
Initial Cost	$25,000,000	$35,000,000	$50,000,000
Annual benefits and costs:			
Power sales	$1,000,000	$1,200,000	$1,800,000
Flood control savings	250,000	350,000	500,000
Irrigation benefits	350,000	450,000	600,000
Recreation benefits	100,000	200,000	350,000
Operating and maintenance costs	200,000	250,000	350,000

The interest rate is 5% and the life of each of the projects is estimated to be 50 years.

(a) By comparing the benefit-cost ratios determine which project should be selected.

(b) Calculate the benefit-cost ratio for each alternative. Is the best alternative selected if the alternative with the maximum benefit-cost ratio is chosen.

(c) If the interest rate is 8% what alternative would be chosen?

21. The federal government is presently considering a number of proposals for improving the speed of mail handling in large urban post offices. The measure of effectiveness to be used to evaluate these mail handling systems is the volume of mail processed per day. The cost of purchasing and installing these various systems, their resulting savings, and their effectiveness are as follows.

System	Initial cost in millions of $	Annual savings in millions of $	Effectiveness in millions of letters processed per day
A	$1,200	$100	5
B	2,000	140	8
C	2,600	230	12
D	4,000	340	13
E	5,100	500	14

If the interest rate is 7% and the life of each system is estimated to be 10 years, plot the cost and effectiveness relationship for each proposal. Which alternatives can be dropped from further consideration? How would you decide between the remaining alternatives?

22. The U.S. government is planning to develop advanced power generation systems so that its coal and uranium resources can be utilized most effectively. The six cases listed below appear to be the most promising concepts to satisfy the available resources and the need for a clean environment.

Cases	Base	I	II	III	IV	V	VI
Energy resources	Petroleum	Uranium	Coal	Coal	Coal	D$_2$O, Li	Solar
Power system	Present technology	Fast breeder-steam	MHD, fuel cell combined GT/ST*		Fuel cell	Fusion	Solar cells

*Comb. GT/ST, combined gas-turbine-steam-turbine.

Based upon the advanced power generation concepts above, the following economic gains of each case as compared to the present technology have been

ECONOMIC BENEFITS OF ADVANCED POWER GENERATION METHODS
(period, 1980–2020)

	Base	Case I	Case II	Case III	Case IV	Case V	Case VI
	P.W.*	P.W.	P.W.	P.W.	P.W.	P.W.	P.W.
Total capital investment	98	114	99	110	121	117	92
Operating cost	212	149	190	158	132	132	196
Benefits	320	340	400	328	335	330	360

*P.W., present worth at 7% in units of 10^9.

identified. For this study, the capital costs of power generating units were based upon 1985 commercial operation after a 5-year R&D period. All present-worth calculations were made at an interest rate of 7%. If benefit-cost analysis is to be considered as a guide to decision making, which project has the highest funding preference? Assume that there is no Do Nothing alternative.

23. For the various alternatives described in Problem 22 consider the following. In the evaluation of advanced power systems the federal government believes that it is necessary to examine the contributions of each power system to the problem of environmental pollution. The level of emissions and water use can vary for the same power demand, depending upon the type of fuel utilized and the efficiency of power conversion. Thus, it is important to consider the cost of these systems and their environmental effects.

	Base	Case I	Case II	Case III	Case IV	Case V	Case VI
Cumulative Total air pollutants (10^9 lb)	4,530	3,080	3,249	3,249	3,080	1,945	1,945
Total water make-up (10^{12} gal)	249	254	194	194	161	253	215
Development cost* (10^{10})	0	4.3	2.0	1.7	1.1	4.7	3.0

*Present worths at 7%.

Show how you can use the cost-effectiveness approach to consider these various factors.

24. A state highway department is considering the location of a new rural highway. Two possible locations are under consideration. For each location, designs for two different numbers of lanes (2 lanes vs. 4 lanes) are studied. It is also

estimated that 2 traffic lanes will be adequate for the next 10 years but after that, 4 traffic lanes will eventually be required. Adding two more lanes after 10 years will cost 1.8 times of current 2-lane construction cost. The following data have been collected for the two locations. In comparing the economic desirability of the highway location, an interest rate of 6% is to be used. For an analysis period of 20 years with zero salvage value, determine which location should be selected by comparing the benefit-cost ratios.

Road Characteristics	Location A		Location B	
	A1 2-Lane	A2 4-Lane	B1 2-Lane	B2 4-Lane
Distance (miles)	12	12	10.5	10.5
Total construction costs	$6,700,000	$10,500,000	$7,500,000	$14,000,000
Annual maintenance and operating costs per mile	$1,000	$1,200	$950	$1,100
Resurfacing cost per mile, end of 10th year	$52,000	$83,000	$52,000	$83,000

Traffic Characteristics	Passenger Car	Commercial Vehicles
Average daily traffic, (all routes) for the first year	3,500	500
Annual growth per year	160	35
Equivalent uniform speed	55 m.p.h.	50 m.p.h.
Value of travel time per vehicle hour	$2.00	$6.50
Vehicle operating cost per mile	$0.05	$0.30

25. The Tennessee Valley Authority (TVA) wants to construct a 150-MW peaking power plant to meet the demand for additional electricity. There are two alternatives under consideration. One alternative is to construct a combustion turbine plant which is particularly well-suited for peaking operation because of its low capital cost. However, the most serious disadvantage of combustion turbine is its poor thermal efficiency which affects the fuel and operating costs. Also under consideration is a fuel cell plant which provides better thermal efficiency even though it is relatively more expensive to construct in comparison with the combustion turbine. Assume the fuel and operating costs are expected to increase at the rate of 6% per year even though the amount of energy produced is expected to be the same each year. The fuel cost in any year n is represented by the following expression:

$$F_n = (C)(H)(U) \left(\frac{8{,}760 \text{ hours/year}}{10^6} \right) P_n$$

where

$$P_n = P_{n-1}(1 + 0.06).$$

The operating cost in any year n is given by:

$$O_n = (C)(H)(U) \left(\frac{8{,}760 \text{ hours/year}}{10^6} \right) Q_n$$

where

$$Q_n = Q_{n-1}(1 + 0.06).$$

In either case, it will be 2 years before either plant would be in full commercial operation. The revenue generated each year by either plant is assumed to be the same. Shown below are the economic and operating data associated with each type of generating plant.

		Combustion Turbine	Fuel Cell Plant
C:	Plant size (kw)	150,000	150,000
H:	Heat rate (Btu/Kw-hr.)	12,700	9,300
U:	Plant utilization factor	15%	15%
N:	Economic service life (years)	25	25
P_0:	Starting fuel cost ($\$/10^6$ Btu)	2.20	2.20
Q_0:	Starting maintenance cost ($\$/10^6$ Btu))	0.19	0.19
	Construction cost ($\$$/kw)	175	240
	Other annual expenses (insurance, tax, depreciation, and maintenance cost) as a fixed % of construction costs	5.5%	5.5%
	Salvage value as a fixed % of construction cost	−1%	−1%

(a) For an interest rate of 12%, determine which alternative is more attractive to undertake. (Annual cost comparison.)

(b) By comparing the incremental BC ratio $\left(\frac{B - C}{I} \right)$, what alternative would be chosen?

(c) By utilizing the incremental BC ratio $\left(\frac{B}{I + C} \right)$, what alternative would be selected? Discuss any difficulty associated with the BC ratio application to this particular problem.

(d) For a higher plant utilization factor at 20%, repeat (a), (b), and (c).

part four

ACCOUNTING, DEPRECIATION,
AND INCOME TAXES

ACCOUNTING AND
COST ACCOUNTING

The accounting system of an enterprise provides a media for recording historical data arising from the essential activities employed in the production of goods and services. Engineering economic analysis provides a means for quantifying the expected future differences in the worth and cost of alternative engineering proposals. As compared with this function, accounting has the objective of providing summaries of the status of an enterprise in terms of assets and liabilities so that the condition of the enterprise may be judged at any point in time. Therefore, it is essential that the function of accounting and the function of engineering economy be clearly distinguished.

Accounting records are one of the most important sources of data for engineering economy studies. In them the analyst will find detailed quantitative data useful in estimating the future outcome of activities similar to those completed. In addition, the outcome of decisions based on economy studies will eventually be revealed in these records and may be used for post audit. For these reasons, it is desirable that the data provided by accounting systems be examined in relation to the requirements of engineering economy studies.

11.1. General Accounting

Two classifications of accounting are recognized: general accounting and cost accounting. Cost accounting is a branch of general accounting and is

11

usually of greater importance in engineering economy studies than is general accounting. Cost accounting will be considered in the next section.

Basic financial statements. The primary purpose of the general accounting system is to make possible the periodic preparation of two basic financial statements for an enterprise. These are:

1. A *balance sheet* setting out the assets, liabilities, and net worth of the enterprise at a stated date.
2. A *profit and loss statement* showing the revenues and expenses of the enterprise for a stated period.

Thus, the balance sheet presents a "snapshot" of the financial condition of an enterprise at a specified point in time. To reflect the level of transactions of the enterprise occurring between the time the previous balance sheet was prepared and the next one is prepared, the enterprise provides the profit and loss statement. Figure 11.1 depicts the relationship between the balance sheet and the profit and loss statement.

The accounts of an enterprise fall into five general classifications—assets, liabilities, net worth, revenue, and expense. Three of these—assets, liabilities, and net worth—serve to give the position of the enterprise at a certain date. The other two accounts—revenue and expense—accumulate profit and loss information for a stated period, which act to change the position of the enterprise at different points in time. Each of these five accounts is a summary of other accounts utilized as part of the total accounting system.

The balance sheet is prepared for the purpose of exhibiting the financial position of an enterprise at a specific point in time. It lists the assets, liabilities, and net worth of the enterprise as of a certain date. The monetary amounts recorded in the accounts must conform to the fundamental accounting equation

$$\text{assets} - \text{liabilities} = \text{net worth.}$$

For example, the balance sheet for Ace Company shows these major accounts as of December 31, 19xx.

ACE COMPANY
BALANCE SHEET
DECEMBER 31, 19xx

Assets		Liabilities	
Cash	$143,300	Notes payable	$ 22,000
Accounts receivable	7,000	Accounts payable	4,700
Raw materials	9,000	Accrued taxes	3,200
Work in process	17,000	Declared dividends	40,000
Finished goods	21,400		$ 69,900
Land	11,000	**Net Worth**	
Factory building	82,000		
Equipment	34,000	Capital stock	$200,000
Prepaid services	1,300	Profit for December	56,100
			256,100
	$326,000		$326,000

Balance sheets are normally drawn up annually, quarterly, monthly, or at other regular intervals. The change of a company's condition during the interval between balance sheets may be determined by comparing successive balance sheets.

Information relative to the change of conditions that have taken place during the interval between successive balance sheets is provided by a profit and loss statement. This statement is a summary of the income and expense for a stated period of time. For example, the profit and loss statement for the Ace Company shows the income, expense, and net profit for the month of December, 19xx.

The balance sheet and the profit and loss statement are summaries in more or less detail, depending upon the purpose they are to serve. They are related to each other; the net profit developed on the profit and loss state-

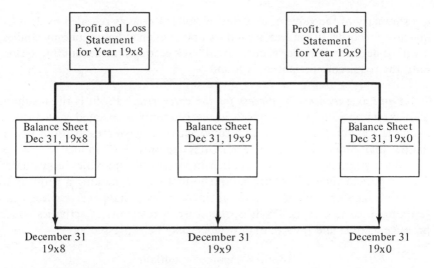

Figure 11.1. Time relationship between balance sheet and profit and loss statement.

ACE COMPANY
PROFIT AND LOSS STATEMENT
MONTH ENDING DECEMBER 31, 19xx

Gross income from sales		$251,200
Cost of goods sold		142,800
Net income from sales		$108,400
Operating expense:		
Rent	$11,700	
Salaries	28,200	
Depreciation	4,800	
Advertising	6,500	
Insurance	1,100	$ 52,300
Net profit from operations before taxes		$ 56,100
Federal and state taxes		12,300
Net profit from operations after taxes		$ 43,800

ment is entered under net worth on the balance sheet as shown above. This net profit is also used as a basis for computing the firm's income tax obligation shown on the balance sheet.

Together, the balance sheet and the profit and loss statement summarize the five major accounts of an enterprise. The major accounts are summaries of other accounts falling within these general classifications as, for example, cash, notes payable, capital stock, cost of goods sold, and rent. Each of these accounts is a summary in itself. For example, the asset item of raw material is a summary of the value of all items of raw material as revealed by detailed inventory records. In the search for data upon which to base economy studies, it will be necessary to trace each account back through the accounting system until the required information is found.

Profit as a measure of success for the enterprise. Profit is the resultant of two components, one of which is the economy associated with income from the activity. It is obvious that some activities have greater profit potentialities than others. In fact, some activities can only result in loss.

When profit is a consideration, it is important that activities be evaluated with respect to their effect on profit. The first step in making a profit is to secure an income. But, to acquire an income necessitates certain activities resulting in certain costs. Profit is, therefore, a resultant of activities which produce income and involve outlay which may be expressed as

$$\text{profit} = \text{income} - \text{outlay}.$$

The total success of an organization is the summation of the successes of all the activities that it has undertaken. Also, the success of a major undertaking is the summation of the successes of the minor activities of which it is constituted. In Figure 11.2 each vertical block represents the income potentialities of a venture, the outlay incurred in seeking it, the outlay incurred in prosecuting it, and the net gain of carrying it on. In this conceptual scheme the several quantities are considered to be measured in a single commensurable term, such as money.

From this figure it is apparent that the extent of the success of an activity

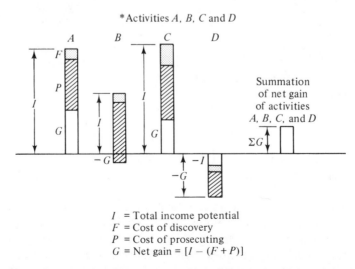

*Activities A, B, C and D

I = Total income potential
F = Cost of discovery
P = Cost of prosecuting
G = Net gain = $[I - (F + P)]$

Figure 11.2. An illustration of the final outcome of several activities.

depends upon its potentialities for income less the sum of the costs of finding it and carrying it on. It is also apaprent that the success of an enterprise is the summation of the net success of the several actitivies undertaken during a period of time.

11.2. Cost Accounting

Cost accounting is a branch of general accounting adapted to registering the costs for labor, material, and overhead on an item-by-item basis as a means of determining the cost of production. The final summary of this information is presented in the form of a cost of goods made and sold statement. It lists the costs of labor, material, and overhead applicable to all goods made and sold during a certain period. For example, the cost of goods made and sold statement for the Ace Company during the month of December, 19xx is as follows:

Direct Material		
In process Dec. 1, 19xx	$ 3,400	
Applied during the month	39,500	
Total	$42,900	
In process Dec. 31, 19xx	4,200	$ 38,700
Direct Labor		
In process Dec. 1, 19xx	$ 4,300	
Applied during the month	51,900	
Total	$56,200	
In process Dec. 31, 19xx	5,700	$ 50,500
Overhead		
In process Dec. 1, 19xx	$ 5,800	
Applied during the month	60,100	
Total	$65,900	
In process Dec. 31, 19xx	7,100	$ 58,800
Cost of Goods Made		$148,000
Finished goods Dec. 1, 19xx		16,200
Total		$164,200
Finished goods Dec. 31, 19xx		21,400
Cost of Goods Sold		$142,800

The cost of goods made and sold statement reflects summary data derived from the four accounts of materials in process, labor in process, overhead in process, and finished goods. The statement is subsidiary to the profit and loss statement since the "cost of goods sold" amount it develops is transferred to the profit statement. Similarly, the profit and loss statement was shown to be subsidiary to the balance sheet since the profit or loss amount developed thereon is transferred to the net worth section of the balance sheet.

The costs that are incurred to produce and sell an item of product are commonly classified as direct material, direct labor, factory overhead, factory cost, administrative cost, and selling cost. The first three are exhibited on the cost of goods made and sold statement, and they give rise to the cost of goods sold entry on the profit and loss statement. Administrative and selling costs appear on the profit and loss statement under operating expense, and they are subtracted from net income to arrive at the net profit amount. Each of these cost classifications will be considered in the paragraphs that follow.

Direct material. The material whose cost is directly charged to a product is termed *direct material.* Ordinarily, the costs of principal items of material required to make a product are charged to it as direct material costs. Charges for direct material are made to the product at the time the material is issued, through the use of forms and procedures designed for that purpose. The sum of charges for materials that accumulate against the product during its passage through the factory constitutes the total direct material cost.

In the manufacture of many products, small amounts of a number of items of material may be consumed which are not directly charged to the product. These items are charged to factory overhead, as will be explained later. They are not directly charged to the product on the premise that the advantage to be gained will not be enough to offset the increased cost of record keeping.

Although perhaps less subject to gross error than records of other elements of cost, records of direct material costs should not be used in engineering economy studies without being questioned. Their accuracy in regard to quantity and price of material should be ascertained. Also, their applicability to the situation being considered should be established before they are used.

Direct labor. *Direct labor* is labor whose cost is charged directly to the product. The source of this charge is time tickets or similar forms used to record the time and wages of workmen whose efforts are applied to a product during its journey through the factory. Unless the allocation of labor costs to products is very closely controlled, records of labor costs charged to specific products are likely to be in error.

As a result of either carelessness or a desire to conceal an undue amount of time spent on a job, some of the time applied to one job may be reported as being applied to another. Thus, direct-labor cost records should be carefully examined for accuracy and applicability to the situations under investigation before being used as data for engineering economy studies.

Various small amounts of labor may not be considered to warrant the record-keeping that is required to charge them as direct labor. Such items of labor became part of the factory overhead. The labor of personnel engaged in such activities as inspection, testing, or moving the product from machine to machine or in pickling, painting, or washing the product is often charged in this way.

Such items as social security, pension, and insurance costs that are nearly proportional to direct wages are sometimes included in arriving at direct labor costs.

Factory overhead. *Factory overhead* is also designated by such terms as factory expense, shop expense, burden, indirect costs, and on-cost. Factory overhead costs embrace all expenses incurred in factory production which are not directly charged to products as direct material or direct labor.

The practice of applying overhead charges arises because prohibitive costly accounting procedures would be required to charge all items of cost directly to the product.

Factory overhead costs embrace costs of material and labor not charged directly to product and fixed costs. Fixed costs embrace charges for such things as taxes, insurance, interest, rental, depreciation and maintenance of buildings, furniture and equipment, and salaries of factory supervision, which are considered to be independent of volume of production.

Indirect material and labor costs embrace costs of all items of material and labor consumed in manufacture which are not charged to the product as direct material or direct labor.

Factory cost. The *factory cost* of a product is the sum of direct material, direct labor, and factory overhead. It is these items that are summarized on the cost of goods made and sold statement. This cost classification separates the manufacturing cost from administrative and selling costs thus giving an indication of production costs over time.

Administrative costs. *Administrative costs* arise from expenditures for such items as salaries of executive, clerical, and technical personnel, office space, office supplies, depreciation of office equipment, travel, and fees for legal, technical, and auditing services that are necessary to direct the enterprise as a whole as distinct from its production and selling activities. Expenses so incurred are often recorded on the basis of the cost of carrying on subdivisions of administrative activities deemed necessary to take appropriate action to improve the effectiveness of administration.

In most cases, it is not practical to relate administrative costs directly to specific products. The usual practice is to allocate administrative costs to the product as a percentage of the product's factory cost. For example, if the annual administrative costs and factory costs of a concern are estimated at $10,000 and $100,000, respectively, for a given year, 10% will be added to the factory cost of products manufactured to absorb the administrative costs.

Selling cost. The *selling cost* of a product arises from expenditures incurred in disposing of the products and services produced. This class of expense includes such items as salaries, commissions, office space, office supplies, rental and depreciation, operation of office equipment and automobiles, travel, market surveys, entertainment of customers, displays, and sales space.

Selling expenses may be allocated to various classes of products, sales territories, sales of individual salesmen, and so forth, as a means of improving the effectiveness of selling activities. In many cases it is considered adequate to allocate selling expense to products as a percentage of their production cost. For example, if the annual selling expense is estimated at $22,000 and the annual production cost is estimated at $110,000, 20% will be added to the production cost of products to obtain the cost of sales.

11.3. Methods for Allocating Overhead Charges

There are four common methods of allocating factory overhead charges to the product being produced. These are the direct-labor-cost method, the direct-labor-hour method, the direct-material-cost method, and the machine-rate method, Each of these bases for allocation will be illustrated by reference to a hypothetical manufacturing concern which will serve to exhibit some aspects of cost accounting pertinent in engineering economy studies.

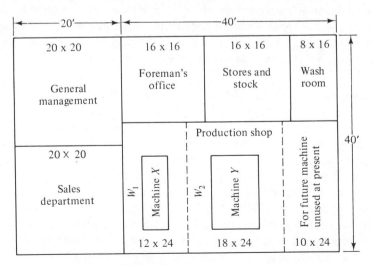

Figure 11.3. Plan of Plasco Company plant facilities.

PLASCO Company is a small plastic manufacturing concern with plant facilities as is shown in Figure 11.3. Land for the plant cost $9,000 and the plant itself, constructed 4 years ago, cost $30,000. Two-thirds of each of these items, or $6,000 and $20,000, are attributed to the production department. Initial cost of Machine X and Machine Y was $18,000 and $30,000, respectively. In addition, other assets of the firm are factory furniture with a first cost of $2,000, small tools and dies which cost $2,100, and stores and stock inventory with a current value of $5,800.

The factory salaries and wages during the current year are estimated to be as follows:

Foreman F supervises factory operations $12,000
Handyman H moves material, takes care of stock and stores, and does
   janitor work 5,200
      Total indirect labor $17,200
Workman W_1 operates Machine X, $4.50/hr. × 2,000 hr. 9,000
Workman W_2 operates Machine Y, $3.60/hr. × 2,000 hr. 7,200
      Total direct labor $16,200

The supplies to be used in the factory during the fifth year are estimated as follows:

Office and general supplies .... $ 600
Water (est. as $\frac{3}{4}$ of bill for entire building)....................... 75
Lighting current (est. as $\frac{2}{3}$ of bill for entire building) 200
Heating fuel (est. as $\frac{2}{3}$ of bill for entire building) 410
Electric power ($260 for Machine X and $400 for Machine Y) 660
Maintenance supplies ($100 for Machine X and $280 for Machine Y).. 380
                                         $2,325

PLASCO Company makes three products L, M, and N. Estimated output, material cost, direct labor hours, and machine hours are given in Table 11.1.

Table 11.1. ESTIMATED ACTIVITY DURING CURRENT YEAR

				Direct-labor hours				Machine hours			
				Workman W_1		Workman W_2		Machine X		Machine Y	
Pro-duct	Esti-mated output	Material cost									
		Each	Total	Each	Total	Each	Total	Each	Total	Each	Total
L	100,000	$0.10	$10,000	0.01	1,000	—	—	0.01	1,000	—	
M	140,000	0.08	11,200	—	—	0.01	1,400	—	—	0.01	1,400
N	80,000	0.12	9,600	0.0125	1,000	0.0075	600	0.0125	1,000	0.0075	600
			$30,800		2,000		2,000		2,000		2,000

The calculation of the PLASCO Company overhead rates during the current year may be summarized as follows:

A. *Overhead Items for Building*

Depreciation, insurance, maintenance on building, ...	$ 2,800
Taxes on building and land, ...	1,000
Interest on present value of building and land, ...	2,400
Water, light, and fuel for factory, ...	685
Total	$ 6,885

B. *Miscellaneous Items of Overhead*

Depreciation, taxes, insurance, and maintenance of factory furniture,..	$ 320
Interest on present value of factory furniture,...............	140
Depreciation, taxes, insurance, and interest on present value of small tools, ...	310
Taxes, insurance, and interest on stores and stock inventory, ..	750
Office and general supplies	600
Total	$ 2,120

C. *Indirect Labor and Labor Overhead*

Salaries of indirect labor of F and H, ...	$17,200
Payroll taxes, ...	3,100
Total	$20,300

D. *Machine X Overhead Items*

Depreciation, taxes, insurance, and maintenance on Machine X,	$ 3,100
Interest on present value of Machine X, ...	800
Supplies for Machine X	100
Power for Machine X	260
Total	$ 4,260

E. *Machine Y Overhead Items*

Depreciation, taxes, insurance, and maintenance on Machine Y,	$ 4,400
Interest on present value of Machine Y, ...	1,300
Supplies for Machine Y	280
Power for Machine Y	400
Total	6,380
Grand total, all factory overhead items	39,945

On the basis of the information given, overhead allocation rates may be calculated as follows:

$$\text{direct-labor-cost rate} = \frac{\text{total factory overhead}}{\text{total direct labor wages}}$$

$$= \frac{\$39,945}{\$16,200} = 2.47.$$

$$\text{direct-labor-hour rate} = \frac{\text{total factory overhead}}{\text{total hours of direct labor}}$$

$$= \frac{\$39,945}{4,000} = \$9.99.$$

$$\text{direct-material-cost rate} = \frac{\text{total factory overhead}}{\text{total direct material cost}}$$

$$= \frac{\$39,945}{\$30,800} = 1.30.$$

Further analysis must be made before machine rates can be established for Machine X and Machine Y. In establishing machine rates, as many items of overhead as possible are directly allocated to each machine before their identity is lost by being charged to an overhead account.

Consider Item A. This item is equal to $6,885 and is equivalent to the rent of the factory building, which has a floor area of 1,600 square feet.

Annual cost per sq. ft. of floor area, $6,885 ÷ 1,600	$ 4.30
Space charge, Machine X, 288 × $4.30	1,238
Space charge, Machine Y, 432 × $4.30	1,858
Total space charged to Machines X and Y	$3,096
Balance of space cost to be allocated, $6,885 − $3,096	$3,789

This balance of Item A, together with Items B and C must be distributed to Machine X and Machine Y on some basis that is estimated to reflect actual conditions. In this example the sums of these items will be allocated equally to the two machines. One-half of the unallocated sum is equal to $\frac{1}{2}(\$3,789 + \$2,120 + \$20,300) = \$13,105$.

Machine X Overhead Charges and Machine Rate

Item D	$ 4,260
Space charge as calculated above	1,238
One-half of unallocated balance of Item A, Item B, and Item C, ..	13,105
Total	$18,603

$$\text{machine rate (Machine X)} = \frac{\text{overhead allocated to Machine X}}{\text{estimated annual hours of operation}}$$

$$= \frac{\$18,603}{2,000} = \$9.30 \text{ per hour.}$$

Machine Y Overhead Charges and Machine Rate

Item E	$ 6,380
Space charge as calculated above	1,858
One-half of unallocated balance of Item A, Item B, and Item C, ..	13,105
Total	$21,343

$$\text{machine rate (Machine } Y) = \frac{\$21,343}{2,000} = \$10.67 \text{ per hour.}$$

The unit cost of Products *L*, *M*, and *N* may now be determined by each of the four methods of allocating factory overhead.

The Factory Cost of Product L

Direct-labor-cost method:

Direct material	$0.100
Direct labor, 0.01 × $4.50	0.045
Overhead, $0.045 × 2.47	0.111
	$0.256

Direct-labor-hour method:

Direct material	$0.100
Direct labor, 0.01 × $4.50	0.045
Overhead, 0.01 × $9.99	0.100
	$0.245

Direct-material-cost method:

Direct material	$0.100
Direct labor, 0.01 × $4.50	0.045
Overhead, $0.10 × 1.30	0.130
	$0.275

Machine-rate method:

Direct material	$0.100
Direct labor, 0.01 × $4.50	0.045
Overhead, 0.01 × $9.30	0.093
	$0.238

The Factory Cost of Product M

Direct-labor-cost method:

Direct material	$0.080
Direct labor, 0.01 × $3.60	0.036
Overhead, $0.036 × 2.47	0.089
	$0.205

Direct-labor-hour method:

Direct material	$0.080
Direct labor, 0.01 × $3.60	0.036
Overhead, 0.01 × $9.99	0.100
	$0.216

Direct-material-cost method:

Direct material	$0.080
Direct labor, 0.01 × $3.60	0.036
Overhead, $0.08 × 1.30	0.104
	$0.220

Machine-rate method:

Direct material	$0.080
Direct labor, 0.01 × $3.60	0.036
Overhead, 0.01 × $10.67	0.107
	$0.223

The Factory Cost of Product N

Direct-labor-cost method:

Direct material	$0.120
Direct labor, 0.0125 × $4.50 + 0.0075 × $3.60	0.083
Overhead, $0.083 × 2.47	0.205
	$0.408

Direct-labor-hour method:

Direct material	$0.120
Direct labor (calculated as above)	0.083
Overhead, 0.02 × $9.99	0.200
	$0.403

Direct-material-cost method:

Direct material	$0.120
Direct labor	0.083
Overhead, $0.12 × 1.30	0.156
	$0.359

Machine-rate method:

Direct material	$0.120
Direct labor	0.083
Overhead, 0.0125 × $9.30 + 0.0075 × $10.67	0.196
	$0.399

The cost of sales for the PLASCO Company is obtained by adding administrative and selling costs to the factory cost. Continuing with the example of the PLASCO Company, suppose that after careful analysis of expenditures, annual administrative costs have been estimated at $28,000 and annual selling costs at $16,500.

Annual direct material costs, direct labor costs, and factory overhead costs for PLASCO have been estimated previously as $30,800, $16,200, and $39,945, respectively. Therefore, the annual estimated cost of sales of PLASCO Company for the current year may be summarized as follows:

Estimated annual direct material cost	$30,800	
Estimated annual direct labor cost	16,200	
Estimated annual factory overhead cost ..	39,945	
Estimated annual factory cost.....................	$86,945	
Estimated annual administrative cost	28,000	
Estimated annual production cost		$114,945
Estimated annual selling cost		16,500
Estimated annual cost of sales		$131,445

Difficulties associated with overhead allocations. An examination of the factory costs obtained for Product *L* in the PLASCO Company example with the four methods of factory overhead employed shows considerable variation. Each of the three costs that make up the total is subject to estimating error. Direct material costs may not be accurate because of variations in pricing, charging a product with more material than is actually used, and the use of estimates. Similarly, and for much the same reasons, direct labor costs as charged will usually be inaccurate to some extent. However, with reasonably good control and accounting procedures, direct material costs and direct labor costs are generally reliable.

If attention is directed to factory overhead costs allocated to Product *L*, it will be observed that amounts allocated by the several methods range from \$0.093 per unit to \$0.130 per unit. The ratio of these amounts is approximately 1.4. The fact that the amounts shown for the several methods differ is evidence that significant differences can result from the choice of a method of overhead allocation.

In actual practice the item of factory overhead is a summation of a great number and variety of costs. It is therefore not surprising that the use of a single, simple method will not allocate factory overhead costs to specific products with precision. Although a particular method may be generally quite satisfactory, wide variations may result in some specific situations.

For example, consider the direct-labor-hour method as it applies to Product *L* and Product *M*. The assumption is that \$9.99 will be incurred for each hour of direct labor, regardless of equipment used. Thus, the amount of overhead allocated to Product *L* and Product *M* is identical, even though it is apparent that the cost of operating the respective machines on which they are processed is quite different. Note that the machine rates of Machine *X* and Machine *Y* are \$9.30 and \$10.67, respectively.

If the direct-material-cost method is used, it is clear that overhead allocations to products will be dependent upon the unit cost as well as upon the quantity of materials used. Suppose, for example, that in the manufacture of tables a certain model may be made from either pine or mahogany. Processing might be identical but the amount charged for overhead might be several times as much for mahogany as for pine because of the difference in the unit cost of the materials used.

Effect of changes in level of activity. In the determination of rates for the allocation of overhead, the activity of the PLASCO Company for the fifth year was estimated in terms of Products *L*, *M*, and *N*. This estimate served as a basis for determining annual material cost, annual direct labor cost, annual direct labor hours, and machine hours. These items then became the denominator of the several allocation rates.

The numerator of the allocation rates was the estimated factory overhead, totaling \$39,945. This numerator quantity will remain relatively constant for changes in activity, as an examination of the items of which it is composed will reveal.

For this reason the several rates for allocating factory overhead will vary in some generally inverse proportion with activity. Thus, if the actual activity is less than the estimated activity, the overhead rate charged will be less than the amount necessary to absorb the estimated total overhead. The reverse is also true. When the under or overabsorbed balance of overhead becomes known at the end of the year, it is usually charged to profit and loss, or surplus.

In engineering economic analyses, the effect of activity on overhead charges and overhead rates is an important consideration. The total overhead charges of the PLASCO Company would remain relatively constant over a range of activity represented, for example, by 1,200 to 2,000 hours of activity for each of the two machines. Thus, after the total overhead has been allocated, the incrementeal cost of producing additional units of product will consist of direct material and direct labor costs.

11.4. Use of Accounting Data in Economy Studies

Since accounting data are the basis for many engineering economy studies, caution should be exercised in their use. An understanding of the relevance of accounting data is essential for its proper utilization. Two examples of the pertinence of cost accounting data are presented in this section.

Cost data must be applicable. It is a common error to infer that a reduction in labor costs will result in a proportionate decrease in overhead costs, particularly if overhead is allocated on a labor cost basis. In one instance a company was manufacturing an oil field specialty. An analysis revealed per item costs as follows:

Direct labor	$ 4.18
Direct material	1.84
Factory overhead, 4.18×2.30	9.61
Factory cost per item	$15.63

The factory cost of $15.36 was slightly less than the price of the item in question. The first suggestion was to cease making the article. But after further analysis it became clear that the overhead of $9.61 would not be saved if the item was discontinued. The overhead rate used was based on heavy equipment required for most of the work in the department and on hourly earnings of workmen who averaged $5.85 per hour. For the job in question only a light drill press and hand tools were used. Little actual reduction in cost would have resulted from not using them in the manufacture of the article.

It has previously been calculated that items of overhead of the PLASCO Company for the building are equal to $4.30 per square foot of factory floor space. Figure 11.3 reveals a currently unused 10 feet by 24 feet space for future machines. The annual cost equivalent of this space is

$$10 \text{ ft.} \times 24 \text{ ft.} \times \$4.30/\text{ft.}^2 = \$1,032.$$

The item need not be included as an item of cost attributable to a machine that may be purchased to occupy the space in question, since no actual additional cost will arise should the space be occupied. The $1,032 item has been entered as an overhead charge to be allocated to products made on the basis of one of several overhead rates. The addition of a new machine will probably result in changes in the overhead allocation rate used, but it will not result in a change in the overhead item pertaining to the building.

Average costs are inadequate for specific analysis. An important function of cost accounting, if not the primary one, is to provide data for decisions relative to the reduction of production costs and the increase of profit from sales. Cost data that are believed to be accurate may lead to costly errors in decisions. Cost data that give true average values and are adequate for overall analyses may be inadequate for specific detailed analyses. Thus, cost data must be carefully scrutinized and their accuracy established before they can be used with confidence in engineering economy studies.

In Table 11.2 actual and estimated cost data relative to the cost of three products have been tabulated. The actual production cost of Products *A*, *B*,

Table 11.2. ACTUAL AND ESTIMATED COST DATA

Product	Direct labor and material costs	Overhead costs, actual	Overhead costs, believed to be	Production cost, actual	Production cost, believed to be
A	$6.50	$2.50	$3.50	$ 9.00	$10.00
B	7.00	3.00	3.00	10.00	10.00
C	7.50	3.50	2.50	11.00	10.00
Average	7.00	3.00	3.00	10.00	10.00

and *C* are $9, $10, and $11, respectively, but because of inaccuracies in overhead costs, the production costs of Products *A*, *B*, and *C* are believed to be equal to $10 for each.

It should be noted that even though the average of a number of costs may be correct, there is no assurance that this average is a good indication of the cost of individual products. For this reason the accuracy of each cost should be ascertained before it is used in an economic analysis.

If, for example, the selling price of the products is based upon their believed production cost, Product *A* will be overpriced and Product *C* will be underpriced. Buyers may be expected to shun Product *A* and to buy large quantities of Product *C*. This may lead to a serious unexplained loss of profit. Average values of cost data are of little value in making decisions relative to specific products.

QUESTIONS AND PROBLEMS

1. What is the relationship between the balance sheet and the profit and loss statement?

2. Describe the difference in viewpoint in respect to time between a balance sheet and an economy study.

3. How does an increase (decrease) in depreciation expense affect the net profit shown on the income statement?

4. What is the effect on federal and state income taxes if the depreciation expense is increased (decreased)?

5. What is the function of general accounting; of cost accounting?

6. Describe the difference between general accounting activities and engineering economy studies?

7. What relationship between accuracy and cost determines the extent of expenditure that can be justified for the maintenance of a cost accounting system?

8. Name several bases for the distribution of overhead costs.

9. What precautions should be exercised in utilizing accounting data in engineering economic analysis?

10. The manufacturing costs of Products X and Y are believed to be $10 per unit. On the basis of this estimate and a desired profit of 10%, the selling price is set at $11 per unit.
    (a) What is the profit if 800 units of Product X and 2,000 units of Product Y are sold?
    (b) If the actual manufacturing costs of Products X and Y are $9 and $11, respectively, what is the actual profit?

11. A small factory is divided into four departments for accounting purposes. The direct labor and direct material expenditures for a given year are as follows:

Department	Direct Labor Hours	Direct Labor Cost	Direct Material Cost
A	1,800	$14,500	$19,000
B	1,890	14,900	6,800
C	2,100	15,050	11,200
D	1,670	13,900	15,000

Distribute an annual overhead charge of $42,000 to Departments A, B, C, and D on the basis of direct labor hours, direct labor cost, and direct material cost.

12. A manufacturer makes 8,500,000 radio tubes per year. An assembly operation is performed on each tube for which a standard piece rate of $1.05 per hundred pieces is paid. The standard cycle time per piece is 0.624 minute. Overhead costs associated with the operation are estimated at $0.36 per operator hour.

If the standard cycle time can be reduced by 0.01 of a minute and if one-half of the overhead cost per operator hour is saved for each hour by which total operator hours are reduced, what will be the maximum amount that can be spent to bring about the 0.01 of a minute reduction in cycle time? Assume that the improvement will be in effect for one year, that piece rates will be reduced in proportion to the reduction in cycle time, and that the average operator's time per piece is equel to the standard time per piece in either case.

13. A factory producing lawn mowers works at 60% of its capacity and produces 21,000 mowers per year. The unit manufacturing cost is computed as follows:

Direct labor cost..	$19.50
Direct material cost	14.50
Overhead cost	10.00
	$44.00

The mowers are marketed through a factory distributor for $52.50 each. It is anticipated that the volume of production can be increased to 28,000 units per year if the price is lowered to $43 per unit. This action would not increase the present total overhead cost. Compute the present profit per year and the profit per year if the volume of production is increased.

14. An automobile parts manufacturer produces batteries and distributor assemblies in his electrical products department. It is believed that the cost of manufacturing the batteries and the distributor assemblies is $18.40 and $23.20 per unit, respectively. These costs were derived on the basis of an equal distribution of overhead charges. A study of the firm's cost structure reveals that overhead would be more equitably distributed if overhead charges against the distributor assembly were 50% more than those against the battery. This conclusion was reached after careful consideration of the nature and source of a $220,000 annual overhead expenditure for these products.

(a) Calculate the unit production cost applicable to the battery and to the distributor assembly if the annual production is made up of 24,000 batteries and 16,000 distributor assemblies.

(b) What is the annual profit if the selling price is $25.80 and $32.00 for the battery and the distributor assembly, respectively?

DEPRECIATION AND DEPRECIATION ACCOUNTING

People satisfy their wants by the consumption of goods and services, the production of which is directly dependent upon the employment of large quantities of producer goods. But producer goods are not acquired without considerable investment. One characteristic of modern civilization is the large investment per worker in production facilities. Although this investment results in high worker productivity it should be recognized that this economy must be sufficient to absorb the reduction in value of these facilities as they are consumed in the production process.

Alternative engineering proposals will affect the type and quantity of producer goods required. As an essential ingredient in the process of want satisfaction, producer goods give rise to capital consumption and investment costs which must be considered in evaluating alternative engineering proposals. An understanding of the depreciation concept is essential if it is to be included as an integral part of engineering economy analysis.

12.1. Classifications of Depreciation

Depreciation may be defined as the lessening in value of a physical asset with the passage of time. With the possible exception of land, this phenome-

12

non is a characteristic of all physical assets. A common classification of the types of depreciation include (1) physical depreciation, (2) functional depreciation, and (3) accidents. The first two of these classifications will be defined and explained in the paragraphs which follow.

Physical depreciation.   Depreciation resulting in physical impairment of an asset is known as *physical depreciation*. Physical depreciation manifests itself in such tangible ways as the wearing of particles of metal from a bearing and the corrosion of the tubes in a heat exchanger. This type of depreciation results in the lowering of the ability of a physical asset to render its intended service. The primary causes of physical depreciation are:

1. Deterioration due to action of the elements including the corrosion of pipe, the rotting of timbers, chemical decomposition, bacterial action, etc. Deterioration is substantially independent of use.
2. Wear and tear from use which subjects the asset to abrasion, shock, vibration, impact, etc. These forces are occasioned primarily by use and result in a loss of value over time.

Functional depreciation.   *Functional depreciation* results not from a deterioration in the asset's ability to serve its intended purpose, but from a change in the demand for the services it can render. The demand for the services of an asset may change because it is more profitable to use a more efficient unit, there is no longer work for the asset to do, or the work to be done exceeds the capacity of the asset. Depreciation resulting from a change in the need for the service of an asset may be the result of:

1. Obsolescence resulting from the discovery of another asset that is sufficiently superior to make it uneconomical to continue using the original asset. Assets also become obsolete when they are no longer needed.
2. Inadequacy or the inability to meet the demand placed upon it. This situation arises from changes in demand not contemplated when the asset was acquired.

Suppose that a manufacturer has a hand riveter in good physical condition, but he has found it more profitable to dispose of it and purchase an automatic riveter because of the reduction in riveting costs that the latter makes possible. The difference between the use value of the hand-operated riveter and the amount received for it on disposal represents a decrease in value due to the availability of the automatic riveter. This cause of loss in value is termed obsolescence. Literally, the hand riveter had become *obsolete* as a result of an improvement in the art of riveting.

As a second example, consider the case of a manufacturer who has ceased producing a certain item and finds that he has machines in good operating condition which he no longer needs. If he disposes of them at a value less than their former use value to him, the difference will be termed depreciation, or loss resulting from obsolescence. The inference is that the machine has become obsolete or out of date as far as the user is concerned.

Inadequacy, a cause of functional depreciation, occurs when changes in demand for the services of an asset result in a demand beyond the scope of the asset. For example, a small electric generating plant has a single 500-kilovolt-amperes (kva) generating unit whose capacity will soon be exceeded by the demand for current. Analyses show that it will cost less in the long run to replace the present unit with a 750-kva unit, which is estimated to meet the need for some time, than to supplement the present unit with a 250-kva unit.

In such a case the original 500-kva unit is said to be *inadequate* and to have been *superseded* by the larger unit. Any loss in the value of the unit below its use value is the result of inadequacy or supersession. The former term is preferred for designating this type of depreciation.

A disposition to replace machines when it becomes profitable to do so instead of when they are worn out has probably been an important factor in the rapid development of this nation. The sailing ship has given way to the steamship; in street transportation, the sequence of obsolescence has been horse cars, cable cars, electric cars, and automotive buses; one generation of computers is rapidly being replaced by the next. In power generation, nuclear plants are becoming an attractive alternative to fossil fuel plants. In manufacturing, improvements in the arts of processing have resulted in widespread obsolescence and inadequacy of equipment. Each technological advance produces improvements that result in the obsolescence of existing assets.

12.2. Accounting for the Consumption of Capital Assets

An asset such as a machine is a unit of capital. Such a unit of capital loses value over a period of time in which it is used in carrying on the productive activities of a business. This loss of value of an asset represents actual piecemeal consumption or expenditure of capital. For instance, a truck tire is a unit of capital. The particles of rubber that wear away with use are actually small physical units of capital consumed in the intended service of the tire. In a like manner, the wear of machine parts and the deterioration of structural elements are physical consumptions of capital. Expenditures of capital in this way are often difficult to observe and are usually difficult to evaluate in monetary terms, but they are nevertheless real.

An understanding of the concept of depreciation is complicated by the fact that there are two aspects to be considered. One is the actual lessening in value of an asset with use and the passage of time, and the other is the accounting for this lessening in value.

The accounting concept of depreciation views the cost of an asset as a prepaid operating expense that is to be charged against profits over the life of the asset. Rather than charging the entire cost as an expense at the time the asset is purchased, the accountant attempts, in a systematic way, to spread the anticipated loss in value over the life of the asset. This concept of amortizing the cost of an asset so that the profit and loss statement is a more accurate reflection of capital consumption is basic to financial reporting and income tax calculation.

A second aim in depreciation accounting is to have, continuously, a monetary measure of the value of an enterprise's unexpended physical capital, both collectively and by individual units such as specific machines. This value can only be approximated with the accuracy with which the future life of the asset and the effect of deterioration can be estimated.

A third aim is to arrive at the physical expenditure of physical capital, in monetary terms, that has been incurred by each unit of goods as it is produced. In any enterprise, physical capital in the form of machines, buildings, and the like is used in carrying on production activities. As machines wear out in productive activities, physical capital is converted to value in the product. Thus, the capital that is lost in wear by machines is recovered in the product processed on them. This lost capital needs to be accounted for in order to determine production costs.

12.3. The Value-Time Function

In considering depreciation for accounting purposes, the pattern of the future value of an asset should be predicted. It is customary to assume that the value of an asset decreases yearly in accordance with one of several mathematical functions. However, choice of the particular model that is to represent the lessening in value of an asset over time is a difficult task. It involves decisions as to the life of the asset, its salvage value, and the form of the mathematical function. Once a value-time function has been chosen, it is used to represent the value of the asset at any point during its life. A general value-time function is shown in Figure 12.1.

Accountants use the term *book value* to represent the original value of an asset less its accumulated depreciation at any point in time. Thus, a function similar to Figure 12.1 can represent book value.

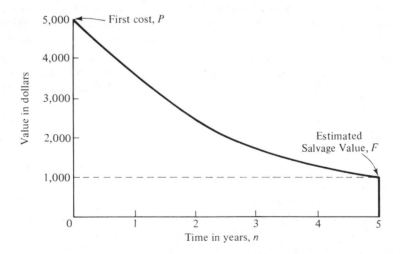

Figure 12.1. A general value-time function.

The book value at the end of any year is equal to the book value at the beginning of the year less the depreciation expense charged during the year. Table 12.1 presents the calculation of book value at the end of each year for an asset with a first cost of $12,000, an estimated life of 5 years, and a salvage value of zero with assumed depreciation charges.

Table 12.1. BOOK VALUE CALCULATION

End of Year	Depreciation charge during year	Book value at end of year
0	——	$12,000
1	$4,000	8,000
2	3,000	5,000
3	2,000	3,000
4	2,000	1,000
5	1,000	0

In determining book value the following notation will be used. Let

P = first cost of the asset;
F = estimated salvage value;
B_t = book value at the end of year t.
D_t = depreciation charge during year t.
n = estimated life of the asset.

A general expression relating book value and the depreciation charge is given by

$$B_t = B_{t-1} - D_t$$

where $B_0 = P$. The sections which follow are concerned with methods for determining D_t for $t = 1, 2, , \ldots, n$.

12.4. Straight-Line Method of Depreciation

The straight-line depreciation model assumes that the value of an asset decreases at a constant rate. Thus, if an asset has a first cost of $5,000 and an estimated salvage value of $1,000, the total depreciation over its life will be $4,000. If the estimated life is 5 years the depreciation per year will be $4,000 ÷ 5 = $800. This is equivalent to a depreciation rate 1 ÷ 5 or 20% per year.

Table 12.2. THE STRAIGHT-LINE METHOD

End of year t	Depreciation charge during year t	Book value at end of year t
0	—	$5,000
1	$800	4,200
2	800	3,400
3	800	2,600
4	800	1,800
5	800	1,000

For this example, the annual depreciation and book value for each year is given in Table 12.2.

General expressions for the calculation of depreciation and book value may be developed for the straight-line method. Table 12.3 shows the depre-

Table 12.3. GENERAL EXPRESSIONS FOR THE STRAIGHT-LINE METHOD

End of year t	Depreciation charge during year t	Book value at end of year t
0	—	P
1	$\dfrac{P - F}{n}$	$P - \left(\dfrac{P - F}{n}\right)$
2	$\dfrac{P - F}{n}$	$P - 2\left(\dfrac{P - F}{n}\right)$
3	$\dfrac{P - F}{n}$	$P - 3\left(\dfrac{P - F}{n}\right)$
t	$\dfrac{P - F}{n}$	$P - t\left(\dfrac{P - F}{n}\right)$
n	$\dfrac{P - F}{n}$	$P - n\left(\dfrac{P - F}{n}\right)$

ciation charge and book value expressions for each year. Thus, the depreciation in any year is

$$D_t = \frac{P - F}{n}$$

the book value is

$$B_t = P - t\left(\frac{P - F}{n}\right)$$

and the depreciation rate per year is $1/n$.

12.5. Declining-Balance Method of Depreciation

The declining-balance mehtod of depreciation assumes that an asset decreases in value at a faster rate in the early portion of its service life than in the latter portion of its life. By this method a fixed percentage is multiplied times the book value of the asset at the beginning of the year to determine the depreciation charge for that year. Thus, as the book value of the asset decreases through time so does the size of the depreciation charge. For an asset with a \$5,000 first cost, a \$1,000 estimated salvage value, an estimated life of 5 years and a depreciation rate of 30% per year the depreciation charge per year is shown in Table 12.4.

Table 12.4. THE DECLINING-BALANCE METHOD

End of year t	Depreciation Charge during year t	Book value at end of year t
0	—	\$5,000
1	(0.30)(\$5,000) = \$1,500	3,500
2	(0.30)(3,500) = 1,050	2,450
3	(0.30)(2,450) = 735	1,715
4	(0.30)(1,715) = 515	1,200
5	(0.30)(1,200) = 360	840

For a depreciation rate R the general relationship expressing the depreciation charge in any year for declining-balance depreciation is

$$D_t = R \cdot B_{t-1}.$$

From the general expression for book value shown in Section 12.3 it is seen that

$$B_t = B_{t-1} - D_t.$$

Therefore, for declining-balance depreciation

$$B_t = B_{t-1} - RB_{t-1} = (1 - R)B_{t-1}.$$

Using this recursive expression it is possible to determine the general expressions for the depreciation charge and the book value for any point in time. These calculations are shown in Table 12.5 for an asset with first cost P. Thus, the depreciation in any year is

$$D_t = R(1 - R)^{t-1}P.$$

Table 12.5. GENERAL EXPRESSIONS FOR THE DECLINING-BALANCE METHOD OF DEPRECIATION

End of year t	Depreciation charge during year t	Book value at end of year t
0	—	P
1	$R \times B_0 = R(P)$	$(1 - R)B_0 = (1 - R)P$
2	$R \times B_1 = R(1 - R)P$	$(1 - R)B_1 = (1 - R)^2 P$
3	$R \times B_2 = R(1 - R)^2 P$	$(1 - R)B_2 = (1 - R)^3 P$
t	$R \times B_{t-1} = R(1 - R)^{t-1} P$	$(1 - R)B_{t-1} = (1 - R)^t P$
n	$R \times B_{n-1} = R(1 - R)^{n-1} P$	$(1 - R)B_{n-1} = (1 - R)^n P$

And the book value is

$$B_t = (1 - R)^t P.$$

If the declining-balance method of depreciation is utilized for income tax purposes the maximum rate that may be used is double the straight-line rate that would be allowed for a particular asset or group of assets being depreciated. Thus, for an asset with an estimated life of n years the maximum rate that may be used with this method is $2(1/n)$. Many firms and individuals choose to depreciate their assets using declining-balance depreciation with the maximum allowable rate. Such a method of depreciation is commonly referred to as the *double-declining-balance* method of depreciation.

In Table 12.4 the book value at the end of year 5 is $840 which is less than the estimated salvage of $1,000. If the salvage value for this example had been zero, it is observed that for this method of depreciation the book value would never reach zero regardless of the length of the time span over which the asset is depreciated. Thus, adjustments are necessary in order to rectify the differences between the estimated and calculated book value of the asset. In most situations these adjustments are made at the time of disposal of the asset when accounting entries are made to account for the difference between the asset's actual value and its calculated book value.

If it is desired to use a depreciation rate that will result in a particular book value at some point in time, it is possible to solve for the rate that should be used. Solving for R from the general expression

$$B_t = (1 - R)^t P$$

gives

$$R = 1 - \sqrt[t]{\frac{B_t}{P}}.$$

For the example previously considered the estimated salvage value is $1,000 at the end of the fifth year and the asset's first cost is \$5,000. Thus,

$$R = 1 - \sqrt[5]{\frac{1,000}{5,000}} = 1 - 0.725 = 27.5\%.$$

This method of determining a depreciation rate for the declining-balance method is rarely used in practice. Usually, the depreciation rate is selected with regard to its effect on income taxes, its effect on the profit and loss statement, and its ease of calculation when assets are grouped for accounting purposes.

12.6. Sum-of-the-Years-Digits Method of Depreciation

The sum-of-the-years-digits depreciation model assumes that the value of an asset decreases at a decreasing rate. If an asset has an estimated life of 5 years, the sum of the years will be $1 + 2 + 3 + 4 + 5 = 15$. Thus, if the first cost of the asset is \$5,000, and the estimated salvage value is \$1,000, the depreciation during the first year will be ($5,000 - $1,000)\frac{5}{15} = \$1,333.33. During the second year, the depreciation will be ($5,000 - $1,000)\frac{4}{15} = \$1,066.67, etc. These values are given in Table 12.6.

Table 12.6. AN EXAMPLE OF THE SUM-OF-THE-YEARS-DIGITS METHOD OF DEPRECIATION

End of year t	Depreciation charge during year t	Book value at the end of year t
0	—	\$5,000
1	$\frac{5}{15}$(\$4,000) = \$1,333	3,667
2	$\frac{4}{15}$(4,000) = 1,067	2,600
3	$\frac{3}{15}$(4,000) = 800	1,800
4	$\frac{2}{15}$(4,000) = 533	1,267
5	$\frac{1}{15}$(4,000) = 267	1,000

The sum-of-the-years for any number of years n can be computed from the expression

$$\sum_{j=1}^{n} j = 1 + 2 + 3 + \ldots + (n - 1) + n = \frac{n(n + 1)}{2}.$$

Thus, for an asset with a 5-year life the sum-of-the-years-digits is $(5)(5 + 1)/2 = 15$ as seen earlier.

The general expressions for the depreciation amount in each year and the book value at the end of each year are presented in Table 12.7. The first cost

Table 12.7. GENERAL EXPRESSIONS FOR THE SUM-OF-THE-YEARS-DIGITS METHOD OF DEPRECIATION

End of year t	Depreciation charge during year t	Book value at the end of year t
0	—	P
1	$\dfrac{n}{n(n+1)/2}(P-F)$	$P - \dfrac{(P-F)}{n(n+1)/2}\,n$
2	$\dfrac{(n-1)}{n(n+1)/2}(P-F)$	$P - \dfrac{(P-F)}{n(n+1)/2}[n+(n-1)]$
3	$\dfrac{n-2}{n(n+1)/2}(P-F)$	$P - \dfrac{(P-F)}{n(n+1)/2}[n+(n-1)+(n-2)]$
t	$\dfrac{n-t+1}{n(n+1)/2}(P-F)$	$P - \dfrac{(P-F)}{n(n+1)/2}\left[\displaystyle\sum_{j=n-t+1}^{n} j\right]$
n	$\dfrac{1}{n(n+1)/2}(P-F)$	$P - \dfrac{(P-F)}{n(n+1)/2}\left[\displaystyle\sum_{j=1}^{n} j\right]$

of the asset is P, while its estimated salvage value and estimated life are F and n, respectively. Thus, the depreciation charge in any year can be expressed as

$$D_t = \frac{n-t+1}{n(n+1)/2}(P-F).$$

And the book value at the end of any year t is

$$B_t = P - \frac{(P-F)}{n(n+1)/2}\left[\sum_{j=n-t+1}^{n} j\right].$$

However, the expression for book value can be simplified when it is seen that

$$\sum_{j=n-t+1}^{n} j = \sum_{j=1}^{n} j - \sum_{j=1}^{n-t} j.$$

In other words, if $n = 6$, $t = 4$, and $n - t + 1 = 3$, then

$$3 + 4 + 5 + 6 = 1 + 2 + 3 + 4 + 5 + 6 - [1 + 2].$$

Since it is known that $\sum_{j=1}^{n} j = n(n+1)/2$ it follows that

$$\sum_{j=n-t+1}^{n} j = \frac{n(n+1)}{2} - \frac{(n-t)[(n-t)+1]}{2}.$$

Substituting back into the equation for book value at any time t,

$$B_t = P - \frac{(P - F)}{n(n+1)/2}\left[\frac{n(n+1)}{2} - \frac{(n-t)(n-t+1)}{2}\right]$$

$$= P - (P - F) + (P - F)\frac{(n-t)(n-t+1)}{n(n+1)}$$

$$= (P - F)\left(\frac{n-t}{n}\right)\left(\frac{n-t+1}{n+1}\right) + F.$$

The sum-of-the-years-digits method of depreciation produces larger depreciation charges in the early life of the asset with the smaller charges occurring later in the asset's life. In this method the depreciation rate decreases through time and this decreasing rate is multiplied times a fixed amount $(P - F)$. This calculation is in contrast to the double-declining method which multiplies a fixed rate times a decreasing book value. Both methods are similar in that the depreciation charges in the early portion of an asset's life are greater than in the later portion of its life.

The sum-of-the-years-digits method of depreciation is not computationally as simple as the double-declining balance method. However, this method is used in practice because the rate at which it charges depreciation does produce a value-time curve that approximates the decrease in value of numerous categories of assets. That is, there are many assets that depreciate more rapidly during the early portion of their life and this fact is reflected by the sum-of-the-years-digits method which depreciates about three-fourths of the depreciable cost of an asset during the first half of its life.

12.7. Sinking-Fund Method of Depreciation

The sinking-fund depreciation model assumes that the value of an asset decreases at an increasing rate. One of a series of equal amounts is assumed to be deposited into a sinking fund at the end of each year of the asset's life. The sinking fund is ordinarily compounded annually and, at the end of the estimated life of the asset, the amount accumulated equals the total depreciation of the asset. Thus, if an asset has a first cost of $5,000, an estimated life of 5 years, an estimated salvage value of $1,000, and if the interest rate is 6%, the amount deposited into the sinking fund at the end of each year is

$$(\$5{,}000 - \$1{,}000)(\overset{A/F\,6,\,5}{0.1774}) = \$709.60.$$

The depreciation charge during any year is the sum of the amount deposited into the sinking fund at the end of the year and the amount of interest earned on the sinking fund during the year. For the conditions assumed, the capital recovered during the first year is \$709.60; during the second year \$709.60 + 0.06 × \$709.60 = \$752.18; during the third year \$709.60 + 0.06(\$709.60 + \$752.18) = \$797.31; etc. These values are given in Table 12.8.

Table 12.8. AN EXAMPLE OF THE SINKING-FUND METHOD OF DEPRECIATION

End of year t	Depreciation charge during year t	Book value at end of year t
0	—	\$5,000
1	\$710	4,290
2	752	3,538
3	797	2,741
4	845	1,896
5	896	1,000

To develop the general expression for the amount of depreciation that will be charged each year for the sinking-fund method of depreciation it is first necessary to understand that the depreciation amount is the sum of two components. The first component is the amount deposited into the sinking fund at the end of each year so that its accumulated value at the end of the asset's life equals the depreciable value of the asset. The second component is the amount of interest earned on the accumulated value of the sinking fund at the beginning of the particular year in question. For an asset with first cost P, estimated salvage value F, life n, and interest rate i, the first component of the depreciation charge each year is $(P - F)(\overset{A/F\,i,\,n}{\quad})$. The second component is different for each year and this is reflected by the second term in the second column of expressions in Table 12.9.

Table 12.9. GENERAL EXPRESSIONS FOR THE SINKING-FUND METHOD OF DEPRECIATION

End of year t	Depreciation charge at end of year t	Book value at end of year t
0	—	P
1	$(P - F)(\overset{A/F\,i,\,n}{\quad}) + i(0)$	$P - (P - F)(\overset{A/F\,i,\,n}{\quad})(\overset{F/A\,i,\,1}{\quad})$
2	$(P - F)(\overset{A/F\,i,\,n}{\quad}) + i\left[(P - F)(\overset{A/F\,i,\,n}{\quad})(\overset{F/A\,i,\,1}{\quad})\right]$	$P - (P - F)(\overset{A/F\,i,\,n}{\quad})(\overset{F/A\,i,\,2}{\quad})$
3	$(P - F)(\overset{A/F\,i,\,n}{\quad}) + i\left[(P - F)(\overset{A/F\,i,\,n}{\quad})(\overset{F/A\,i,\,2}{\quad})\right]$	$P - (P - F)(\overset{A/F\,i,\,n}{\quad})(\overset{F/A\,i,\,3}{\quad})$
t	$(P - F)(\overset{A/F\,i,\,n}{\quad}) + i\left[(P - F)(\overset{A/F\,i,\,n}{\quad})(\overset{F/A\,i,\,t-1}{\quad})\right]$	$P - (P - F)(\overset{A/F\,i,\,n}{\quad})(\overset{F/A\,i,\,t}{\quad})$
n	$(P - F)(\overset{A/F\,i,\,n}{\quad}) + i\left[(P - F)(\overset{A/F\,i,\,n}{\quad})(\overset{F/A\,i,\,n-1}{\quad})\right]$	$P - (P - F)(\overset{A/F\,i,\,n}{\quad})(\overset{F/A\,i,\,n}{\quad})$

Thus, for sinking-fund depreciation the depreciation amount for any year t can be determined from the general expression

$$D_t = (P - F)(\overset{A/Fi,\,n}{}) + i\left[(P - F)(\overset{A/Fi,\,n}{})(\overset{F/Ai,\,t-1}{})\right].$$

By rewriting this expression

$$D_t = (P - F)(\overset{A/Fi,\,n}{})\left[1 + i(\overset{F/Ai,\,t-1}{})\right]$$

but $(\overset{F/Ai,\,t-1}{})$ is $[(1 + i)^{t-1} - 1]/i$ so that

$$D_t = (P - F)(\overset{A/Fi,\,n}{})[1 + (1 + i)^{t-1} - 1] = (P - F)(\overset{A/Fi,\,n}{})(1 + i)^{t-1}$$

$$D_t = (P - F)(\overset{A/Fi,\,n}{})(\overset{F/Pi,\,t-1}{}).$$

The general book value expression for sinking-fund depreciation is

$$B_t = P - (P - F)(\overset{A/Fi,\,n}{})(\overset{F/Ai,\,t}{}).$$

Although these expressions may appear complicated, the purpose of depreciation accounting is to charge the depreciable portion of the asset over the estimated life of the asset. The sinking-fund method of depreciation is just another systematic method of accomplishing this objective.

12.8. Service Output Method of Depreciation

In some cases, it may not be advisable to assume that capital is recovered in accordance with a theoretical value-time model such as those considered previously. An alternative is to assume that depreciation occurs on the basis of service performed without regard to the duration of the asset's life. Thus, a trencher might be depreciated on the basis of pipeline trench completed. If the trencher has a first cost of $11,000 and a salvage value of $600, and if it is estimated that the trencher would dig 1,500,000 linear feet of pipeline trench in its life, the depreciation charge per foot of trench dug may be calculated as

$$\frac{\$11,000 - \$600}{1,500,000} = 0.006933 \text{ per foot.}$$

The undepreciated capital at the end of each year is a function of the number of feet of trench dug during the year. If, for instance, 300,000 feet of pipeline trench were dug during the first year, the undepreciated balance at the end of the year would be $11,000 - 300,000($0.006933) = $8,920.10. This analysis would be repeated at the end of each year.

12.9. Depletion

Depletion differs in theory from depreciation in that the latter is the result of use and the passage of time while the former is the result of the intentional, *piecemeal removal* of certain types of assets. Depletion refers to an activity that tends to exhaust a supply and the word literally means emptying. When natural resources are exploited in production, depletion indicates a lessening in value with the passage of time. Examples of depletion are the removal of coal from a mine, timber from a forest, stone from a quarry, and oil from a reservoir.

In the case of depletion, it is clear that a portion of the asset is disposed of with each sale. But when a machine tool is used to produce goods for sale, a portion of its productive capacity is a part of each unit produced and thus, is disposed of with each sale. A mineral resource has value only because the mineral may be sold and, similarly, the machine tool has value because what it can produce may be sold. Both depletion and depreciation represent decreases in value through the using up of the value of the asset under consideration.

There is a difference in the manner in which the capital recovered through depletion and depreciation must be handled. In the case of depreciation, the asset involved usually may be replaced with a like asset, but in the case of depletion such replacement is usually not possible. In manufacturing, the amounts charged for depreciation are reinvested in new equipment to continue operation. However, in mining, the amounts charged to depletion cannot be used to replace the ore deposit and the venture may sell itself out of business. The return in such a case must consist of two portions—the profit earned on the venture and the owners' capital which was invested. In the actual operation of ventures dealing with the piecemeal removal of resources, it is common to acquire new properties, thus enabling the venture to continue.

The *cost* method of depletion is similar to the service method of depreciation discussed in Section 12.8. That is, the depletion charge is based on the amount of the resource that is consumed and the initial cost of the resource. Suppose a reservoir containing an estimated 1,000,000 barrels of oil required an initial investment of $1,700,000 to develop. For this reservoir the unit depletion rate is $1,700,000/1,000,000 bbls. = $1.70/bbl. If 50,000 bbls. of oil are produced from this reservoir during a year the depletion charge is (50,000 bbls.) ($1.70/bbl.) = $85,000. Since the estimates of the number of units of the resource remaining vary from year to year the unit depletion rate is recalculated by dividing the unrecovered cost of the resource by the estimated units of the resource remaining.

For certain resources an optional method of calculating the depletion charge is provided by the United States income tax laws. This optional method is sometimes referred to as the *percentage* method of depletion. Percentage depletion allows a fixed percentage of the gross income produced by the sale of the resource to be the depletion charge. Thus, over the life of an asset the total depletion charges may exceed the initial cost of the asset. However, it is required that for any period the depletion charge may not exceed 50% of the net taxable income for that period computed without the depletion allowance. Some of the fixed percentages for depletion allowed by the tax laws for certain natural resources are oil and gas, uranium and sulfur, 22%; coal and salt, 10%; and gravel 5%.

Using the oil reservoir example again, suppose that the price of oil is $9/bbl., transportation charges are $0.80/bbl., and all other expenses associated with the operation of the oil reservoir are $70,000. To find the allowed depletion charge, the following calculations are required:

Sales 50,000 bbls. @ $9/bbl.	$450,000
Less transportation (50,000 bbls.)($0.80/bbl.)	40,000
Gross depletion income	410,000
Depletion rate	22%
Percentage depletion charge	$ 90,200

Now it is necessary to determine if the figure of $90,200 exceeds the maximum depletion charge allowed by the tax laws for this method of calculating depletion charges. The following calculations specify the maximum depletion charge that is permitted:

Gross depletion income	$410,000
Less expenses	70,000
	340,000
Deduction limitation	50%
Maximum depletion charge	$170,000

Since $90,200 is less than $170,000, the full depletion charge of $90,200 is allowable in this circumstance. By comparing this amount to the $85,000 permitted under the *cost* method it would be advantageous to apply the *percentage* method in this situation.

12.10. Capital Recovery With Return Is Equivalent for All Methods of Depreciation

It has been seen that the straight-line, declining-balance, sum-of-the-years-digits, and sinking-fund methods of depreciation all lead to different value-time functions for book value. It is therefore interesting to note that if the retirement of any asset takes place at the age predicted and the book value equals the estimated salvage value, the depreciation amount and interest on the undepreciated balance can be shown to be equivalent to the capital recovery with a return for *any* method of depreciation. Recall that capital recovery with a return is defined as $(P - F)(\overset{A/P\,i,\,n}{\quad}) + Fi$.

If $(P - F) = A + B + C + \ldots + N$, where $A, B, C, \ldots, N$ are capital recovered amounts for the successive years, we get the following table:

End of Year	Depreciation at End of Year	Interest on Undepreciated Balance at End of Year
0	0	0
1	A	$Ai + Bi + \ldots + Ni + Fi$
2	B	$Bi + \ldots + Ni + Fi$
n	N	$Ni + Fi$

Interest, Fi, on the salvage value, F, will be equal for all methods of depreciation and need not be given further consideration. The quantity $B + Bi$ as of the end of year 2 is equivalent to

$$(B + Bi) \times \frac{1}{(1 + i)} = B$$

as of the end of year 1. Addition of this amount, B, to Bi results in a total $(B + Bi)$ as of the end of year 1. This sum is in turn equivalent to B as of the end of year 0. By similar calculations, quantities involving symbols A to N inclusive will be found to have a worth of $A, B, \ldots, N$ as of the end of year 0, respectively. Since $(A + B + \ldots + N)$ equal $(P - F)$, and since $A, B, \ldots, N$ may be chosen to represent depreciation by any method, it may be concluded that the present-worth for i of the depreciation calculated by any method plus the interest on the undepreciated balance is equivalent to the present-worth of $(P - F)(\overset{A/P\,i,\,n}{\quad})$ at the beginning of the depreciation period.

In the illustrative example on straight-line depreciation, an asset having a first cost of $5,000 was depreciated to a salvage value of $1,000 in a period of 5 years. The sum of *depreciation* and *interest on unrecovered balance* is shown in the second column of the following table. For example for year 3, $1,004 = $800 + 0.06($3,400).

End of Year	Sum of Depreciation and Interest on Undepreciated Balance		Single Payment Present-Worth Factor		
1	$1,100	×	$P/F\,6,1$ (0.9434)	=	$1,037.74
2	1,052	×	$P/F\,6,2$ (0.8900)	=	936.28
3	1,004	×	$P/F\,6,3$ (0.8396)	=	842.96
4	956	×	$P/F\,6,4$ (0.7921)	=	757.25
5	908	×	$P/F\,6,5$ (0.7473)	=	678.55
			Total Present Worth	=	$4,252.78

The comparable figures calculated for sinking-fund depreciation are given in the second column of the following table.

Year No.	Sum of Depreciation and Interest on Undepreciated Balance		Single Payment Present-Worth Factor		
1	$1,009.60	×	$P/F\,6,1$ (0.9434)	=	$ 952.46
2	1,009.60	×	$P/F\,6,2$ (0.8900)	=	898.54
3	1,009.60	×	$P/F\,6,3$ (0.8396)	=	847.66
4	1,009.60	×	$P/F\,6,4$ (0.7921)	=	799.70
5	1,009.60	×	$P/F\,6,5$ (0.7473)	=	754.47
			Total Present Worth	=	$4,252.83

The slight difference between the two resulting values, $4,252.78 and $4,252.83, results from using tables of too few decimal places.

There are ordinarily only two real transactions in an asset's depreciation. These are its purchase and its sale as salvage. In the above example the asset was purchased for $5,000 and its salvage 5 years later was presumed to have a value of $1,000. The present worth of these two amounts as of the time of purchase follows.

Present worth of $5,000 disbursement at time of purchase 5,000.00
Present worth of $1,000 received from sale of salvage value (a receipt

is a negative disbursement), $1,000($\overset{P/F\,6,\,5}{0.7473}$) .... −747.30

Total present worth $4,252.70

Compare this with the two previous results. Also, note that capital recovery with return is

$$CR(6\%) = (\$5,000.00 - \$1,000.00)(\overset{A/P\,6,\,5}{0.2374}) + 1,000.00(0.06) = \$1,009.60$$

and that

$$\$4,252.70(\overset{A/P\,6,\,5}{0.2374}) = \$1,009.59.$$

Although this relationship between depreciation plus the interest on the undepreciated balances and capital recovery with a return does exist it is important for decision-making purposes that only the actual cash flows be considered. Depreciation charges by themselves are only accounting entries and therefore they do not represent an actual disbursement of cash.

12.11. Depreciation and Engineering Economy Studies

As was indicated earlier, depreciation is a cost of production. An asset is actually consumed in producing goods and thus its depreciation is a production cost. If the cost of capital consumption is neglected, profits will appear to be higher than they are by an amount equal to the depreciation that has taken place during the production period.

In economic analysis dealing with physical assets, it is necessary to compute the equivalent annual cost of capital recovered plus return, so that alternatives involving competing assets may be compared on an equivalent basis. Regardless of the depreciation model chosen to represent the value of the asset over time, the equivalent annual cost of capital recovered and return will be

$$CR(i) = (P - F)(\overset{A/P\,i,\,n}{\hspace{1em}}) + Fi.$$

Therefore, in making annual cost comparisons of alternative engineering proposals involving physical assets, this expression may be used to represent the equivalent annual costs of capital recovery for interest rate i.

In engineering economic analysis the primary importance of depreciation is its effect on estimated cash flows resulting from the payment of income taxes. Depreciation as an amortized cost influences profits as shown on a company's profit and loss statement since depreciation appears as an expense to be deducted from gross income. Income taxes are paid on the resulting net income figure and these taxes do represent actual cash flows although the depreciation charges are bookkeeping entries.

Depreciation is based upon estimates. It would be desirable to know the amount and pattern of an asset's depreciation at any point during its life in order that exact charges could be made against products as they are produced. Unfortunately, the depreciation of an asset cannot be known with certainty until after the asset has been retired from service.

Usually, it is impractical to defer calculation of depreciation costs until after an asset has been disposed of at the end of its life. In fact, depreciation costs of an asset should be taken into account prior to the purchase of the asset, as one of the factors to be considered in arriving at the desirability of purchasing it.

Since information on depreciation is needed on a current basis for making decisions, it has become a practice to estimate the amount of depreciation an asset will suffer and the pattern in which this depreciation will occur. This involves estimates of the service life, salvage value, and depreciation method. Usually, the first cost of an asset is known with considerable accuracy. However, since the service life, salvage value, and pattern of depreciation refer to events in the future, they cannot be known with certainty. These estimates are usually based upon experience with similar assets and the judgment of the estimator.

Usually, the most important factor in figuring the annual depreciation charge for an asset is the estimate of its useful life. The useful life can be determined from the firm's experience with particular assets, or if the estimator's experience is inadequate, the general experience of the industry can provide a basis for this estimate. As long as the lives utilized in depreciation calculations are reasonable and there is no clear and convincing basis for change, adjustments in these estimated lives will not be required.

An alternative to determining useful life on the basis of operating conditions and experience is the more liberal Class Life Asset Depreciation Range (ADR) system. The ADR system allows taxpayers to take as a reasonable depreciation charge an amount based on any life selected within a range specified for designated classes of assets. Some examples of these asset classes and the ranges of lives permitted are presented in Table 12.10.

By using the Class Life ADR system disputes should be minimized between the taxpayer and the Internal Revenue Service concerning the useful life of property. Otherwise, the taxpayer must be able to provide justification for the lives he utilizes in his depreciation calculations.

Table 12.10. CLASSES OF ASSETS AND RANGES OF LIVES PERMITTED FOR THE CLASS LIFE ASSET DEPRECIATION RANGE (ADR) SYSTEM

Asset Description	Lower Limit	Asset Guideline Period	Upper Limit
Depreciable Assets Used in all Business Activities			
Aircraft	5	6	7
Automobiles, taxis	2.5	3	3.5
Railroad cars and locomotives	12	15	18
Vessels, barges, tugs	14.5	18	21.5
Depreciable Assets Used in the Following Activities			
Agriculture equipment	8	10	12
Computers	5	6	7
Construction	4	5	6
Electric production plant			
Hydraulic	40	50	60
Nuclear	16	20	24
Steam	22.5	28	33.5
Manufacturing			
Electrical equipment	9.5	12	14.5
Electronic products	6.5	8	9.5
Ferrous metals	14.5	18	21.5
Furniture	8	10	12
Motor vehicles	9.5	12	14.5
Paper pulps	13	16	19
Petroleum			
Drilling	5	6	7
Exploration	11	14	17
Refining and marketing	13	16	19
Rubber products	11	14	17
Textile mill products	11	14	17
Recreation and amusement	8	10	12

QUESTIONS AND PROBLEMS

1. What is the difference between the accountant's concept of depreciation and depreciation as it is commonly understood by the general public?

2. Describe the value-time function and name its essential components?

3. Does a depreciation charge represent an actual disbursement of funds? Explain.

4. Discuss the difference between physical and functional depreciation?

5. How does depreciation affect the profit and loss statement of a company?

6. Explain the difference between capital recovery with return as used by the engineering economist and the accountant's concept of depreciation. Under what

conditions will the annual depreciation charge equal capital recovery with return?

7. A man, who purchased a car for $2,500, was offered amounts for his car in succeeding years as follows:

Year	1	2	3	4	5	6
Offer	$1,900	$1,350	$925	$650	$425	$300

On the basis of the offers, calculate the depreciation and the undepreciated value of the car for each year of its life. Is this the accountant's view of depreciation?

8. A truck which cost $23,000 had an estimated useful life of 8 years and estimated salvage value of $4,800. How much depreciation should be allocated on the new truck in each year and what is the undepreciated value of the truck at the end of the fifth year assuming the use of straight-line depreciation?

9. An asset has a first cost of $60,000 with a salvage of $5,000 after 11 years. If it is depreciated by declining-balance method using a rate of 10% what will be the
   (a) depreciation charge in the second year?
   (b) depreciation charge in the eighth year?
   (c) book value in the eighth year?

10. An asset was purchased 10 years ago for $4,800. It is being depreciated in accordance with the straight-line method for an estimated total life of 20 years and salvage value of $800. What is the difference in its current book value and the book value that would have resulted if declining-balance depreciation at a rate of 10% had been used?

11. A machine acquired at a cost of $32,000 had an estimated life of 5 years. Residual salvage value was estimated to be $3,500.
    (a) Determine the annual depreciation charges during the useful life of the machine under the sum-of-the-years-digits method of depreciation.
    (b) What depreciation rate makes the book value at the end of 5 years equal to the estimated salvage value if the declining-balance method of depreciation is used?

12. Calculate the book value at the end of each year using straight-line, sum-of-the-years-digits, double-declining balance and sinking-fund methods of depreciation for an asset with an initial cost of $50,000 and an estimated salvage value of $10,000 after 4 years. The interest rate is 15% and the results should be presented in tabular form.

13. A central air-conditioning unit is purchased for $25,000 and it has an expected life of 10 years. The salvage value for the unit at that time is expected to be $4,000. What will be the book value at the end of 7 years for (a) straight-line depreciation, (b) sum-of-the-years-digits depreciation, (c) double-declining balance depreciation, and (d) sinking-fund depreciation for 8%?

14. An electronic calculator has a first cost of $1,000 with an estimated salvage value of $250 at the end of 5 years. The interest rate is 7%.
    (a) Graph the annual depreciation by the straight-line, sinking-fund, sum-of-the-years-digits, and double-declining methods for each year.
    (b) Graph the book values obtained by the four methods for the end of each year.

15. An apartment complex is purchased for 1.8 million dollars. It has an estimated life of 25 years and the salvage value at that time is expected to be $500,000. After 5 years what will be the total depreciation charged for (a) straight-line depreciation, (b) sum-of-the-years-digits depreciation, and (c) double-declining balance depreciation?

16. An asset has a first cost of $45,000 with an estimated life of 10 years. The salvage value at that time is estimated to be $5,000. What is the total accumulated depreciation charged during the first 4 years of the asset's life for (a) straight-line depreciation, (b) double-declining-balance depreciation, (c) sum-of-the-years-digits depreciation, and (d) sinking-fund depreciation for an interest rate of 12%?

17. A special purpose machine is purchased for $200,000 with an expected life of 12 years. If its salvage value is expected to be zero, what will be the depreciation charge in the first and fifth years and what will be the book value at the end of the third year for (a) straight-line depreciation, (b) double-declining-balance depreciation, and (c) sum-of-the-years digits depreciation?

18. A company has purchased a numerically controlled machine for $150,000. It is estimated that it will have a salvage value of $50,000 4 years from now. What rate must be used with the declining-balance method of depreciation so that the book value of the machine will be equal to its salvage value at the end of its life? (a) Using the rate just calculated with declining-balance depreciation find the depreciation and book value for each year of the machine's life. (b) Compare those figures with similar figures for straight-line and sum-of-the-years-digits depreciation.

19. Filters costing $1,000 per set are used to separate particles from hot gases being discharged into the air. These filters must be replaced every 2,500 hours. If the plant operates 24 hours per day, 30 days per month, what is the depreciation charge per month for these filters?

20. An automatic control mechanism is estimated to provide 3,000 hours of service during its life. The mechanism costs $5,200 and will have a salvage value of zero after 3,000 hours of use. What is the depreciation charge if the number of hours the mechanism is used per year is (a) 600 (b) 1,300?

21. A gold mine that is expected to produce 850,000 grams of gold is purchased for $280,000. Gold can be sold for $0.52 per gram to the federal government. If 99,000 grams are produced this year what will be the depletion allowance for (a) cost depletion and (b) percentage depletion where the fixed percentage for gold is 22%?

22. An oil reservoir is estimated to have 200,000 barrels of oil. The cost of pur-

chasing the lease and well is $880,000. If crude oil is selling for $7.50 per barrel and 30,000 barrels of oil per year are produced, what is the amount of the depletion allowance for (a) cost depletion and (b) percentage depletion at a fixed percentage of 22%?

23. A sulfur mining operation has a gross income of $500,000 in February from the production of sulfur. All expenses with the exception of depletion expenses for this month amount to $390,000. The percentage depletion rate for sulfur is 22%. What is the depletion allowance for this month?

24. A drilling rig was purchased for $4,000. One year later an offer of $4,200 was received for it. Two years later, the offer was $3,000 and 3 years later, a $2,000 offer was received.
    (a) Determine the depreciation of the rig for each year based on the offers received.
    (b) Determine the interest on the undepreciated balance for each of the 3 years using an interest rate of 8%.
    (c) Determine the uniform year-end amount for the 3-year period which is equivalent to the sum of the depreciation and interest on the undepreciated balance found above.
    (d) Determine the annual cost of capital recovery with a return for the 3-year period using the initial cost and the last offer. Compare with (c).

25. An asset has a first cost of $6,000 and an estimated salvage of $1,000 at the end of 4 years. The interest rate is 10%. Repeat (a), (b), and (c) of Problem 24 using 10% instead of 8% for straight-line and sum-of-the-years-digits methods of depreciation. Now compare the results of (c) for both methods. Are they equal?

INCOME TAXES IN
ECONOMIC ANALYSIS

Income tax laws are the result of legislation over a period of time. Since they are man made, they incorporate many diverse ideas, some of which appear to be in conflict. These laws are expressed through a number of provisions and rules intended to meet current conditions. The provisions and rules are not absolute in application, but rather are subject to interpretation when applied.

Income taxes are levied by the federal government as well as by many individual states. State income tax laws will not be considered here because the principles involved are similar to those for federal tax laws, because state income rates are relatively small, and because there is a great diversity of state income tax law provisions.

13.1. Relation of Income Taxes to Profit

In most cases, the desirability of a venture is measured in terms of differences between income and cost, receipts and disbursements, or some other measure of profit. It is the specific function of economic analysis to determine

13

future profit potential that may be expected from prospective engineering proposals being examined. But income taxes are levies on profit that result in a reduction of its magnitude.

Regardless of how public-spirited a person may be or how clearly he may understand the government's need for income taxes, and even if he should place a high value upon the services he may receive in return for his income tax payments, income taxes are disbursements that differ from other disbursements associated with undertakings only in the manner in which their magnitudes are determined. Thus, in relation to engineering economy studies, income taxes are merely another class of expenditure, which require special treatment. Such taxes must be taken into account along with other classes of costs in arriving at the fruits accruing to sponsors of an undertaking.

Of the aggregate profit earned in the United States from such productive activities as manufacturing, construction, mining, lumbering and others in which engineering analysis is important, the aggregate disbursement for income taxes will be approximately 30 to 40% of net income. Their importance becomes clear when it is realized that in some cases income taxes may result in disbursements of as much as 70% of the profit to be expected from a venture.

13.2. Individual Federal Income Tax

With few exceptions, every individual under 65 having a gross income of $2,350 or more during the year must file a tax return. For married persons under 65 years of age a gross income of $3,400 or more is required before a tax return must be submitted. Failure to pay the tax when it is due results in an interest charge at the rate of 9% per year from the due date until the tax is paid. In addition, if the failure to pay is due to willful neglect rather than some reasonable cause the penalty is $\frac{1}{2}$% of the unpaid tax for each month of delinquency.

An individual's income tax obligation is determined by applying a graduated tax rate to his net income from salaries, fees, commissions, and business activities less certain exemptions and deductions. For example, consider Mr. Doe who is unmarried, qualifies as the head of a household, and has four dependents. An outline of the steps necessary to compute Mr. Doe's tax obligation is given below.

*Individual Federal Income Tax Outline**

1. Salary, wages, commission or other compensation received $12,000
2. Net income from business activities:
   This item consists of income received from the conduct of trade or business activities less the expense of carrying on such activities. These expenses include such items as rents, wages, fees, raw material, supplies, services, depreciation of facilities and losses incurred from the sale and exchange of other than capital assets. Business income less business expense   8,000
3. Income from capital gains:
   This item consists of 50% of the amount that net long-term (assets held more than 6 months) capital gains exceed net short-term (assets not held more than six months) capital losses. ($900 long-term gain less $300 short-term loss) × 0.5   300
4. Total adjusted gross income $20,300
5. Less deductions as follows:
   a. Contributions not generally allowed in excess of 50% of adjusted gross income (for contributions directly to religious, educational organizations and organizations providing medical research) $1,000
   b. Interest paid   200
   c. Taxes paid   1,600
   d. Medical expenses in excess of 3% of adjusted gross income   0
   e. Loss from fire, storms, or other casualty not compensated by insurance, etc.   0
   f. Exemption. The taxpayer is permitted an exemption of $750 ($1,500 if 65 years old or older) (Blindness entitles a taxpayer an additional exemption of $750)

for himself and $750 for each dependent. ($750 for a dependent wife 65 years old or older and $750 additional if she is blind.) A dependent is a person, 50% or more of whose support is paid for by the taxpayer, and whose annual earnings are less than $750 per year. The $750 limitation on income does not apply to any child who is under 19 years of age or who is a student (Doe, and his four dependents, five at $750) $3,750

       Total deductions and exemptions 6,550   6,550

6. Taxable income $13,750

*Applicable to tax year 1976.

The amount of income tax due on the taxable income may be calculated from tabulated values such as those given in Tables 13.1 and 13.2. Table 13.2 is applicable to Mr. Doe's situation; therefore, his tax obligation will be $2,440 + 0.27 \times ($13,750 − $12,000), or $2,912.

For a second illustration, suppose that Mr. Blue has a wife and three children, and he had items of income identical with those of Mr. Doe. The number of exemptions in both cases is five. Mr. and Mrs. Blue elect to file

Table 13.1. TAX RATES FOR MARRIED TAXPAYERS FILING JOINT RETURNS AND CERTAIN WIDOWS AND WIDOWERS (1976)

If the taxable income is:	The income is:
Not over $1,000	14% of taxable income
$ 1,000 to $ 2,000	$140 + 15% of excess over $ 1,000
2,000 to 3,000	290 + 16% of excess over 2,000
3,000 to 4,000	450 + 17% of excess over 3,000
4,000 to 8,000	620 + 19% of excess over 4,000
8,000 to 12,000	1,380 + 22% of excess over 8,000
12,000 to 16,000	2,260 + 25% of excess over 12,000
16,000 to 20,000	3,260 + 28% of excess over 16,000
20,000 to 24,000	4,380 + 32% of excess over 20,000
24,000 to 28,000	5,660 + 36% of excess over 24,000
28,000 to 32,000	7,100 + 39% of excess over 28,000
32,000 to 36,000	8,660 + 42% of excess over 32,000
36,000 to 40,000	10,340 + 45% of excess over 36,000
40,000 to 44,000	12,140 + 48% of excess over 40,000
44,000 to 52,000	14,060 + 50% of excess over 44,000
52,000 to 64,000	18,060 + 53% of excess over 52,000
64,000 to 76,000	24,420 + 55% of excess over 64,000
76,000 to 88,000	31,020 + 58% of excess over 76,000
88,000 to 100,000	37,980 + 60% of excess over 88,000
100,000 to 120,000	45,180 + 62% of excess over 100,000
120,000 to 140,000	57,580 + 64% of excess over 120,000
140,000 to 160,000	70,380 + 66% of excess over 140,000
160,000 to 180,000	83,580 + 68% of excess over 160,000
180,000 to 200,000	97,180 + 69% of excess over 130,000
over $200,000	110,930 + 70% of excess over 200,000

Table 13.2. TAX RATES FOR TAXPAYERS WHO QUALIFY AS HEAD OF HOUSEHOLD (1976)

If the taxable income is:	The income tax is:
Not over $1,000	14% of the taxable income.
$ 1,000 to $ 2,000	$ 140 + 16% of excess over $ 1,000
2,000 to 4,000	300 + 18% of excess over 2,000
4,000 to 6,000	660 + 19% of excess over 4,000
6,000 to 8,000	1,040 + 22% of excess over 6,000
8,000 to 10,000	1,480 + 23% of excess over 8,000
10,000 to 12,000	1,940 + 25% of excess over 10,000
12,000 to 14,000	2,440 + 27% of excess over 12,000
14,000 to 16,000	2,980 + 28% of excess over 14,000
16,000 to 18,000	3,540 + 31% of excess over 16,000
18,000 to 20,000	4,160 + 32% of excess over 18,000
20,000 to 22,000	4,800 + 35% of excess over 20,000
22,000 to 24,000	5,500 + 36% of excess over 22,000
24,000 to 26,000	6,220 + 38% of excess over 24,000
26,000 to 28,000	6,980 + 41% of excess over 26,000
28,000 to 32,000	7,800 + 42% of excess over 28,000
32,000 to 36,000	9,480 + 45% of excess over 32,000
36,000 to 38,000	11,280 + 48% of excess over 36,000
38,000 to 40,000	12,240 + 51% of excess over 38,000
40,000 to 44,000	13,260 + 52% of excess over 40,000
44,000 to 50,000	15,340 + 55% of excess over 44,000
50,000 to 52,000	18,640 + 56% of excess over 50,000
52,000 to 64,000	19,760 + 58% of excess over 52,000
64,000 to 70,000	26,720 + 59% of excess over 64,000
70,000 to 76,000	30,260 + 61% of excess over 70,000
76,000 to 80,000	33,920 + 62% of excess over 76,000
80,000 to 88,000	36,400 + 63% of excess over 80,000
88,000 to 100,000	41,440 + 64% of excess over 88,000
100,000 to 120,000	49,120 + 66% of excess over 100,000
120,000 to 140,000	62,320 + 67% of excess over 120,000
140,000 to 160,000	75,720 + 68% of excess over 140,000
160,000 to 180,000	89,320 + 69% of excess over 160,000
Over $180,000	103,120 + 70% of excess over 180,000

a joint return. The taxable income will remain at $13,750, and Mr. and Mrs. Blue may file a joint return. If this is done, rates in Table 13.1 apply and the tax to be paid is $2,260 + 0.25 × ($13,750 − $12,000) or $2,697.

If Mr. and Mrs. Blue had divided the $13,750 between them and filed separate returns in accordance with the law, their total tax would have been $2,697, or the same as resulted from a joint return. However, if they had divided the income between them unequally, the tax to be paid would have been greater.

Item 2 in the outline above, *Net income from business activities*, consists essentially of business income less business expenses. Some items such as depreciation and allowed amortization must be determined by calculation.

Such items are treated in subsequent sections as is the matter of capital gains appearing in Item 3 of the outline.

Adjusted gross income. The term adjusted gross income means net earnings as determined in accordance with the provisions of federal income tax law. The starting point is the taxpayer's total income. Certain items of income are excluded in whole or in part from the total income to determine gross income. Examples of excluded income are interest on certain federal, state or municipal obligations, health and accident insurance benefits, and annuities.

From the gross income certain deductions are made to arrive at the adjusted gross income. These deductions consist essentially of expenses incurred in carrying on a trade, profession, or business, reimbursement for expenses incurred in employment and certain longterm capital gains explained elsewhere.

Taxable income. The individual's taxable income is his adjusted income less deductions he is permitted to take. Deductions may be made for contributions to qualified individuals, fraternal organizations, governmental units of the United States, religious and educational organizations, and organizations providing medical care and hospitalization.

Interest paid by a taxpayer in carrying on business activities is taken into account in arriving at adjusted gross income. But interest paid on borrowed funds made for personal purposes is deductible under Item 5b in the above outline.

Practically all taxes, except federal income taxes, incurred and necessary to carry on business activity are taken into account in arriving at the adjusted gross income of an individual. In addition, individuals are permitted to make deductions for payment of most state and local taxes except estate, inheritance, and gift taxes, and for a few federal taxes with the exception of federal income, estate, and gift taxes.

Medical expenses in excess of 3% of adjusted gross income, incurred by a taxpayer for himself or his dependents, are deductible under Item 5d, if these expenses are not compensated for by insurance or other means. Cost of medicines and drugs in excess of 1% of adjusted gross income is included as a part of medical expense.

Losses sustained from fire, storm, theft, or other casualties are deductible to the extent they are not compensated by insurance or similar means as shown under Item 5e.

The exemptions listed in Item 5f of the outline cover most of the provisions relative to this classification.

In lieu of itemized deductions 5a to 5c, the taxpayer may elect to take a standard deduction. The standard deduction on the joint return of a husband and wife is $2,600 or 16% of the adjusted income, whichever is the least. The maximum standard deduction on a separate return of a single person is $2,300.

13.3. Corporation Federal Income Tax

The term *corporation*, as used in income tax law, is not limited to the artificial entity usually known as a corporation but may include joint stock associations or companies, some types of trusts, and some limited partnerships. In general, all business entities whose activities or purposes are the same as those of corporations organized for profit are taxed as such. This discussion will be based upon the tax requirements that apply to the usual business corporation.

The business corporation is subject to two taxes—the normal tax and the surtax. The normal tax is equal to 20% on the first $25,000 of taxable income plus 22% on taxable income over $25,000. The surtax is equal to 26% of taxable income in excess of $50,000. In addition, a corporation may be liable for an accumulated earnings tax.

As a general rule, the taxable income of a corporation is computed in the same manner as that of an individual. There are, however, several provisions peculiar to corporations. Although corporations are in general entitled to the same deductions as individuals, deductions of a personal nature such as medical expenses, child care, alimony, or exemptions for the taxpayer and his dependents are excluded. Corporations are entitled to deductions for partially exempt interest received as income on certain government obligations, for dividends received, and for certain organizational expenses to which individuals are not entitled.

In general, a corporation's income tax is a levy on its net earnings, the difference between the income derived from and the expense incurred in business activity, with some exceptions. Because of the continuing nature of business activity and the fact of annual tax periods, net income must usually be a calculated amount.

The following is an outline of steps to be taken in computing the normal income tax and the surtax of a corportation, illustrated by entering amounts applicable to a particular corporation:

1. Gross Income: This item embraces gross sales less cost of sales, dividends received on stocks, interest received on loans and bonds, rents, royalties, gains and losses from capital or other property .. $700,000
2. Deductions: This item embraces expense not deducted elsewhere as, for example, in cost of sales. Includes compensation of officers,

wages, and salaries, rent, repairs, bad debts, interest, taxes, contributions (not in excess of 5% of taxable income), losses by fire, storms, and theft, depreciation, depletion, advertising, contributions to employee benefit plans, and special deductions for partially exempt bond interest and partially exempt dividends 600,000

3. Taxable income $100,000
4. First $25,000 of taxable income, $25,000 × 0.20 .... $ 5,000
5. Taxable income in excess of $25,000 .... $ 75,000
6. Normal tax on balance, $75,000 × 0.22 .... $ 16,500
7. Balance subject to surtax, $100,000 − $50,000 .... $ 50,000
8. Balance multiplied by surtax rate $50,000 × 0.26 .... $ 13,000
9. Total normal tax plus surtax (Items 4, 6, and 8) .... $ 34,500

Corporate taxable income with some exceptions may be expressed as an equation applicable for taxable incomes from $50,000 and up. This equation is

$$T = \$5,000 + (E_b - \$25,000) \times 22\% + (E_b - \$50,000) \times 26\%$$

where

T = total income tax;
E_b = earnings before taxes (taxable income).

Effective income tax rates. Engineering economy studies involving income taxes can be simplified by the determination of applicable *effective income tax rates*. An effective income tax rate, as the term is used here, is a single rate which when multiplied by the taxable income of a venture under consideration will result in the income tax attributable to the venture. Effective income tax rates are essentially average rates that are applicable over increments of income.

The tax rate on any increment of taxable corporate income between $0 and $25,000 is of course 20%, and the tax rate for any increment over $25,000 is 22%. With the surtax of 26% on taxable income above $50,000, the corporate taxpayer is paying a rate of 48% on all earnings above that figure. The previous statement does not hold to the extent that there are capital gains, exempt interest, and dividends to which special rates apply.

The average tax rate, t_a, of any increment of taxable income may be calculated as follows: Find the difference of the tax payable on incomes corresponding to the upper and lower limits of the increments and divide this difference by the amount of the increment. For example, suppose a corporation wishes to find the average tax rate on the increment of income between $45,000 and $60,000. The tax payable on $45,000 is equal to $5,000 + ($45,000 − $25,000) × 0.22 = $9,400, and the tax payable on $60,000 is equal to $5,000 + ($60,000 − $25,000) × 0.22 + ($60,000 − $50,000) × 0.26 = $15,300. The difference is $15,300 − $9,400 = $5,900. The average tax rate over the increment is

$$t_a = \frac{\$5,900}{\$60,000 - \$45,000} = \frac{\$5,900}{\$15,000} = 0.393, \text{ or } 39.3\%.$$

When there are capital gains, losses to be carried back or over, actual depreciation and depletion rates different from those allowed by tax schedules, and other conditions, many factors may be involved in the determination of effective income tax rates. These determinations require careful study by individuals proficient in tax matters.

Once they are determined for a particular purpose, effective rates serve as a means for transmitting the thought and analyses that went into the determination in a single figure. By the use of effective rates the consideration of income taxes in economy studies is reduced to two distinct factors, namely, (1) the determination of effective tax rates applicable to particular activities or particular classes of activities and (2) the application of the effective tax rates in economy studies.

The investment credit. To encourage the use of modern plants and equipment, the federal government provides tax incentives that favor the acquisition of capital assets. The investment credit is an allowed amount that directly reduces a taxpayer's liability, and it is determined by the dollar amount of investment for "qualified property."[1] No credit is allowed for property with a useful life of less than 3 years while the credit applies for depreciable real property (except buildings) used in manufacturing, production, extraction, or providing utility services.

The investment credit is calculated by multiplying the investment cost of qualified assets by a rate that is determined by the useful life of the asset and whether the corporation is considered to be industrial or a utility. If the life is estimated to be 7 years or more, the rates used for industrials and utilities are 7% and 4%, respectively. For an asset with an expected life of between 5 and 7 years, the rates are $\frac{2}{3}$ of 7% for industrials and $\frac{2}{3}$ of 4% for utilities. Where the estimated life is between 3 and 5 years, the industrial rate is $\frac{1}{3}$ of 7% while the utility rate is $\frac{1}{3}$ of 4%. There is a limit to the total investment tax credit that may be taken in any year. Thus, the excess of investment credit over $25,000 cannot exceed 50% of the precredit tax liability less $25,000.

Suppose a utility has invested $4,500,000 this year in new plants and

---

[1] *Prentice-Hall Federal Tax Handbook*, (Englewood Cliffs, New Jersey: Prentice-Hall, Inc.).

equipment that qualify for the investment credit. The company's tax liability is $400,000 before considering the investment credit. If the useful lives of the qualified assets purchased are more than 7 years, the investment credit rate is 4%. In this situation the maximum investment credit that would be allowed is

$$\$25,000 + 50\%(\$400,000 - \$25,000) = \$212,500$$

For the $4,500,000 of new acquisitions, the investment credit is expressed as 4%($4,500,000) = $180,000. Since this amount is less than $212,500, the full $180,000 credit is then deducted directly from the precredit tax liability of $400,000 yielding a tax obligation of $220,000. It is important to note that the credit is allowed only for the year the property is placed in service and it does not affect the depreciation base. If the investment credit exceeds the maximum amount allowed for the year, the excess credit may be carried forward or backward for a prescribed number of years.

The effect of interest on income taxes. Interest paid for funds borrowed by an individual or a corporation to carry on a profession, trade, or business is deductible from income as expense of carrying on such activity. In addition, interest paid on funds borrowed for purposes not connected with a profession, trade, or business is deductible from the adjusted income in computing an individual's income tax.

Because interest paid is deductible as an expense the amount of borrowed funds may have a marked effect upon the amount of income tax that must be paid. For example, consider two corporations designated as Firm *A* and Firm *B* that are essentially identical except that in Firm *A* no borrowed funds are used and in Firm *B* an average of $200,000 is borrowed at the rate of 9% during a year. Assume that taxable income before interest payments in both cases is $80,000. Shown below are the taxable income and the income taxes that each firm would be required to pay.

	Firm A	*Firm B*
Income	$80,000	$80,000
Interest expense	0	18,000
Taxable income	$80,000	$62,000
Taxes	$24,900	$16,260

The difference in taxes to be paid represents a savings in taxes of $24,900 − $16,260 = $8,640 to be realized if the $200,000 is borrowed. With the interest costs on the loan being $18,000 and a tax saving of $8,640, Firm *B* is effectively paying only $18,000 − $8,640 = $9,360 to borrow the money. This is equivalent to an interest rate of 4.68%.

Capital gains and losses. Capital gains and losses are recognized as being short-term if they apply to assets held not more than 6 months. Long-term gains and losses apply to assets held more than 6 months.

If the aggregate of short-term and long-term activities results in a net loss for a year, a corporation may not deduct such loss from current income. But such loss may be carried forward for 3 subsequent years, being considered as a short-term capital loss, and offset against capital gains during that period.

If the aggregate of short-term and long-term activities results in a net gain, the tax may be calculated by either of two methods. The method adopted is the one which produces the smaller tax. These two methods are referred to as the *regular* and *alternative* methods. For the regular method, the net capital gain is included with the other taxable income, and the total tax obligation is calculated. Thus, the tax rate for this method is subject to the normal rate and the surtax rate.

The alternative method applies a 30% rate to the net gain, and this amount is then added to the tax figured on the taxable income less the net gain. This corporate tax rate can be contrasted with the maximum 25% rate on long-term capital gains of $50,000 or less that would be paid by individuals using the so-called *alternative tax* method.

Research and experimental expenditures. Research and experimental activities result in new knowledge which has little value in itself, but is valuable only in application. If knowledge is used it may be presumed to result in increased taxable income of the activity in which it is used, and, therefore, in increased income taxes from this source. This viewpoint seems to be borne out by a provision which became effective for tax years beginning after December 31, 1953, and permits expenditures for research and experimentation to be deducted in the year in which they were incurred. The taxpayer also has the option of treating such expenditures as deferred expenses chargeable to capital account and deducting it evenly over a period of 60 or more months. In the latter case, the activity is treated essentially the same as an asset that depreciates.

The provision permitting expenditures for research and experimental activities to be deducted in the year in which they are incurred is indeed an encouragement to carry on such activities when the income tax rate is high. For example, suppose that the effective tax rate of a taxpayer applicable to the consideration of research expenditures is 0.45 and that such expenditures amount to $100,000 during a year. If the research results in no increase in income, its cost to the taxpayer will be $100,000(1 − 0.45), or $55,000. Thus, in a sense, research expenditures are in part underwritten by the government until they result in increased income, if ever.

13.4. Depreciation and Income Taxes

Since the amount of taxes to be paid during any one year is dependent upon deductions made for depreciation, the latter is a matter of consideration for the Internal Revenue Service of the United States Treasury Department and state taxing agencies. Directives are issued by governmental agencies as guides to the taxpayers in properly handling depreciation for tax purposes.

In general, an asset must be used for the purpose of producing an income, whether or not an income actually results from its use, in order that a deduction may be made for its depreciation. In cases where an asset such as an automobile is used both as a means for earning income that is taxable and for personal use, a proportional deduction is allowable for depreciation. Intangible property such as patents, designs, drawings, models, copyright, licenses, and franchises may be depreciated.

There are restrictions which limit the percentage of the first cost of an asset that may be depreciated during the initial years of life. Depreciation models that yield depreciation amounts in early years which are in excess of that permitted by the Internal Revenue Service may not be used for tax purposes. For example, when the declining-balance model is used the rate of depreciation must not exceed twice the allowable straight-line rate. If a 5-year life is considered reasonable for an asset whose installed cost is $1,000 and there is no salvage value, its straight-line depreciation rate would be 20% and its annual depreciation would be $200 per year. The corresponding maximum allowable declining balance rate would be 40%. This would result in a deduction for depreciation of $1,000 \times 0.40 = $400 the first year, ($1,000 - $400) \times 0.40 = $240 the second year, etc.

Effect of method of depreciation on income taxes. A taxpayer has considerable choice in the method of depreciation he may use for tax computation. If the effective tax rate remains constant and the operating expense is constant over the life of the equipment, the depreciation method used will not alter the total of the taxes payable over the life of the equipment. But methods providing for high depreciation and consequent low taxes in the first years of life will be of advantage to the taxpayer because of the time value of money.

For illustration of the comparative effect of the straight-line and the fixed percentage on a diminishing balance method, consider the following example. Assume that a taxpayer has just installed a machine whose first cost is $1,000, whose estimated life is 10 years, and whose salvage value is nil. The machine is estimated to have a constant operating income before depreciation and income taxes of $200 per year. The taxpayer estimates the applicable effective income tax rate for the life of the machine to be 40%. He considers money to be worth 6% and wishes to compare the effect of the straight-line method and the fixed percentage on a declining balance method. Let the first method be represented by Alternative A and the second method be represented by Alternative B. These alternatives are summarized in Tables 13.3 and 13.4.

The total income tax paid during the life of the equipment is equal to $400 with either alternative, but the present worths of the taxes paid differ. For Alternative A, the present worth of taxes paid as of the beginning of year 1 is equal to $294.40. The corresponding figure for Alternative B is $275.60. The difference in favor of the declining balance method is $18.80, or 6.4% of the Alternative A tax.

By examining the book value of an asset over its life for the various methods of depreciation, one can easily see which methods produce larger depreciation deductions in the early portion of the asset's life. The time-value functions of the four depreciation methods discussed in Chapter 12 are presented in Figure 13.1. Since the book value of an asset is the first cost of the asset less accumulated depreciation charges, the curves with the lowest values in their early life are those which are most favorable to the taxpayer.

From Figure 13.1 it is seen that double-declining balance and sum-of-the-years-digits depreciation have faster depreciation write-offs than straight-

Table 13.3. ALTERNATIVE A

Income Taxes for Straight-Line Method of Depreciation and 10 Year Life							
Year end no. A	First cost B	Income before depr. and income tax C	Annual book depr. D	Total book depr. to date ΣD E	Income less depr. (taxable income) $C - D$ F	Income tax rate G	Income tax $F \times G$ H
0	$1,000						
1		$ 200	$100	$ 100	$100	0.4	$ 40
2		200	100	200	100	0.4	40
3		200	100	300	100	0.4	40
4		200	100	400	100	0.4	40
5		200	100	500	100	0.4	40
6		200	100	600	100	0.4	40
7		200	100	700	100	0.4	40
8		200	100	800	100	0.4	40
9		200	100	900	100	0.4	40
10		200	100	1,000	100	0.4	40
		$2,000					$400
Present worth of income taxes = $40($\overset{P/A\ 6,\ 10}{7.3601}$) = $294.40							

Table 13.4. ALTERNATIVE B

		Income before depr. and income tax	Annual book depr.	Total book depr. to date	Income less depr. (taxable income)	Income tax rate	Income tax
Year end no.	First cost						
A	B	C	D	E ΣD	F $C-D$	G	H $F \times G$
0	$1,000						
1		$ 200	$200	$ 200	$ 0	0.4	$ 0
2		200	160	360	40	0.4	16
3		200	128	488	72	0.4	29
4		200	102	590	98	0.4	39
5		200	82	672	118	0.4	47
6		200	66	738	134	0.4	54
7		200	52	790	148	0.4	59
8		200	42	832	158	0.4	63
9		200	34	866	166	0.4	66
10		200	134	1,000	66	0.4	27
		$2,000					$400

Present worth of income taxes $= \sum_{n=1}^{10} [\text{Col. } H \times (\overset{P/F\,6,\,n}{\ })] = \$275.60$

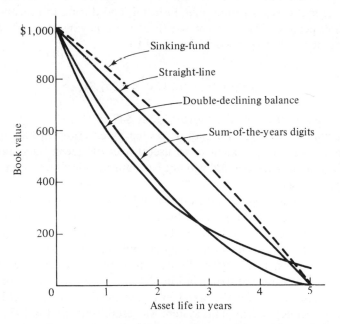

Figure 13.1. Book value computed by four methods of depreciation.

363

line or sinking-fund depreciation. Thus, for tax purposes double-declining-balance and sum-of-the-years-digits depreciation have found wide acceptance since they effectively postpone the payment of taxes to a later date. Such a postponement is favorable to the taxpayer because money has a time value and a dollar paid in taxes at the present is worth more than a dollar paid in taxes at some future time. For the same reasons the sinking-fund method of depreciation is the least appealing of the four methods shown in Figure 13.1.

Effect of estimated life on income taxes. There is often little connection between the life of an asset that will be realized and that which a taxpayer may use for tax purposes. If the applicable effective tax rate remains constant through the realized life of an asset, the use of a shorter life for tax purposes will usually be favorable to a taxpayer. The reason is that use of short estimated lives for tax purposes results in relatively high annual depreciation, low annual taxable income and, consequently, low annual income taxes during the early years of an asset's life. Even though early low annual income taxes will result in correspondingly higher annual income taxes during the later years of an asset's life, the present worth of all income taxes during the asset's life will be less.

If the conditions in Alternative *A* in the previous section had permitted the use of an estimated life of 5 years for tax purposes, the annual income tax for the first 5 years would have been nil, and during the second 5 years of life, $80 per year. The present worth of these payments at 6% as of the beginning of the first year would be equal to

$$\overset{P/A\,6,5}{\$80(\ 4.212)}\ \overset{P/F\,6,5}{(0.7473)} = \$251.81.$$

The difference in favor of the shorter life is $42.59.

The matter of useful life may be settled by entering into an agreement with the Treasury Department as to useful lives of assets. Such agreements are binding on both parties and will not be changed unless facts not previously contemplated arise.

If challenged in regard to the estimated life selected for assets, the burden of proof rests upon the taxpayer. Changes of rate during the life of an asset must be based upon facts related to the use of the asset. For example, markedly increased use of a machine over that on which a depreciation rate was initially established would generally be considered an acceptable reason for using an increased rate of depreciation. An alternative to determining useful life as just described is for the taxpayer to use the Class Life Asset Depreciation Range (ADR) system. This system was established by the 1971 Revenue Act and it allows the selection of useful life by industry classifications. (See Table 12.10.)

The lessening in value due to extraordinary obsolescence ordinarily cannot be anticipated; consequently, it will usually be taken into account by an adjustment in the remaining life of an asset. When it becomes apparent that the value of an asset is being adversely affected by revolutionary inventions, radical economic changes, and abnormal needs that will result in retirement of the asset at a shorter life than originally estimated, the estimated life may be decreased to reflect the new conditions.

Special provisions are made for the rapid amortization of facilities that have been certified by proper authority as necessary in the national interest. Deductions for tax purposes may be made for such assets on the basis of amortization over a 60-month period regardless of their estimated useful life. This provision has essentially the same effect as permission to use an estimated life of 5 years for the assets involved.

Effect of a gain or loss on disposal on income taxes. When a depreciable asset used in business is exchanged or sold for an amount greater or smaller than its book value, this "gain" or "loss" on disposal has an important effect on income taxes. If the asset has been held for more than 6 months and it is disposed of for an amount greater than the asset's book value at the time of disposal the "gain" is treated as a long-term capital gain. If the asset is sold for less than its book value the "loss" on disposal is treated as a long-term capital loss. The disposal of an asset for an amount equal to the asset's estimated salvage value or book value at the time of disposal incurs no taxes.

For example, suppose an asset purchased for $7,000 has a life of 3 years and produces revenue of $4,000 per year. If the asset is depreciated by the straight-line method and it is estimated that the salvage value of the asset will be $1,000, the annual depreciation charge is ($7,000 − $1,000)/3 = $2,000. The calculations of taxes for an effective tax of 40% and long-term capital gain tax of 30% are presented in Table 13.5 in which it is assumed that

Table 13.5. TAXES FOR AN ASSET WHEN ITS ACTUAL SALVAGE VALUE EXCEEDS ITS ESTIMATED SALVAGE VALUE

End of Year	Before-Tax Cash Flow	Deprecia- tion	Book Value	Taxable Income		Taxes
				Ordinary Taxable Income	Capital Gains or Loss	
0	− $7,000	—	$7,000	—		—
1	4,000	$2,000	5,000	$2,000		$800
2	4,000	2,000	3,000	2,000		800
3	4,000+ 1,500	2,000	1,000	2,000	$500	800 + 150

$1,500 is received from the disposal of the asset. The taxable income for year 3 consists of two elements. The first element consists of the ordinary revenue received less the annual depreciation or $4,000 − $2,000 = $2,000. The second element reflects the fact that the actual amount received on disposal exceeds the estimated salvage value at the time of disposal. Thus, a capital gain of $1,500 − $1,000 = $500 is realized. For corporations this capital gain is taxed at a rate of 30% while the other income is taxed at the effective rate. If the amount realized on the sale of the asset had been less than the estimated salvage value, the difference would appear as a capital loss. This loss could then be used to offset other capital gains resulting from other activities in year 3. If there are no other capital gains during year 3, the loss can be carried forward for 5 years and offset against capital gains during that period.

13.5. Depletion and Income Taxes

Natural resources such as minerals, oil, gas, timber, and certain others are found in deposits or stands of definite quantities. They are exploited by removal. As quantities are removed to produce income, the amount remaining in a deposit or stand is reduced.

In arriving at taxable income, depletion is deducted from gross income. The basis for determining the amount of depletion for a given year is the total cost of the property, the total number of units in the property, and the number of units sold during the year in question. This may be expressed by the equation:

$$\text{depletion for year} = \frac{\text{cost of property}}{\text{total units in property}} \times \text{units sold during year.}$$

Oil, gas, and mineral properties are also depleted for tax purposes on the basis of a percentage of gross income, provided that the amount allowed for depletion is not greater than 50% of the taxable income of the property before depletion allowances. In general, percentage depletion will be elected if it results in less tax than does depletion based on cost.

Typical of percentage depletion allowances for mineral and similar resorces are the following:

Oil and gas wells 22%
Sulphur, uranium, asbestos, bauxite, graphite, mica, antimony, bismuth,
  cadmium, cobalt, lead, manganese, nickel, tin, tungsten, vanadium,
  zinc 22%
Various clays, diatomaceous earth, dolomite, feldspar, and metal mines
  if not in 22% group 14%
Coal, lignite, sodium chloride 10%
Brick and tile clay, gravel, mollusk shells, peat, pumice, and sand 5%

Timber may not be depleted on a percentage basis. In general, timber is depleted for tax purposes on the basis of its cost at a specific date, the total number of units of timber on the property, and the units of timber removed during the tax year. Special rules are applicable for determining cost and for revising the number of units on the property from time to time.

Effect of depletion method on income taxes. Most resources subject to depletion may be depleted on either a cost basis or a percentage basis in computing income taxes. Consider an example of a mineral deposit estimated to contain 60,000 units of ore and purchased at a cost of $24,000 including equipment. The depletion rate is $24,000 ÷ 60,000, or $0.40 per unit of ore.

This property is also subject to percentage depletion at rate of 22% per year, but not in excess of 50% of the net taxable income of the property before depletion. An analysis of this situation is presented in Table 13.6.

Table 13.6. APPLICATION OF COST AND PERCENTAGE DEPLETION TO A MINERAL DEPOSIT

Year *A*	Units produced *B*	Gross income $(B \times \$2)$ *C*	Operating cost *D*	Net income before depletion and income tax *E*	50% of net income $(E \times 0.5)$ *F*	Allowable cost depletion at $0.40 per unit *H*	Allowable percentage depletion at 22% $(C \times 0.22)$ *I*	Taxable income $(E \text{ less } F, H, \text{ or } I)$ *J*	Income tax (tax rate of 0.3)$(J \times 0.3)$ *K*
1	20,000	$40,000	$20,000	$20,000	$10,000	$8,000	$8,800‡	$11,200	$3,360
2	16,000	32,000	16,000	16,000	8,000	6,400	7,040‡	8,960	2,688
3	12,000	24,000	14,000	10,000	5,000*	4,800	5,280	5,000	1,500
4	8,000	16,000	8,000	8,000	4,000	3,200	3,960‡	4,040	1,212
5	4,000	8,000	6,000	2,000	1,000	1,600†	1,760	400	120
									$8,880

*Deduct this amount for depletion as full percentage depletion will not be allowed and because it is greater than cost depletion.

†Deduct this amount for depletion because it is greater than the allowable percentage depletion that may be used.

‡Deduct this amount for depletion because it is full percentage depletion allowances and it is greater than allowable cost depletion.

By taking advantage of the two methods of depletion, the income taxes on the property above total $8,880. If only cost depletion had been used, income taxes in successive years would have been $3,600, $2,880, $1,560, $1,440, and $120 for a total of $9,600.

13.6. Economy Calculations Involving Income Taxes

The introduction of income taxes into economy studies requires special consideration because of three factors in the nature of income taxes. One factor is that income taxes are dependent upon net income. What interests a taxpayer is the net return of a venture after income taxes. Net earning before income taxes, which determines the amount of income taxes, is a difference of income and cost which makes net earning subject to the joint effect of errors in estimates of the latter two quantities. A second factor is that in computing income taxes only interest on borrowed money may be considered a cost. A third factor is that the method of depreciation used in computing income taxes must be considered in analyses.

A basic tax equation. The following symbolism will be used and will apply to annual quantities:

E_b = net profit before income taxes;
E_a = net profit after income taxes;
G = estimated gross income from the activity under consideration;
C = estimated annual costs, all not included elsewhere;
D = estimated annual depreciation;
D_t = estimated annual depreciation allowed for tax purposes;
I = interest paid on borrowed funds;
t = effective applicable income tax rate;
T = income tax payable.

A general equation for annual profit before income taxes is derived as follows:

$$E_b = G - (C + D + I) \text{ and } E_a = E_b - T$$

and

$$T = [G - (C + D_t + I)]t.$$

Then

$$E_a = [G - (C + D + I)] - [G - (C + D_t + I)]t.$$

This equation may be simplified by assuming that $D_t = D$ and substituting D for D_t. This may be justified by the fact that although actual and allotted annual depreciations may differ, the total actual depreciation should equal the total allowed depreciations over the life of an asset. If this substitution is made

$$E_a = E_b(1 - t) = [G - (C + D + I)](1 - t).$$

Notice should be taken of the fact demonstrated in this equation that income after income taxes of a venture is equal to its gross income times $(1 - t)$ less its costs times $(1 - t)$.

The effect of taxes on alternatives with equal income. When two alternatives have equal income, their comparison may be based only upon costs. In many cases, gross incomes of alternatives can be assumed to be identical even though their amount may not be known. For example, when either of two machines can satisfactorily handle an equal flow of production, their contribution to the total income of the enterprise may be assumed to be equal even though the amount of the contribution of each to income may not be known. Assume that Machine A and Machine B result in gross incomes of G and G', respectively. Then income taxes of the alternatives can be taken into account as follows:

Machine A: $\quad E_a = G(1 - t) - (C + D + I)(1 - t)$

Machine B: $\quad E'_a = G'(1 - t) - (C' + D' + I')(1 - t).$

But by the conditions given, $G = G'$, and

$$E_a - E'_a = [-(C + D + I) + (C' + D' + I')](1 - t).$$

If the costs of Machine A are less than the costs of Machine B, the annual "saving" in operating costs before income taxes will amount to $(C' + D' + I') - (C + D + I)$. Thus, the advantage of Machine A over Machine B after income taxes will be equal to this "saving" multiplied by $(1 - t)$.

After-tax cash flows for comparison of alternatives. The cash flows that represent the actual receipts and disbursements associated with an investment alternative can be either before or after-tax cash flows. Since taxes constitute a substantial portion of the disbursements that are related to an alternative it is sound decision making to compare the after-tax cash flows of investment alternatives.

To determine the after-tax cash flows of an investment alternative a tabular method may be used. Tabular methods have the advantage that they can be made to reflect complex situations with simple mathematics. They are also easy for the layman to understand.

Suppose a firm has $30,000 available for investment. Three possible alternatives have been suggested and they are considered to be mutually exclusive. Alternative $A1$ and Alternative $A2$ differ only in the method of depreciation used in the calculation of their after-tax cash flows. Alternative

$A1$ requires sum-of-the-years-digits depreciation while Alternative $A2$ uses the straight-line method. The net before-tax cash flows for Alternative $A1$ and Alternative $A2$ are presented in Column B of Tables 13.7 and 13.8, respectively. The lives of these alternatives are 5 years and the estimated salvage value is zero. The effective rate is 40% and the MARR *after taxes* is considered to be 10%.

Table 13.7. TABULAR METHOD FOR ALTERNATIVE $A1$

End of Year	Before-Tax Cash Flow	Depreciation Charges	Taxable Income* $B + C$	Taxes $-0.4 \times D$	After-Tax Cash Flow $B + E$
A	B	C	D	E	F
0	−$30,000				−$30,000
1	10,000	− $10,000	$ 0	$ 0	10,000
2	10,000	− 8,000	2,000	− 800	9,200
3	10,000	− 6,000	4,000	−1,600	8,400
4	10,000	− 4,000	6,000	−2,400	7,600
5	10,000	− 2,000	8,000	−3,200	6,800

*Consider only those revenues and costs that affect taxable income. (The initial cost of investment has no effect on taxable income.)

Table 13.8. TABULAR METHOD FOR ALTERNATIVE $A2$

End of Year	Before-Tax Cash Flow	Depreciation Charges	Taxable Income $B + C$	Taxes $-0.4 \times D$	After-Tax Cash Flow $B + E$
A	B	C	D	E	F
0	−$30,000				−$30,000
1	10,000	−$6,000	$4,000	− $1,600	8,400
2	10,000	− 6,000	4,000	− 1,600	8,400
3	10,000	− 6,000	4,000	− 1,600	8,400
4	10,000	− 6,000	4,000	− 1,600	8,400
5	10,000	− 6,000	4,000	− 1,600	8,400

Table 13.9. TABULAR METHOD FOR ALTERNATIVE $A3$

End of Year	Before-Tax Cash Flow	Cash Flow for Loan	Cash Flow After Loan Payment	Deprecia-tion Charges	Taxable Income $D + E$	Taxes $-0.4 \times F$	After-Tax Cash Flow $D + G$
A	B	C	D	E	F	G	H
0	−$50,000	$20,000	−$30,000				−$30,000
1	12,000	−1,600	10,400	−$5,000	$5,400	−$2,160	8,240
2	12,000	−1,600	10,400	− 5,000	5,400	− 2,160	8,240
3	12,000	−1,600	10,400	− 5,000	5,400	− 2,160	8,240
4	12,000	−1,600	10,400	− 5,000	5,400	− 2,160	8,240
5	12,000	−1,600	10,400	− 5,000	5,400	− 2,160	8,240
5	25,000*	−20,000	5,000				5,000

*Salvage value.

Alternative $A3$ is somewhat different from $A1$ and $A2$ in that it requires an initial investment of $50,000. Since the firm has only $30,000 available it must borrow the additional $20,000 required at an interest rate of 8% per year in order to undertake Alternative $A3$. This alternative has a life of 5 years, similar to $A1$ and $A2$, but the estimated salvage at the end of its life is expected to be $25,000. Column B of Table 13.9 presents the before-tax cash flow of the alternative exclusive of the loan and its associated interest of payments. The cash flow related to the loan is shown in Column C of Table 13.9. Now that the after-tax cash flows are available for each of the three alternatives the computation of the present-worth amounts for these alternatives is straightforward. For an after-tax MARR of 10% the present-worth on total investment for each alternative is computed.

$$PW(10)_{A1} = -\$30,000 + \left[\$10,000 - \$800(\overset{A/G\,10,\,5}{1.8101})\right](\overset{P/A\,10,\,5}{3.7908}) = \$2,419$$

$$PW(10)_{A2} = -\$30,000 + \$8,400(\overset{P/A\,10,\,5}{3.7908}) = \$1,843$$

$$PW(10)_{A3} = -\$30,000 + \$8,240(\overset{P/A\,10,\,5}{3.7908}) + \$5,000(\overset{P/F\,10,\,5}{0.6209}) = \$4,341.$$

Economically, the most desirable alternative is $A3$. Although Alternative $A1$ and Alternative $A2$ are economically the same on a before-tax basis, Alternative $A1$ is favored over Alternative $A2$ on an after-tax basis. This difference between these two alternatives arises because of the effect the difference methods of depreciation have on income taxes. This subject was discussed earlier in Section 13.4.

Suppose that Alternative $A3$ is implemented and it is now 5 years later. An examination of the receipts and disbursements produced by Alternative $A3$ over the last 5 years reveals that the predictions about the future regarding this alternative were accurate with one exception. Instead of the salvage value being $25,000 as predicted the actual amount received was $30,000. Thus, the after-tax cash flow resulting from the salvage value is the estimated salvage value plus the after-tax gain realized on disposal. That is, the after-tax contribution from the $30,000 received on disposal is

$$\$25,000 + (\$30,000 - \$25,000)(1 - 0.30) = \$28,500.$$

However, since $20,000 of that amount is used to repay the loan, the net benefit derived from the $30,000 salvage is $8,500. That amount now replaces the $5,000 amount shown for year 5 in Column H of Table 13.9. Thus, if one is interested in the after-tax rate of return actually earned on the investment made in Alternative $A3$ all that is required is to solve for i in the following equation:

$$0 = -\$30,000 + \$8,240(\overset{P/A\,i,\,5}{\quad}) + \$8,500(\overset{P/F\,i,\,5}{\quad})$$

By trial and error

$$i^*_{A3} = 17.3\%.$$

Thus, in retrospect, Alternative $A3$ can be considered a good investment since it has produced a return greater than the MARR of 10%.

There are many instances in the evaluation of economic alternatives where only the costs of the alternatives are considered. Most often this occurs where alternatives that are intended to provide the same service are to be compared. Since the benefits to be derived are equal for each alternative, it is common practice to eliminate the estimation of associated revenues to reduce the evaluation effort.

It might appear that ignoring revenues for alternatives would prevent the comparison of these alternatives on an after-tax basis. Fortunately, this is not the case. It can be seen that the direct application of the procedures presented in Tables 13.7, 13.8, and 13.9 will permit the incorporation of tax effects in the comparison cash flows having equal revenue streams. By following the sign convention just used, (+) revenues, (−) costs, the tabular method will produce tax-adjusted cash flows that can be directly compared as long as (1) their revenues are assumed to be equal, and (2) the firm on the whole is realizing a profit. To see how the procedure operates, examine the cost-only cash flow in Table 13.10.

Table 13.10. TAX EFFECTS FOR A COST CASH FLOW

End of Year A	Before-Tax Cash Flow B	Depreciation Charges C	Taxable Income $B + C$ D	Taxes (Savings) $-0.42 \times D$ E	After-Tax Cash Flow $B + E$ F
0	− $45,000				− $45,000
1	− 15,000	− $10,000	− $25,000	$10,500	− 4,500
2	− 15,000	− 10,000	− 25,000	10,500	− 4,500
3	− 15,000	− 10,000	− 25,000	10,500	− 4,500
4	− 15,000	− 10,000	− 25,000	10,500	− 4,500
4	5,000*				5,000

*Salvage value.

For this example, straight line depreciation is used. The estimated salvage for the investment is $5,000, and the effective tax rate is 42%. Since there are no annual operating revenues being considered, the taxable income appears as $25,000 in costs each year. That is, the effect of this project is to reduce the firm's profits by $25,000 per year (if revenues aren't considered). Assuming that the firm has sufficient earnings from other activities to offset these costs, there will be a "savings" in taxes of $10,500. By following the sign convention utilized in the previous tables, the taxes in Column E are positive indicating that $10,500 in taxes would be avoided yearly if the costs in Column B and C are incurred.

The after-tax cash flow is found by adding these tax savings to the before-tax cash flow shown in Column B. By comparing this after-tax cash flow with a similarly calculated after-tax cash flow for a competing alternative, the alternative with the minimum equivalent after-tax cost can be identified. Thus, without any change in procedure, an after-tax comparison can be made for projects having the same revenue stream. Since this comparison does not require consideration of the possible revenue effects of the alternatives, much time and effort may be saved in performing the analysis.

PROBLEMS

1. A self-employed book salesman is 32 years old, married, and has two children. Last year he earned $2,000 in commissions in connection with which he had $600 traveling and other expenses. From the operation of a technical book shop he had $66,000 gross income and had expenses of $43,400, exclusive of those expenses listed below which should properly be charged to the book shop. In addition, he sold a trailer for $1,800 that he purchased 9 months ago for $1,400. During the year he made contributions of $400 to his church, Red Cross, etc., paid $500 and $300 in interest, respectively, on mortgages on his home and his book shop. To keep abreast of the fields in which he sells books he is a member of several technical societies and pays annual dues of $80 per year. Last year he paid $1,400 in federal income tax. On his home and bookshop, he paid, respectively, $200 and $400 in state, county, and municipal taxes. Medical expenses were $120. Water from a leaky pipe caused $140 damage to books in his shop of which $120 was compensated by insurance.
   (a) Determine adjusted income.
   (b) Determine taxable income.
   (c) Determine the income tax to be paid; use the proper table in the text and assume that the salesman and his wife file a joint return.

2. From data in Problem 1 determine the effective income tax rate applicable to the salesman's (a) taxable income, and (b) adjusted income.

3. The salesman is considering giving up selling on commission.
   (a) What would have been his net income after taxes if he had done so in the year covered in Problem 1?
   (b) What is the effective income tax rate pertinent to the increment of income from the activity he is considering giving up?

4. Mr. Brown, who is in business for himself, is interested in an activity which will necessitate the borrowing of $3,000 for 9 months at 6% interest. The new activity is expected to increase his taxable income by $600, bringing his total taxable

income to $17,000. If Mr. Brown is single and qualifies as head of a household, what will be his net income after income taxes?
(a) If he undertakes the new activity and it turns out as estimated?
(b) If he does not undertake the new activity?

5. A corporation's taxable income for next year is estimated at $1,000,000. It is considering an activity for next year which it is estimated will result in an additional taxable income of $100,000. Calculate the effective income tax rate applicable.
(a) To the present estimated $1,000,000 taxable income.
(b) To the $100,000 estimated increment of income.
(c) If the profit to the new venture is $100,000 before income taxes, what will it be after income taxes are paid?

6. The same conditions exist as in Problem 5, except that a new activity has been undertaken but instead of resulting in an increase of $100,000 in taxable income as estimated, it has resulted in a decrease in the taxable income of $50,000. Calculate
(a) Net income after income taxes on a taxable income of $50,000.
(b) Net income after income taxes if the new venture results in a loss of $50,000.
(c) The loss in income after income taxes that would be caused by the $50,000 loss.

7. To develop a new product a corporation is considering an investment of $1,500,000 in research and development. It is estimated that the increase in taxable income from marketing the new product in the coming year will be $6,000,000. Currently, the taxable income for the firm is $30,000,000 per year.
(a) What will be the increase in the net income after taxes if the research and marketing program is successful?
(b) What will be the decrease in the net income after taxes if the research and marketing program is unsuccessful?

8. A corporation is considering a study to develop a new process to replace a present method of manufacturing transistors which, if successful, is estimated to result in a saving of $300,000 in fulfilling a contract that will be completed during the current tax year. The study is estimated to cost $25,000 during the same tax year. If the new process is not successful, the corporation taxable income will be $5,000,000.
(a) If the study is successful, what will be the increase in income after taxes?
(b) If the study is unsuccessful, what will be the decrease in income after taxes?

9. A corporation has a taxable income of $50,000 during a year. It is considering a venture that will result in an additional income of $20,000 during next year.
(a) What is the effective income tax rate if earnings remain at $50,000?
(b) What is the effective income tax rate if the new venture is undertaken and turns out as estimated?
(c) What is the effective income tax rate applicable to the increment of income due to the new venture?

10. Corporation A has a total investment of $1,000,000, uses no borrowed funds, and has a taxable income of $100,000.

(a) How much will it have for payment of dividends after income taxes?

(b) If it is capitalized at $1,000,000 what would be the rate of return it earned for its stockholders?

Corporation B is identical with A except that it is capitalized at $800,000, uses $200,000 of funds borrowed at 9%, and its taxable income before payment of interest on borrowed funds is $100,000.

(c) How much will it have for payment of dividends after income taxes?

(d) What rate of return did it earn for its stockholders?

11. An asset with an estimated life of 10 years and a salvage value of $5,000 is purchased by a firm for $60,000. The firm's effective tax rate is 50% and the minimum attractive rate of return is 10%. The gross income from the asset before depreciation and taxes will be $8,500 per year. Calculate the present worth of the taxes for the straight-line and sum-of-the-years-digits methods of depreciation.

(a) Assume that the firm is profitable in its other activities.

(b) Assume that the firm has no profits in its other activities.

12. The applicable effective tax rate of a taxpayer is 0.45. He has purchased a machine for $2,000 whose life and salvage value for tax purposes is 4 years and zero, respectively. The estimated annual income of the machine is $1,200 per year before income taxes. Compare the present worth of the income taxes that will be payable if the straight-line method, the sum-of-the-years-digits depreciation, and the double-declining-balance methods are elected, and if the interest rate is 15%.

13. A corporation purchases an asset for $60,000 with an estimated life of 5 years and a salvage value of zero. The gross income per year will be $20,000 before depreciation and taxes. If the tax rate applicable to this activity is 40% and if the interest rate is 12%, calculate the present worth of the income taxes if the straight-line method of depreciation is used; if the sinking-fund method of depreciation is used; if the double-declining-balance method of depreciation is used; if the sum-of-the-years-digits method is used.

14. A corporation purchases a $60,000 asset with an estimated life of 4 years and expected salvage value of $30,000. If after 4 years the asset is sold for $20,000, what are the capital gains or losses for the straight-line depreciation method and the double-declining-balance depreciation method in the fourth year? Assume all accounting adjustments between actual and estimated salvage value occur at the time the asset is sold. What are the capital gain tax effects of these two alternative methods of depreciation?

15. The income tax rate of a corporation is 45%. Four months ago it purchased a warehouse for $44,000. It has just received an offer to sell the warehouse for $50,000 for a short-term gain of $6,000. What selling price 2 to 8 months later will be equivalent to the present $50,000 offer in terms of income after taxes if the corporation has no other short-term gains in prospect?

16. A contractor purchased $400,000 worth of equipment to build a dam. For tax purposes, he was permitted to depreciate the equipment by the straight-line

method on the basis of no salvage value and a 5-year life. After 2 years and the completion of the dam, the contractor sold the equipment for $300,000. The contractor attributed sale of the equipment at an amount higher than the "book value" to the excellence of maintenance. He estimated that he spent $16,000 per year more for maintenance than necessary "just to keep the equipment running," and by so doing reduced his operating costs exclusive of maintenance on the contract by $12,000 per year. The contractor's income tax rate is 42% and the long-term capital gain tax rate is 30%. The contractor averaged $48,000 taxable income for the 2 years required to complete the contract. There were no short-term capital losses.

(a) How much capital gains tax would the contractor pay as a result of selling his equipment?

(b) What was his total income tax including capital gains tax for the 2-year period, and what was his total income after income taxes?

(c) What would have been his total income tax for the 2-year period if he had spent $12,000 per year less maintenance and as a result his operating cost had been increased by $16,000 per year and his equipment had been sold for its book value? What was his total income after income taxes for the 2-year period?

(d) How much was his total income after income taxes for the 2-year period increased by his maintenance policy if his estimates were correct?

17. An oil lease is purchased for $200,000 and it is estimated that there are 400,000 barrels of oil on this lease. It costs $60,000 per year exclusive of depletion charges to recover the oil and the oil can be sold for $7.50 per barrel. The effective tax rate is 40% and 40,000 barrels of oil are produced each year. If the interest rate is 12% and the lease will produce for 10 more years, what is the present worth of the after-tax cash flow for this lease for (a) unit depletion and (b) percentage depletion?

18. A mineral deposit discovered by a mining company is estimated to contain 100,000 tons of ore. The company has made an initial investment of $5,000,000 to recover the ore which sells for $210 per ton. The company has a minimum attractive rate of return of 15% and an effective tax rate of 46%. The fixed percentage depletion rate for this mineral is 14%. If the ore is produced and sold at a rate of 20,000 tons per year and the operating expenses exclusive of depletion expenses are $1,000,000 per year, what is the present worth of the after-tax cash flow for this company if (a) unit depletion is used and (b) percentage depletion is used?

19. A prospector acquired a mine containing 300,000 tons of tungsten ore for $580,000. Annual operating costs of the mine are $132,000, and 40,000 tons of ore are being sold at the rate of $14.50 per ton. The prospector wishes to compare the results of using cost depletion and percentage depletion (22%). Determine the income tax for each method assuming that the applicable income tax rate is 0.35.

20. Consider the following two investment alternatives:

	Alternative A	Alternative B
Initial investment	$10,000	$5,000
Service life	5 years	5 years
Salvage value	$1,000	0
Depreciation method	Sum-of-digits	Straight-line

ESTIMATED OPERATING COSTS AND REVENUES

	End of Year	1	2	3	4	5
Alternative A	Operating cost	$10,000	$10,500	$12,000	$14,000	$16,000
	Revenue	17,000	16,900	17,800	19,200	20,600
Alternative B	Operating cost	$1,200	$1,000	$1,500	$1,300	$1,200
	Revenue	6,400	6,200	6,700	6,500	6,400

(a) Determine the rate of return on the after-tax cash flow for each alternative (income tax rate is 50%).

(b) For a given rate of interest of 6%, which alternative is more attractive to undertake?

21. An investment proposal has the following estimated cash flow before taxes:

End of Year	0	1	2	3	4
	−$50,000	$30,000	$10,000	$5,000	$2,000 + $20,000 (salvage value)

The effective tax rate is 40%, the tax rate on capital gains is 30%, and the interest rate is 10%. Assume the company considering this proposal is profitable in its other activities. Find the after-tax cash flows and the present worth of those cash flows for (a) the straight-line method of depreciation, (b) the double-declining method of depreciation, and (c) the sum-of-the-years-digits depreciation.

22. An investment proposal is described by the following tabulation:

Year	1	2	3
Gross income at end of year ..	$ 8,000	$16,000	$20,000
Investment at beginning year ..	14,000	10,000	0
Operating cost	2,000	3,000	4,000
Depreciation charge	8,000	8,000	8,000

The company considering this proposal is profitable in its other activities. Thus, any depreciation not used to offset income for this proposal can be used to reduce taxes on the income from the other activities. Find the rate of return earned by this proposal (a) before taxes and (b) after taxes. The effective income tax rate is 40%.

23. A $20,000 investment in machinery is under consideration. The project is expected to have a life of 5 years and no salvage value. The estimated annual income from the project is $10,000 with annual operating expenses of $4,000. The investment will be depreciated by the straight-line method. If a 40% income tax rate is applied, compute the rate of return on the proposed investment after-tax cash flow under the following financing policies:
    (a) The investment is provided from the firm's equity funds.
    (b) The initial investment is borrowed at 10% with repayment of interest at the end of each period and the loan principal repaid at the end of 5 years.

24. A firm is considering buying a new machine which costs $40,000. The machine has an economic life of 8 years, an estimated salvage value of $4,000, and annual operating disbursements of $3,000. The machine is expected to generate a revenue of $18,000 annually and will be depreciated by the sum-of-the-years-digits method. Assuming the effective income tax rate is 40%, compute the annual equivalent at 10% for the after-tax cash flow under the following financing:
    (a) The initial investment is financed from equity funds.
    (b) The initial investment is borrowed at 8% with repayment of principal and interest in eight equal annual amounts.
    (c) The initial investment is borrowed at 8% with repayment of interest at the end of each period and the loan principal repaid at the end of 8 years.

25. An asset is being considered whose first cost, life, salvage value, and annual operating expenses, respectively, are estimated at $15,000, 10 years, zero, and $800. The asset will be depreciated by the straight-line method of depreciation. The effective income tax is 40%. Determine the equivalent annual cost at 12% for the after-tax cash flow if
    (a) The initial investment is made from equity funds.
    (b) The initial investment is borrowed at 8% with repayment of principal and interest in 10 equal annual amounts.
    (c) The initial investment is borrowed at 8% with repayment of interest at the end of each period and the loan principal repaid at the end of 10 years.

26. A firm is considering the purchase of an asset. The asset requiring an investment of $100,000 is expected to have a service life of 5 years with a salvage value of $2,000. The firm has an opportunity to finance the asset from borrowed funds at an interest rate of 8%. The loan should be repaid with 5 equal annual payments (principal and interest). The estimated net annual income before tax is $50,000. The depreciation policy adopted by the firm is the *double declining balance method assuming switch to straight-line depreciation where advantageous*. The income tax rate is 50%, and the long-term capital gains tax, if applicable, is 30%.
    (a) Determine the optimal time to switch from the double declining balance to straight-line depreciation for the tax accounting purpose.
    (b) Compute the cash flow after taxes, and present in a tabular format.

27. A new machine costing $140,000 is to be installed in a certain manufacturing

plant. Its useful life is predicted to be 5 years with a salvage value of $20,000. Annual net operating income before tax is estimated to be $40,000. The machine will be depreciated by the sum-of-years-digit method over five years. The investment tax credit applicable to this asset is 7% on its first cost. For an effective tax rate of 48%, determine the equivalent annual worth at 10% for the after-tax cash flow. Assume that the firm purchasing this machine is profitable in its other activities.

28. A telephone company is faced with the prospect of having to replace one of its central switching systems. The new computerized switching system will cost $2,000,000. The system is estimated to have a 15-year useful life with no salvage value and would be depreciated on a straight-line basis. The new system will result in a net annual savings of $300,000. Assuming that the installation of the new switching system would result in an investment tax credit of 4% of its first cost,
    (a) Find the rate of return on the project before taxes.
    (b) Find the rate of return on the project after taxes if the effective tax rate is 48%.

29. A firm has an opportunity to expand one of its production facilities at a cost of $200,000. The asset has an expected economic life of 10 years, after which the asset is not expected to have any salvage value. The double-declining-balance depreciation will be used for tax purposes for the new asset. A 7% investment tax credit will be taken if the asset is purchased. The annual operating cost is estimated at $5,000. If the asset is purchased, the firm is capable of financing the acquisition entirely with an 8%, 10-year unsecured term loan. The loan is payable over this period in equal annual payments which include both principal and interest. The income tax rate applicable for this firm is 40% and any capital gains tax rate is assumed to be 30%. Determine the after-tax annual equivalent cost at 10% for this investment.

part five

ESTIMATES, RISK,
AND UNCERTAINTY

ESTIMATING
ECONOMIC ELEMENTS

Estimating in the physical environment approximates certainty in most applications. Examples are the pressure that a confined gas will develop under a given temperature, the current flowing in a conductor as a function of the voltage and resistance, and the velocity of a falling body at a given point in time. Much less is known with certainty about the economic environment with which the engineering process is concerned. Economic laws depend upon the behavior of people and are unlike physical laws which depend upon well-ordered cause and effect relationships.

A large portion of creative engineering activity has as its objective the search for activities with profit potential that is high in relation to the risk involved. This search necessitates the estimation of pertinent facets of the anticipated economic outcome. In this chapter attention is directed to the methods and techniques for estimating economic elements.

14.1. The Elements To Be Estimated

The success of an activity as a whole may be estimated. Thus, if the establishment of a construction department is under consideration by a corporation, it might be directly estimated that the department will yield a

14

return to the firm equivalent to a certain per cent per year on the amount invested. This return will be a resultant of a number of prospective receipts and disbursements, which may be classified as income, operating expenses, depreciation, interest, and taxes. It is a rare individual who can combine accurately four complex items to obtain their resultant without resort to paper and pencil, even when the items are clearly known. Thus, for best results in estimating the final outcome of an undertaking, it will almost always be found advantageous to begin with detailed estimates of income and costs as they may be expected to originate in the future. The detailed estimates are then combined mathematically to obtain their results.

Success of a venture in terms of economy is determined by considering the relationship between the input and output of the venture through time, with the time value of money taken into consideration. But anticipated success can be imaginary if certain elements are omitted. Thus, an important task in an economy study is the delineation of inputs and outputs associated with the proposed activity.

Outputs. Structures, processes, systems, and activities are normally proposed in response to a need or requirement. Outputs, therefore, should be considered first and in conjunction with the need. Benefit, worth, effectiveness, and other terms are used to describe outputs in relationship to requirements.

The outputs of commercial organizations and governmental agencies are endless in variety. Commercial outputs are differentiated from governmental outputs by the fact that it is usually possible to evaluate the former accurately but not the latter. A commercial organization offers its products to the public. Each item of output is evaluated by its purchaser at the point of exchange. Thus, the monetary values of past and present outputs of commercial concerns are accurately known item by item.

Since engineering is concerned with the future, it is concerned with future output. Generally, information on two subjects is needed to come to a sound conclusion. One of these is the physical output that may be expected from a certain input. This is a matter for engineering analysis. The second is a measure of output that may be expressed in terms of monetary income.

Monetary income is dependent upon two factors: one is the volume of output, in other words, the amount that will be sold; the other is the monetary value of the output per unit. The determination of each of these items for the future must of necessity be based upon estimates. Market surveys and similar techniques are widely used for estimating the future output of commercial concerns. In the case of large scale systems or projects the monetary income to the contractor is determined through contractual agreement.

Under some circumstances—for example, if the income is represented by the saving resulting from an improvement in a process for manufacturing a product made at a constant rate—an estimate of income is easily made. But estimating income for new products with reasonable accuracy may be very difficult. Extensive market surveys and even trial sales campaigns over experimental areas may be necessary to determine volume. When work is done on contract, as is the case with much construction work, for example, the necessity for estimating income is eliminated. Under these circumstances the income to be received is known in advance with certainty from the terms of the contract.

Internal intermediate outputs of commercial organizations are determined with great difficulty and are usually estimated as dictated by judgment. For example, the value of the contribution of an engineer, a production clerk, or a foreman to a final output is rarely known with reasonable accuracy to either the employee or to his superior. Similarly, it is difficult to determine the value of the contributions of most intermediate activities to the final result.

The outputs of many governmental activities are distributed without regard to the amount of taxes paid by the recipient. When there is no evaluation at the point of exchange, it appears utterly impossible to evaluate many governmental activities. John Doe may recognize the desirability of the Department of Defense, the U.S. Forest Service, or Public Health activities, but he will find it impossible to demonstrate their worth in monetary terms. However, some government outputs, particularly those which are localized, such as highways, drainage, irrigation, and power projects, may be fairly accurately evaluated in monetary terms by caluculating the reduction in cost or the increase in income they bring about for the user.

Input. Any activity that is undertaken requires an input of thought, effort, material, and other elements for its performance. In a purposeful activity, an input of some value is surrendered in the hope of securing an output of greater value. The terms input and output as used here have the same meaning as when they are used to designate, for instance, the number of heat units that are supplied to an engine and the number of energy units it contributes for a defined purpose.

A most important item of input is services of people for which salaries and wages are paid. In a commercial organization the total input of human services, as measured by the cost for a given period of time, is ordinarily reflected quite accurately. Where involuntary service is secured, as in military services of governments, it will be clear that the input of human services is not necessarily reflected by amounts paid out as wages. In a commercial organization the input of human services may be classified under the headings of direct labor, indirect labor, and investigation and research. Of these, direct labor is the only item whose amount is known with reasonable accuracy and whose identity is preserved until it becomes a part of output.

Input devoted to investigation and reasearch is particularly hard to relate to particular units of output. Much research is conducted with no particular specified goal in mind and much of it results in no appreciable benefit that can be associated with a particular output. Expenditures for people for investigation and research may be made for some period of time before this type of service has a concrete effect upon output. Successful research of the past may continue to affect output for a long time in the future. The fact that expenditures for this type of service do not parallel in time the benefits provided makes it very difficult to relate them to an organization's output.

Indirect labor, supervision, and management have characteristics falling between those of direct labor and of investigation and research. The input of indirect labor and, to a lesser extent, that of supervision, parallel the output fairly closely in time, but their effects can ordinarily be identified only with broad classes of output items.

Management is associated with the operations of an organization as a whole. Its important function of seeking out desirable opportunities is similar in character to research. Input in the form of management effort is difficult to associate with output either in relation to time or to classes of product.

As difficult as it may be to associate certain inputs with a final measurable output, such as of products sold on the market, input can ordinarily be identified closely with intermediate ends that may or may not be measurable in concrete terms. For example, the cost of the input of human effort assigned to the engineering department, the legal department, the labor relations department, and the production department are reflected with a high degree of accuracy by the payrolls of each. However, the worth of the output of the personnel assigned to these departments may and usually does defy even reasonably accurate measurement.

A second major category of input is that of material. Many items of material are acquired to meet the objectives of commercial and governmental enterprise. For convenience, material items may be classified as direct material, indirect material, equipment, land, and buildings.

Inputs of direct material are directly allocated to final and measurable outputs. The measure of material items of input is their purchase price plus costs for purchasing, storage, and the like. This class of input is subject to reasonably accurate measurement and may be quite definitely related to final output, which in the case of commercial organizations is easily measurable.

Indirect material and power inputs are measurable in much the same way and with essentially the same accuracy as are direct material and power inputs. One of the important functions of accounting is to allocate this class of input in concrete terms to items of output or classes of output. This may ordinarily be done with reasonable accuracy.

An input in the form of an item of equipment requires that an immediate expenditure be made, but its contribution to output takes place piecemeal over a period of time in the future which may vary from a short time to many years, depending upon the use life of the equipment. Inputs of equipment are accurately measurable and can often be accurately allocated to definite output items except in amount. The latter limitation is imposed by the fact that the number and kinds of output to which any equipment may contribute are often not known until years after many units of the product have been distributed. The function of depreciation accounting is to allocate equipment inputs to outputs.

Inputs of land and buildings are treated in essentially the same manner as inputs of equipment. They are somewhat more difficult to allocate to output because of their longer life and because a single item, such as a building, may contribute simultaneously to a great many output items. Allocation is made with the aid of depreciation and cost accounting techniques and practices.

Allocation of inputs of indirect materials, equipment, buildings, and land rests finally upon estimates or judgments. Although this fact is often obscured by the complexities of and the necessary reliance upon accounting practices for day-to-day operations, it should not be lost sight of when economy studies are to be made.

Capital in the form of money is a very necessary input, although it must ordinarily be exchanged for producer goods in order for it to make a contribution to output. Interest on money used is usually considered to be a cost of production and so may be considered to be an input. Its allocation to output will necessarily be related to the allocation of human effort, services, material, and equipment in which money has been invested.

Taxes are essentially the purchase of governmental service required by private enterprise. Since business activity cannot be carried on without the payment of taxes, they comprise a necessary input. There are many types of taxes, such as ad valorem, excise, sales, and income. The amounts may be precisely known and, therefore, are accurately called inputs. However, it is often difficult to allocate taxes to outputs especially in the case of income taxes which are levied after the profit is derived.

14.2. Cost Estimating Methods

A cost estimate is an opinion based upon analysis and judgment of the cost of a product, system, or service. This opinion may be arrived at in either a formal manner or an informal manner by several methods, all of which assume that experience is a good basis for predicting the future. In many cases the relationship between past experience and future outcome is fairly direct and obvious.

In other cases the relationship between past experience and future outcome is unclear, because the proposed product, system, or service differs in some significant way from its predecessors. The challenge is to project from the known to the unknown by using experience with existing items. The techniques used for cost estimating range from intuition at one extreme to detailed mathematical analysis at the other.

Estimating by engineering procedures. Estimating by engineering procedures can be described as an examination of separate segments of work at a low level of detail. The engineering estimator begins with a set of drawings and specifies each engineering task, equipment and tool need, and material requirement. Costs are assigned to each element at the lowest level of detail and these are then synthesized into a total for the product or project.

Time standards for construction or production operations exist for many common tasks. These are usually developed by industrial engineers and constitute the minimum time required to complete a given task with normal worker skill and tools. Standards are best applied in engineering estimating procedures when a long, stable production run of identical items is contemplated. They are normally not useful in estimating required for complex systems in which one-of-a-kind is to be fabricated. For example, production runs of advanced military and space systems are normally short with both design configurations and production processes continuing to evolve rapidly over time.

Engineering estimating procedures require more hours of effort and data than are likely to be available in the development of some systems or projects. One large aerospace firm judges that the engineering approach in estimating the cost of an airframe would require about 5,000 estimates. The cost of an effort such as this makes other estimating methods attractive, especially if the alternative method gives comparable results.

Combining thousands of detailed estimates into an overall estimate can lead to an erroneous result, for the whole often turns out to be greater than the sum of its parts. The engineering estimator works from sketches, blueprints, or descriptions for some items that have not been completely designed. He can only assign costs to work that he knows about. The effect of a low estimate may be compounded because detail estimating is attempted on only a portion of the labor hours. A number of construction or production labor elements, such as planning, rework, coordination and testing, are usually factored in as a percentage of the detail estimates. Other cost elements, such as maintenance, inspection, and production control, are factored in as a percentage of the production labor required. Thus, small errors in detailed estimates can result in large errors in the total cost estimate.

Another source of error in estimates made by the engineering method is the significant variability that occurs in the fabrication of successive units. Production runs of like models may be of limited length and are often subject to design changes. In the case of defense systems, production rates vary frequently and unexpectedly. The proportion of new components may be significant from model year to model year as the manufacturer seeks to adapt a product to market demand. The effect of these forces can often be represented best by mathematical or statistical functions which describe technological progress.

Estimating by analogy. When a firm is venturing into a new area, estimating by analogy can be very effective. For example, aircraft companies bidding on missile programs in the 1950s drew analogies between aircraft and missiles as a basis for estimating. Appropriate adjustments were made for differences in size, number of engines, and performance. This is an instance of estimating by analogy at the macrolevel.

Estimating by analogy may also occur at the microlevel. The direct labor hours required to make a part may be estimated by referring to that required on similar jobs. The basis for the estimate is the similarity that exists between the known item and the proposed part. Some estimators with backgrounds as mechanics, tool makers, or foremen are able to estimate the required times very closely. They are often consulted when an estimate is needed quickly.

The cost of direct labor is often estimated in relation to the cost of direct material. These relationships are known with reasonable accuracy for different kinds of activities. For example, the cost of labor to lay 1,000 bricks is approximately equal to the cost of the bricks and necessary mortar.

At all levels of aggregation, much estimating is performed by analogy. For example, Project A required 12,000 direct labor hours and 5,000 equipment hours. Given the similarities and differences between Project A and proposed Project B, the direct labor hours and equipment hours might be estimated to be 8,000 and 3,200, respectively. By applying current labor and equipment hour rates, and an applicable overhead rate, a total project cost can be estimated. Or, within Project A the estimator may find elements that are analogous to elements in Project B. From this the cost of Project B may be estimated. In this instance, analogy becomes part of the engineering method of estimating.

A major disadvantage of estimating by analogy is the high degree of judgment required. Considerable experience and expertise are required to identify and deal with appropriate analogies and to make adjustments for perceived differences. However, because the cost of estimating by analogy is low, it can be used as a check on other methods. Often it is the only method that can be used because the product, system, or service is in a preliminary stage of development.

Statistical estimating method. The statistical method of cost estimating may utilize statistical techniques ranging from simple graphical curve fitting to complex multiple correlation analysis. In either case the objective is to find a functional relationship between changes in costs and the factor or factors upon which the cost depends, such as output rate, lot size, weight, and so forth.

Figure 14.1 illustrates a simple curve representing the cost per horsepower as a function of the horsepower for a class of electric motors. The data points were taken from the manufacturer's list price and provide a statistical base for the curve. Estimates of the cost of motors of any horsepower within the range illustrated can be made from the curve. It would be dangerous, however, to attempt an estimate much beyond the range for which the cost data are available.

Although statistical cost estimating techniques are preferred in most situations, there are cases in which engineering methods or estimating by

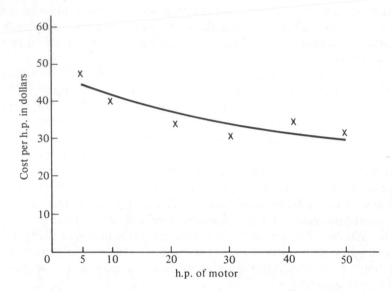

Figure 14.1. Cost of electric motors per horsepower, based on a manufacturer's list price for squirrel-cage induction, 3-phase, 60-cycle, 230-volt, 1,200-rpm motors.

analogy are required because the data for a systematic historical base do not exist. The product may utilize some new unfamiliar manufacturing method, thus invalidating data from previous items as a statistical base.

There will always be situations in which analogy or engineering methods are required, but the statistical approach is useful and sufficient in long-range planning. Total cost may be estimated directly as a function of power output, weight, square feet, volume, etc. Industry-wide data or maintenance experience and energy consumed can be treated statistically and added to statistically estimated costs associated with the item itself.

Statistical estimating techniques will vary according to the purpose of the study and the information available. In a conceptual study it is desirable to have a procedure that gives the total expected cost of the product or system. Allowances for contingencies are to compensate for unforseen changes that will have to be made. Later, as the product or system moves closer to final design, it is desirable to have a procedure that will yield estimates of its component parts. Additional engineering effort can then be applied to reduce the cost of those components which are high contributors to overall cost.

In planning, industry-wide labor and overhead rates can be obtained from statistical publications and used to give a rough cost estimate for a given item. As the item nears the time for bidding in its design cycle, data that are specific for a particular contractor and particular location can be used. As more is known, more specific statistical data can be used as is the case of a product currently in production. In this stage of the product, accounting records can be used to produce a relatively precise cost figure.

14.3. Adjustment of Cost Data

Data must be consistent and comparable if they are to be useful in an estimating procedure. Inconsistency is often inherent in cost data because there are differences in definitions, production quantity differences, missing cost elements, inflation, and so forth. In this section some approaches found useful in coping with this inherent inconsistency are presented.

Cost data categories. Different accounting practices make it necessary to adjust basic cost data. Companies record their costs in different ways; often they are required to report costs to governmental agencies in categories that differ from those used internally. These categories also change from time to time. Because of these definitional differences, the first step in cost estimating should be to adjust all data to the definition being used.

Consider the following example. A manufacturing firm has a planning department that is responsible for both production planning and planning for plant expansion. Production planning costs are those associated with scheduling, sequencing, machine loading, etc., for a specific product. Plant expansion planning leads to costs associated with adapting the manufacturing facility to the production demands created by the numerous products in process and the products anticipated to be in process.

Planning costs in the firm may be charged several ways. Some may be allocated to specific product lines, such as those associated with production planning. Plant planning costs may be charged to overhead, or they may be prorated to product lines on the basis of some formula. Also, all costs arising from the department could be charged to overhead and then apportioned back to the product lines depending upon the labor or machine hours utilized by the product. Therefore, a precise definition applied consistently over time is required so that the planning cost element will have value in estimating for a new product line.

Consistency is also needed in defining physical and performance characteristics. The weight of an item depends upon what is included. Gross weight, empty weight, and airframe weight apply to aircraft with each term differing in exact meaning. Speed can be defined in many ways: maximum speed, cruising speed, etc. Differences such as these can lead to erroneous cost estimates in which the estimate is statistically derived as a function of a physical characteristic. When cost data are collected from several sources, an understanding of the definitions of physical and performance characteristics is as important as an understanding of the cost elements.

Another requiring clear definition is the question of nonrecurring and recurring costs. Recurring costs are a function of the number of units produced but nonrecurring costs are not. If research and development work on new products is charged off as an expense against current production, it does not appear against the new product. In this case, the cost of initiating a new product would be understated and the cost of existing products overstated. Separation of nonrecurring and recurring costs should be made by means of a downward adjustment of the production costs for existing products and an account to collect research and development costs for new products.

In addition to the first cost needed to construct or produce a new structure or system, there are recurring costs of a considerable magnitude associated with operations, maintenance, and disposal. Power, fuel, lubricants, spare parts, operating supplies, operator training, maintenance labor, and logistics support are some of these recurring costs. The life of many complex systems when multiplied by recurring costs such as these leads to an aggregate expenditure which may be large when compared with the first cost. Limited cost estimating effort is often applied to the category of costs associated with operations and maintenance. This may be unfortunate, for the life-cycle cost of the structure or system is the only correct basis for judging its worth.

Price level changes. The price of goods and services, together with the labor, material, and energy required in their production, changes over time. Down through history, inflationary pressures in the economic system have acted to increase the price of most items. These increases have been very significant in recent years. Prices have decreased for only very brief periods.

Figure 14.2 shows the change in average weekly earnings of production workers in manufacturing over the last two decades. The weekly wage rate has increased by a factor of about 3 from the mid-1950s to the mid-1970s. Thus, if the labor cost component of an automobile was $1,000 in the mid-1950s, it would be almost $3,000 in the mid-1970s. Fortunately, increased productivity has kept the labor cost of the automobile at about the same percentage over this 20-year period.

Adjustments in cost data are often made by means of an index con-

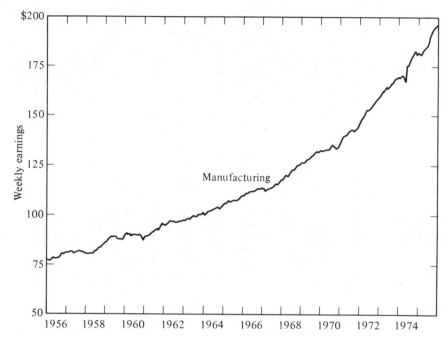

Figure 14.2. Changes in weekly earnings.

structed from data in which one year is selected as the base. The value of that year is expressed as 100, and the other years in the series are then expressed as a percentage of this base. Hourly earnings from 1960 through 1975 for production workers could be converted to an index by using any of the years as a base. Table 14.1 uses both 1960 and 1970 as a base.

The information needed to construct a labor index such as is given in Table 14.1 is available in the Bureau of Labor Statistics' monthly publication entitled *Employment and Earnings*. Indexes for labor categories can be developed from this source, as is shown in Table 14.2.

Average hourly earnings are also a function of location as shown in Table 14.3. The differences are quite significant.

Cost indexes are also available for commodities. Figure 14.3 gives the wholesale price index, and Figure 14.4 gives the consumer price index from 1965 through 1975.

Figure 14.5 gives the cost index for industrial buildings as computed and published by the Austin Company. This index is based on prices in major industrial areas for a 116,760 square-foot flat-roofed, steel-framed structure and an 8,325 square-foot office building. Estimates include site-work, fluores-

Table 14.1. AVERAGE HOURLY EARNINGS INDEX

Year	Average Hourly Earnings	Index with 1960 as Base Year	Index with 1970 as Base Year
1960	$2.20	100	68
1961	2.25	102	69
1962	2.31	105	71
1963	2.37	108	73
1964	2.44	111	75
1965	2.51	114	77
1966	2.59	118	80
1967	2.72	124	84
1968	2.88	131	89
1969	3.06	139	94
1970	3.24	147	100
1971	3.44	156	106
1972	3.66	166	113
1973	3.89	177	120
1974	4.23	192	131
1975	4.65	211	144

Table 14.2. AVERAGE HOURLY EARNINGS INDEX FOR PRODUCTION LABOR CATEGORIES $(1965 = 100)$

Year	Electrical Equipment and Supplies	Transportation Equipment	Instruments and Related Products	Miscellaneous Manufacturing Industries
1960	089	086	091	091
1961	092	089	093	093
1962	093	091	093	092
1963	095	094	095	094
1964	098	097	097	097
1965	100	100	100	100
1966	102	104	103	103
1967	108	108	109	109
1968	113	114	113	116
1969	120	122	120	123
1970	129	128	128	132
1971	136	138	136	137
1972	142	146	142	144
1973	149	159	147	152
1974	162	171	161	163
1975	180	190	175	178

Table 14.3. AVERAGE HOURLY EARNINGS OF
PRODUCTION WORKERS BY LOCATION IN 1975

State	Avg. Hourly Earnings
Alabama	$4.15
California	5.23
Florida	4.10
Kentucky	4.63
Maryland	5.06
Massachusetts	4.47
Michigan	6.12
Nevada	5.28
New Mexico	3.69
New York	4.94
Pennsylvania	4.92
Texas	4.65
Washington	5.86
Wisconsin	5.19
Wyoming	5.39

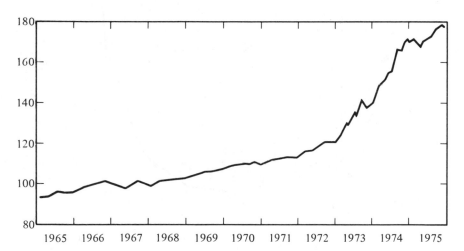

Figure 14.3. Wholesale price index (1967 = 100).

cent lighting with uniform illumination, mechanical ventilation, office air
conditioning, sprinklers for fire protection, toilet facilities, process services,
sanitary plumbing, and storm drainage. This index is an example of how the
costs for labor, materials, etc., can be combined into one composite index.

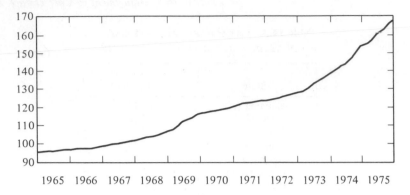

Figure 14.4. Consumer price index (1967 = 100).

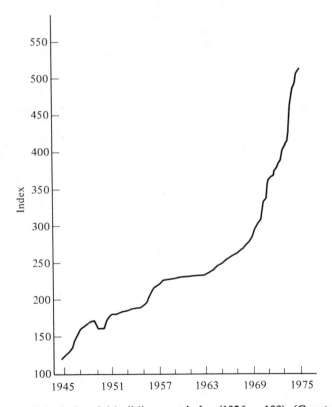

Figure 14.5. Industrial building cost index (1926 = 100). (Courtesy The Austin Company)

The adjustment of costs for yearly price increases based on indexes is not always easy. While the average labor rate may increase by 6% in a given year, the labor rate of a particular firm may be either more or less. Also, indexes may not be available for specific material items or for certain purchased parts. A third problem arises when expenditures are made over a number of years for a long duration project. In this latter case, the costs in early years will require less adjustment than those in later years.

Whenever price level changes are contemplated, one should consider the fact that increasing productivity tends to offset increased labor costs. The increase in productivity is not uniform from firm to firm or even within a specific company. However, it is a well-known fact of economic life that the upper limit on wage increases is set by the attainable productivity increase. Wage increases which exceed increases in productivity tend to depress profits and/or increase prices which, in turn, adds to inflation.

Adjustments because of learning. Learning takes place within an individual or within an organization as a function of the number of units produced. It is commonly accepted that the amount of time required to complete a given task or unit of product will be less each time the task is undertaken. The unit time will decrease at a decreasing rate and this time reduction will follow a predictable pattern.

The empirical evidence supporting the concept of learning was first noted in the aircraft industry. The reduction in direct labor man-hours required to build an aircraft was observed and found to be predictable. Since then the learning curve has found applications in other industries as a means for adjusting costs for items produced beyond the first one.

Most learning curves are based upon the assumption that the direct labor man-hours needed to complete a unit of product will decrease by a constant percentage each time the production quantity is doubled. A typical rate of improvement in the aircraft industry is 20% between doubled quantities. This establishes an 80% learning function and means that the direct labor man-hours needed to build the second aircraft will be 80% of the hours required to build the first aircraft. The fourth aircraft will require 80% of the man-hours that the second required, and the eighth aircraft will require 80% of the fourth, etc. This relationship is given in Table 14.4.

An analytical expression for the learning curve may be developed. Let

x = the unit number;
Y_x = the number of direct labor hours required to produce the xth unit;
K = the number of direct labor hours required to produce the first unit;
ϕ = the slope parameter of the learning curve.

Table 14.4. UNIT, CUMULATIVE, AND CUMULATIVE AVERAGE DIRECT LABOR
MAN-HOURS FOR AN 80% FUNCTION WITH UNIT 1 AT 100,000 HOURS

Unit Number	Unit Direct Labor Man-hours	Cumulative Direct Labor Man-hours	Cumulative Average Direct Labor Man-hours
1	100,000	100,000	100,000
2	80,000	180,000	90,000
4	64,000	314,210	78,553
8	51,200	534,591	66,824
16	40,960	892,014	55,751
32	32,768	1,467,862	45,871
64	26,214	2,392,453	37,382

From the assumption of a constant percentage reduction in direct labor
hours for doubled production units,

$$Y_x = K\phi^1 \quad \text{where} \quad x = 2^0 = 1;$$
$$Y_x = K\phi^2 \quad \text{where} \quad x = 2^1 = 2;$$
$$Y_x = K\phi^3 \quad \text{where} \quad x = 2^2 = 4;$$
$$Y_x = K\phi^4 \quad \text{where} \quad x = 2^3 = 8.$$

Therefore,

$$Y_x = K\phi^d$$

where

$$x = 2^d.$$

Taking the common logarithm gives

$$\log Y_x = \log K + d \log \phi$$

where

$$\log x = d \log 2.$$

Solving for d gives

$$d = \frac{\log Y_x - \log K}{\log \phi} \quad \text{and} \quad d = \frac{\log x}{\log 2}$$

from which

$$\frac{\log Y_x - \log K}{\log \phi} = \frac{\log x}{\log 2}$$

$$\log Y_x - \log K = \frac{\log x(\log \phi)}{\log 2}.$$

Let

$$n = \frac{\log \phi}{\log 2}.$$

Therefore,

$$\log Y_x - \log K = n \log x.$$

Taking the antilog of both sides gives

$$\frac{Y_x}{K} = x^n$$

$$Y_x = Kx^n.$$

Application of the above equation can be illustrated by reference to the example of an 80% progress function with unit 1 at 100 direct labor man-hours. Solving for Y_8, we see that the number of direct labor man-hours required to build the eighth unit gives

$$Y_8 = 100(8)^{\log 0.8/\log 2}$$
$$= 100(8)^{-0.322}$$
$$= \frac{100}{(8)^{0.322}}$$
$$= \frac{100}{1.9535} = 51.2.$$

The man-hour information from the learning curve can be extended to cost estimates for labor by multiplying by the labor rate which applies. In doing this the analyst must take into consideration that the subsequent units may be completed months or years after the initial units. Adjustments for labor rate increases might have to be made along with the adjustment for learning which is inherent in the application of the learning curve.

Adjustments for the degree of perfection. Figure 14.6 shows the approximate general relationship between tolerance in inches and the cost of production. According to Figure 14.6, the specification of a tolerance of 0.001 inch would increase the cost over a specification of 0.003 inches by

$$\frac{32 - 12}{12} = 1.67, \text{ or } 167\%.$$

Adjustments in production cost data can be made for future items that exhibit either more or less perfection from relationships like that exhibited in

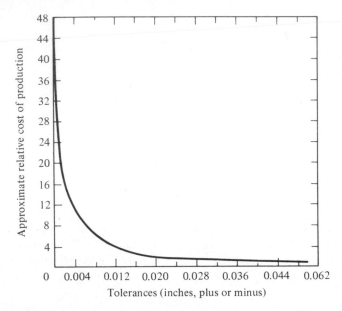

Figure 14.6. General cost relationship of various degrees of accuracy.

Figure 14.6. However, perfection is often a false ideal unless perfection is taken to mean that which is most appropriate.

To secure a desirable result even in the physical world, compromises from perfection must often be made. For example, increasing the compression ratio of an internal combustion engine, ideally, will increase its efficiency but at the same time necessitate use of fuel of higher antiknock characteristics. In order to produce a smooth-running engine, a compromise between an ideal compression ratio and characteristics of available fuel will have to be made. Cost estimating for an appropriate degree of perfection suggests that consideration be given to the determination of the degree of perfection that, if specified, will be beyond the capability of the process to which it applies.

14.4. Cost Estimating Relationships

Estimates of a result will usually be more accurate if they are based upon estimates of the factors having a bearing on the result than if the result is estimated directly. For example, in estimating the volume of a room it will usually prove more accurate to estimate the several dimensions of the room and calculate the volume than to estimate the volume directly. Similar reasoning applies to estimates of cost and other economic elements.

Statistical estimating methods often lead to mathematically fitted functions called *cost estimating relationships* (CERs). A cost estimating relationship as a functional model that mathematically describes the cost of a structure, system, or service as a function of one or more independent variables. Of course, there must be a logical or theoretical relationship of the variables to cost, a statistical significance of the contribution of the variables, and independence among the variables.

Determining item cost involving learning. As one example of the development of a cost estimating relationship consider the production of a lot of N units. The cumulative number of direct labor man-hours required to produce N units may be expressed as

$$T_N = Y_1 + Y_2 + \ldots + Y_N = \sum_{x=1}^{N} Y_x$$

where Y_x is the number of direct labor man-hours required to produce the xth unit. This was derived in the previous section as

$$Y_x = Kx^n.$$

An approximation for the cumulative direct labor man-hours is given by

$$T_N \cong \int_0^N Y_x \, dx$$

$$\cong \int_0^N Kx^n \, dx$$

$$\cong K\left(\frac{N^{n+1}}{n+1}\right).$$

Dividing by N gives an approximation for the cumulative average number of direct labor man-hours expressed as

$$V_N \cong \frac{1}{n+1}(KN^n).$$

Item cost per unit may be expressed in terms of the direct labor cost, the direct material cost, and the overhead cost. Let

LR = direct labor hourly rate;
DM = direct material cost per unit;
OH = overhead rate expressed as a decimal fraction of the direct labor hourly rate.

Item cost per unit, C_i, can be expressed as

$$C_i = V_N(LR) + DM + V_N(LR)(OH).$$

Or, by substituting for V_N

$$C_i = \frac{KN^n}{n+1}(LR)(1 + OH) + DM.$$

As an example application of this cost estimating relationship, consider a situation in which 8 units are to be produced. The direct labor hourly rate is \$7 per hour, the direct material cost per unit is \$600, and the overhead rate is 0.90. It is estimated that the first unit will require 100 direct labor man-hours and that an 80% learning curve is applicable. The item cost per unit is then estimated to be

$$C_i = \left[\frac{100(8)^{-0.322}}{-0.322 + 1}\right](\$7)(1.90) + \$600$$

$$= \left[\frac{51.2}{0.678}\right](\$7)(1.90) + \$600$$

$$= \$1,004 + \$600 = \$1,604.$$

This cost estimating relationship gives the cost per unit as a function of the direct labor cost, the direct material cost, and the overhead cost. It also adjusts the direct labor man-hours for learning in accordance with an estimated rate.

Power law and sizing model. Equipment cost estimating can often be accomplished by using the power law and sizing model. Equipment to which this cost estimating relationship applies must be similar in type and vary only in size. The economies of scale are in terms of size expressed in the relationship which is given as

$$C = C_r\left(\frac{Q_c}{Q_r}\right)^m$$

where

C_c = cost for design size Q_c;
C_r = known cost for reference size Q_r;
Q_c = design size;
Q_r = reference design size;
m = correlating exponent, $0 < m < 1$.

If $m = 1$, a linear relationship exists and the economies of scale do not apply. For most equipment m will be approximately 0.5, and for chemical processing equipment it is approximately 0.6.

As an example application of power law and sizing model, assume that a 200-gallon reactor with glass lining and jacket cost $9,500 in 1970. An estimate is required for a 300-gallon reactor to be purchased and installed in 1978. The price index in 1970 was 180 and is anticipated to be 210 in 1978. An estimate of the correlating exponent for this type of equipment is 0.50. Therefore, the estimated cost in 1970 dollars is

$$C_{1970} = \$9,500 \left(\frac{300}{200}\right)^{0.5} = \$11,635.$$

And the cost in 1978 dollars is

$$C_{1978} = \$11,635 \left(\frac{210}{180}\right) = \$13,574.$$

This cost estimating relationship relates a known cost to a future cost by means of an exponential relationship. Determination of the exponent is essential to the estimation process.

Other estimating relationships. The widespread use of estimating relationships speaks of their value in a wide range of situations. These relationships are available in the form of equations, sets of curves, nomograms, and tables.

In the aircraft and defense industry it is common to express the cost of airframe, power plant, avionics, and so forth as a function of weight, impulse, speed, etc. It is normally the next generation of aircraft or missile that is of interest. The estimated cost is determined by an extrapolation from known values for a sample group to an unknown proposed vehicle. This information is then used in cost-effectiveness studies and in budget requests.

Pollution control equipment such as sewage plants can be estimated from functions which relate the investment cost to the capacity in gallons per day. A municipality considering a plant with primary and secondary treatment capability may refer to estimating relationships based upon plants already built. The estimator then would make two adjustments. One would be for the difference in construction cost due to the passage of time since the relationship was derived. The other would be a correction for cost differences based on location.

Cost estimating relationships are probably most plentiful for estimating the cost of buildings. These relationships are normally based upon square feet and/or volume. They take into consideration costs due to labor differentials for location. An individual beginning discussions about the construction of a new home will almost always begin thinking in terms of the cost per square foot.

Engineering costs for complex projects are available as a function of the installed cost of the structure or system being designed. These costs are expressed as a percentage of the installed cost of such projects as office buildings and laboratories, power plants, water systems, and chemical plants. In each case the engineering cost is a decreasing percentage as the installed project cost increases.

14.5. Service Life Estimation

The cost of owning and operating a machine, system, or vehicle depends upon its anticipated service life. Much has been written about the service life of equipment. Compilations summarizing the depreciation of many types of equipment in many different situations are available.

Unfortunately, such data are only of limited value as a basis for estimating the service life of a particular item of equipment. For the most part, the information that is available consists of tables giving the average life of various types of structures, machines, and so forth. These have been prepared by people of various degrees of competence and ability and are largely based on judgments. One difficulty with such tables is that the conditions under which the facilities were used are not sufficiently described to enable application to be made to a particular situation.

When an asset depreciates through use, a prediction must be made of the extent to which it will be used. If depreciation is caused by the elements, the rate at which deterioration progresses must be established. Even more difficult are the predictions that seek to determine when a machine will become obsolete because of new inventions and new needs, or inadequate because of unanticipated demand.

When a number of units are used under similar conditions, mortality data can be useful in determining the probable life. Electronic components, telephone poles, light bulbs, railroad ties, and similar items fall into this category.

Mortality data for 30,000 wooden telephone poles treated with a certain preservative are given in Table 14.5. It is assumed that all the poles were installed at the same time. Thus, the age interval in Column 1 will also be the elapsed time since installation. From these data several useful curves may be drawn as shown in Figure 14.7.

Examination of the mortality frequency curve shows that some poles do not survive as long and some poles survive much longer than the average of

Table 14.5. COMPILATION OF MORTALITY DATA FOR TELEPHONE POLES*

Age interval, years (1)	Units retired during age interval, % (2)	Survivors at beginning of age interval, % (3)	Service during age interval, %-years (4)	Remaining service at beginning of age interval, %-years (5)	Expectancy at beginning of age interval, years (6)	Probable life at beginning of age interval, years (7)
0– ½	0.00	100.00	50.00	1,067.88	10.68	10.68
½– 1½	0.35	100.00	99.82	1,017.88	10.18	10.68
1½– 2½	0.74	99.65	99.28	918.06	9.21	10.71
2½– 3½	1.45	98.91	98.19	818.78	8.28	10.78
3½– 4½	3.19	97.46	95.86	720.59	7.39	10.89
4½– 5½	2.96	94.27	92.79	624.73	6.63	11.13
5½– 6½	5.68	91.31	88.47	531.94	5.83	11.33
6½– 7½	6.11	85.63	82.58	443.47	5.18	11.68
7½– 8½	7.28	79.52	75.88	360.89	4.54	12.04
8½– 9½	9.63	72.24	67.42	285.01	3.95	12.45
9½–10½	10.37	62.61	57.43	217.59	3.48	12.98
10½–11½	10.33	52.24	47.07	160.16	3.07	13.57
11½–12½	9.54	41.91	37.14	113.09	2.70	14.20
12½–13½	9.06	32.37	27.84	75.95	2.34	14.84
13½–14½	7.26	23.31	19.68	48.11	2.06	15.56
14½–15½	6.44	16.05	12.83	28.43	1.77	16.27
15½–16½	3.77	9.61	7.73	15.60	1.62	17.12
16½–17½	3.01	5.84	4.33	7.87	1.35	17.85
17½–18½	1.84	2.83	1.91	3.54	1.25	18.75
18½–19½	0.50	0.99	0.74	1.63	1.65	20.15
19½–20½	0.12	0.49	0.43	0.89	1.82	21.32
20½–21½	0.13	0.37	0.31	0.46	1.24	21.74
21½–22½	0.21	0.24	0.13	0.15	0.63	22.13
22½–23½	0.03	0.03	0.02	0.02	0.67	23.17
23½–24½	0.00	0.00	0.00	0.00	0.00	23.50
Total..	100.00	..	1,067.88	Avg. life = 1,067.88 ÷ 100 = 10.68 yrs.		

*From Marston, Anson, and Agg, *Engineering Valuation*, McGraw-Hill Book Co., 1936.

the group, but that the greatest rate of retirement centers around the average age. In order to maintain service, poles that are retired are immediately replaced. These replacements are subject to the same mortality frequency curve as the original group. Thus, after a few years there are renewals of original poles, renewals of renewals, renewals of renewals of renewals, and so forth. A summation of these renewals is given by the renewals curve in Figure 14.7.

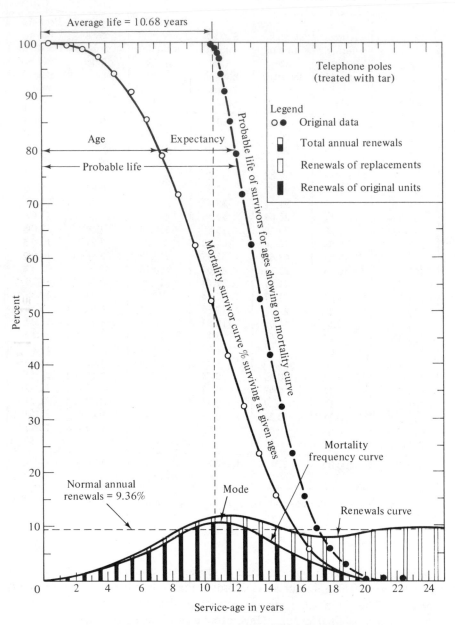

Figure 14.7. Mortality curves for 30,000 telephone poles.

The mortality survivor curve shows the number of original poles surviving at any service age. This curve is very similar to the survivorship curves for humans which are used in life insurance actuarial work.

The curve showing probable life of survivors is useful in estimating the life of those of the original groups that remain in use. Note that the probable life of the group at zero service age is equal to the average life of the group.

When depreciation is dependent upon wear and tear, the extent of use will be the determining factor in its service life. Experience with like or similar machines will have to be used to provide the needed service life estimate. Mortality curves as presented in this section are only applicable to certain classes of assets under certain conditions, and they cannot be applied to the problem of service life estimation in general.

14.6. Judgment in Estimating

The need for judgment in the estimating process is often discussed. Although this need is self-evident, one of the problems in the past has been too much reliance on judgment and too little on quantitative approaches.

People are involved in applying judgment. Because of this there is the problem of personal bias. A person's assigned role or position seems to influence his forecasts. Thus, a tendency toward low estimates appears among those whose interests are served by low estimates, principally proponents of a new structure, project, or product. Similarly, there are people whose interests are served by caution. Their estimates are likely to run higher than would be the case if they were not subject to pressures for caution.

The primary use of judgment should be to decide whether or not an estimating relationship can be used for an advanced system. Second, judgment must determine what adjustments will be necessary to take into account the effect of a technology that is not present in the sample. Judgment is also required to decide whether or not the results obtained from an estimating relationship are reasonable in comparison with the past cost of similar items. Any estimate that implies that new, higher performance equipment can be produced for less than the cost of existing units should be subject to critical examination.

When a proposed project contains considerable uncertainty, it is often wise to hold capital equipment investment to a minimum until outcomes become clearer, even though such a decision may result in higher maintenance and operation costs for the present system. Such action amounts to a decision to incur higher costs temporarily in order to reserve the privilege of making a second decision when the situation becomes clearer.

The accuracy of estimates with respect to events in the future is at least to some extent inversely proportional to the span of time between the estimate and the event. It is often appropriate to incur expense for the privilege of deferring the decision until better estimates can be made.

QUESTIONS AND PROBLEMS

1. Explain why engineering economy studies must rely heavily upon estimates.

2. Explain why an estimate of a result will probably be more accurate if it is based upon estimates of the factors that have a bearing on the result than if the result is estimated directly.

3. Contrast outputs expected from commercial activities with those from governmental activities.

4. List the common items of input that are utilized in the production of goods and services.

5. Briefly describe the advantages and disadvantages of the three cost estimating methods.

6. Use the data in Table 14.1 to compute a new set of indexes with 1965 as a base year.

7. Plot the data in Table 14.1 and estimate the average hourly earnings in 1980 by extrapolation.

8. If the total cost for manufacturing labor was $628,000 in 1962, estimate the total cost in 1975 using the information in Figure 14.2.

9. Estimate the annual rate of inflation in the wholesale price index from 1965 to 1975 using the information in Figure 14.3.

10. Estimate the annual rate of inflation in the consumer price index from 1965 through 1975 using the information in Figure 14.4.

11. Suppose that 15,000 man-hours are required to build a certain system for which the rate of learning for subsequent systems will be 15% on doubled quantities.
    (a) Calculate the labor man-hours to build the second, fourth, eighth, and the sixteenth systems, and plot these points.
    (b) Estimate the labor man-hours required for the tenth and twentieth units and compare this estimate to the result obtained from the formula.

12. Use the information in Problem 11 to compute the cumulative labor man-hours for a group of 16 systems by both the exact and the approximate methods.

13. Estimate the cumulative average number of direct labor man-hours for a product requiring 4 man-hours for the first unit if 16 units are to be produced with an improvement of 20% between doubled production quantities.

14. Estimate the item cost per unit for a production lot of 1,000 units if the first unit requires 8 labor man-hours, the 80% learning curve applies, the labor rate is $4.80 per hour, the direct material cost is $40 per unit, and the overhead rate is $1.60.

15. A large heat exchanger cost \$7,500 in 1967 and must be replaced soon with a larger unit. The present unit has an effective area of 250 feet and its replacement should have an area of 350 feet. Replacement is anticipated in 1980 when the price index is estimated to be 270 with 1967 as the base year. Estimate the cost of the unit in 1980 if the correlating coefficient for this type of equipment is 0.6.

16. Use the data in Table 14.5 to estimate the average annual equivalent cost of a telephone pole which has an installed cost of \$260 and no salvage value. The interest rate is 8% compounded continuously.

17. Describe a situation in which a person's role or position has influenced his estimate of an undertaking.

ESTIMATES AND
DECISION MAKING

The future outcome of a venture can be predicted more accurately if sufficient knowledge is available. Considerable emphasis is placed on securing appropriate data and using them carefully to arrive at estimates that are representative of future outcomes to the greatest possible extent. When all facts about alternative courses of action are known in accurate quantitative terms, the relative merit of each may be expressed in terms of a single decision number. Decision making in this case is simple.

Decisions must always be made in spite of the fact that quantitative considerations are based upon estimates that are subject to error. It must be remembered that the final calculated amounts embody the errors in the estimated quantities. Qualitative knowledge must be used to fill gaps in what is known about an undertaking. In this chapter several methods for recognizing and compensating for errors in estimates are presented.

15.1. An Example Decision Based Upon Estimates

An example will be used to illustrate some aspects of estimating, treating estimated data, and arriving at an economic decision. To simplify the discussion of these subjects, the example presented in this section will be used in the next two sections of this chapter. Income taxes will not be considered.

15

The purchase of a machine for producing a microcircuit, now performed in another manner, is considered likely to result in a saving. It is known with absolute certainty that the machine will cost $8,000 installed. All other factors pertinent to the decision are unknown and must be estimated.

Income estimate. From a study of available data and the result of judgment, it has been estimated that a total of 6,000 units of the circuit will be made during the next 6 years. The number to be made each year is not known; but, since it is believed that production will be fairly well distributed over the 6-year period, it is believed that the annual production should be taken as 1,000 units. A detailed consideration of materials used, time studies of the methods employed, wage rates, and the like, have resulted in an estimated saving of $5.50 per unit, exclusive of the costs incident to the operation of the machine if the machine is used. Combining the estimated production and the estimated unit saving results in an estimated saving (income) of 1,000 × $5.50 = $5,500 per year.

Capital recovery with return estimate. The machine is a single-purpose machine and no use is seen for it except in processing the product under consideration. Its service life has been taken to be 6 years to coincide with the estimated production period of 6 years. It is believed that the salvage value of the machine will be offset by the cost of removal at retirement. Thus, the estimated net receipts at retirement will be zero.

Interest is considered to be an expense in this evaluation, and the rate has been estimated at 8%. The next step is to combine the estimates of first cost, service life, and salvage value to determine the estimated annual capital recovery with a return. For the first cost of $8,000, a service life of 6 years, a salvage value of zero, and an interest rate of 8%, the resultant estimate of the annual cost of capital recovery and return will be:

$$(\$8,000 - \$0)(\overset{A/P\,8,\,6}{0.2163}) + \$0(0.08) = \$1,730.$$

Operation and maintenance costs estimate. Operation and maintenance costs will ordinarily be made up of several items such as power, supplies, spare parts, and labor. For simplicity, consider that the operation and maintenance expenses of the equipment in this example consist of four items. Each of these items is estimated on the basis of the number of units of product that it is estimated are to be processed per year as follows:

Power	$ 600
Maintenance labor	400
Operating and maintenance supplies	200
Operating labor	1,400
	$2,600

The estimated net annual profit for the project may be summarized as follows:

Estimated total annual income		$5,500
Estimated annual capital recovery with return	$1,730	
Estimated annual operation and maintenance cost	2,600	
Estimated total annual cost	$4,330	
Estimated net annual profit		$1,170

This final statement means that the project will result in an equivalent annual profit of $1,170 per year for a period of 6 years if the several estimates prove to be accurate.

The resultant equivalent annual profit is itself an estimate, and experience teaches that the most certain characteristic of estimates is that they nearly always prove to be inaccurate, sometimes in small degree and often in large degree. Once the best possible estimates have been made, however, whether they eventually prove to be good or bad, they remain the most objective basis upon which to base decision. It should be realized that decision making can never be an entirely objective process.

15.2. Allowances for Errors in Estimates

The better the estimate, the less allowance need be made for error. It should be realized at the outset that allowances for variances do not make up for a deficiency of knowledge in the sense that allowances correct errors. The allowances presented in this section are merely rule of thumb methods of eliminating some consequences of error.

As an example of the effect of an allowance in an economic undertaking, consider the following illustration. A contractor has estimated the cost of a project on which he has been asked to bid at $100,000. If he undertakes the job, he wishes to profit by 10% or $10,000. How shall he make allowance for errors in his cost estimate? If he makes an allowance of 10% for errors in his estimates, comparing to the very low factor of safety of 1.10, his estimated cost becomes $110,000. To allow for his profit margin, he will have to enter a bid of $121,000. But the higher his bid, the less the chance that he will be the successful bidder. If he is not the successful bidder, his allowance for errors in his estimate may have served to insure not only that he did not profit from the venture, but that he was left with a loss equal to the cost of making the bid. This illustration serves to emphasize the necessity for considering the cost of making allowances for errors in estimates.

Allowing for error in estimates by high interest rates. A policy common to many industrial concerns is to require that prospective undertakings be justified on the basis of a high minimum acceptable rate of return, say 30%. One basis for this practice is that there are so many opportunities which will result in a return of 30% or more that those yielding less can be ignored. But since this is a much greater return than most concerns make on the average, the high rate of return represents an allowance for error. It is hoped that if undertaking of ventures is limited to those that promise a high rate of return, none or few will be undertaken that will result in a loss.

Returning to the example of the machine given previously, suppose that the estimated income and the estimated cost of carrying on the venture when the interest rate is taken at 30% are as follows:

Estimated total annual income	$5,500
Estimated annual capital recovery with return, $8,000 \left(\overset{A/P\ 30,\ 6}{0.3784} \right)$	$3,027
Estimated annual operating and maintenance cost	$2,600
Estimated total annual cost	$5,627
Estimated net annual profit	−$127

If the calculated loss based on the high interest rate is the deciding factor, the venture will not be undertaken. Though the estimates as given above, except for the interest rate of 30%, might have been correct, the venture would have been rejected because of the arbitrary high interest rate taken, even though the resulting rate of return would be almost 30%

Suppose that the total operating costs had been estimated as above but that annual income had been estimated at $6,000. On the basis of a policy to accept ventures promising a return of 30% on investment, the venture would be accepted. But if it turned out that the annual income was, say, only $5,000, the venture would result in loss regardless of the calculated income with the high interest rate. In other words, an allowance for error embodied in a high rate of return does not prevent a loss that stems from incorrect estimates if a venture is undertaken that will result in loss.

Allowing for error in estimates by rapid payout. The effect of allowing for error in estimates by rapid payout is essentially the same as that of using high interest rates for the same purpose. Let it be assumed that a policy exists that equipment purchases must be based upon a 3-year payout period when the interest rate is taken at 8%.

Returning to the example of past paragraphs, suppose that the estimated income and the estimated cost of carrying on the venture when a 3-year payout period is taken is as follows:

Estimated annual income (estimated for 6 years but taken as 3 years to
  conform to policy) $5,500
Estimated annual capital recovery with return,
  $A/P\,8,3$
  $8,000 (0.3880)$ $3,104
Estimated annual operating and maintenance cost $2,600
Estimated total annual $5,704
Estimated net annual profit .... −$204

Under these conditions the venture would not have been undertaken. The effect of choosing conservative values for the components making up an estimate is to improve the certainty of a favorable result, if the outcome results in values that are more favorable than those chosen.

15.3. Considering a Range of Estimates

A plan for the treatment of estimates considered to have some merit is to make a least favorable estimate, a fair estimate, and a most favorable estimate of each situation. The *fair estimate* is the estimate that appears most reasonable to the estimator after a diligent search for and a careful analysis of data. This estimate might also be termed the most likely estimate.

The *least favorable estimate* is the estimate that results when each item of data is given the least favorable interpretation that the estimator feels may reasonably be realized. The least favorable estimate is definitely not the very worst that could happen. This is a difficult estimate to make. Each element of each item should be considered independently in so far as this is possible. The least favorable estimate should definitely not be determined from the fair estimate by multiplying the latter by a factor.

The *most favorable estimate* is the estimate that results when each item of data is given the most favorable interpretation that the estimator feels may reasonably be realized. Comments similar to those made in reference to the least favorable estimate, but of reverse effect, apply to the most favorable estimate. The use of the three estimates will be illustrated by application to the example of previous sections.

Items Estimated	Least Favorable Estimate	Fair Estimate	Most Favorable Estimate
Annual number of units	900	1,000	1,100
Savings per unit	$5.00	$5.50	$6.00
Annual saving	$4,500	$5,500	$6,600
Period of annual savings, n	4	6	8
Capital recovery with return, $8,000($ $A/P\,8,n$ $)$	$2,415	$1,730	$1,392
Operating cost and maintenance			
Item a	$700	$600	$500
Item b	500	400	300
Item c	300	200	150
Item d	$1,600	$1,400	$1,200
Estimated total of capital recovery, return and operating items	$5,515	$4,330	$3,542
Estimated net annual saving in prospect for n years............................	−$1,015	$1,170	$3,058

An important feature of the least favorable, fair, and most favorable estimate plan of comparison is that it provides for bringing additional information to bear upon the situation under consideration. Additional information results from the estimator's analysis and judgment in answering two questions relative to each item: "What is the least favorable value that this item may reasonably be expected to have?" and the reverse, "What is the most favorable value that this item may reasonably be expected to have?"

Judgment should be made item by item, for a summation of judgments can be expected to be more accurate than a single judgment of the whole.

A second advantage of the three-estimate plan is that it reveals the consequences of deviations from the fair or most likely estimate. Even though the calculated consequences are themselves estimated, they show what is in prospect for different sets of conditions. It will be found that the small deviations in the direction of unfavorableness may have disastrous consequences in some situations. In others even a considerable deviation may not result in serious consequences.

The results above can be put on other bases for comparison. The present-worth basis has merit in this instance because of the variation in the number of years embraced by the above three estimates. The estimated present-worth of savings is calculated as follows:

Items Estimated	Least Favorable Estimate	Fair Estimate	Most Favorable Estimate
Net annual saving	−$1,015	$1,170	$3,058
Period of annual savings, n	4	6	8
Present worth of saving(P/A 8, n)	−$3,337	$5,355	$17,356

Another viewpoint may be useful, for example, the net estimated savings per unit of product. This is computed as follows:

Items Estimated	Least Favorable Estimate	Fair Estimate	Most Favorable Estimate
Net annual saving	−$1,015	$1,170	$3,058
Annual number of units	900	1,000	1,100
Net saving per unit of product	−$1.13	$1.17	$2.78

There are many who feel that it is an aid to judgment to have several bases on which to compare a single situation. Since the cost of making extra calculations is usually insignificant in comparison with the worth of even a small improvement in decision, the practice should be followed by all who feel they benefit from the additional information. But there are limits beyond which further calculations can serve no useful purpose. This occurs when calculations are made that are beyond the scope of the data used. In the example above it would seem that no useful purpose would be served, for instance, by averaging the results calculated from the least favorable, fair, and most favorable estimates.

15.4. Sensitivity Analysis

The crude techniques of the previous sections are limited in their usefulness. Decision makers are typically interested in the full range of possible outcomes that would result from the variances in estimates. Sensitivity analysis permits a determination of how sensitive the final results are to changes in the values of the estimates.

To illustrate the technique of sensitivity analysis, the variation in the annual equivalent amount for three investment alternatives to changes in their gradient, the interest rate used, and their expected service lives will be examined. The annual equivalent profit expected from each of these alternatives is the basis for judging their economic desirability.

Alternative A requires an initial investment of $1,000 followed by receipts that are a decreasing gradient series as shown in Figure 15.1. The annual equivalent profit for Alternative A is expressed as

$$AE(i)_A = -\$1,000(\overset{A/P\,i,n}{\quad}) + \$1,000 - G(\overset{A/G\,i,n}{\quad}).$$

Alternative B produces $1,300 per year for n years from an initial investment of $4,000 as shown in Figure 15.1. This alternative's receipts are an equal payment series, and its annual equivalent profit is expressed as

$$AE(i)_B = -\$4,000(\overset{A/P\,i,n}{\quad}) + \$1,300.$$

Alternative C requires an investment of $5,000 which will produce revenues that are expected to increase uniformly over the investment's life as shown in Figure 15.1. The equivalent annual profit for this alternative is expressed as

$$AE(i)_C = -\$5,000(\overset{A/P\,i,n}{\quad}) + \$1,000 + G(\overset{A/G\,i,n}{\quad}).$$

Consider the situation in which it is expected that the lives for each of the three alternatives will be 10 years and that the appropriate interest rate is 15%. What are the effects on profit for these three alternatives when the values of G range from $0 to $200? From the expressions for the three alternatives' annual equivalent profit, it is observed that Alternative A and Alternative C are directly affected by changes in G, while Alternative B remains unaffected. These observations are clearly evident in Figure 15.2.

If, on the other hand, it is believed that the greatest variances will be associated with the estimated interest rate to be used, the changes in the annual equivalent profit as a function of interest rate can be investigated. Suppose

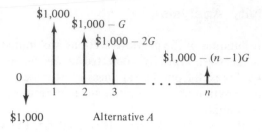

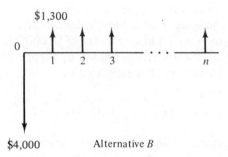

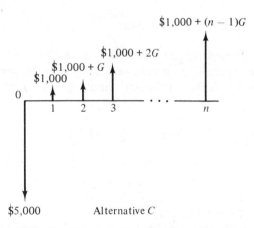

Figure 15.1. Cash flows for three investment alternatives.

that the gradient amount is $100 for Alternatives A and C and that the expected life for each of the three alternatives is 10 years. If these two parameters are held fixed, the changes in equivalent annual profit can be computed for interest rates ranging from 0% to 30%. The results of such an analysis are presented in Figure 15.3.

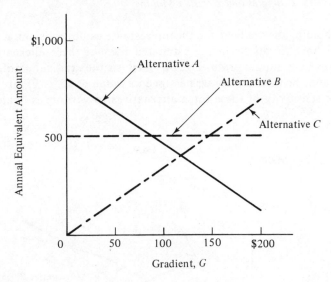

Figure 15.2. Sensitivity of the annual equivalent amount to the gradient.

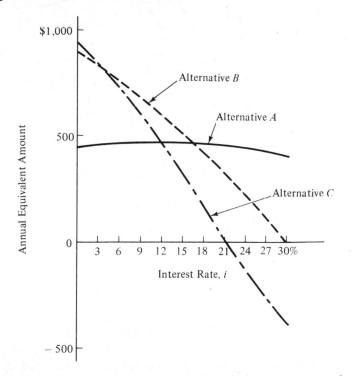

Figure 15.3. Sensitivity of the annual equivalent amount to the interest rate.

By holding the gradient and the interest rate values constant, it is possible to study how variations in the estimated lives of these alternatives affect their equivalent annual profit. For this analysis the gradient amount was set at $100 and the interest rate was assigned a value of 15%. Figure 15.4 indicates the sensitivity of these three alternatives to various estimates of their lives.

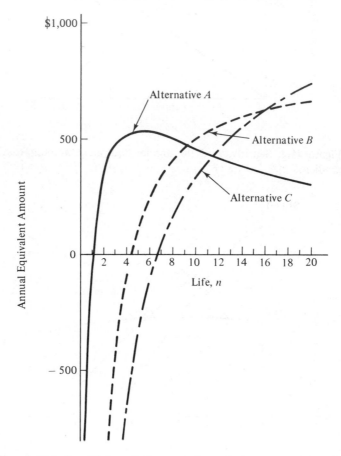

Figure 15.4. Sensitivity of the annual equivalent amount to the alternative's life.

An examination of Figure 15.2 reveals that the profitability of Alternative *A* and Alternative *C* is significantly affected by the gradient amount that is estimated. Obviously, Alternative *B*'s profitability is unaffected by such an estimate. It is also seen that for gradient amounts ranging from $0 to $200 all the alternatives will produce some positive contribution to profit. Thus, with regard to the estimate of the gradient amount, there seems to be little chance that financial losses will occur from any of these undertakings.

However, when comparing these three alternatives it is possible to determine for a particular estimate which alternative would be preferred. From Figure 15.2 it is seen that Alternative A is preferred to the others as long as the estimated gradient amount is less than $88, while Alternative B is preferred for the range of values from $88 to $147. For values estimated over $147, Alternative C is preferred. Use of this information provides insight regarding the sensitivity of these alternatives to the estimates of the gradient amount.

Similar interpretations can be made from Figure 15.3 and 15.4. For example, from Figure 15.3 it is seen that for the range of interest rates considered Alternative A will always be profitable. However, if the interest rate exceeds 22%, Alternative C will become unprofitable. Alternative B becomes unprofitable if the interest rate exceeds 30%. From Figure 15.3 it is possible to specify the range of interest rates for which one alternative is more profitable than the others. These ranges are as given in Table 15.1. Thus, if it is

Table 15.1. PREFERRED ALTERNATIVES FOR INTEREST RATE RANGES

Preferred Alternative	Range of Interest Rates
Alternative C	$0 \leq i \leq 4\%$
Alternative B	$4\% \leq i \leq 16\%$
Alternative A	$16\% \leq i \leq 30\%$

believed that the interest rate that should be utilized in this analysis would be in the vicinity of 10%, Alternative B is the obvious choice. Relatively large deviations from this 10% rate would not change the decision, and therefore the decision can be used with confidence.

The effect of various estimates of the alternatives' lives on profitability is presented in Figure 15.4. Here it is observed that Alternatives A, B, and C must have lives that exceed 2, 5, and 7 years, respectively, in order to assure a profit. In addition, the range of lives for which one alternative dominates the others can be easily determined. These ranges are given in Table 15.2.

Table 15.2. PREFERRED ALTERNATIVES FOR LIFE RANGES

Preferred Alternative	Range of Life
Alternative A	$0 \leq n \leq 9$
Alternative B	$9 \leq n \leq 15$
Alternative C	$15 \leq n \leq 20$

Unless Alternative *C* is expected to have a life that exceeds 15 years, it should not be adopted. Thus, if it is anticipated that these undertakings would span approximately 10 years, Alternative *B* is favored. If the actual time span exceeds the estimated life of 10 years by a substantial amount, Alternative *B* is still preferred. However, if the actual life realized is only slightly less than the 10 years estimate, Alternative *A* will be preferred to Alternative *B*. The impact of overestimating or underestimating the actual lives for these investments becomes evident. It is this type of information that can provide the decision maker with a better understanding of the effect of estimates of future outcomes.

QUESTIONS AND PROBLEMS

1. Relate the "factor of safety" concept in engineering design to allowances for estimating error in engineering economy studies.

2. Additional equipment must be purchased for use in a warehouse to meet an increased work load. It is estimated that the equipment would handle an additional 4,000 tons of goods per year for which the estimated income would be $7 per ton. The equipment will cost $20,000, its service life is estimated to be 10 years, and the estimated annual maintenance and operation costs are $15,000. If the interest rate is 8 %, compute the estimated net annual profit.

3. Would the equipment in Problem 2 be purchased if the additional work load per year is estimated to be 3,000 tons?

4. Would the equipment in Problem 3 be purchased if the minimum acceptable rate of return is 15 %?

5. Suppose that the following most favorable estimates apply to the equipment in Problem 2:

Additional tonnage handled	5,500
Income per ton	$ 7.50
Service life in years	12
Annual maintenance and operation costs......................	$12,000

What is the annual net profit for this most favorable estimate?

6. Compute the least favorable, fair, and most favorable estimates of the cost per mile for an automobile you would personally use from estimates of the least favorable, fair, and most favorable first costs, service life, salvage value, interest rate, maintenance and operating costs, and miles driven per year. Your estimates should be those which you might personally experience.

7. A special metal-working machine was purchased for $75,000. It was estimated that the machine would result in a saving on production cost of $10,500 per

year for 20 years. With a zero salvage value at the end of 20 years, what is the anticipated rate of return? It now appears that the machine will soon become inadequate. If the machine is sold after 6 years of use for $20,000, what is the change in the anticipated rate of return resulting from the incorrect service life estimate?

8. A machine design department is attempting to determine the annual equivalent cost which would be experienced by a machine currently being designed. It is estimated that it will cost $72,000 to design and build the machine. A service life of 7 years with a salvage value of $7,000 is anticipated, but these estimates are subject to error. A service life of 6 years with a salvage value of $8,000 and a service life of 8 years with a salvage value of $6,000 are also possible. If the interest rate is 10%, analyze the sensitivity of the annual equivalent cost to changes from the anticipated values.

9. Three investment alternatives are under consideration with the following economic parameters:

Investment Alternative	Initial Investment	Annual Income	Service Life	Salvage Value	Interest Rate
A	X	$700	n	x	i
B	Y	$2,000 - aG$	n	0	i
C	$3,000	$500 + bG$	n	y	i

where $a, b = 0, 1, 2, \ldots, n - 1$. Analyze the sensitivity of the present worth if
(a) G ranges from $50 to $200 with other parameters fixed at $X = \$2,000$, $Y = \$1,500$, $n = 10$, $x = \$200$, $y = \$300$, and $i = 10\%$.
(b) n ranges from 0 to 20 years with other parameters fixed as in (a) with $G = \$100$.
(c) i ranges from 0% to 25% with other parameters fixed as in (a) with $G = \$100$.

10. Analyze the sensitivity of the annual equivalent cost for the data given in Problem 9 if
(a) X and Y vary between $1,000 and $3,000 with $x = \$200$, $y = \$300$, $n = 10$, $i = 10\%$, and $G = \$100$.
(b) x and y vary between $0 and $300 with $X = \$2,000$, $Y = \$1,500$, $n = 10$, $i = 15\%$, and $G = \$100$.

DECISION MAKING
INVOLVING RISK

There is usually little assurance that predicted outcomes will coincide with actual outcomes. The economic elements upon which a course of action depends may vary from their estimated values because there is a multitude of chance causes. Not only are the estimates of future economic effects problematical, but in addition the anticipated future worth of most ventures is known only with a degree of assurance. It is this lack of certainty about the future that makes economic decision making one of the most challenging tasks faced by individuals, industry, and government.

This chapter begins with an introduction to probability theory as a quantitative tool for dealing with risk in decision making. The techniques for using probability concepts presented in this chapter assume that probabilities can be ascribed to future events. Decisions based on this assumption are classified as decisions under risk.

16.1. Basic Probability Theory

To formally incorporate uncertainty about future events into a logical decision process it is necessary to utilize probability theory. Probability theory consists of an extensive body of knowledge concerned with the quanti-

16

tative treatment of uncertainty. Since probability theory is well developed and rigorously defined it is appropriate to apply it to decision problems involving uncertainty.

By using probability theory it is possible to uniquely define events so that no ambiguities exist and so that each statement made within the theory is explicit and clearly understood. Probability theory allows uncertainty to be represented by a number so that the uncertainty of different events can be directly compared. In addition, the structure of probability theory prevents the introduction of extraneous notions without full knowledge of the decision maker.

Some probability notions. The probability that an event will occur may be expressed by a number that represents the likelihood of the occurrence. This likelihood may be determined by examining all available evidence related to the occurrence of the event. Thus, probability can be viewed as a state of mind since it represents belief about the likelihood of an event's occurring.

To illustrate, suppose you were presented with a normal looking coin and you were asked to guess the probability of the coin landing with heads up after it is tossed. Since you have never tossed this particular coin you might answer that the probability of heads occurring is approximately one-half. This answer is based upon all your past experience of tossing two-sided coins which has led you to believe that the probability of heads occurring is one-half. Now suppose you were allowed to toss the coin in question 100 times and heads occurred 60% of the time. Your belief that this coin will yield heads 50% of the time is somewhat shaken. If you tossed the coin one million times and heads occurred 70% of the time you would be convinced that this coin is not "fair" and that the likelihood of heads occurring is approximately 70%. Because your knowledge about the frequency of heads for this coin has increased, your feeling about the likelihood of getting a heads from a toss of this coin has been substantially modified. Thus, your information leads you to believe that the probability of a heads occurring when *this* coin is tossed is approximately 0.7.

Accordingly, any decision involving a wager on the basis of a toss of this coin will certainly be affected by your feeling about the likelihood of heads occurring. Since our beliefs about the occurrence of future events are the bases for making investment decisions, the concept of probability as a state of mind is quite useful in the analysis of economic alternatives.

The idea that probabilities can be subjective is disputed in some circles. However, because the usefulness of this concept for decision problems is well established, subjective probabilities have become an integral part of economic decision making. To take advantage of existing probability theory it will be assumed, with justification, that subjective probabilities obey the laws of classical probability theory.

Probability axioms. The three probability axioms are:

1. For any event A, the probability of A, $P(A) \geq 0$.
2. $P(S) = 1$, where S is the set of all possible events or outcomes under consideration.
3. If $AB = \varnothing$, then $P(A + B) = P(A) + P(B)$.

That is, for two mutually exclusive events the probability that event A or B occurs is equal to the sum of their probabilities.

The conditional probability of event A occurring given that event B has occurred is described as the probability of A given B, $P(A/B)$

$$P(A/B) = \frac{P(AB)}{P(B)}$$

where $P(AB)$ equals the probability of both event A and event B occurring.

When events A and B are independent,

$$P(AB) = P(A)P(B).$$

Thus, the probability of the event (A and B) equals the probability of event A multiplied by the probability of event B. When events A and B are independent, the conditional probability of event A occurring given that event B has occurred equals the probability that event A occurs expressed as

$$P(A/B) = \frac{P(AB)}{P(B)} = \frac{P(A)P(B)}{P(B)} = P(A).$$

Probability distribution functions. A *random variable* is a function which assigns a value to each event included in the set of all possible events. For instance, if a coin is to be tossed twice a random variable describing the number of heads occurring can have the values 0, 1, or 2. When a random variable is discrete as in the coin tossing example a *probability mass function* is used to describe the probability of the random variable being equal to a particular value.

If the probability of a coin showing heads is considered to be 0.5 the probability mass function for the random variable "number of heads" for two tosses of the coin is found in the following manner. The possible outcomes of the two tosses can be described by the events H_1, heads on first toss; H_2, heads on second toss; T_1, tails on first toss; and T_2, tails on second toss. The possible outcomes of tossing the coin twice are shown in Column A of Table 16.1. The probability of each outcome is shown in Column B of Table

Table 16.1. RANDOM VARIABLE "NUMBER OF HEADS" FOR TWO TOSSES OF A COIN

Possible Outcomes A	Probability of Occurrence B	Number of Heads C
$H_1 \ H_2$	$P(H_1)P(H_2) = (0.5)(0.5) = 0.25$	2
$H_1 \ T_2$	$P(H_1)P(T_2) = (0.5)(0.5) = 0.25$	1
$T_1 \ H_2$	$P(T_1)P(H_2) = (0.5)(0.5) = 0.25$	1
$T_1 \ T_2$	$P(T_1)P(T_2) = (0.5)(0.5) = 0.25$	0

16.1. These probabilities are easy to calculate since the outcome on the first toss is considered to be independent of the outcome on the second toss. Thus, the probabilities in Column B consist of the probability of the given event occurring on the first toss multiplied times the probability of the given event occurring on the second toss. The value of the random variable "number of heads" is shown in Column C.

The *probability mass function* can be directly described from the information in Table 16.1. The probability that the random variable "number of heads" has the value 2 is 0.25. The probability of having no heads occurring for two tosses is also 0.25. Since the events H_1T_2 and T_1H_2 are mutually exclusive it is necessary to sum the probabilities of those events to determine the probability that the random variable "number of heads" is equal to 1. Thus, the probability of getting one head on two tosses of the coin is 0.5. The probability mass function describing the probability of there being a particular number of heads after two tosses of a coin is shown in Figure 16.1. For any probability mass function the sum of the probabilities for all possible outcomes must total to one and the probability of each possible outcome must be greater than or equal to zero and less than or equal to one. Thus, for the random variable x

$$0 \le P(x) \le 1 \qquad \sum_x P(x) = 1$$

where $\sum_x$ indicates summing over all possible values of x.

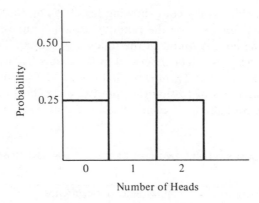

Figure 16.1. Probability mass function for the number of heads in the toss of two coins.

When a random variable is continuous a *probability density function* is used to relate the probability of an event to a value or range of values for the random variable. A probability density function for a random variable x is shown in Figure 16.2.

The probability of an event occurring is described by the area under the probability density function for those values of x included in the event. For example, in Figure 16.2 the dark area under the curve for $1 \le x \le 3$ represents the probability that the random variable x will occur within that range. In order to satisfy the axioms of probability a probability density function must have the following properties:

$$0 \le f(x) < \infty \quad \text{and} \quad \int_{-\infty}^{\infty} f(x)\, dx = 1.$$

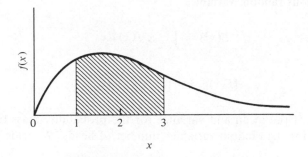

Figure 16.2. Probability density function.

The probability of an event in the range a to b is defined as

$$P(a \leq x \leq b) = \int_a^b f(x)\,dx.$$

Mean and variance. There are two parameters of a probability distribution function that help characterize the function. The first of these parameters is the *mean* or *expected value* of the random variable. For a discrete random variable x the expected value is defined as

$$E(x) = \sum_x xP(x).$$

For a continuous random variable x the expected value is defined as

$$E(x) = \int_{-\infty}^{\infty} xf(x)\,dx.$$

The mean of a probability distribution is a measure of central tendency or concentration of mass for the distribution.

The second parameter is a measure of the spread of the probability distribution and it is called the *variance*. The variance for any random variable x is defined as

$$\text{Var}\,(x) = E\{[x - E(x)]^2\} = E(x^2) - [E(x)]^2.$$

For discrete random variables

$$E(x^2) = \sum_x x^2 P(x)$$

and

$$[E(x)]^2 = \left[\sum_x xP(x)\right]^2.$$

For continuous random variables

$$E(x^2) = \int_{-\infty}^{\infty} x^2 f(x)\, dx$$

and

$$[E(x)]^2 = \left[\int_{-\infty}^{\infty} xf(x)\, dx\right]^2.$$

Calculation of the mean and variance for the probability mass function in Figure 16.1 for the random variable "number of heads" N yields

$$E(N) = \sum_{N} NP(N) = 0(0.25) + 1(0.50) + 2(0.25) = 1$$

$$\text{Var}\,(N) = E(N^2) - [E(N)]^2 = 0^2(0.25) + 1^2(0.50) + 2^2(0.25) - 1^2 = 0.5.$$

16.2. Expected Value Decision Making

If probability distributions are used to describe the economic elements that make up an investment alternative, the expected value of the cost or profit can provide a reasonable basis for comparing alternatives. The expected profit or cost of a proposal reflects the long-term profit or cost that would be realized if the investment were repeated a large number of times and if its probability distribution remained unchanged. Thus, when large numbers of investments are made it may be reasonable to make decisions based upon the average or long-term effects of each proposal. Of course, it is necessary to recognize the limitations of using the expected value as a basis for comparison on unique or unusual projects when the long-term effects are less meaningful.

To see how these ideas can be useful for decision making, consider the following game. A coin will be tossed twice. If no heads occur, $100 will be lost. If one head occurs, $40 will be won. If two heads occur, $80 will be won. What is the expected value of the random variable G, "profit from game"?

$$E(G) = -\$100(0.25) + \$40(0.50) + \$80(0.25) = \$15.$$

If one were able to participate in a large number of such games the expected or average winnings per bet over the long run would be approximately $15.

Because most industries and governments are generally long-lived the expected value as a basis for comparison seems to be a sensible method for evaluating investment alternatives under risk. The long-term objectives of such organizations may include the maximization of expected profits or minimization of expected costs. If it is desired to include the effect of the time value of money where risk is involved all that is required is to state expected profits or costs as expected present-worths, expected annual equivalents, or expected future-worths.

Designing against flood damage.   As a first example of decision making using expected values, suppose that a community has a water treatment facility located on the flood plain of a river. The construction of a levee to protect the facility during periods of flooding is under consideration. Data on the costs of construction and expected flood damages are shown in Table 16.2.

Table 16.2. PROBABILITY AND COST INFORMATION FOR DETERMINING OPTIMUM LEVEE SIZE

Feet (x) A	Number of years river maximum level was x feet above normal B	Probability of river being x feet above normal C	Loss if river level is x feet above levee D	Initial Cost building levee x feet high E
0	24	0.48	$ 0	$ 0
5	12	0.24	100,000	100,000
10	8	0.16	150,000	210,000
15	3	0.06	200,000	330,000
20	2	0.04	300,000	450,000
25	1	0.02	400,000	550,000
	$\overline{50}$	$\overline{1.00}$		

Using historical records which describe the maximum height reached by the river during each of the last 50 years, the frequencies shown in Column B of Table 16.2 were ascertained. From these frequencies are calculated the probabilities that the river will reach a particular level in any one year. The probability for each height is determined by dividing the number of years for which each particular height was the maximum by 50, the total number of years.

The damages that are expected if the river exceeds the height of the levee are related to the amount by which the river height exceeds the levee. These costs are shown in Column D. It is observed that they increase in relation to the amount the flood crest exceeds the levee height. If the flood crest is 15 feet and the levee is 10 feet the anticipated damages will be $100,000, whereas a flood crest of 20 feet for a levee 10 feet high would create damages of $150,000.

The estimated costs of constructing levees of various heights are shown in Column E. The company considers 12% to be their minimum attractive rate of return and it is felt that after 15 years the treatment plant will be re-located away from the floodplain. The company wants to select the alternative that minimizes its total expected costs. Since the probabilities are defined as the likelihood of a particular flood level in any one year, the expected equivalent annual costs is an appropriate choice for the basis for comparison.

An example of the calculations required for each levee height is demonstrated for two of the alternatives.

5-Foot Levee

Annual investment cost $= \$100,000(\overset{A/P\,12,\,15}{0.1468}) = \$14,682$

Expected annual damage $= (0.16)(\$100,000) + (0.06)(\$150,000)$
$+ (0.04)(\$200,000) + (0.02)(\$300,000) = \underline{39,000}$

Total expected annual cost $\qquad \$53,682$

10-Foot Levee

Annual investment cost $= \$210,000(\overset{A/P\,12,\,15}{0.1468}) = \$30,828$

Expected annual damage $= (0.06)(\$100,000) + (0.04)(\$150,000)$
$+ (0.02)(\$200,000) = \underline{16,000}$

Total expected annual cost $\qquad \$46,828$

The costs associated with the alternative levee heights are summarized in Table 16.3. The levee height that minimizes the total expected annual costs is

Table 16.3. SUMMARY OF ANNUAL CONSTRUCTION AND FLOOD DAMAGE COSTS

Levee Height (Feet)	Annual Investment Cost	Expected Annual Damage	Total Expected Annual Costs
0	$ 0	$80,000	$80,000
5	14,682	39,000	53,682
10	30,828	16,000	46,828
15	48,450	7,000	55,450
20	66,069	2,000	68,069
25	80,751	0	80,751

the levee which is 10 feet in height. The selection of a smaller levee would not provide enough protection to offset the reduced construction costs while a levee higher than 10 feet requires more investment without providing proportionate savings from expected flood damage. The use of expected value in determining the cost of flood damage is reasonable in this case since the 15-year period under consideration allows time for long-term effects to appear.

Introducing a new product. Suppose a firm is planning to introduce a new product which is similar to an existing product. After a detailed study by the marketing and production departments estimates are made of possible future cash flows as related to varying market conditions. The estimates shown in Table 16.4 indicate that this new product will produce cash flow *A*

Table 16.4. PROBABILITY OF OCCURRENCE FOR NEW PRODUCT PROPOSAL CASH FLOWS

	Probability of Occurrence for Cash Flow		
	A	*B*	*C*
Year	$P(A) = 0.1$	$P(B) = 0.3$	$P(C) = 0.6$
0	−$30,000	−$30,000	−$30,000
1	11,000	11,000	4,000
2	10,000	11,000	7,000
3	9,000	11,000	10,000
4	8,000	11,000	13,000

as demand decreases, cash flow *B* if demand remains constant, and cash flow *C* if demand increases. Based upon past experience and projections of future economic activity it is believed that the probabilities of decreasing, constant, and increasing demand will be 0.1, 0.3, and 0.6, respectively.

This firm generally uses the present-worth criterion to make such decisions and the minimum attractive rate of return is 10%. By calculating the present-worth for each level of demand the firm develops a probability mass function for the present-worth amount. Because this firm makes numerous decisions when risk is involved it uses the expected value of the present-worth amount to determine the acceptability of such ventures. From Table 16.4 the expected present-worth may be calculated as

$$E[PW(10)] = (0.1)PW(10)_A + (0.3)PW(10)_B + (0.6)PW(10)_C$$

$$= (0.1)\left[-\$30{,}000 + \$11{,}000(\overset{P/A\,10,\,4}{3.170}) \right.$$

$$\left. - \$1{,}000(\overset{A/G\,10,\,4}{1.3812})(\overset{P/A\,10,\,4}{3.170}) \right]$$

$$+ (0.3)\left[-\$30{,}000 + \$11{,}000(\overset{P/A\,10,\,4}{3.170}) \right]$$

$$+ (0.6)\left[-\$30{,}000 + \$4{,}000(\overset{P/A\,10,\,4}{3.170}) \right.$$

$$\left. + \$3{,}000(\overset{A/G\,10,\,4}{1.3812})(\overset{P/A\,10,\,4}{3.170}) \right]$$

$$= (0.1)(\$492) + (0.3)(\$4{,}870) + (0.6)(-\$4{,}185)$$

$$= -\$1{,}001.$$

The expected value for the present-worth of this proposal is $-\$1,001$ and therefore the venture is rejected. If the expected value had been positive, this new product would have been introduced.

16.3. Expectation-Variance in Decision Making

In many decision situations it is desirable not only to know the expected value of the basis for comparison but to also have a measure of the dispersion of its probability distribution. The variance of a probability distribution provides such a measure and its value in decision making is illustrated in the following example.

Suppose a firm has a set of four mutually exclusive alternatives from which one is to be selected. The probability mass functions describing the likelihood of occurrence of the present-worth amounts for each alternative are described in Table 16.5. For example, the probability that Alternative

Table 16.5. PROBABILITY DISTRIBUTIONS OF PRESENT-WORTH AMOUNTS FOR FOUR ALTERNATIVES

	Present-Worth					Expected Present-Worth	Variance (000,000)
Alternatives	$-\$40,000$ A	$\$10,000$ B	$\$60,000$ C	$\$110,000$ D	$\$160,000$ E	F	G
Alternative $A1$	0.2	0.2	0.2	0.2	0.2	$60,000	$5,000
Alternative $A2$	0.1	0.2	0.4	0.2	0.1	$60,000	$3,000
Alternative $A3$	0.0	0.4	0.3	0.2	0.1	$60,000	$2,500
Alternative $A4$	0.1	0.2	0.3	0.3	0.1	$65,000	$3,850

$A2$ will realize a cash flow that has a present-worth equal to $60,000 is 0.4 while the probability that it will have a $110,000 present-worth is 0.2. The expected present-worth and the variance of each probability distribution is shown in Table 16.5. These values are calculated by using the definitions in Section 16.1. For example, the expected present-worth for Alternative $A2$ is

$$E(PW_{A2}) = (0.1)(-\$40,000) + (0.2)(\$10,000) + (0.4)(\$60,000)$$
$$+ (0.2)(\$110,000) + (0.1)(\$160,000) = \$60,000$$

and the variance is

$$\text{Var}\ (PW_{A2}) = E(PW_{A2}^2) - [E(PW_{A2})]^2$$
$$= (0.1)(-\$40,000)^2 + (0.2)(\$10,000)^2 + (0.4)(\$60,000)^2$$
$$+ (0.2)(\$110,000)^2 + (0.1)(\$160,000)^2$$
$$- (\$60,000)^2$$
$$= \$3,000 \times 10^6.$$

An examination of the expected present-worth values in Table 16.5 indicates that there is relatively little difference among the alternatives. However, an examination of the distribution of possible present-worths for each alternative gives additional insight into the desirability of each alternative. First, it is important to determine the probability that the present-worth of each alternative will be less than zero. This probability represents the likelihood of the investment yielding a rate of return less than the minimum attractive rate of return. From Table 16.5, Column A

$$P(PW_{A1} \le 0) = 0.2$$
$$P(PW_{A2} \le 0) = 0.1$$
$$P(PW_{A3} \le 0) = 0.0$$
$$P(PW_{A4} \le 0) = 0.1$$

Since it is desirable to minimize these probabilities, it appears that on this basis Alternative $A3$ is most desirable.

The second important consideration is the variance of these probability distributions. Because the variance indicates the dispersion of the distribution it is usually desirable to try to minimize the variance since the smaller the variance the less the variability or uncertainty associated with the random variable. Alternative $A3$ has the minimum variance as shown in Table 16.5.

It is clear that Alternative $A3$ is more desirable than $A1$ or $A2$ since their expected present-worths are the same and $A3$ is preferred on the basis of the two criteria just discussed. However, the decision is not so obvious when comparing Alternatives $A3$ and $A4$. Alternative $A4$ has a larger expected present-worth but it is not as desirable as $A3$ on the basis of minimum variance and minimum chance of the present-worth being less than zero. In cases such as these the decision maker must weight the importance of each factor and decide if he would prefer more variability in the possible outcomes in order to achieve a higher expected value or less chance of the present-worth being negative. It is possible the relative importance of these three factors could be quantified and then a single basis for comparison could be developed for each alternative.

By having the additional information that can be derived from probability distributions it is likely that a more intelligent decision can be made. Of course, the astute decision maker must balance the economic trade-off between the cost of developing better information for decision making and the saving that he hopes to realize from better selection of alternative. Thus, it may not be economic to use elaborate techniques to consider small projects while on the other hand the use of more sophisticated analyses may provide substantial payoffs when very large expenditures are being considered.

16.4. Decision Trees in Decision Making

In many decision-making problems it is desirable to recognize that future decisions are affected by actions that are taken at the present. Too often decisions are made without consideration of their long-term effects. As a result decisions which initially appeared sound may place the decision maker in an unfavorable position with respect to future decisions. For decision problems where consideration of sequences of decisions is important and probabilities of future events are known, the use of *decision-flow diagrams* or *decision trees* for analysis is usually a very effective technique.

The application of decision trees to investment problems is illustrated by the analysis of the following problem.[1] Suppose a firm is planning to produce a product that has never been marketed previously. Because this product is somewhat different from the firm's existing products it will be necessary to construct a separate production facility to manufacture this new product.

Based on information supplied by the firms marketing group it is believed that the demand for this new product will be significant over the next 10 years. If the product is to be a good seller it is believed that over the next 10 years there will be a high demand for the product. If the product becomes a fad then it is anticipated that the demand will be high for the first 2 years followed by a low demand for the remaining 8 years. If the product is a poor seller then the demand is expected to be low for the next 10 years. Thus, the demand for this new product is expected to follow one of three demand patterns.

	Demand	Period*	Demand	Period	Designation
Good Seller:	High	1st;	High	2nd	(H_1, H_2)
Fad:	High	1st;	Low	2nd	(H_1, L_2)
Poor Seller:	Low	1st;	Low	2nd	(L_1, L_2)

*In this example the first 2 years are designated as the first period while the remaining 8 years are considered to be the second period.

---

[1] This problem suggested by an example in "Decision Trees for Decision Making" by J. F. Magee, *Harvard Business Review*, July-August, 1964.

The first alternative is to build a large plant that would suffice for the full 10 years of the product's life. The second alternative is to build a small plant and after 2 years of observing the product sales make a decision whether to expand the small plant. Which alternative to select is the decision problem under consideration.

Structuring the decision tree. The use of a decision tree to display the alternatives and the possible chance events (demand) that can occur is illustrated in Figure 16.3. Beginning at the left and moving to the right, the decision tree spans 10 years, the life of the product. The nodes of the tree from which the tree's branches emanate are either decision nodes, □, or chance nodes, ○. The branches that emanate from a decision node represent alternative courses of actions about which the decision maker must make a choice. On the other hand, the branches leaving the chance nodes represent chance events which represent outcomes of Nature. The occurrence of a chance event can be considered to be a random variable over which the decision maker has no control. Chance events are usually controlled by exogenous forces such as weather, sun spots, the marketplace, etc. In our example, the chance events represent the demand for the product over the next 10 years.

Starting at the left, the first node represents the choice of building a big plant or a small plant. If a big plant is built, then the next possible events are the chance events which represent the possiblity of experiencing a high demand or a low demand for the first 2 years. If a big plant is built and a high demand is experienced for the first 2 years, the next chance node represents the possibility of a high demand occurring or low demand occurring for the remaining 8 years. Thus, one sequence of branches beginning at the left and ending at the right represents one of the possible sequence of events that can result from actions by the decision maker and Nature.

Notice that if the small plant is built the decision about whether to expand or not to expand is made after 2 years only if there is a high demand. If demand is low in the first 2 years it is known that demand will be low for the remaining 8 years. Therefore, the decision not to expand is obvious and no decision node is required following low demand in the first 2 years.

Once the structure of the decision tree is determined the next task is to ascertain the costs and revenues that are associated with each of the decision alternatives and the possible chance outcomes. These amounts should then be written on the appropriate branches of the tree. For our example the costs of plant construction and the net profits expected from sale of the product for the various market conditions and plant sizes are shown in Figure 16.4. Thus, the investment required to build the big plant is seen to be $4,000,000 while the net profit for the first 2 years is $861,000 per year if the big plant is built. If the big plant is built and there is high demand in the first 2 year

Decision point 1   Decision point 2

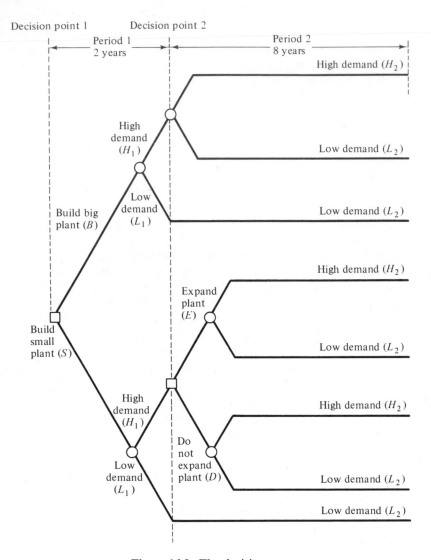

Figure 16.3. The decision tree.

period, the net profit is $1,650,000 per year for the remaining 8 years, if there is high demand in Period 2.

The cost to build the small plant is $2,600,000 and its expansion cost is $3,173,000 if the expansion is undertaken. All the costs are shown as negative values on the decision tree while the revenues are positive amounts. The amounts that are shown on a per year basis represent the annual net profit received during the 2 years of Period 1 or the 8 years of Period 2.

Since the costs and revenues occur at different points in time over the

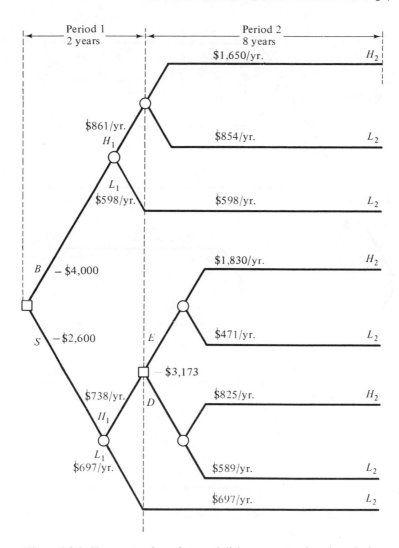

Figure 16.4. Forecasts of receipts and disbursements placed on decision tree (000s).

10-year study period it is appropriate to convert the various amounts on the tree's branches to their equivalent amounts. For the problem being considered the MARR is 15% and Figure 16.5 shows the receipts and disbursements on the branches transformed to their present-worth equivalents.

To illustrate, if the big plant is built, the present-worth of the net revenues that result from a high demand for the first 2 years is

$$PW(15) = \$861{,}000(\overset{P/A\ 15,\ 2}{1.626}) = \$1{,}400{,}000.$$

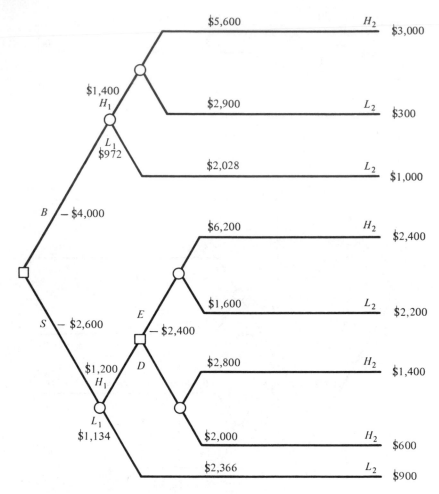

Figure 16.5. Present-worth amounts of receipts and disbursements and for each outcome (000s).

If the big plant is built and high demand is experienced for the remaining 8 years the present-worth of those revenues is

$$PW(15) = \$1,650,000(\overset{P/A\,15,\,8}{4.487})(\overset{P/F\,15,\,2}{0.7562}) = \$5,600,000.$$

Now that the present-worths of costs and revenues have been determined it is possible to sum these figures along each possible sequence of branches, starting each time at the left-most node. Thus, if the big plant is built, the demand is high in Period 1 and Period 2, the total present-worth of that sequence of events is $-\$4,000,000 + \$1,400,000 + \$5,600,000 = \$3,000,000$. This amount is placed at the tip of the right-most branch representing such an outcome. This procedure is repeated for each possible sequence of branches and the resulting amounts are shown in Figure 16.7. If the small plant is built, demand is high for Period 1, the plant is expanded and demand is low in Period 2 the net present-worth of such a sequence of events is

$$-\$2,600,000 + \$1,200,000 - \$2,400,000 + \$1,600,000 = -\$2,200,000.$$

Nature's tree. Once the cost and revenue information is in the form shown in Figure 16.6, it is necessary to place the probabilities of the chance events occurring at the chance nodes on the decision tree. Suppose that the probabilities of the product being a good seller, a fad, or a poor seller are estimated by the marketing department. These estimated probabilities are

$$P(\text{good seller}) = P(H_1 H_2) = \tfrac{2}{5}$$
$$P(\text{fad}) \qquad\quad = P(H_1 L_2) = \tfrac{1}{5}$$
$$P(\text{poor seller}) = P(L_1 L_2) = \tfrac{2}{5}.$$

Whenever it is necessary to compute the probabilities for a decision tree it is usually very helpful to first construct Nature's tree, that is, construct a tree that indicates Nature's options as shown in Figure 16.6. In Nature's tree all the nodes are chance nodes. At the beginning of each branch is placed

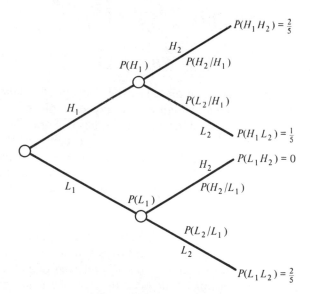

Figure 16.6. Nature's tree.

the probability of following that branch given that you are at the node preceding that branch. The probability that branch H_2 is selected given that there has been high demand in Period 1 (H_1) is the conditional probability $P[H_2 \mid H_1]$ and it is placed at the beginning of the H_2 branch that radiates from the H_1 branch. At each tip of Nature's tree is placed the probability that the sequence of events represented by that tip will occur. These probabilities are calculated by multiplying the probabilities on all the branches that lead from the initial node on Nature's tree to each of the tips. Thus, to find the probability that H_1 and L_2 occur, the probability on branch H_1 is multiplied times the probability on branch L_2 following H_1. This calculation gives $P(H_1L_2) = P(H_1)P(L_2 \mid H_1)$.

The probabilites that are required for the decision tree are $P(H_1), P(L_1),$ $P(H_2 \mid H_1)$, and $P(L_2 \mid H_1)$. By having Nature's tree and the probabilities for the tips of the tree, all that is required to find $P(H_1)$ is to *add* each of the probabilities at the tips that contains an event H_1. Thus,

$$P(H_1) = P(H_1H_2) + P(H_1L_2) = \tfrac{2}{5} + \tfrac{1}{5} = \tfrac{3}{5}.$$

Similarly,

$$P(L_1) = P(L_1H_2) + P(L_1L_2) = 0 + \tfrac{2}{5} = \tfrac{2}{5}.$$

With the $P(H_1)$ now known the conditional probabilities are calculated in the following manner:

$$P(H_2 \mid H_1) = \frac{P(H_1H_2)}{P(H_1)} = \left(\frac{\frac{2}{5}}{\frac{3}{5}}\right) = \frac{2}{3}$$

$$P(L_2 \mid H_1) = \frac{P(H_1L_2)}{P(H_1)} = \left(\frac{\frac{1}{5}}{\frac{3}{5}}\right) = \frac{1}{3}.$$

These probabilities are then placed at the appropriate chance nodes on the decision tree as is shown in Figure 16.7. For example, the chance node following the decision to expand the small plant has two branches H_2 and L_2. The probability of taking the H_2 branch is the probability that H_2 occurs *given* that H_1 has already occurred. Therefore, $P(H_2 \mid H_1) = \tfrac{2}{3}$ is placed on the H_2 branch and $P(L_2 \mid H_1) = \tfrac{1}{3}$ is placed on the L_2 branch. For the case where the demand is low in Period 1 the probability $P(L_1) = P(L_1L_2) = \tfrac{2}{3}$ since $P(L_2 \mid L_1) = 1$ and $P(H_2 \mid L_1) = 0$.

The "rollback" procedure. At this point it is now possible to solve the decision tree in order to see which alternative should be undertaken. The solution technique is relatively simple and it is referred to as the "rollback" procedure. Starting at the tips of the decision tree's branches and working back toward the initial node of the tree the following two rules are used:

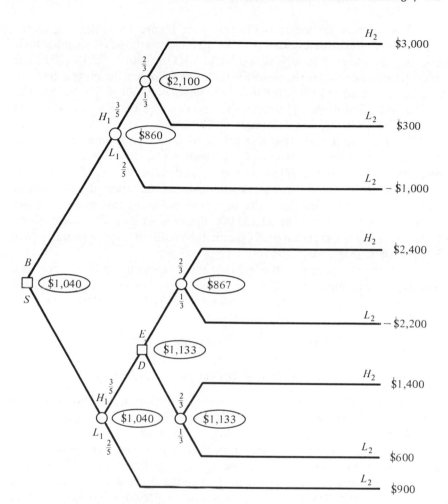

Figure 16.7. Rollback solution of decision tree (000s).

1. If the node is a chance node calculate the *expected* value of that node based on the "rolled backed" values on the adjacent nodes to the right of the node being considered.
2. If the node is a decision node *select* the maximum profit or minimum cost from the adjacent nodes to the right of the node being considered.

As each node is considered starting at the right of the decision tree the values calculated for Rules 1 and 2 should be placed just to the right of the node and circled. By working backward through the decision tree certain alternatives can be eliminated from further consideration and thus the "rollback" procedure is efficient for large decision trees.

The rollback technique is illustrated in Figure 16.7. Starting at the top of the decision tree it is seen that the first node to the left is a chance node and the expected value is $\frac{2}{3}(\$3,000,000) + \frac{1}{3}(\$300,000) = \$2,100,000$. The next chance node that is right-most is the node following the plant expansion branch. The expected value for this node is $\frac{2}{3}(2,400,000) + \frac{1}{3}(-\$2,200,000) = \$867,000$. The other right-most node is a chance node and its expected value is $\frac{2}{3}(\$1,400,000) + \frac{1}{3}(\$600,000) = \$1,133,000$.

The right-most node that has yet to be evaluated is now the decision node concerning the expansion of the small plant. Based on the previous calculations it is seen that if the plant is expanded the expected future profits will be $867,000 while if the plant is not expanded the expected future profits will be $1,133,000. Therefore, the optimum policy at this node is to not expand the small plant. The $1,133,000 figure is written after the decision node indicating the expected profit that will be realized if the optimum policy is followed at that node.

Now the next nodes to the left that must be evaluated by the rollback procedure are chance nodes. The expected values at those nodes are calculated from the values on the adjacent nodes to the right that were previously calculated. These two expected values are

$$\frac{3}{5}(\$2,100,000) + \frac{2}{5}(-\$1,000,000) = \$860,000$$

and

$$\frac{3}{5}(\$1,133,000) + \frac{2}{5}(\$900,000) = \$1,040,000.$$

The last node to be evaluated is the initial decision node. If the big plant is built the expected profits will be $860,000 while if the small plant is built the expected profit is $1,040,000. (This expected profit figure assumes the optimum policy is followed in the future). Therefore, the maximum figure is selected for the decision node and the optimum policy is to build the small plant. If the demand is high in Period 1, then the decision not to expand the plant should be followed. If demand is low in Period 1, no expansion of the small plant is the only course of action. If this policy is followed then the expected profit from this venture is $1,040,000. Since the present-worth amount is positive, it assures the firm an expected return of better than 15% in addition to being the best of the alternatives under consideration.

The expected value of perfect information. With the information that is available to the firm in our example it is seen that by following an optimal policy the expected present-worth of the profit is $1,040,000. If the firm could obtain additional information about the occurrence of future demand levels it may be that the profit figure could be improved. Usually, such additional information costs money since the firm must allocate its resources for further research or it must purchase the information from sources external to the firm (i.e., market research consultants).

It is foolish to pay for additional information if it will not provide additional profits that exceed the cost of the information. In order to evaluate the possible benefits that can be derived from additional information it is necessary to calculate the *expected value of perfect information* (EVPI).

Suppose there is a market research group that does have perfect information about the future demand levels for the product being considered in this example. What would be the most money that the firm would be willing to pay for this additional information? Assume the firm knew with certainty that there would be high demand in both periods (H_1, H_2). Then the optimum policy to follow is to build the big plant since that payoff ($3,000,000) exceeds the payoffs possible from building the small plant and expanding ($2,400,000) or builing the small plant and not expanding ($1,400,000). If the demand were known to be (H_1, L_2) then the best strategy is to build the small plant and not expand since that payoff ($600,000) exceeds the returns received if the big plant is built ($300,000) or if the small plant is constructed and then expanded (−$2,200,000). Thus, for each possible state of nature it is necessary to determine the strategy that will maximize the payoff. For our example the three states of nature are (H_1, H_2), (H_1, L_2), and (L_1, L_2) and the maximum payoffs for each state of nature are indicated in Table 16.6.

Table 16.6. MAXIMUM PAYOFF

State of Nature	Maximum Payoff
Good seller (H_1, H_2)	$3,000
Fad seller (H_1, L_2)	$ 600
Poor seller (L_1, L_2)	$ 900

Before receiving the perfect information from the market research group the firm can calculate the *expected profit with perfect information* (EPPI). This is accomplished by summing for each possible state of nature the probability that a particular state will occur multiplied by the maximum payoff achievable for that state of nature. These calculations are shown in Table 16.7 and it is seen that the expected profit for perfect information is $1,680,000. Since the firm can achieve expected profits of $1,040,000 without the additional information, the expected value of perfect information is $1,680,000 − $1,040,000 = $640,000. The firm should never pay over $640,000 for additional information even if that information predicts the future with certainty. Thus, the EVPI gives an upper limit reflecting the firm's expected improvement in profits if a source of perfect information were available. Information that is less than perfect can only provide a profit improvement that is less than the expected value of perfect information.

Table 16.7. CALCULATING THE EXPECTED PROFIT FOR PERFECT INFORMATION

State of Nature	Best strategy	Maximum payoff	Probability the state of nature occurs	Expected payoff for each state
A	B	C	D	$C \times D$
H_1, H_2	Big plant	\$3,000,000	$(3/5)(2/3) = 2/5$	\$1,200,000
H_1, L_2	Small plant; no expansion	600,000	$(3/5)(1/3) = 1/5$	120,000
L_1, L_2	Small plant with expansion	900,000	$(2/5)(1) = 2/5$	360,000
Expected profit with perfect information (EPPI) = \$1,680,000					

16.5. Monte Carlo Methods

Monte Carlo is the name given to a class of simulation approaches to decision making in which probability distributions describe certain system parameters. In many of these cases, an analytical solution is not possible because of the way in which the probabilities must be manipulated. In other cases, the Monte Carlo approach is preferred because of the level of detail that it exhibits.

Decision situations to which Monte Carlo methods may be applied are characterized by empirical or theoretical distributions. The Monte Carlo approach utilizes these distributions to generate random outcomes. These outcomes are then combined in accordance with the economic analysis technique being applied to find the distribution of the present-worth, the annual equivalent cost distribution, etc.

It is necessary to generate values at random from the distributions representing system parameters. There are many ways of doing this, including mechanical, mathematical, digital computer, etc. In this section, a simple example based upon the mechanical approach for discrete distributions is presented.

A defense contractor wishes to enter a bid on a defense project that requires special instrumentation. Two alternatives are being considered. The first has a high first cost and low operation and maintenance (O&M) costs, but the second has a low first cost and high operation and maintenance costs. Although the first cost of each instrumentation alternative is known with certainty, the annual operation and maintenance costs are uncertain.

Table 16.8 gives the probabilities associated with the operation and maintenance costs for each instrumentation alternative. Alternative *A* has a first cost of \$70,000 because of its high degree of automation. Alternative *B*

Table 16.8. PROBABILITIES FOR OPERATION AND
MAINTENANCE COST

Alternative *A*		Alternative *B*	
O&M Cost	Probability	O&M Cost	Probability
$2,000	1/6	$12,000	1/6
$3,000	1/2	$25,000	1/3
$5,000	1/3	$30,000	1/3
		$40,000	1/6

requires considerable operator attention and has a first cost of only $20,000.
Neither instrumentation alternative will have any salvage value at the end of
the contract period. The interest rate is 10%.

The contract duration is uncertain and is estimated to be either 1 year,
2 years, or 3 years. Table 16.9 gives the probability distribution describing

Table 16.9. CONTRACT DURATION
PROBABILITIES

Contract Duration in Years, *n*	Probability the Duration is *n*
1	0.25
2	0.50
3	0.25

this uncertainty. It will be recognized as the same distribution as was given
in Figure 16.1. Since this is the case, values can be generated for contract
duration by tossing two coins.

The method for generating a sequence of simulated contract durations is
mechanical in nature. This method can also be used to generate simulated
operation and maintenance costs for each alternative. For Alternative *A*
one die can be tossed with certain outcomes used to represent certain cost
occurrences. This same approach can be used for Alternative *B*. Table 16.10
summarizes the three mechanical means for generating the data pertinent to
the instrumentation choice described in this example.

The process for generating simulated contract durations and operation
and maintenance costs for the alternatives presented in Table 16.10 can now
be applied with the techniques of engineering economy over several trials.
This is shown in Table 16.11 for 100 trials with a summary at 10 trials. After
10 trials the annual equivalent cost for Alternative *A* is $48,630. For Alterna-
tive *B* the annual equivalent cost is $41,623 after 10 trials. The annual equiva-

Table 16.10. GENERATION OF SIMULATED VALUES

Simulated Value	Possible Outcomes	Probability of Outcome	Simulation Technique	Assignment of Outcome
Contract Duration, *n*	1	1/4	Tossing Two Coins	*HH*
	2	1/2		*HT* or *TH*
	3	1/4		*TT*
O&M Costs for Alternative *A*	$ 2,000	1/6	Tossing One Die	1
	$ 3,000	1/2		2, 3, or 4
	$ 5,000	1/3		5 or 6
O&M Costs for Alternative *B*	$12,000	1/6	Tossing One Die	1
	$25,000	1/3		2 or 3
	$30,000	1/3		4 or 5
	$40,000	1/6		6

lent cost difference is $7,007 as shown in the last column of Table 16.11.

Although Alternative *B* appears to be best, it must be remembered that only 10 trials lead to this conclusion. It is entirely possible for a larger sample to yield different results. Consider the behavior of the mean annual equivalent cost for Alternative *A* as the number of trials increases to 100 as shown in Figure 16.8. Note how the average annual equivalent cost fluctuates early

Table 16.11. MONTE CARLO COMPARISON OF TWO PLANS

Trial Number	Outcome of Coin Tossing	Contract Duration *n*	(a) Capital Recovery Cost, Plan *A*; $70,000($A/P$ 10, n$)	(b) Capital Recovery Cost, Plan *B*; $20,000($A/P$ 10, n$)	Outcome of Die Toss for Plan *A*
1	*HH*	1	$77,000	$22,000	2
2	*HT*	2	40,334	11,524	4
3	*TH*	2	40,334	11,524	1
4	*TH*	2	40,334	11,524	5
5	*TT*	3	28,147	8,042	2
6	*HII*	1	77,000	22,000	3
7	*TH*	2	40,334	11,524	6
8	*HT*	2	40,334	11,524	1
9	*TT*	3	28,147	8,042	3
10	*TH*	2	40,334	11,524	6
.	.	.	.	.	.
.	.	.	.	.	.
.	.	.	.	.	.
100	*HT*	2	40,334	11,524	2

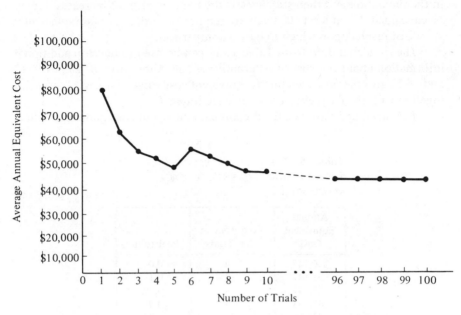

Figure 16.8. Convergence of the annual equivalent cost with increasing trials.

Table 16.11. (CONTINUED)

(c) Annual O&M Costs for Plan *A*	(d) = (a) + (c) Annual Equiva- lent Cost for Plan *A*	Outcome of Die Toss for Plan *B*	(e) Annual O&M Costs for Plan *B*	(f) = (b) + (e) Annual Equiva- lent Cost for Plan *B*	(d) − (f) Difference in Annual Equiva- lent Cost Plan *A*-Plan *B*
$3,000	$80,000	3	$25,000	$47,000	$33,000
3,000	43,334	3	25,000	36,524	6,810
2,000	42,334	6	40,000	51,524	−9,190
5,000	45,334	2	25,000	36,524	8,810
3,000	31,147	1	12,000	20,042	11,105
3,000	80,000	5	30,000	52,000	28,000
5,000	45,334	4	30,000	41,524	3,810
2,000	42,334	5	30,000	41,524	810
3,000	31,147	6	40,000	48,042	−16,895
5,000	45,334	4	30,000	41,524	3,810
.	486,300	.	.	$416,228	$70,070
.	Mean AEC = 48,630	.	.	Mean AEC = $41,623	Mean Diff. = $7,007
.	.	.	.	.	.
3,000	43,334	4	30,000	41,524	1,810

in the simulation and then stabilizes as the number of trials increases. It may be concluded that at least 100 trials are required in order to obtain sufficiently stabilized results upon which to base a comparison.

The simulated data from Table 16.11 can be used to develop additional information about the cost of Alternative *A* and Alternative *B*. Tables 16.12 and 16.13 give the possible annual equivalent cost values for each alternative together with the frequency of occurrence for each.

The mean and variance for the annual equivalent cost under each alter-

Table 16.12. FREQUENCY DISTRIBUTION OF THE ANNUAL EQUIVALENT COST FOR ALTERNATIVE A

Annual Equivalent Cost	Frequency in 100 Trials	Probability
$30,147	4	0.04
31,147	13	0.13
33,147	9	0.09
42,334	8	0.08
43,334	22	0.22
45,334	18	0.18
79,000	5	0.05
80,000	13	0.13
82,000	8	0.08
	100	1.00

Table 16.13. FREQUENCY DISTRIBUTION OF THE ANNUAL EQUIVALENT COST FOR ALTERNATIVE B

Annual Equivalent Cost	Frequency in 100 Trials	Probability
$20,042	4	0.04
23,524	9	0.09
33,042	9	0.09
34,000	4	0.04
36,524	17	0.17
38,042	8	0.08
41,524	17	0.17
47,000	7	0.07
48,042	4	0.04
51,524	8	0.08
52,000	8	0.08
62,000	5	0.05
	100	1.00

native can be estimated from the frequencies given in Tables 16.12 and 16.13. For Alternative A the mean is

$$0.04(\$30,147) + 0.13(\$31,147) + \ldots + 0.08(\$82,000) = \$50,229.$$

And the variance is

$$0.04(\$30,147)^2 + 0.13(\$31,147)^2 + \ldots +$$
$$0.08(\$82,000)^2 - (\$50,229)^2 = \$346,802,050.$$

For Alternative B the mean is

$$0.04(\$20,042) + 0.09(\$23,524) + \ldots +$$
$$0.05(\$62,000) = \$40,158.$$

And the variance is

$$0.04(\$20,042)^2 + 0.13(\$23,524)^2 + \ldots +$$
$$0.05(\$62,000)^2 - (\$40,158)^2 = \$101,232,300.$$

The annual equivalent cost difference between Alternative A and Alternative B can now be estimated more precisely. It is $\$50,229 - \$40,158 = \$10,071$. This compares with a difference of $\$7,007$ after 10 trials. Thus, Alternative B is favored on the basis of the expected value approach.

It should be noted that Alternative B also leads to a smaller variance in the annual equivalent cost than the variance for Alternative A. Thus, from an expectation-variance viewpoint Alternative B is clearly superior to Alternative A. It is unlikely that this conclusion would be altered by pursuing the Monte Carlo simulation beyond 100 trials.

PROBLEMS

1. Because the geological structure of a certain mountain is unknown, the cost of constructing a highway tunnel is a random variable described as follows:

Cost (X)	Probability Cost is X
$ 8,000,000	0.2
10,000,000	0.4
12,000,000	0.3
14,000,000	0.1

What is the expected value of the cost of the tunnel? Is the expected cost approach appropriate in this situation? What bid should be submitted if the contractor wishes to be 90% sure that the cost will not exceed the income?

2. Suppose that an asset costing $7,500 will result in an annual saving of $2,000 for as long as the asset remains serviceable. The probabilities that the asset will remain serviceable for a certain number of years are as follows:

Number of Years Asset Functions (n)						
	1	2	3	4	5	6
Probability Asset is Productive Exactly n Years	0.1	0.2	0.2	0.3	0.1	0.1

What is the expected net present-worth of the venture if the interest rate is 10%? Should the investment be made?

3. A company is considering the purchase of a concrete plant for $4,000,000. The success of the plant depends on the amount of highway construction undertaken over the next 4 years. It is known that there are three possible levels (A, B, and C) of federal support for the construction of new highways. Shown below are the receipts and disbursements (in millions of dollars) the company expects from the concrete plant for each level of government support.

	End of Year				
	0	1	2	3	4
Support Level:					
A	− $4.0	$2.0	$2.0	$2.0	$2.0
B	− 4.0	1.0	1.0	1.0	1.0
C	− 4.0	0.5	0.5	1.0	2.5

The probabilities that support levels A, B, and C are realized over the next 4 years are 0.3, 0.1, 0.6, respectively. If the minimum attractive rate of return is 12%, what is the expected present-worth of this investment opportunity?

4. Uncertainty as to the rate of technological innovation means that a proposed computer system will have a service life which is unknown. If the initial cost of developing a computer system is $10 million and each year of service life pro-

duces net revenues of $4.5 million, what is the expected present-worth for an interest rate of 10% if the lifetime of the system is described by the following probabilities?

	Number of Years of Life (n)				
	1	2	3	4	5
Probability Life is Exactly *n* Years	0.1	0.2	0.4	0.2	0.1

5. A company is considering the introduction of a "new" umbrella although the future success of this product is uncertain. The product will be sold for only 5 years, and if the 5 years are "wet" years, the following cash flows (in hundred thousands) are anticipated:

					Year		
Probability	*Cash Flow*	0	1	2	3	4	5
0.6	W1	− $4	$2	$3	$2	$1	$0
0.4	W2	− 4	1	2	4	3	2

On the other hand, if the next 5 years are "dry" years, the cash flows are likely to be as follows:

					Year		
Probability	*Cash Flow*	0	1	2	3	4	5
0.5	D1	− $4	$0	$1	$1	$1	$1
0.5	D2	− 4	1	1	2	1	1

The interest rate is 10%.
(a) What is the expected present-worth if the next 5 years are going to be "wet"?
(b) What is the expected present-worth if the next 5 years are going to be "dry"?
(c) If the best available information indicates that the next 5 years will be "wet" with probability 0.7,
   (i) What is the probability that the next 5 years will be "dry"?
   (ii) What is the expected present-worth of the proposal?

6. A retail firm has experienced three basic responses to its previous advertising campaigns. Presently, the firm has two advertising programs under consideration. Because each advertising program has a different emphasis it is expected that the percentage of people responding in a particular manner will vary according to the program undertaken. Shown below are the cash flows that are expected for each of the three possible customer responses. Each cash flow shown assumes 100% response.

PROGRAM A

Response	Percentage Responding	End of Year			
		0	1	2	3
1	20%	− $500,000	$300,000	$200,000	$100,000
2	60%	− 500,000	200,000	200,000	300,000
3	20%	− 500,000	200,000	200,000	200,000

PROGRAM B

Response	Percentage Responding	End of Year			
		0	1	2	3
1	30%	− $600,000	300,000	200,000	200,000
2	40%	− 600,000	200,000	300,000	300,000
3	30%	− 600,000	300,000	300,000	300,000

For an interest rate of 10% which advertising campaign has the largest expected present-worth?

7. A toy manufacturer must decide whether to market a new doll or to update a doll that is currently being marketed. He has the following information to use in making his decision:

	Initial Cost	Net Return per Year	Probability of Duration of Sales (in years)				
			1	2	3	4	5
New doll	$50,000	$25,000	0.1	0.2	0.3	0.2	0.2
Updated old doll	20,000	12,000	0.4	0.3	0.2	0.1	0.0

Using the expected present-worth, which alternative should the manufacturer choose if the interest rate is 15%?

8. A plant is to be built to produce blasting devices for construction jobs, and the decision must be made as to the extent of automation in the plant. Additional automatic equipment increases the investment costs but lowers the probability of shipping a defective device to the field which must then be shipped back to the factory and dismantled at a cost of $10. The operating costs are identical for the different levels of automation. It is estimated that the plant will operate 10 years, the interest rate is 20%, and the rate of production is 100,000 devices per year for all levels of automation.

Find the level of automation that will minimize the expected annual cost for the investment costs and probabilities given below.

Level of Automation	Probability of Producing a Defective	Cost of Investment
1	0.10	$100,000
2	0.05	150,000
3	0.02	200,000
4	0.01	275,000
5	0.005	325,000
6	0.002	350,000
7	0.001	400,000

9. A dam is being planned for a certain river of erratic flow. It has been determined by past experience that a dam of sufficient capacity to withstand various flow rates where the probability of these rates being exceeded in any one year is 0.10, 0.05, 0.025, 0.0125, and 0.00625 will cost $142,000, $154,000, $170,000, $196,000, and $220,000, respectively; will require annual maintenance amounting to $4,600, $4,900, $5,400, $6,500, and $7,200, respectively; and will suffer damage of $122,000, $133,000, $145,000, $170,000, and $190,000, respectively, if subjected to flows exceeding its capacity. The life of the dam will be 40 years with no salvage value. For an interest rate of 10% calculate the annual cost of the dam including probable damage for each of the five proposed plans and determine the dam size that will result in a minimum cost.

10. It has been proposed to build a drive-in car wash and the decision must be made as to the number of individual facilities to be provided. A greater number of facilities means fewer customers turned away (and thus more customers serviced) but each additional facility requires additional investment. Assume the interest rate is 10%, an investment life of 10 years, and that each car serviced produces an operating surplus of $1. Using the following data, determine the number of facilities that will result in the highest expected profit.

No. of facilities	Total required Invest- ment	Probability of Averaging n cars per year								
		2,000	4,000	6,000	8,000	10,000	12,000	14,000	16,000	18,000
1	$10,000	0.6	0.4	—	—	—	—	—	—	—
2	18,000	0.2	0.6	0.2	—	—	—	—	—	—
3	25,000	0.1	0.3	0.4	0.2	—	—	—	—	—
4	32,000	0.1	0.1	0.2	0.3	0.2	0.1	—	—	—
5	38,000	0.05	0.05	0.1	0.2	0.4	0.1	0.1	—	—
6	43,500	0.05	0.05	0.1	0.1	0.2	0.3	0.1	0.1	—
7	50,000	0.05	0.05	0.1	0.1	0.2	0.3	0.1	0.05	0.05
8	55,000	0.05	0.05	0.1	0.1	0.2	0.3	0.1	0.05	0.05

11. A company has developed probability distributions representing the probabilities that various annual equivalent amounts will be realized from the three mutually exclusive projects that are under consideration. These distributions are given below.

Annual equivalent profit:	Alternative		
	A1	A2	A3
$ 5,000	0.10	0.00	0.00
10,000	0.10	0.20	0.00
15,000	0.20	0.20	0.20
20,000	0.30	0.20	0.50
25,000	0.30	0.20	0.20
30,000	0.00	0.20	0.10

Calculate the expected annual equivalent profit and the variance for each probability distribution. Which alternative would you consider to be the most attractive?

12. An advertising agency has developed four alternative advertising compaigns for one of its clients. The client has studied the alternatives and has developed probability distributions describing the present-worth of the net profits expected if it invests in a particular advertising program. The probability distributions for the four ad programs are as follows:

Ad program	Net present-worth of profits				
	−$50,000	−$10,000	$10,000	$50,000	$100,000
A	0.10	0.20	0.30	0.30	0.10
B	0.05	0.15	0.40	0.40	0.00
C	0.40	0.00	0.00	0.00	0.60
D	0.00	0.10	0.40	0.50	0.00

Calculate the mean, the variance, and the probability that the net present-worth of profits is less than zero. Which advertising program would you undertake? Explain your choice.

13. To buy a numerically controlled machine a manufacturer must pay $200,000. If the machine is purchased, there are five different manufacturing processes that can utilize this machine. By using the machine exclusively in process A, B, C, D or E, respectively, the annual income that will be realized is $120,000, $130,000, $150,000, $160,000, or $200,000. The life of the machine is expected to vary according to where the machine is used. The probability that the machine will provide service for exactly 1, 2, 3, or 4 years is shown below for each process.

	Life of Machine (years)			
Process	1	2	3	4
A	0.25	0.25	0.25	0.25
B	0.30	0.30	0.30	0.10
C	0.30	0.40	0.25	0.05
D	0.20	0.60	0.20	0.00
E	0.50	0.50	0.00	0.00

Calculate the means and variance of the present-worth amounts for each of these five processes. The interest rate is 10%. Which process could use this machine most effectively?

14. An inventor has developed an electronic instrument to monitor the impurities in the metal produced by a smelting process. A company that markets this type of equipment is considering the purchase of the patent rights to this instrument for $40,000. The company feels that there is one chance in 5 of the device becoming a successful seller. It is estimated that if the device is successful it will produce net revenues of $150,000 a year for the next 5 years. If the product is not a success no revenues will be received. The company's interest rate is 15%.
  (a) Draw a decision tree describing the decision options and determine the best decision policy.
  (b) If there is a market research group that can provide perfect information about the success of this product, what would be the most the company should be willing to pay for their service?
  (c) Suppose the market research group can make a market survey that with probability 0.7 will give a favorable result if the device will be a success and with probability 0.9 will give an unfavorable result if the device will not sell. How much would this survey be worth to the company?

15. An electronics firm is trying to decide whether or not to manufacture a new communications device. The decision to produce the device means an investment of $5 million, and the demand for such a device is not known. If demand is *high*, the company expects a return of $2.0 million each year for 5 years. If the demand is *moderate*, the return will be $1.6 million each year for 4 years, and a *light* demand means a return of $.8 million each year for 4 years.

It is estimated that the probability of a light demand is 0.1 and the probability of a high demand is 0.5. Interest is 10%.

(a) On the basis of expected present-worth, should the company make the investment?

(b) Someone has proposed that a survey be taken to establish the actual demand to be experienced. What is the maximum value of such a survey?

(c) A survey is available at a cost of $75,000 which has the following characteristics:

		Survey results say favorable or unfavorable with following probabilities:	
		Favorable	*Unfavorable*
If demand is:	High	0.9	0.1
	Moderate	0.5	0.5
	Low	0.0	1.0

Should the company make the survey? If so, should the company produce the device if the survey says unfavorable? What is the expected profit in this case?

16. A wholesaler is studying his warehouse needs for the next 8 years. At present three alternatives are under consideration: A new warehouse can be built to replace the existing facility, the existing facility can be expanded, or the decision to expand can be postponed. If the decision is postponed the wholesaler will wait 4 years and then decide whether to expand the existing facility or to leave it as is. To expand the present facility now will cost $400,000 while to build a new facility will cost $700,000. Consumer demand for the wholesaler's products is expected to be either high for all 8 years (H_1, H_2), high for 4 years and low for the remaining 4 years (H_1, L_2) low for the first 4 years and high for the last 4 years (L_1, H_2) or low for all 8 years (L_1, L_2). Depending upon these demand levels the following annual receipts are expected to be received from the two alternatives that required an investment now:

	Demand			
	H_1H_2	H_1L_2	L_1H_2	L_1L_2
Alternatives:				
Build new warehouse now	$320,000	$160,000	$110,000	$80,000
Expand existing warehouse now	200,000	150,000	100,000	50,000

If the decision to construct additional warehouse capacity is postponed, the cost of expanding the warehouse 4 years from now is expected to be $600,000. In this case it is anticipated that the annual revenues for high and low demand during the first 4 years will be $50,000 and $20,000, respectively. The annual revenues for the remaining 4 years are shown below.

	Demand	
	H_2	L_2
Decision:		
Expand after 4 years	$400,000	$100,000
Do not expand after 4 years	80,000	40,000

The interest rate is 12% and the probabilities that demand will be at a particular level through the 8-year period are

	Demand			
	H_1H_2	H_1L_2	L_1H_2	L_1L_2
Probability	0.3	0.2	0.1	0.4

Using decision tree analysis determine the decision policy that should be followed so that the wholesaler's expected profits are maximized.

17. Urn 1 contains 6 white balls and 4 black balls while Urn 2 contains 7 white balls and 3 black balls. A sample of 2 balls is going to be taken and on this basis you must decide from which urn the sample was selected. Although you do not see the drawing of the sample it is known that the sample is equally likely to come from either urn. If you make the correct decision you will win $10 and if you make an incorrect decision you win nothing. Using decision tree analysis find the decision policy that will maximize your expected winnings and indicate what the expected winnings will be. (a) Assume that you may choose whether the

sample is drawn with or without replacement. (With replacement, the first ball drawn is returned to the urn before the second ball is selected.)

(b) Assume that you may choose whether the second ball shall be drawn with or without replacement after you have seen the result of drawing the first ball.

18. Apply the Monte Carlo method with 100 trials to the situation described in Problem 1.
    (a) Plot the average cost after 10, 20, 30, ... , 100 trials to illustrate convergence.
    (b) Plot the statistical distribution of cost and find the ratio of costs above $12,000,000 to the total number of cost elements generated.
    (c) Discuss the findings of this exercise in comparison with the theoretical results found in the solution to Problem 1.

19. The salvage value of a certain asset depends upon its service life as follows:

Service Life	Salvage Value
3 years	$18,000
4 years	$12,000
5 years	$ 6,000

If the asset has a first cost of $120,000 and the interest rate is 12%, find its expected equivalent annual cost if each service life is equally likely to occur.

20. Two equipment investment alternatives are under consideration. Alternative A requires an initial investment of $15,000 with an annual operating cost of $2,000. The service life–salvage value possibilities are:

Service Life	Salvage Value
3 years	$4,000
4 years	$2,500

Each service life is equally likely to occur. Alternative B requires an initial investment of $30,000 with a negligible annual operating cost. The salvage value as a function of the service life is $8,000-$500(n), where n is the service life with the following probability distribution:

Service Life	Probability
7 years	0.4
8 years	0.3
9 years	0.3

If the interest rate is 10%, find the average annual equivalent cost for each alternative by 10 Monte Carlo trials. Discuss your result as a basis for choosing one alternative over the other.

21. Develop theoretical distributions from the information given in Section 16.5 which may be compared with the frequency distributions developed by Monte Carlo and exhibited in Tables 16.12 and 16.13. What is the expected cost for each alternative? Why does it differ from that found after 100 Monte Carlo trials?

DECISION MAKING
UNDER UNCERTAINTY

It may be impossible to assign probabilities to the occurrence of future events associated with some decision situations. Often no meaningful data are available from which probabilities may be derived. In other instances the decision maker may be unwilling to assign a probability, as is sometimes the case when the event is unpleasant.

When probabilities are not available for the assignment to future events, the situation is called *decision making under uncertainty*. As compared with decision making under certainty and under risk, this decision situation is more abstract. In this chapter decisions under uncertainty are structured in a formal manner and some useful decision rules are applied.

17.1. The Payoff Matrix

A particular decision can result in one of several outcomes depending upon which of several future events takes place. For example, a decision to go on a picnic can result in a high degree of satisfaction if the day turns out to be sunny or in a low degree of satisfaction if it rains. These levels of satis-

17

faction would be reversed if the decision were made to stay home. Thus, for the two states of nature, sun and rain, there are different payoffs depending upon the alternative chosen.

A *payoff matrix* is a formal way of exhibiting the interaction of decision alternatives and the states of nature. In this usage alternatives have the same meaning as before, that is, courses of action between which choice is contemplated. The states of nature need not be natural events such as sun and rain. This phrase is used to describe a wide variety of future events over which the decision maker has no control. The payoff matrix gives a qualitative or quantitative payoff for each possible future state and for each alternative under consideration.

As an example of the structuring of a payoff matrix, consider the following situation. An engineering and construction firm has the opportunity to bid on two contracts. The first contract pertains to the design and construction of a plant to convert solid waste into steam for heating purposes in a city. The second contract pertains to the design and construction of a steam distribution system within the city. The firm may be awarded either contract X or contract Y or both contract X and contract Y. Thus, there are three possible outcomes or "states of nature."

In considering the opportunities afforded by these contracts, the firm identifies five alternatives. Alternative $A1$ is for the firm to serve as project manager, with all of the work to be subcontracted. Alternative $A2$ is for the firm to subcontract the design but to do the construction. Alternative $A3$ is for the firm to subcontract the construction but to do the design. Alternative $A4$ is for the firm to do both the design and construction. Alternative $A5$ calls for the firm to bid jointly with another organization which has the capability to undertake an innovative project of this type.

Once the states of nature and the alternatives are identified, the next step is to derive payoff values. In this example, 15 payoff values must be developed. By listing anticipated disbursements and receipts over time identified with each alternative, for each state of nature, the present value of profit is found. Suppose that these present values are in thousands of dollars as exhibited in Table 17.1.

Table 17.1. PAYOFF MATRIX FOR PROFIT IN THOUSANDS OF DOLLARS

		States of Nature		
		X	Y	X and Y
	$A1$	−4,000	1,000	2,000
	$A2$	1,000	1,000	4,000
Alternatives	$A3$	−2,000	1,500	6,000
	$A4$	0	2,000	5,000
	$A5$	1,000	3,000	2,000

From the payoff matrix it can be seen that the firm could incur a present loss of $4 million if Alternative $A1$ is chosen and contract X is awarded. If contract Y is awarded, the present profit would be $1 million. The present profit would be $2 million if both contracts are awarded. Thus, each row of the payoff matrix represents the outcomes expected for each state of nature (column) for a particular alternative (row).

Individual payoff values in a payoff matrix need not be monetary in character. They may be qualitative or quantitative expressions of the utility expected from each of the several alternatives. It is essential, however, that the payoff values be expressed in some common and directly comparable measure such as present worth or annual equivalent amount. In Table 17.1, the payoff values are present-worth amounts.

Before proceeding, the payoff matrix should be examined for dominance. If for two alternatives one would always be preferred no matter which future occurs, the preferred alternative dominates and the other alternative may be discarded.

In Table 17.1, Alternative $A1$ may be discarded since it is dominated by other alternatives. Therefore, the payoff matrix can be reduced to the form shown in Table 17.2. This reduced payoff matrix completely rules out the

Table 17.2. REDUCED PAYOFF MATRIX IN THOUSANDS OF DOLLARS

		States of Nature		
		X	Y	X and Y
Alternatives	$A2$	1,000	1,000	4,000
	$A3$	−2,000	1,500	6,000
	$A4$	0	2,000	5,000
	$A5$	1,000	3,000	2,000

alternative of the firm's serving as project manager with all of the design and construction work to be subcontracted. The rules presented in the following sections may be used to assist in the selection of one of the four remaining alternatives.

17.2. The Laplace Rule

If the firm were willing to assign probabilities to the states of nature in Table 17.2, the decision situation would be classified as decision making involving risk. The techniques of the previous chapter could then be applied, and the best alternative would be chosen by applying the proper criteria.

Suppose, however, that the firm is unwilling to assess the states of nature in terms of their probabilities of occurrence. In the absence of these probabilities one might reason that each possible state of nature is as likely to occur as any other. The rationale of this assumption is that there is no stated basis for one state of nature to be more likely than any other. This is called the *Laplace principle* or the *principle of insufficient reason* based upon the philosophy that nature is assumed to be indifferent.

Table 17.3. COMPUTATION OF AVERAGE PAYOFF IN MILLIONS OF DOLLARS

Alternative	Average Payoff
$A2$	($1,000 + $1,000 + $4,000) ÷ 3 = $2,000
$A3$	(−$2,000 + $1,500 + $6,000) ÷ 3 = $1,833
$A4$	($0 + $2,000 + $5,000) ÷ 3 = $2,333
$A5$	($1,000 + $3,000 + $2,000) ÷ 3 = $2,000

Under the Laplace principle the probability of the occurrence of each future state of nature is assumed to be $1/n$, where n is the number of possible future states. To select the best alternative one would compute the arithmetic average for each. For the payoff matrix of Table 17.2 this is accomplished as shown in Table 17.3. Alternative $A4$ results in a maximum profit of $2,333,000 and would be selected by this procedure.

17.3. Maximin and Maximax Rules

Two very simple decision rules are available for dealing with decisions under uncertainty. The first is the *maximin* rule based on an extremely pessimistic view of the outcome of nature. The use of this rule would be justified if it is judged that nature will do her worst. The second is the *maximax* rule based upon an extremely optimistic view of the outcome of nature. Use of this rule is justified if it is judged that nature will do her best.

Because of the pessimism embraced by the maximin rule, its application will choose the alternative that assures the best of the worst possible outcomes. If P_{ij} is used to represent the payoff for the ith alternative and the jth state of nature, then the required computation is

$$\max_{i} \, [\min_{j} P_{ij}].$$

Consider the decision situation described by the payoff matrix of Table 17.2. The application of the maximin rule requires that the minimum value in each row be selected. Then the maximum value is identified from these and associated with the alternative which would produce it. This procedure is illustrated in Table 17.4. Selection of either Alternative $A2$ or $A5$ assures the firm of a payoff of at least $1,000,000 regardless of the outcome of nature.

Table 17.4. PAYOFF IN THOUSANDS OF DOLLARS BY THE MAXIMIN RULE

Alternative	$\min_{j} P_{ij}$
$A2$	$1,000
$A3$	$-$2,000
$A4$	$0
$A5$	$1,000

The optimism of the maximax rule is in sharp contrast to the pessimism of the minimax rule. Its application will choose the alternative that assures the best of the best possible outcomes. As before, if P_{ij} represents the payoff for the ith alternative and the jth state of nature, then the required computation is

$$\max_i \, [\max_j P_{ij}].$$

Consider the decision situation of Table 17.2. The application of the maximax rule requires that the maximum value in each row be selected. Then the maximum value is identified from these and associated with the alternative which would produce it. This procedure is illustrated in Table 17.5. Selection of Alternative $A3$ is indicated. Thus, the decision maker may receive a payoff of $6,000,000 if nature is benevolent.

Table 17.5. PAYOFF IN THOUSANDS OF DOLLARS BY THE MAXIMAX RULE

Alternative	Max P_{ij}
A2	$4,000
A3	$6,000
A4	$5,000
A5	$3,000

A decision maker who chooses the maximin rule considers only the worst possible occurrence for each alternative and selects that alternative which promises the best of the worst possible outcomes. In the example where $A2$ was chosen, the firm would be assured of a payoff of at least $1,000,000, but it could not receive a payoff any greater than $4,000,000. Or, if $A5$ were chosen, the firm could not receive a payoff any greater than $3,000,000. Conversely, the firm that chooses the maximax rule is an optimist one that decides solely on the basis of the highest payoff offered for each alternative. Accordingly, in the example in which $A3$ was chosen, the firm faces the possibility of a loss of $2,000,000 in the quest of a payoff of $6,000,000.

17.4. The Hurwicz Rule

Because of the extreme nature of the decision rules presented in the previous section, they are alien to many decision makers. Most people possess a degree of optimism or pessimism somewhere between the extremes. A third approach to decision making under uncertainty involves an index of relative optimism and pessimism. It is called the *Hurwicz rule*.

A compromise between optimism and pessimism is embraced in the

Hurwicz rule by allowing the decision maker to select an index of optimism, α, such that $0 \le \alpha \le 1$. When $\alpha = 0$ the decision maker is pessimistic about the outcome of nature, while an $\alpha = 1$ indicates optimism about nature. Once α is selected the Hurwicz rule requires the computation of

$$\max_i \{\alpha[\max_j P_{ij}] + (1 - \alpha)[\min_j P_{ij}]\}$$

where P_{ij} is the payoff for the ith alternative and the jth state of nature.

As an example of the Hurwicz rule, consider the payoff matrix of Table 17.2 with $\alpha = 0.2$. The required computations are shown in Table 17.6 and Alternative $A2$ would be chosen by the firm.

Table 17.6. PAYOFF IN THOUSANDS OF DOLLARS BY THE HURWICZ RULE WITH $\alpha = 0.2$

Alternative	$\alpha[\max_j P_{ij}] + (1 - \alpha)[\min_j P_{ij}]$	
$A2$	$0.2(\$4,000) + 0.8(\$1,000)$	$=$ $\$1,600$
$A3$	$0.2(\$6,000) + 0.8(-\$2,000)$	$=$ $-\$400$
$A4$	$0.2(\$5,000) + 0.8(0)$	$=$ $\$1,000$
$A5$	$0.2(\$3,000) + 0.8(\$1,000)$	$=$ $\$1,400$

Additional insight into the Hurwicz rule can be obtained by graphing each alternative for all values of α between zero and one. This makes it possible to identify the values of α for which each alternative would be favored. Such a graph is shown in Figure 17.1. It may be observed that Alternative $A2$ yields a maximum expected payoff for all values of $\alpha \le 1/2$. Alternative $A4$ exhibits a maximum for $1/2 \le \alpha \le 2/3$ and Alternative $A3$ gives a maximum for $2/3 \le \bar{\alpha} \le \bar{1}$. There is no value of α for which Alternative $A5$ would be best except at $\alpha = 0$ where it is as good an alternative as $A2$.

When $\alpha = 0$, the Hurwicz rule gives the same result as the maximin rule, and when $\alpha = 1$, it is the same as the maximax rule. This may be shown for the case where $\alpha = 0$ as

$$\max_i \{0[\max_j P_{ij}] + (1 - 0)[\min_j P_{ij}]\} = \max_i [\min_j P_{ij}].$$

And, for the case where $\alpha = 1$

$$\max_i \{1[\max_j P_{ij}] + (1 - 1)[\min_j P_{ij}]\} = \max_i [\max_j P_{ij}].$$

Thus, the maximin rule and the maximax rule are special cases of the Hurwicz rule.

The philosophy behind the Hurwicz rule is that many people focus on the most extreme outcomes or consequences in arriving at a decision. By

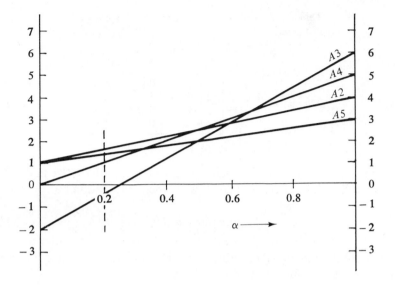

Figure 17.1. Values for the Hurwicz rule representing four alternatives.

use of this rule, the decision maker may weight the extremes in such a manner as to reflect the relative importance attached to them.

17.5. The Minimax Regret Rule

If a decision maker selects an alternative and a state of nature occurs such that he could have done better by selecting another alternative, he "regrets" his original selection. This regret is the difference between the payoff which could have been achieved with perfect knowledge of the future and the payoff which was actually received from the alternative chosen. The *minimax regret rule* is based on the premise that a decision maker wishes to avoid any regret or at least to minimize his maximum regret about a decision.

Application of the minimax regret rule requires the formulation of a regret matrix. This is accomplished by identifying the maximum payoff for each state (column). Next, each payoff in the column is subtracted from the maximum payoff identified and this is repeated for each column. For the payoff matrix of Table 17.2 the maximum payoffs are $1,000, $3,000, and $6,000 for X, Y, and X and Y, respectively. Thus, the requests for X, applicable to Alternatives $A2$ through $A5$, are $1,000 − $1,000 = 0; $1,000 − (−$2,000) = $3,000; $1,000 − $0 = $1,000; $1,000 − $1,000 = $0. Repeating this computation for each state results in the regret matrix shown in Table 17.7.

Table 17.7. REGRET MATRIX IN THOUSANDS OF DOLLARS

| | | States of Nature | | |
		X	Y	X and Y
Alternative	$A2$	0	2,000	2,000
	$A3$	3,000	1,500	0
	$A4$	1,000	1,000	1,000
	$A5$	0	0	4,000

If the regret values are designated R_{ij} for the ith alternative and the jth state, then the minimax regret rule requires the computation of

$$\min_{i} [\max_{j} R_{ij}].$$

This computation is shown in Table 17.8. Selection of Alternative $A4$ assures the firm of a maximum regret of $1,000,000.

Table 17.8. PAYOFF IN THOUSANDS OF DOLLARS BY THE MINIMAX REGRET RULE

Alternative	$\max_{j} P_{ij}$
$A2$	$2,000
$A3$	$3,000
$A4$	$1,000
$A5$	$4,000

A decision maker who uses the minimax regret rule as a decision criterion will make that decision which will result in the least possible opportunity loss. Individuals who have a strong aversion to criticism would be tempted to apply this rule because it puts them in a relatively safe position with respect to the future states of nature which might occur. In this regard this criterion has a conservative underlying philosophy.

17.6. Summary of Decision Rules

The alternatives selected by the decision rules presented in this chapter are summarized in Table 17.9. It will be noted that the rules do not give consistent results. They are developed to give insight into those decision situations in which probabilities are not or cannot be assigned to the occurrence of future events.

Table 17.9. COMPARISON OF RULES

Decision Rule	Alternative Selected
Laplace	A4
Maximin	A2 or A5
Maximax	A3
Hurwicz ($\alpha = 0.2$)	A2
Minimax regret	A4

Examination of the courses of action recommended by the five decision rules indicates that each has its own merit. Several factors may influence a decision maker's choice of a rule in a given decision situation. The decision maker's attitude toward uncertainty (pessimistic or optimistic) and his personal utility function are important influences. Thus, the choice of a particular decision rule for a given decision situation must be based upon the subjective judgment of the decision maker. This is what one would expect, for in the absence of probabilities concerning future events it is not possible to derive a completely objective decision procedure.

PROBLEMS

1. The following matrix gives the payoffs in utiles for three alternatives and three possible states of nature:

		States of Nature		
		S1	*S2*	*S3*
	A1	50	80	80
Alternatives	A2	60	70	20
	A3	90	30	60

Which alternative would be chosen under the Laplace principle? The maximin rule? The maximax rule? The Hurwicz rule with $\alpha = 0.75$? The minimax regret rule?

2. The following matrix gives the payoffs in utiles for four alternatives and four possible states of nature:

| | States of Nature | | | |
	S1	S2	S3	S4
A1	2	6	0	0
A2	2	2	2	2
A3	0	8	0	0
A4	4	4	0	2

Alternatives (labels at left of A1–A4)

Which alternative would be chosen under the Laplace principle? The maximin rule? The maximax rule? The Hurwicz rule with $\alpha = 0.60$? The minimax regret rule?

3. Graph the Hurwicz rule for all values of α using the payoff matrix of Problem 2.

4. The following matrix gives the dollar profits expected for five investments and four different levels of sales:

| | Levels of Sales | | | |
	L1	L2	L3	L4
I1	15	11	12	9
I2	7	9	12	20
I3	8	8	14	17
I4	17	5	5	5
I5	6	14	8	19

Investments (labels at left of I1–I5)

Which investment would be chosen under the maximin rule? The maximax rule? The Hurwicz rule with $\alpha = 0.7$? The minimax regret rule?

5. The following matrix gives the expected profit in thousands of dollars for five marketing strategies and five potential levels of sales:

| | Levels of Sales | | | | |
	L1	L2	L3	L4	L5
M1	10	20	30	40	50
M2	20	25	25	30	35
M3	50	40	5	15	20
M4	40	35	30	25	25
M5	10	20	25	30	20

Strategies (labels at left of M1–M5)

Which marketing strategy would be chosen under the maximin rule? The maximax rule? The Hurwicz rule with $\alpha = 0.4$? The minimax regret rule?

6. Graph the Hurwicz rule for all values of α using the payoff matrix of Problem 5.

7. A construction firm is considering the purchase of a number of different pieces

of equipment. The firm knows that, depending upon future projects won by bidding, certain types of equipment will have varying costs. Shown below is a cost matrix indicating the equivalent annual costs associated with a particular piece of equipment and its use on a particular project. The equipment alternatives are mutually exclusive and this firm anticipates that they will win the contract on only one of the projects. What piece of equipment would they purchase if they based their decision on the (a) minimax rule, (b) minimin rule, (c) Hurwicz rule for $\alpha = 0.3$, (d) minimax regret rule, and (e) the Laplace rule? Graph the Hurwicz rule for each alternative for all values of α.

			Projects	
		A	*B*	*C*
	1	$100	$90	$ 60
Equipment	2	70	80	90
	3	30	30	140
	4	100	20	120

8. Assume that you have a sum of money you wish to invest. After some thought and preliminary analysis, you have narrowed the possibilities to the following:

    *A*1: Invest in speculative stocks.
    *A*2: Invest in blue chip stocks.
    *A*3: Invest in government bonds.

You have also considered the following three future states:

    *S*1: War.
    *S*2: Peace without economic recession.
    *S*3: Peace with economic recession.

Your preliminary analysis yields the following payoff matrix in terms of rate of return:

	*S*1	*S*2	*S*3
*A*1	20	1	−6
*A*2	9	8	0
*A*3	4	4	4

  (a) What course of action do the following decision criteria indicate: Laplace, maximin, maximax, and Hurwicz with $\alpha = 0.5$?

(b) Which investment do you prefer? Why?

9. Consider the example of an engineering and construction firm used in this chapter. Suppose that the firm restructures the states of nature as follows:

> $S1$ = the firm receives no contract.
> $S2$ = the firm gets at least one contract.
> $S3$ = the firm gets both contracts.

The following payoff matrix applies in which the values are present profits in thousands of dollars with the values under $S2$ being the averages of the payoffs in columns X and Y of Table 17.1:

		States of Nature		
		$S1$	$S2$	$S3$
	$A1$	−8,000	−1,500	2,000
	$A2$	−4,000	1,000	4,000
Alternatives	$A3$	−5,000	−250	6,000
	$A4$	−3,000	1,000	5,000
	$A5$	0	2,000	2,000

(a) Would the firm make a different decision under the restructured payoff matrix if the Laplace rule is applied?

(b) Does the decision depend upon the order in which the states of nature are listed if the Laplace rule is applied? The maximin rule? The minimax rule?

part six

ECONOMIC ANALYSIS
OF OPERATIONS

ANALYSIS OF CONSTRUCTION
AND PRODUCTION OPERATIONS

Engineering proposals cannot meet human needs if they remain in the form of plans and specifications. Once economic feasibility is assured, these proposals should be converted into structures, systems, and products through the process of construction and/or production. The physical environment is altered through these operations, resulting in the creation of utility and the satisfaction of human needs.

Construction and production operations require a sequence of activities and the employment of producer goods of various types. Producer goods in the form of construction and production equipment, coupled with organized activity, are the instruments used by mankind to alter the physical environment. In this chapter some economic aspects of construction and production operations are examined from the standpoint of the activities and equipments utilized.

18.1. Critical Path Methods

Many engineering projects involving design, development, and construction operations are nonrepetitive in nature. Although a particular project is to be executed only once, many interdependent activities are re-

18

quired. Some must be executed simultaneously. Each will require personnel and equipment, the cost of which will depend upon the resource commitment. The total project duration and its overall cost depend upon the activities on the *critical path* and the resources allocated to them. Critical path methods (CPM) for dealing with this situation are presented in this section.

Activity and event networks. A network is the basic structural entity behind all critical path methods. It is used to portray the interrelationships among the activities associated with a project. Figure 18.1 illustrates a network that represents the activities and events in connection with the preparation of a foundation. Four events and three activities are shown.

An event in CPM is represented by a circle. It indicates the completion of an activity. In the case of an initial event it indicates project start or the initiation of the associated activities. Events A and C in Figure 18.1 indicate the start of excavation and the start of equipment move activities. Event D represents the completion of foundation preparation.

Activities in CPM are represented by arrows that interconnect events. These arrows symbolize the effort required to complete an event in units of time. Their direction specifies the order in which the events must occur. For example, activities BD (set forms) and CD (equipment move-in) are required

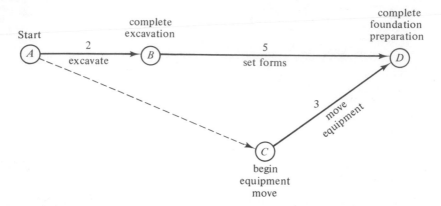

Figure 18.1. Activity-event network for preparing a foundation.

in order for event D (complete preparation) to be achieved. The precedence relationships between event D and event B and between event D and event C are established by the arrows. Events B and C must be realized before the foundation is prepared (event D).

In some network representations it is necessary or useful to create dummy activities to clarify precedence relationships. A dummy activity is illustrated in Figure 18.1 by a broken arrow that connects activity A with activity C, establishing that event C is preceded by event A. Even though dummy activities do not require time and do not consume resources, they may alter the completion time of a project by establishing the order in which events are performed.

Associated with each activity is an elapsed time estimate. This estimate is usually shown above the arrow in the network. The length of the arrows need not be proportional to the duration of the activity. In Figure 18.1 the time estimates are weeks.

Two important absolute times for each event must be identified. The first is the earliest time, T_E, which indicates the calendar time at which an event can take place, provided all previous activities were completed at their earliest possible times. The second is the latest time, T_L, for each event, defined as the latest calendar time at which an event can be completed without delaying initiating the following activities and completion of the project.

The difference between the latest and the earliest times for each event is called the *slack time* for that event. Positive slack for an event indicates the maximum amount of time that it can be delayed without delaying subsequent events and the overall project. In the example, event B has no slack since the latest time for completion of excavation and the earliest time for setting the forms is the same point in time. Any delay in realizing event B would cause a delay in the project. From this it is evident that the activities of excavation and setting the forms are "critical." The activity of equipment move-in is "noncritical" to the realization of event D since the start of equipment move-in can occur at any time on or before 3 weeks before the completion of the project. Thus, 4 weeks is the slack time for event C.

Finding the critical path. In CPM all critical activities must be identified and given special attention. This is necessary because any delay in performing these activities may lead to a delay in subsequent activities and the project as a whole. Often it is possible to identify the critical path by tracing the activities along a sequence of events which have zero slack time. The sum of the activity times along this critical path gives the shortest possible elapsed time for project completion. This is 7 weeks in the foundation example.

As an example of a systematic procedure for finding the critical path, consider a project requiring the construction of a certain structure. Five major events must be realized through seven activities as illustrated in Figure 18.2. The number appearing above each arrow is the estimated duration of the activity in months, with the arrow indicating the precedence relationship.

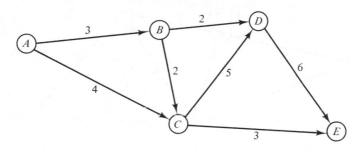

Figure 18.2. A network for the construction of a certain structure.

The procedure for finding the critical path starts by determining the earliest time and the latest time for each event. By subtracting T_E from T_L for each event, the slack time is found. To facilitate this process each event can be labeled (T_E, T_L, S), where S represents the slack time for the event. First, T_E's are determined for each event and are labeled $(T_E, \ , \ )$. Next, T_L's are found and the labels become $(T_E, T_L, \ )$. Finally, S is found by subtracting T_E from T_L and completing the label (T_E, T_L, S) for each event.

The earliest time for each event is determined by adding the earliest time for the event immediately preceding to the activity duration connecting the two events. When there is more than one preceding event, the largest value of the sum is chosen as the earliest time for the event under consideration. The process for determining these earliest times proceeds forward from the initial event, which is assumed to have an earliest time of zero. This process is illustrated in Figure 18.3 and Table 18.1 for the example under consideration.

Table 18.1. COMPUTATION OF EARLIEST TIMES

Event	Immediate Predecessor	T_E for Immediate Predecessor Plus Activity Time	Largest Sum = The earliest Time
A	None	$0 + 0 = 0$	0
B	A	$0 + 3 = 3$	3
C	A	$0 + 4 = 4$	
	B	$3 + 2 = 5 \leftarrow$	5
D	B	$3 + 2 = 5$	
	C	$5 + 5 = 10 \leftarrow$	10
E	C	$5 + 3 = 8$	
	D	$10 + 6 = 16 \leftarrow$	16

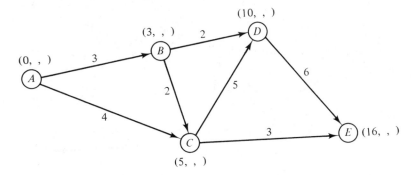

Figure 18.3. The earliest time for each event.

The latest time for each event can be determined in a manner similar to that used in finding the earliest time except that the process starts with the final event and proceeds backward. The latest time for this final event is the same as its earliest time, for, if it were not, the project would be delayed. For each event the latest time is determined by subtracting the duration time of an activity connecting the event and an event immediately succeeding from the latest time of the event immediately succeeding. If there is more than one event immediately succeeding, the smallest value of the difference is chosen to be the latest time for the event under consideration. A slack time for each event is then found by computing the difference between T_L and T_E as illustrated in Figure 18.4 and Table 18.2.

The critical path can now be identified with the T_L, T_E, and S values entered in Figure 18.4. This is done by tracing all zero slack events and view-

Table 18.2. COMPUTATION OF LATEST TIMES

Event	Immediate Successor	T_L for Immediate Successor Minus Activity Time	Smallest Difference = The Latest Time
E	None	$16 - 0 = 16$	16
D	E	$16 - 6 = 10$	10
C	D	$10 - 5 = 5 \longleftarrow$	5
	E	$16 - 3 = 13$	
B	C	$5 - 2 = 3 \longleftarrow$	3
	D	$10 - 2 = 8$	
A	B	$3 - 3 = 0 \longleftarrow$	0
	C	$5 - 4 = 1$	

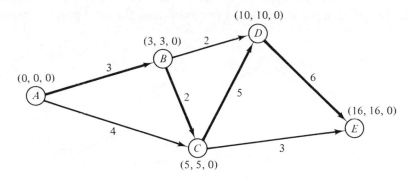

Figure 18.4. Labeling of events and identification of critical path.

ing them as a path. In this example the critical path is $A \to B \to C \to D \to E$ with a minimum project duration of 16 months.

Economic aspects of CPM: The CPM example above was presented under the assumption that the activities are performed in normal time with a normal allocation of resources. This is called a *normal schedule*. When one or more activities are performed with additional resources to shorten their time durations, a *crash schedule* is said to exist.

Two costs are associated with a project. Direct costs exist for each activity. These costs increase with an increase in the resources allocated for the purpose of decreasing the activity time. Indirect costs exist which are associated with the overall project duration. These increase in direct proportion to an increase in the total time required for project completion. This relationship should be self-evident, for overhead costs continue independent of activity levels and they depend upon the passage of time.

In most projects it is of interest to investigate the economic aspects of increasing direct costs as activities are expedited in the light of decreasing overhead costs. An optimum allocation of resources to each activity is sought so that the sum of direct and indirect costs will be a minimum.

The first step in finding the optimum project schedule is to find the critical path under normal resource conditions. This is the starting point. Next, the activity on the critical path which has the least impact on direct cost is shortened. When shortening this activity, other critical paths may appear. If they do, it is necessary to reduce the duration of activities on these paths an equal number of time units for each path. No further activity time reduction should be attempted either when a limit has been reached or when the indirect costs saved are not greater than the extra expenditure of direct resources. This process is repeated for all activities on the critical path or paths.

Consider the example project shown in Figure 18.5. Indirect costs of $1,000 can be saved for each day removed from the total project duration.

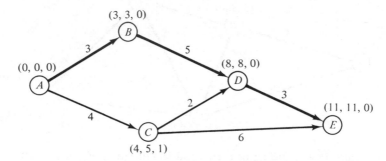

Figure 18.5. A CPM network with normal activities.

Event times shown are those that would occur under normal conditions. The label for each event represents T_E, T_L, and S as defined earlier. The critical path is $A \longrightarrow B \longrightarrow D \longrightarrow E$.

Data on activity durations under normal and crash conditions are given in Table 18.3. These data are daily costs computed from the assumption that each additional worker adds a direct cost of $50 per day. Additional equipment costs are those estimated to arise for each day reduced from the normal activity duration.

The project depicted in Figure 18.5 will take 11 days to complete under the normal activity times shown. No crash cost is incurred and therefore there is no saving in indirect cost.

A 10-day schedule can be achieved by reducing *BD* by one day. From Table 18.3 a crash cost of $700 is found. A saving of $1,000 in indirect cost will result. No further reduction in activity *BD* should be attempted because the critical path will change. This 10-day schedule is shown in Figure 18.6.

Two critical paths now exist: the original path and a new path $A \longrightarrow C$

Table 18.3. COMPUTATION OF CRASH COSTS

Activity	Number of Days Reduction	Additional Labor Cost			Additional Equipment Cost	Total Crash Cost
		Additional Workers	Working Days	Additional Costs		
AB	1	2	2	$200	$ 600	$ 800
AC	1	3	3	450	300	750
BD	1	1	4	200	500	700
	2	3	3	450	450	900
	3	3	2	300	1,700	2,000
CD	0	0	2	0	0	0
CE	1	0	5	0	950	950
	2	4	4	800	1,100	1,900
DE	0	0	3	0	0	0

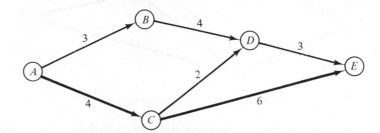

Figure 18.6. A ten-day schedule.

→ *E*. Reducing activity *BD* one more day yields a 9-day schedule. This leads to a crash cost of $900 which is less than the crash cost for shortening *BD* one day and *AB* one day. The duration of one activity in the critical path *A* → *C* → *E* should also be shortened by one day at the same time. A one-day reduction of *AC* costs $750 which is less than shortening *CE* by one day. A saving of $2,000 in indirect cost will result. This 9-day schedule is shown in Figure 18.7.

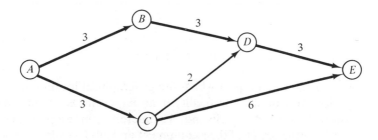

Figure 18.7. A nine-day schedule.

In the 9-day schedule, path $A \rightarrow B \rightarrow D \rightarrow E$ and path $A \rightarrow C \rightarrow E$ remain critical as in the 10-day schedule. Further equal time reduction should be sought in these critical paths.

An 8-day schedule is obtained by reducing AB by one day at a crash cost of $800 and BD by 2 days at a crash cost of $900 for a total of $1,700. This is better than a 3-day reduction in activity BD. It is necessary to reduce CE and AC by one day each at a crash cost of $950 plus $750 or $1,700. This is better than a 2-day reduction in CE. A saving of $3,000 is possible by this schedule. Three critical paths result: $A \rightarrow B \rightarrow D \rightarrow E$, $A \rightarrow C \rightarrow D \rightarrow E$, and $A \rightarrow C \rightarrow E$, each with a duration of 8 days. No further reductions are possible. The resulting schedule is shown in Figure 18.8.

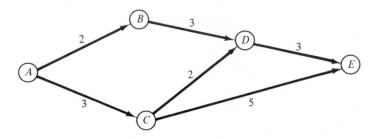

Figure 18.8. An eight-day schedule.

Finally, it is necessary to find the optimum schedule. This is accomplished by summarizing the crash costs and the savings in Table 18.4. A maximum net saving occurs for a schedule of 9 days.

Table 18.4. NET SAVINGS FOR CRASH SCHEDULES

Crash Schedule	Crash Cost	Indirect Cost Saving	Net Saving
10-day	$ 700	$1,000	$ 300
9-day	1,650	2,000	350
8-day	3,400	3,000	−400

18.2. Economic Aspects of Location

There are many situations in which the geographical location of a construction project or a production facility is an important economic consideration. Contractors will normally not be interested in projects that are located away from the center of their equipment and labor pools. Production managers normally seek to locate new plants in close proximity to markets, materials, and labor.

SSStartet

Locating temporary facilities. It is not uncommon for temporary service facilities to be set up in the area in which temporary activities are to be undertaken. The economy of setting up such facilities and the economy of various locations within the area must be determined.

Consider an example of contractor who is to build a dam requiring 300,000 cubic yards of gravel. Two feasible locations have been identified and their characteristics are shown in Table 18.5.

Table 18.5. COMPARISON OF TWO MATERIAL SOURCES

Location and Cost Data	Source *A*	Source *B*
Distance, pit to dam site in miles	1.8	0.6
Cost of gravel per cubic yard at pit	—	$0.08
Purchase price of pit	$7,200	—
Road construction necessary	$2,400	None
Overburden to be removed at $0.24 per cubic yard	—	60,000
Hauling cost per cubic yard per mile	$0.08	$0.10

In order to make a selection on the basis of economy, the cost of securing the required gravel from either source should be determined. The cost of 300,000 cubic yards of gravel from Source A is calculated as follows:

```
Purchase price of pit  ........................................   $ 7,200
Road construction  .........................................      2,400
Hauling cost, 300,000 cu. yd. × 1.8 miles × $0.08 per cu. yd. per mile   43,200
                                                               ─────────
                                                                $52,800
```

The cost of 300,000 cubic yards of gravel from Source B is:

```
Cost of gravel at pit, 300,000 yd. @ $0.08  .....................   $24,000
Removal of overburden, 60,000 yd. @ $0.24  ....................     14,400
Hauling cost, 300,000 cu. yd. × 0.6 mile × $0.10 per cu. yd. per mile   18,000
                                                               ─────────
                                                                $56,400
```

Locating permanent facilities. The selection of a location for a plant is a long-term commitment. Once a plant has been constructed, the expense and disruption necessary to move it to a more favorable location are generally so great that they are impractical, even though failure may result from the unfavorable characteristics of the original location. Therefore, the search for and evaluation of plant sites justify very careful consideration.

When profit is the measure of success, the best location is the one in which production and marketing effort will result in the greatest profit. Location may affect the cost at which raw materials are gathered, the cost of production, the cost of marketing, and the volume of product that can be sold.

Evaluation of a location begins with research to determine the volume of sales and income promised by the given location. Then research is directed along the lines suggested by the factors listed above to gain data with which to calculate the cost of gathering raw material, processing it, and delivering finished products to the consumer in the volume that can be sold. Evaluation of plant locations consists essentially of operating the enterprise under consideration "on paper" at each location studied. The result of this approach in evaluating three locations for a small glassware plant requiring an investment of approximately $180,000 are given in Table 18.6.

Table 18.6. COMPARATIVE EVALUATION OF THREE PLANT LOCATIONS

	Location A	Location B	Location C
Market:			
Annual sales ....(A)	$260,000	$260,000	$260,000
Selling expense ....(B)	44,000	43,000	46,000
Net income from sales	$216,000	$217,000	$214,000
Production costs:			
Supplies and raw materials	$ 69,000	$ 71,000	$ 62,000
Transportation—in and out	27,000	26,000	23,000
Fuel, power, and water	13,000	17,000	18,000
Wages and salaries	64,000	63,000	61,000
Miscellaneous items	8,000	8,000	10,000
Fixed costs other than interest	11,000	11,000	11,000
Net production costs ....(C)	$192,000	$196,000	$185,000
Selling plus production costs,			
$B + C$ ....(D)	236,000	239,000	231,000
Annual profit, $A - D$ ....(E)	24,000	21,000	29,000
Rate of return, $E \div \$180,000$	13.3%	11.7%	16.1%

Consider the competitive advantage of a plant at Location C over one at Location B. A plant at Location C has an advantage in net income of over 4% by virtue of its location to offset price reductions and operating and selling efficiency of a competing plant at Location B.

18.3. Economic Operation of Equipment

Most production facilities may be operated at rates below, equal to, or above their normal capacity. Facilities are usually inefficient and costly to operate when utilized at a rate of output below their normal capacity. For example, if a production department is allotted more space than is needed for its efficient operation, a number of losses may be expected to arise. The fixed costs such as maintenance, taxes, insurance, and interest will be higher than necessary. Heat, light, and janitor service will be wasted in the unused space, and the cost of supervision and material handling may be expected to be higher than necessary.

The load that is considered proper to impose on equipment is usually indicated by its makers. Such indicated normal loads are often not the optimum for economy. Operation of equipment above its normal capacity will result in increased production at the expense of increased power consumption and shortened life. But, in many cases the overall result is a reduced cost per unit produced. If such is the case, the equipment in question should be overloaded so that the resulting economy may be realized.

The economy of loading to normal capacity. A mining company that is expanding its operations is in need of direct current for an electrolytic process. During the next year an average of 12 kw of direct current energy will be needed. Planned expansion of the existing operations is expected to increase the rate at which direct current is needed by 2 kw per year until 30 kw are needed. At this point, the demand for current will remain constant. Energy will be needed 2,000 hours per year regardless of the rate at which is it used.

Investigation reveals that the need for direct current can best be provided by an a-c to d-c motor-generator set. These sets can be purchased in a variety of capacities. Plan A involves the purchase of a 20-kw set now for 5 years of use at which time the demand will have reached 20 kw. At this time, a 30-kw set will be purchased.

Efficiency-load curves provided by the vendor show that the 20-kw set has efficiencies of 48, 69, 78, and 76% at $\frac{1}{4}$, $\frac{1}{2}$, $\frac{3}{4}$, and full load respectively. The 30-kw set is slightly more efficient at its $\frac{1}{4}$, $\frac{1}{2}$, $\frac{3}{4}$, and full load. The purchase rate of a-c current is $0.02 per kw-hr. Interest is taken at 10%. Taxes, insurance, maintenance, and operating costs will be neglected in the interest of simplicity. The 20-kw set will cost $1,230 installed. It is estimated that $800 will be allowed for the 20-kw set on the installed purchase price of $1,580 of the 30-kw set 5 years hence. An analysis of the present-worth cost of the power consumed for each of the 5 years is given in Table 18.7.

The present-worth of providing 5 years of service under Plan A is calculated as follows:

Total present worth of power bill for 5 years (ΣE in table)	$3,091
Present-worth cost of 20-kw set installed	1,230
Present worth of 20-kw set trade-in receipt, $800($\overset{P/F\,10,\,5}{0.6209}$)..........	−497
Present worth of 30-kw set, $1,580($\overset{P/F\,10,\,5}{0.6209}$)	981
	$4,805

Table 18.7. PRESENT-WORTH COST OF CONSUMED POWER FOR PLAN *A*

Year	Output rate d-c current, in kw (A)	Efficiency at d-c output rate (B)	Input rate a-c current, A ÷ B (C)	Annual power bill C × $0.02 × 2,000 hr. (D)	Present worth of annual power bill $D \times (\overset{P/F\,10,\,n}{\quad\quad})$ (E)
1	12	74	16.2	$ 648	$589
2	14	77	18.2	728	602
3	16	78	20.5	820	616
4	18	78	23.1	924	631
5	20	76	26.3	1,052	653

Plan *B* involves the purchase of the 30-kw set now so that it will be available to meet the anticipated demand. The present-worth cost of the power consumed for each of the first 5 years is given in Table 18.8.

Table 18.8. PRESENT-WORTH COST OF CONSUMED POWER FOR PLAN *B*

Year	Output rate d-c current, in kw (A)	Efficiency at d-c output rate (B)	Input rate a-c current, A ÷ B (C)	Annual power bill C × $0.32 × 2,000 hr. (D)	Present worth of annual power bill, $D \times (\overset{P/F\,10,\,n}{\quad\quad})$ (E)
1	12	65	18.5	$ 740	$673
2	14	70	20.0	800	661
3	16	74	21.6	864	649
4	18	77	23.4	936	639
5	20	79	25.3	1,012	628

The present-worth cost of providing five years of service under Plan *B* is calculated as follows:

Total present worth of power bill for 5 years (ΣE in table) $3,250
Present-worth cost of 30-kw set installed $1,580
$4,830

Power costs beyond the first 5 years have not been taken into account in the analysis above because they will be equal in subsequent years since a 30-kw set will be used regardless of whether Plan *A* or *B* is adopted. The assumption that a 30-kw unit 5 years old is equivalent to a new set was made to simplify the analyses and because electrical equipment ordinarily has a long service life.

The economy of loading above normal capacity. The life of a piece of equipment is usually inversely proportional to the load imposed upon it. But the output is directly proportional to the load imposed. When such is the case there exists a least cost load that will determine the level of operation for maximum economy.

In some cases, particularly with short-lived assets, the economic load may be determined by experiment. If the effect of various loadings upon the life, maintenance, and other operating costs of a unit of equipment are known or can be estimated with reasonable accuracy, it is practical to determine the load that will result in the greatest economy.

Consider the following example. A concern has an automatic plastic moulding machine that produces one piece for each revolution it makes. It is now being operated at the rate of 150 r.p.m. and produces 18.6 pounds of product per hour. The machine costs $6,150 and is estimated to have an operating life of 10,000 hours. Direct wages plus labor overhead applicable to the operation amount to $1.37 per hour. Average present maintenance and power are $0.072 and $0.064 per hour, respectively. Output of product and cost of power used per hour are estimated to be directly proportional to the r.p.m. of the machine. Since centrifugal forces on machine parts increase with the square of the speed, it is estimated that maintenance costs will increase in proportion to the square of the speed, and that the useful life of the machine will be inversely proportional to the square of the speed. Table 18.9 is constructed on the basis of these assumptions.

On the basis of the estimated use, it would be desirable to increase the speed of the machine to 200 r.p.m. The expected saving per pound of prod-

Table 18.9. RELATIONSHIP OF COST TO OPERATING SPEED

Operating speed in r.p.m. A	Estimated life in hr., $10,000 \times (150)^2/A^2$ B	Average output in pounds per hr. $18.6 \times A/150$ C	Labor cost per hour D	Average maintenance cost per hr., $\$0.072 \times A^2/(150)^2$ E	Power cost per hr., $\$0.064 \times A/150$ F	Average depreciation cost per hr., $\$6,150 \div B$ G	Total cost of operation per hr., $D+E+F+G$ H	Cost per pound, $H \div C$ I
150	10,000	18.6	$1.37	$0.072	$0.064	$0.615	$2.121	$0.114
200	5,630	24.8	1.37	0.128	0.085	1.092	2.675	0.108
250	3,600	31.0	1.37	0.200	0.107	1.708	3.385	0.109

uct would be

$$(\$0.114 - \$0.108) \div \$0.114 = 5.3\%.$$

This is a worthwhile saving, particularly since it shortens the capital recovery period and so lessens the possibilities of losses from obsolescence and inadequacy.

Economic load distribution between machines. In many cases two or more machines are available for the same kind of production. Thus, two or more boilers may be available to produce steam or two or more turbine generators may be on hand to produce power. If the total load is less than the combined capacity of the machines that are available, the economy of operation will, in some measure, depend upon the portion of the total load that is carried by each machine.

As an example of a simple method of determining the load distribution between two machines suppose that two diesel-engine-driven d-c generators, one of 1,000-kw capacity and the other of 500-kw capacity, are available. The efficiencies for different outputs and the corresponding inputs for each are given in Table 18.10.

Table 18.10. EFFICIENCIES FOR DIFFERENT OUTPUTS

Output, in kw	Machine A, 1,000 kw Output		Machine B, 500 kw Output	
	Efficiency in per cent	Input in kw	Efficiency in per cent	Input in kw
100	15.1	663	20.7	483
200	22.0	909	27.9	719
300	26.3	1,141	31.8	943
400	29.5	1,356	32.5	1,231
500	31.2	1,603	32.0	1,563
600	32.3	1,858		
700	32.5	2,154		
800	32.6	2,454		
900	32.4	2,778		
1,000	32.3	3,096		

Suppose that the total load to be met at a certain time is 1,200 kw. This load may be distributed in several ways as is shown in Table 18.11. It is observed that the desired output of 1,200 kw can be produced with a minimum input when the larger unit carries 800 kw and the smaller unit 400 kw.

In the example above input can readily be converted to units other than kilowatts. Suppose that it is desired to express input in terms of dollars. Let it be assumed that fuel oil having an energy content of 18,800 Btu weighs 7.48 pounds per gallon and costs $0.46 per gallon delivered. One kw-hr. is equivalent to 3,410 Btu. From these data the cost per kilowatt hour is calculated to be

$$\frac{\$0.46 \times 3,410}{7.48 \times 18,800} = \$0.0112.$$

Table 18.11. DISTRIBUTION OF LOAD BETWEEN MACHINES

Output in kw			Input in kw		
Machine A	Machine B	Total	Machine A	Machine B	Total
1,000	200	1,200	3,096	719	3,815
900	300	1,200	2,778	943	3,721
800	400	1,200	2,454	1,231	3,685
700	500	1,200	2,154	1,563	3,717

At this rate of cost for fuel, 1,200 kw-hr. output of energy can be produced for 3,685 × $0.0112 = $41.27 if the loads on the two machines are 800 kw and 400 kw, respectively. If the load distribution had been 1,000 kw and 200 kw on the machines, the cost would have been $42.73, an increase over the better method of operation of 3.5%. This is a rather high rate of saving considering the ease with which it is made. The saving is a result of a correct decision based on knowledge of load distribution.

18.4. Producing to a Variable Demand

The demand for many manufactured goods is seasonal. Such variation in demand reflects changes in human wants caused by seasonal changes such as temperature, rainfall, and hours of sunshine. Some items may vary in demand because of social customs as is the case with the sale of fireworks.

The seasonal goods manufacturer may make his product at a relatively low rate throughout the year and store it until it is needed. Or he may acquire sufficient facilities to manufacture the product at a rate equal to the demand during the period in which it is sold.

The disadvantage of the first plan is that storage cost is relatively high, and the disadvantage of the second method is that a rather large equipment investment will be required. The most desirable plan may be a compromise of the two plans above.

The method of solution of this and similar situations may be illustrated by an example. Let it be assumed that 36,000 units of a product are sold during a 4-month period each year as follows:

Month of Year	Number Sold
9th....................	2,000
10th....................	10,000
11th....................	15,000
12th....................	9,000

One machine can make 36,000 units of this product during a year. The fixed charges on the required machine for such items as interest, taxes, insurance, space to house machine, and depreciation due to causes exclusive of usage amounts to $2,000 per year. The cost for depreciation due exclusively to usage, power, supplies, maintenance, and so forth amounts to $0.25 per hour or $40 per month of operation on the basis of a 160-hour month.

Although one machine can meet the demand for the product, the accumulation of finished products throughout the year will result in considerable expense for storage. The expense for storage can be reduced by using more machines to shorten the period required to make the year's needs.

The costs associated with plans of production based on using one, two, and three machines will be determined. For an output of 36,000 units per machine, 12, 6, and 4 months, respectively, will be required to manufacture the annual output of 36,000 units with one, two, and three machines. The number sold each month and the number in storage at the end of each month during the year are given in Table 18.12 and are shown graphically in Figure 18.9.

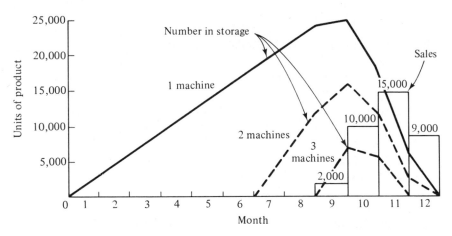

Figure 18.9. Graphical presentation of storage requirements.

The average number in storage for one, two, and three machines is 13,083, 4,083, and 1,083, respectively. These results are obtained by adding the average of the number of units of product in storage at the beginning

Table 18.12. STORAGE REQUIREMENTS FOR MANUFACTURING TO A SEASONAL
DEMAND

Month	Sales During Month	Number in Storage at End of Month for		
		1 Machine	2 Machines	3 Machines
1		3,000		
2		6,000		
3		9,000		
4		12,000		
5		15,000		
6		18,000		
7		21,000	6,000	
8		24,000	12,000	
9	2,000	25,000	16,000	7,000
10	10,000	18,000	12,000	6,000
11	15,000	6,000	3,000	0
12	9,000	0	0	0

and end of each month in the year for each plan and dividing the resulting
sum by 12. The product is valued at $3 per unit. The sum of interest, taxes,
and insurance is taken as 10% of the unit cost and storage costs as $0.20 per
unit per year of storage.

On the basis of the above data the cost with one, two, and three machines
will be as follows.

One Machine

Fixed charges on machine, $1 \times \$2,000$	$2,000
Variable charge on machine, $12 \times \$40$	480
Interest, taxes, and insurance, $13,083 \times \$3 \times 0.10$	3,925
Storage cost, $13,083 \times \$0.20$	2,617
Total cost with one machine	$9,022

Two Machines

Fixed charges on machines, $2 \times \$2,000$	$4,000
Variable charge on machines, $6 \times 2 \times \$40$	480
Interest, taxes, and insurance, $4,083 \times \$3 \times 0.10$	1,225
Storage cost, $4,083 \times \$0.20$	817
Total cost with two machines	$6,522

Three Machines

Fixed charges on machines, $3 \times \$2,000$	$6,000
Variable charge on machines, $4 \times 3 \times \$40$ ...;...................	480
Interest, taxes, and insurance, $1,083 \times \$3 \times 0.10$	325
Storage cost, $1,083 \times \$0.20$	217
Total cost with three machines	$7,022

On the basis of the analysis above, a saving of $500 per year would result
from using two machines in place of three and a saving of $2,500 would result

from using two machines in place of one. It should be realized that there may be considerable hazard in the use of any of the three plans. If only one machine is used, there is a possibility that the product may become outmoded before the sales period or that sales may not be up to expectations. If three machines are used, there is a large investment that may not be recovered before the machines are rendered obsolete or inadequate.

18.5. Equipment Selection for Expanding Operations

In the early stages of the enterprise, when production volume is low, it will usually prove to be economical to purchase equipment whose fixed costs are low. In the latter stages, when sales are approaching the ultimate level, high fixed cost equipment permitting low variable production costs may be most economical. In this connection consider the following example.

It is estimated that annual sales of a new product will begin at 1,000 units the first year and increase by increments of 1,000 units per year until 4,000 units are sold during the fourth and subsequent years. Two proposals for equipment to manufacture the product are under consideration.

Proposal *A* involves equipment requiring an investment of approximately $10,000. Annual fixed cost with this equipment is calculated to be $2,000 and the variable cost per unit of product will be $0.90. The life of the equipment is estimated at 4 years.

Proposal *B* involves equipment requiring an investment of approximately $20,000. Fixed cost of this equipment is estimated at $3,800 per year and variable cost per unit of product will be $0.30. The life of this equipment is also estimated at 4 years.

On the basis of the ultimate annual production of 4,000 units, cost per unit will be as follows:

Proposal A

Annual cost for 4,000 units of product,
$2,000 + (4,000 × $0.90) $5,600.00
Cost per unit, $5,600 ÷ 4,000 .... $1.40

Proposal B

Annual cost for 4,000 units of product,
$3,800 + (4,000 × $0.30) $5,000.00
Cost per unit, $5,000 ÷ 4,000 .... $1.25

On the basis of the ultimate rate of production, Proposal *B* is superior to Proposal *A*. On the basis of the total production during the life of the equipment, the analyses in Table 18.13 apply.

Table 18.13. UNIT COSTS OVER EQUIPMENT LIFE

Year of Life	No. of Units Made	Proposal A Fixed Cost	Proposal A Variable Cost	Proposal B Fixed Cost	Proposal B Variable Cost
1	1,000	$2,000	1,000 × $0.90 = $ 900	$ 3,800	1,000 × $0.30 = $ 300
2	2,000	2,000	2,000 × $0.90 = 1,800	3,800	2,000 × $0.30 = 600
3	3,000	2,000	3,000 × $0.90 = 2,700	3,800	3,000 × $0.30 = 900
4	4,000	2,000	4,000 × $0.90 = 3,600	3,800	4,000 × $0.30 = 1,200
	10,000	$8,000	$9,000	$15,200	$3,000

The cost per unit under Proposal A is $8,000 + $9,000 divided by 10,000 or $1.70. Under Proposal B it is $15,200 + $3,000 divided by 10,000 or $1.82. The calculated advantage of Proposal A over Proposal B would have been increased by considering the time value of money. But perhaps even more important than the difference in cost per unit, particularly for a new enterprise that must conserve its funds or when there is considerable doubt that production schedules will be reached, is the lessser investment required by Proposal A.

At rates of operation approaching capacity, single-purpose machines may be expected to produce more economically than general-purpose machines. Specialized facilities usually result in higher fixed costs and lower variable costs than general-purpose facilities. Consequently, specialized equipment is generally advantageous for high volumes of work and general-purpose equipment is advantageous for low volumes of work.

Single-purpose machines particularly are subject to obsolescence because of change in design or demand for the product on which they are specialized. Soundness of analyses involving specialized facilities rests heavily upon the accuracy of estimates of the volume of work to be performed by them.

PROBLEMS

1. Eleven activities and eight events constitute a certain research and development project. The activities and their expected completion times are as follows:

Activity	*Completion Time in Weeks*
$A \longrightarrow B$	5
$A \longrightarrow C$	6
$A \longrightarrow D$	3
$B \longrightarrow E$	10
$B \longrightarrow F$	7
$C \longrightarrow E$	8

Activity	Completion Time in Weeks
$D \rightarrow E$	2
$E \rightarrow F$	1
$E \rightarrow G$	2
$F \rightarrow H$	5
$G \rightarrow H$	6

(a) Represent the project in the form of an activity-event network.
(b) Calculate T_E and T_L for each event and label each with T_E, T_L, and S.
(c) Identify the critical path and calculate the shortest possible time for project completion.

2. In Problem 1 assume that event C must occur before event D occurs. Modify the activity-event network for this assumption and show how the critical path would be changed.

3. Nine activities and six events are required to execute and complete a certain construction program. The activities and their completion times in weeks under normal and under expedited conditions are as follows:

Activity	Normal		Expedited	
	Duration	Cost	Duration	Cost
AB	10	$3,000	8	$ 5,000
AC	5	2,500	4	3,600
AD	2	1,100	1	1,200
BC	6	3,000	3	12,000
BE	4	8,500	4	8,500
CE	7	9,800	5	10,200
CF	3	2,700	1	3,500
DF	2	9,200	2	9,200
EF	4	300	1	4,600

A linear crash cost-time relationship exists between the normal and expedited conditions. Overhead of $11,000 per week can be saved if the program is completed earlier than the normal schedule.
(a) Represent the program in the form of an activity-event network.
(b) Find the earliest time within which the program can be completed under normal conditions.
(c) Find the minimum cost schedule for the program.

4. Eight maintenance activities, their normal durations in weeks, and the crew sizes for the normal condition are as follows:

Activity	Duration in Weeks	Crew Size
AC	6	8
BC	6	7
BE	2	6
CD	3	2
CE	4	1
DF	7	4
EF	4	6
FG	5	10

(a) Represent the maintenance project in the form of an activity-event network.
(b) Identify the critical path and find the minimum time for project completion.
(c) Extra crew members can be used to expedite activities BC, CE, DF, and EF at a cost given below:

Extra Crew	Weeks Saved	Crash Cost per Week
1	1	$100
2	2	120
3	2	200
4	3	250

If there is a penalty cost of $1,250 per week beyond the minimum maintenance time of 17 weeks, recommend the minimum cost schedule.

5. A contractor has been awarded the contract to construct an earth fill dam requiring 500,000 cubic yards of fill. Two possible sources of fill have been found: Source A, above the dam site at a distance of 4 miles at which the pit will be operated and maintained by the contractor; Source B which is below the dam site at a distance of 2.6 miles from which the contractor must purchase the fill by the cubic yard. The haul cost per cubic yard will be different because in one instance the loaded trucks will travel down slope and in the other they will travel up slope. The following estimates are available:

	A	B
Purchase price of pit	$10,000	....
Cost of fill per cu. yd. at pit	....	$0.22
Construction of haul road	6,200	....
Maintenance of haul road	2,400	....
Hauling cost per cu. yd. per mile	0.12	0.18

Which source should the contractor choose? What savings will result from choosing this source?

6. Two locations are under consideration for a new brick plant. A cost analysis applicable to each location is given below.

	Location A	Location B
Capital cost per 1,000 bricks	$ 6.50	$ 4.70
Material cost per 1,000 bricks	8.75	10.10
Labor cost per 1,000 bricks	20.20	16.90
Distribution cost per 1,000 bricks	18.00	19.25
Total cost per 1,000 bricks	$53.45	$50.95

(a) Calculate the ratio of the profit per 1,000 bricks produced at Location *A* to the profit per 1,000 bricks produced at Location *B* for selling prices of $52, $54, $56, and $58 per 1,000 bricks.

(b) What is the minimum selling price per 1,000 bricks at each location if the profit must be at least 10%?

7. Electric lamps rated at 200 watts and 110, 115, or 120 volts can be purchased for $0.68 each. When the lamps are placed in a 115-volt circuit, the following data apply:

	110 v Lamp	115 v Lamp	120 v Lamp
Average watts input	209.1	195.2	182.8
Average lumens output per watt of input..	18.30	16.84	15.48
Average life of lamp in hours	420	750	1,340

Energy costs $0.035 per 1,000 watt-hours. Determine the cost per million lumen hours for each lamp under the three conditions.

8. The normal operating speed of wire weaving looms is 300 r.p.m. At this speed the looms have a life expectancy of 5 years, an annual maintenance cost of $280, and an annual power cost of $120. The machines cost $4,000 each and have negligible salvage value. Annual space and overhead charges are estimated at $340 per machine. The operator's rate is $4.45 per hour and each operator runs three machines regardless of their speed. The interest rate is 8%. Experimental work on the operation of these machines has resulted in the following data:

Speed r.p.m.	Life in Years	Annual Maintenance Cost	Annual Power Cost
250	7.2	$180	$100
300	5.0	280	120
350	4.2	370	145
400	2.8	660	185
450	2.2	880	240

Summarize the items of cost in a table and determine the most economical revolutions per minute at which to run the machines. Assume that a loom operating at a normal operating speed of 300 r.p.m. will produce 300 units of product per year (each year consists of 40 hours a week for 50 weeks) and that production is directly proportional to the speed of the loom.

9. An order for 8 tons of a certain insulating material has been received. The material is packaged in 1-ton lots for shipment. Two trucks are available, one with a capacity of 7 tons and the other with a capacity of 4 tons. The material must be hauled 25 miles to a job site. Find the economic load distribution between the trucks if the following costs apply:

	Fuel Cost per Mile	*Labor Cost per Ton*	*Other Costs per Ton-Mile*
7-Ton truck	$0.10	$5	$1.20
4-Ton truck	0.07	7	0.80

10. In the operation of a certain type of equipment, repairs average $60 each and the cost of a shutdown for emergency repairs entails an additional loss of $140. In the past, there has been no routine periodic inspection of the equipment. It is believed that periodic inspections would result in a worthwhile saving.

In order to get data on which to determine the frequency of periodic inspections for greatest economy, five groups of an equal number of machines A, B, C, D, and E were inspected 0, 1, 2, 3, and 5 times per week, respectively. The cost per inspection was found to be $12.50. The plant operates 5 days per week, 50 weeks per year. The results of a year's run are shown below:

Machine Group	*Inspections per Week*	*Emergency Repairs*	*Non-emergency Repairs*
A	0	31	27
B	1	20	36
C	2	13	41
D	3	9	43
E	5	4	46

How many inspections should be made per week and what will be the total yearly cost for inspection, loss due to emergency repairs, and cost of repairs for this many inspections per week?

11. A manufacturer of a seasonal item finds that his sales are at the rate of 10,000 units per month, each month of the year except September and October when the rate of sales is 40,000 units per month. This product is made on single-purpose machines that have an initial cost of $24,000 each, an estimated life of

6 years, and no salvage value. Fixed charges other than interest and depreciation on each of these machines amount to $810 and the variable cost amounts to $0.80 per hour of operation. Machine operators are paid $4.60 per hour. Each machine normally produces 25 units of product per hour, and the plant operates 160 hours per month. The material entering the product is received as used and costs $5.10 per unit of product. Charges for interest, taxes, and insurance on the average inventory of finished goods are estimated to be 9% of the variable cost of manufacture and materials. Charges for storage of inventory amount to $0.26 per unit per year of average inventory. How many machines should be employed in the production of the item for minimum cost if the interest rate is 8%?

12. The expected demand for a certain item during the months of January through June is 10,000 per month. It is expected that the demand will be 30,000 per month during the months of July through September and 20,000 per month during the months of October through December. The item can be ordered for a January 1 delivery at $3.95 per unit, for an April 1 delivery at $4 per unit, and for an August 1 delivery at $4.05 per unit. The item cannot be delivered at other times during the year. The cost to initiate and complete a purchase is $150 and the storage cost is $0.10 per unit per year. Interest, insurance, taxes, and other storage costs are estimated to be 9% of the investment in average inventory.

As an alternative, the item can be manufactured on single-purpose machines with the following characteristics:

Machine cost	$18,000
Service life in years	8
Salvage value	0
Annual fixed cost excluding depreciation	$800
Variable cost per hour of operation	$0.42
Hours available per month	160
Production in units per hour	25
Interest rate	8%

The direct labor cost for the operator is $4.50 per hour. The direct material cost per unit is $3. Calculate the number of machines for minimum cost if the manufacturing alternative is chosen and compare this with the least cost purchase policy.

13. A firm is considering the expansion of its operations to include a plating department that will require 5,000,000 kw-hr. of electrical energy per year. The maximum rate of consumption is estimated to be 1,400 kilowatts. The power can be purchased from a power company for $0.022 per kw-hr. providing the firm will construct the required 25 miles of transmission line and furnish necessary transformers. The cost of the high tension line will be $14,500 per mile and the transformers required will cost $28,000. The life of these assets will be 30 years with a salvage value of $12,000.

As an alternative, the firm can construct and operate its own power plant. The plant can be located on a river which would furnish adequate cooling water for condensing purposes if a steam plant is selected or for cooling if a diesel plant is installed. A good location is also available for a hydrogenerating plant 36 miles from the firm's plant site. Costs involved in the construction and maintenance of the three types of power plants are as follows:

	Steam	*Diesel*	*Hydro*
Total investment per kw of capacity	$210.00	$260.00	$310.00
Transmission equipment cost	0	0	$364,000
Fuel required in pounds per kw-hr	1.85	0.10	0
Cost of fuel per pound	$ 0.007	$ 0.021	—
Annual cost of maintenance	$54,000	$35,000	$28,000
Taxes and insurance as a % of investment	2	2	2
Life of plant and equipment in years	20	16	30
Salvage value as a % of original cost....	10	12	5

Determine the cost of each of the four methods of supplying the required power if the interest rate is 8 % compounded continuously.

MATHEMATICAL AND GRAPHICAL
MODELS FOR OPERATIONS

Mathematical models are finding common use in connection with many operational systems. Such models are quantitative abstractions of those aspects of the system important to its description and control. By formulating a mathematical model it is possible to determine values for policy variables which lead to least cost or maximum profit operation.

The delay in developing models for operational systems may be attributed to an early preoccupation with the physical environment. During much of history the limiting factors were predominantly physical. But, with the accumulation of knowledge about physical phenomena, complex operational systems have been developed to produce goods and services. Decision makers are becoming increasingly aware that experience, intuition, and judgment must be augmented with analysis techniques using mathematical models if such systems are to be operated economically.

19.1. Linear Break-Even Models

There are two aspects of an industrial enterprise. One consists of assembling labor, facilities, and material for the production of goods or services. The other consists of the distribution of the goods or services that have been

19

produced. The success of an enterprise depends upon its ability to carry on these activities to the end that there may be a net difference between receipts for goods and services sold and the input necessary to produce and distribute them.

If receipts and costs are assumed to be linear functions of the quantity of product to be made and sold, analysis of their relationships to profit is greatly simplified. If this assumption is made, the patterns of income and costs will appear as in Figure 19.1. Fixed production costs are represented by the line *HL*.

The sum of variable production costs and distribution costs is represented by the line *HK*. Income from sales is represented by the line *OJ*. It must be remembered that this linear representation is only an approximation of actual operating conditions.

Analysis of existing or proposed operations that are represented with a break-even model can be made mathematically or graphically. For an illustration of the mathematical methods, let

N = number of units of product made and sold per year;

R = the amount received per unit of product in dollars; R is equal to the slope of *OJ*;

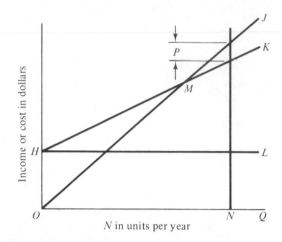

Figure 19.1. General graphical representation for income, cost, units of output, and profit.

$I = RN$, the annual income from sales in dollars; $I = RN$ is the equation of line OJ;
$F =$ fixed cost in dollars per year, represented by OH and HL;
$V =$ variable cost per unit of product; V is equal to the slope of HK;
$TC =$ the sum of fixed and variable cost of N units of product, $F + VN$; $TC = F + VN$ is the equation of line HK;
$P =$ annual profit in dollars per year; $P = I - TC$; negative values of P represent loss;
$M =$ break-even point; at this point $P = 0$;
$Q =$ capacity of plant in units per year.

The break-even point occurs when income is equal to cost. In Figure 19.1 this occurs where lines OJ and HK intersect. At this point $I = TC$ and $RN = F + VN$. Solving for N,

$$N = \frac{F}{R - V}$$

which is the abscissa value of the break-even point.

If $F/(R - V)$ is substituted for N in $I = RN$ or $TC = F + VN$, the ordinate of the break-even point may be found. The value of the ordinate in terms of dollars of income or cost will be

$$I = R\left(\frac{F}{R - V}\right) \quad \text{and} \quad TC = F + \frac{VF}{R - V}.$$

As an example, suppose it is desired to find the break-even point when $R = \$11$, $F = \$4,000$, and $V = \$5$. Then

$$N = \frac{F}{R - V} = \frac{\$4,000}{\$11 - \$5} = 667 \text{ units per year}$$

and

$$I = TC = RN = \$11 \times 667 = \$7,337.$$

Since P is the annual profit it is often desirable to have a relationship that expresses P as a function of the number of units made and sold, N. This relationship may be derived as follows:

$$\begin{aligned} P &= I - TC \\ &= RN - (F + VN) \\ &= (R - V)N - F. \end{aligned}$$

For example, if the annual profit is required when $R = \$11$, $F = \$4,000$, $V = \$5$, and $N = 800$, the following calculations may be made:

$$\begin{aligned} P &= (R - V)N - F \\ &= (\$11 - \$5)800 - \$4,000 \\ &= \$6(800) - \$4,000 = \$800. \end{aligned}$$

Even though the analysis presented was based upon the assumption that fixed cost, variable cost, and income are linear functions of the quantity made and sold, it is very useful in evaluating the effect on profit of proposals for new operations not yet implemented and for which no data exist. Consider a proposed activity consisting of the manufacturing and marketing of a certain plastic drafting template for which the sale price per unit is estimated to be $0.30. The machine required in the operation will cost $1,400 and will have an estimated life of 8 years. It is estimated that the cost of production including power, labor, space, and selling expense will be $0.11 per unit sold. Material will cost $0.095 per unit. An interest rate of 8% is considered necessary to justify the required investment. The costs associated with this activity are:

	Fixed Costs (Annual)	Variable Costs (Per Unit)
Capital recovery with return $1,400(0.1740) .... [A/P 8,8]	$244	
Insurance and taxes	34	
Repairs and maintenance	22	$0.005
Material		0.095
Labor, electricity, and space...............		0.110
Total......................................	$300	$0.210

The difficulty of making a clear-cut separation between fixed and variable costs becomes apparent when attention is focused on the item for repair and maintenance in the above classification. In practice, it is very difficult to distinguish between repairs that are a result of deterioration that takes place with the passage of time and those that result from the wear and tear of use. However, in theory, the separation can be made as shown in this example and is in accord with fact, with the exception, perhaps, of the assumption that repairs from wear and tear will be in direct proportion to the number of units manufactured. To be in accord with actualities, depreciation also undoubtedly should have been separated so that a part would appear as variable cost.

In this example, $F = \$300$, $TC = \$300 + \$0.21\ N$, and $I = \$0.30\ N$. If F, TC, and I are plotted as N varies from 0 to 6,000 units, the result will be as shown in Figure 19.2.

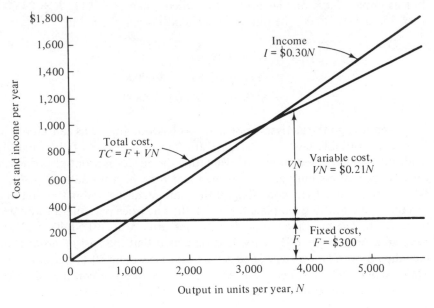

Figure 19.2. Fixed cost, variable cost, and income per year.

The cost of producing templates will vary with the number made per year. The production cost per unit is given by $F/N + V$. If production cost per unit, variable cost per unit, and income per unit are plotted as N varies from 0 to 6,000 the results will be as shown in Figure 19.3. It will be noted that the fixed cost per unit may be infinite. Thus, in determining unit costs, fixed cost has little meaning unless the number of units to which it applies are known.

Income for most enterprises is directly proportional to the number of units sold. However, the income per unit may easily be exceeded by the sum of the fixed and the variable cost per unit for low production volumes. This is shown by comparing the total cost per unit curve and the income per unit curve shown in Figure 19.3.

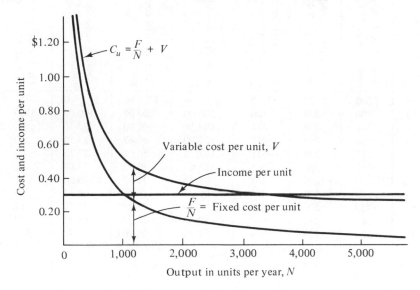

Figure 19.3. Fixed cost, variable cost, and income per year.

19.2. Nonlinear Break-Even Analysis

Table 19.1 gives cost data for a hypothetical plant which has a maximum capacity of 10 units of product per year and rates of production ranging from 0 to 10 units per year. These data and those presented in Tables 19.2 and 19.3 will be used as a framework for an analysis of relationships between production costs, distribution cost, income, and profit where the functions are nonlinear.

The annual total production cost of the hypothetical plant as its rate of production varies from 0 to 10 units per year is given in column 2 of Table 19.1. It should be specifically noted that no distribution costs are embraced in these data.

Annual fixed cost is given as $200 per year in column 3. The annual fixed cost in this example should be considered to be the annual cost of

Table 19.1. RELATIONSHIP OF PRODUCTION COST, NET INCOME FROM SALES, PROFIT, AND THE NUMBER OF UNITS MADE AND SOLD PER YEAR

(1) Annual output, number of units	(2) Annual total production cost, A, Fig. 19.4	(3) Annual fixed production cost, B, Fig. 19.4	(4) Annual variable production cost, C, Fig. 19.4	(5) Average production cost per unit, D, Fig. 19.4	(6) Average fixed cost per unit	(7) Average variable cost per unit, E, Fig. 19.4	(8) Incremental production cost per unit, F, Fig. 19.4
0	$200	$200	$ 0	$ ∞	$ ∞	$ 0	—
1	300	200	100	300.00	200.00	100.00	$100
2	381	200	181	190.50	100.00	90.50	81
3	450	200	250	150.00	66.67	83.30	69
4	511	200	311	127.75	50.00	77.75	61
5	568	200	368	113.60	40.00	73.60	57
6	623	200	423	103.83	33.33	70.50	55
7	679	200	479	97.00	28.57	68.48	56
8	740	200	540	92.50	25.00	67.50	61
9	814	200	614	90.44	22.22	68.22	74
10	924	200	724	92.40	20.00	72.40	110

Table 19.2. RELATIONSHIP OF GROSS INCOME FROM SALES, NET INCOME FROM SALES, DISTRIBUTION COST, AND NUMBER OF UNITS SOLD PER YEAR

(1) Annual output, number of units	(2) Annual gross income from sales, J, Fig. 19.5	(3) Annual distribution cost, K, Fig. 19.5	(4) Annual net income from sales, G, Fig. 19.5	(5) Incremental distribution cost per unit, L, Fig. 19.5	(6) Incremental net annual income from sales per unit, H, Fig. 19.5
0	$ 0	$ 0	$ 0	—	—
1	140	38	102	$ 38	$102
2	280	72	208	34	106
3	420	104	316	32	108
4	560	136	424	32	108
5	700	170	530	34	109
6	840	211	629	41	99
7	980	261	719	50	90
8	1,120	325	795	64	76
9	1,260	405	855	80	60
10	1,400	505	895	100	40

Table 19.3. RELATIONSHIP OF GROSS INCOME, PRODUCTION COST, DISTRIBU-
TION COST, PROFIT, AND NUMBER OF UNITS MADE AND SOLD PER YEAR

Annual output, number of units	Annual total production cost, A, Fig. 19.6	Annual distribution cost, K, Fig. 19.6	Annual total of production and distribution cost, M, Fig. 19.6	Annual gross income from sales, J., Fig., 19.6	Annual profit N, Figs. 19.4 and 19.6	Incremental total production and distribution costs, O, Fig. 19.6
0	$200	$ 0	$ 200	$ 0	−$200	—
1	300	38	338	140	−198	$138
2	381	72	453	280	−173	115
3	450	104	554	420	−134	101
4	511	136	647	560	−87	93
5	568	170	738	700	−38	91
6	623	211	834	840	6	96
7	679	261	940	980	40	106
8	740	325	1,065	1,120	55	125
9	814	405	1,219	1,260	41	154
10	924	505	1,429	1,400	−29	210

maintaining the plant whose capacity is 10 units per year in operating condition at a production rate of zero units per year. The fixed cost of a plant is
analogous to the standby costs of a steam boiler whose fire is banked but
which is ready to furnish steam on short notice.

The difference between the annual total production cost in column 2
and the fixed production cost in column 3 constitutes the annual variable
cost and appears in column 4. Average unit production, fixed cost, and variable cost appear in columns 5, 6, and 7, respectively.

The increment production costs per unit given in column 8 were obtained by dividing the differences between successive values of annual total
production cost as given in column 2 by the corresponding difference between
successive values of annual output given in column 1. In this case the divisor
will be equal to unity, but it should be understood that an increment of production of several units may be convenient in some analyses.

The values in all columns of Table 19.2 except those in column 6 and
the values of all columns of Tables 19.2 and 19.3 have been plotted with
respect to the annual output in number of units in Figures 19.4, 19.5 and
19.6 and have been keyed for identification.

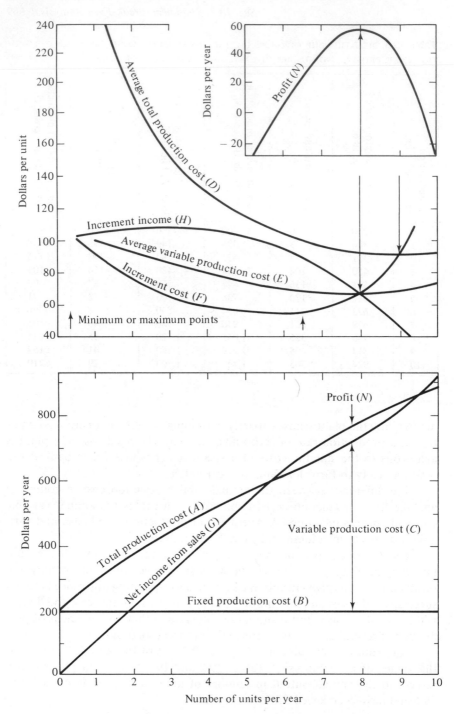

Figure 19.4. Graphical presentation of the data given in Table 19.1.

Consider curve *A* in Figure 19.4. Total production cost curves of actual operations take a great variety of forms. Ordinarily, however, a producing unit will produce at minimum average cost at a rate of production between zero output and its maximum rate of output. Thus, the average unit cost of production will decrease with an increase in the rate of production from zero until a minimum average cost of production is reached and then the average cost per unit will rise.

The average total production cost is given by curve *D* in Figure 19.4. It will be noted that it reaches a minimum at a rate of nine units per year. The average variable production cost is given in curve *E*; its minimum occurs at eight units per year. Curves *D* and *E* should be considered in relation to column 6 of Table 19.1. It should be noted that fixed cost per unit is inversely proportional to the number of units per year. Reduction of fixed cost per unit is an important factor tending toward lower average cost with increases in rates of production.

Incremental production costs are given in curve *F*; their minimum value is reached at six units per year. The form of this curve shows that the incremental cost of production per unit declines until six units per year are produced and then it rises.

An interesting fact to observe is that the incremental cost curve *F* intersects the average total production cost curve *D* and the average variable production costs curve *E* at their minimum points. The incremental cost curve *F* is a measure of the slope of curve *A*. It may be noted that the slope of curve *A* decreases until six units per year is reached and then it increases. The same could have been said of a variable production cost curve *C* if one had been plotted.

Pattern of income and cost of distribution. Further consideration will be given to the curves of Figure 19.4 after income from sales of product and the cost of distribution have been explained. For simplicity, the cost of distribution will be considered to be a summation of an enterprise's expenditures to influence the sale of its products and services. Such items as advertising, sales administration, salesmen's salaries, and expenditures for packaging and decoration of products done primarily for sales appeal will be included in the cost of distribution. Gross annual income from sales is the total income received from customers as payment for products. In columns 2 and 3 of Table 19.2 gross annual income from sales nd the annual cost of distribution are given for various rates of product sales of the firm for which cost-of-production data were given in Table 19.1 and Figure 19.4.

The difference between the gross annual income from sales and the annual cost of distribution is equal to the annual net income from sales given in column 4. Incremental distribution costs per unit are given in column 5 and incremental net income from sales per unit appears in column 6. The

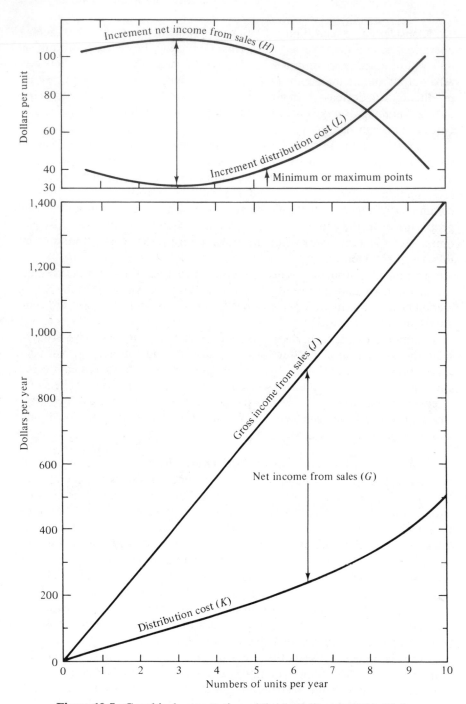

Figure 19.5. Graphical presentation of the data given in Table 19.2.

values of each of these columns have been plotted with respect to annual output in Figure 19.5. The gross annual income from sales, curve *J* in Figure 19.5, is a straight line and is typical of situation's in which products are sold at a fixed price. The annual incremental distribution cost, curve *L*, first falls slightly until a minimum is reached and then it rises. This is typical of situations in which sales effort is relatively inefficient at low levels and in which sales resistance increases with increased number of units sold.

A pattern of distribution cost, made up of fixed and variable costs similar to those of production, might have been taken. Ordinarily, fixed distribution costs are a minor consideration; they have been omitted from this analysis in the interest of simplicity.

A representation of the net income from sales is given as *G* in Figure 19.5. The positive difference between net income from sales and the total cost of production, curve *A* in Figure 19.4, represents profit. Profit is shown as *N* in Figure 19.4, it should be noted that the maximum profit, or minimum loss if no profit is made, occurs at a rate of production at which the incremental income curve *H* intersects the incremental cost curve *F*, or at eight units per year. From column 8, Table 19.1, and column 6, Table 19.2, it may be observed that the incremental product on cost and incremental net income for a ninth unit are $74 and $60, respectively. Thus, a loss of $14 profit would be sustained if the activity were increased to nine units. When the profit motive governs, there is no point in producing beyond the point where the incremental cost of the next unit will exceed the incremental income from it.

If the sales effort in the example above is resulting in sales of nine or ten units per year, a number of steps might be taken. The sales effort could be reduced, causing sales to drop. The price could be increased, causing sales to decrease and the income per unit to increase. The plant could be expanded or other changes could be made to alter the pattern of production costs. Consideration of any of the steps above would require a new analysis embracing the altered factors.

Consolidation of production and distribution cost. It is common practice to consolidate production and distribution cost for analysis of operations. To illustrate this practice an analysis of the example above will be made in this manner. The method is illustrated in Table 19.3, whose values have been plotted as several curves in Figure 19.6.

Profit in this case will be equal to the difference between gross income *J* and total production and distribution cost *M*. The point of maximum profit will occur when the incremental gross income curve *P* is intersected by a rising incremental production and distribution cost curve *O*. This occurs at a rate of 8 units per year. It would be noted that if production reaches 10 units, a loss of $29 will be sustained from the year's operation.

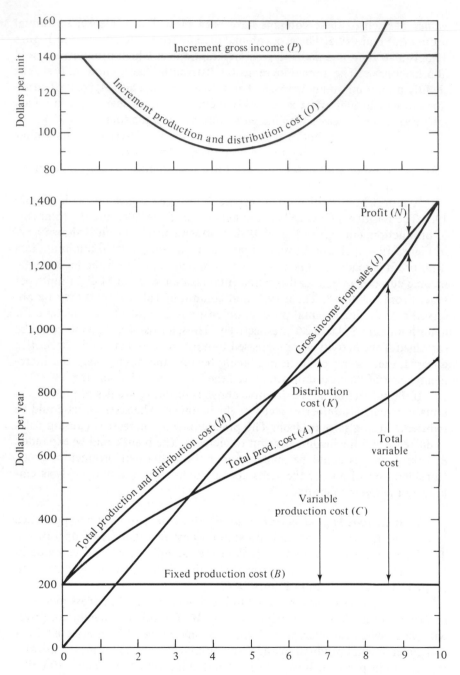

Figure 19.6. Graphical presentation of the data given in Table 19.3.

19.3. Linear Programming Models

Operational situations involving many activities which compete for scarce resources can often be optimized by use of the general linear programming model. Usually, any operation involving a linear effectiveness function and linear resource constraints can be expressed in the format of the linear programming model formally stated as optimize the effectiveness function

$$E = \sum_{j=1}^{n} e_j x_j$$

subject to the constraints

$$\sum_{j=1}^{n} a_{ij} x_j = b_i \qquad i = 1, 2, \ldots, m$$

$$x_j \geq 0 \qquad j = 1, 2, \ldots, n.$$

Optimization requires either maximization or minimization of E depending upon the measure of effectiveness involved. The decision maker has control of the variables, x_j. Not directly under his control are the effectiveness coefficients, e_j, the constants a_{ij}, and the constants, b_j.

Consider the following production example. Two products compete for scarce machine time during the production process. A single unit of product A requires 2.0 minutes of punch press time and 4.0 minutes of assembly time. The profit for product A is $0.60 per unit. A single unit of product B requires 3.0 minutes of punch press time and 2.5 minutes of welding time. The profit for product B is $0.75 per unit. The available capacity of the punch press department is 1,200 minutes per week. The welding department has an available capacity of 500 minutes per week and the assembly department can supply 1,600 minutes of capacity per week. The manufacturing and other data for this production situation are summarized in Table 19.4.

In this example, two products compete for scarce production time. The objective is to determine the quantity of product A and the quantity of prod-

Table 19.4. PRODUCTION DATA FOR TWO PRODUCTS

Department	Product A	Product B	Capacity
Punch press	2.0	3.0	1,200
Welding	0	2.5	500
Assembly	4.0	0	1,600
Profit	$0.60	$0.75	

uct B to produce so that total profit will be maximized. This will require maximizing the linear effectiveness function

$$TP = \$0.60A + \$0.75B$$

subject to the linear constraints

$$2.0A + 3.0B \le 1,200$$
$$0A + 2.5B \le 500$$
$$4.0A + 0B \le 1,600$$

where both A and B must be ≥ 0.

The graphical equivalent of the algebraic statement of this two-product problem is shown on Figure 19.7. The set of linear restrictions defines a region of feasible solutions. This region lies below $2.0A + 3.0B = 1,200$ and is restricted further by the requirements that $B \le 200$, $A \le 400$, and that both A and B be non-negative. Thus, the scarce production time (resources) determine which combinations of these activities are feasible and which are not feasible.

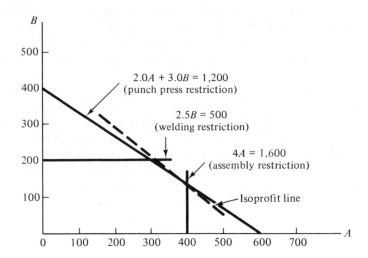

Figure 19.7. Graphical solution for a two-product production problem.

The production quantity combinations of A and B that fall within the region of feasible solutions constitute feasible production programs. That combination of A and B which maximizes profit is sought. The relationship of A and B is $A = 1.250 B$ and is based on the relative profit of each product. The total profit will depend upon the production quantity combination chosen. Thus, there is a family of isoprofit lines, one of which will have at least one point in the region of feasible production quantity combinations and have a maximum distance from the origin. The member that satisfies this condition intersects the region of feasible solutions at the extreme point $A = 400$, $B = 133$. This is shown as a broken line in Figure 19.7, and represents a total profit of $\$0.60(400) + \$0.75(133) = \$340$. No other production quantity combination would result in a higher profit.

This illustration dealt with only two activities making the graphical solution method practical. Usually, however, several activities are involved in a linear programming application. In this case an optimization technique known as the simplex method may be used.

19.4. Models for Inventory Operations

When the decision has been made to procure a certain item, it becomes necessary to determine the procurement quantity that will result in a minimum cost. The demand for the item may be met by producing a year's supply at the beginning of each year or by procuring a day's supply at the beginning of each day. Neither of these extremes may be the most economical in terms of the sum of costs associated with procurement and holding the item in inventory. Models for inventory operations are used to determine the most economical procurement quantity.

The economic purchase quantity. If the demand for an item is met by purchasing once per year, the cost incident to purchasing will occur once, but the large quantity received will result in a relatively high inventory holding cost for the year. Conversely, if orders are placed several times per year, the cost incident to purchasing will be incurred several times per year, but since small quantities will be received, the cost of holding the item in inventory will be relatively small. If the decision is to be based on economy of the total operation the purchase quantity that will result in a minimum annual cost must be determined. Let

TC = total yearly cost of providing the item;
D = yearly demand for the item;
N = number of purchases per year;
t = time between purchases;
Q = purchase quantity;
C_i = item cost per unit (purchase price);
C_p = purchase cost per purchase order;
C_h = holding cost per unit per year made up of such items as interest, insurance, taxes, storage space, and handling.

The total yearly cost will be the sum of the item cost for the year, the purchase cost for the year, and the holding cost for the year. That is,

$$TC = IC + PC + HC.$$

The item cost for the year will be the time cost per unit times the yearly demand in units, or

$$IC = C_i(D).$$

If it is assumed that the demand for the item is constant throughout the year, the purchase lead time is zero and no shortages are allowed; the resulting inventory system may be represented graphically as shown in Figure 19.8.

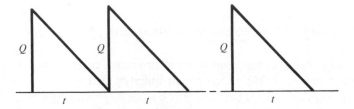

Figure 19.8. Graphical representation of an inventory process for purchasing.

The purchase cost for the year will be the cost per purchase times the number of purchases per year, or

$$PC = C_p(N).$$

But since N is the yearly demand divided by the purchase quantity

$$PC = \frac{C_p(D)}{Q}.$$

Since the interval, t, begins with Q units in stock and ends with none, the average inventory during the cycle will be $Q/2$. Therefore, the holding cost for the year will be the holding cost per unit times the average number of units in stock for the year, or

$$HC = \frac{C_h(Q)}{2}.$$

The total yearly cost of providing the required item is the sum of the item cost, purchase cost, and holding cost, or

$$TC = C_i(D) + \frac{C_p(D)}{Q} + \frac{C_h(Q)}{2}.$$

The purchase qunatity resulting in a minimum yearly cost may be found by differentiating with respect to Q, setting the result equal to zero, and solving for Q as follows:

$$\frac{dTC}{dQ} = -\frac{C_p(D)}{Q^2} + \frac{C_h}{2} = 0$$

$$Q^2 = \frac{2C_p(D)}{C_h}$$

$$Q = \sqrt{\frac{2C_p(D)}{C_h}}.$$

As an example of the use of this model, assume that the annual demand for a certain item is 1,000 units. The cost per unit is $6 delivered. Purchasing cost per purchase order is $10 and the cost of holding one unit in inventory for one year is estimated to be $1.32.

The economic purchase quantity may be found by substituting the appropriate values in the derived relationship as follows:

$$Q = \sqrt{\frac{2(\$10)(1,000)}{\$1.32}}$$

$$= 123 \text{ units.}$$

Total cost may be expressed as a function of Q by substituting the costs and various values of Q into the total cost equation. The result is shown in Table 19.5. The tabulated total cost value for $Q = 123$ is the minimum cost purchase quantity for the condition specified. Total cost as a function of Q is illustrated in Figure 19.9.

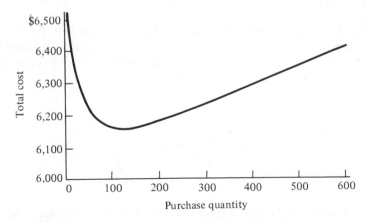

Figure 19.9. Total cost as a function of purchase quantity.

Table 19.5. TABULATED VALUES OF TOTAL COST AS
A FUNCTION OF PURCHASE QUANTITY

Purchase Quantity	Total Cost
50	$6,233
100	$6,166
123	$6,162
150	$6,165
200	$6,182
300	$6,231
400	$6,289
600	$6,413

The economic production quantity. When the decision has been made to produce a certain item, it becomes necessary to determine the production quantities which are determined in a manner similar to determining economic purchase quantities. The difference in analysis is because a purchased lot is received at one time while a production lot accumulates as it is made. Let

TC = total yearly cost of providing the item;
D = yearly demand for the item;
N = number of production runs per year;
t = time between production runs;
Q = production quantity;
C_i = item cost per unit (production cost);
C_s = set-up cost per production run;
C_h = holding cost per unit per year made up of such items as interest, insurance, taxes, storage space, and handling;
R = production rate.

If it is assumed that the demand for the item is constant, the production rate is constant during the production period, the production lead time is zero, and no shortages are allowed; the resulting inventory system may be represented graphically as in Figure 19.10.

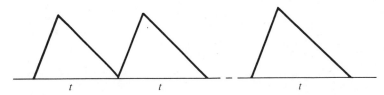

Figure 19.10. Graphical representation of an inventory process for production.

The total yearly cost will be the sum of the item cost for the year, the set-up cost for the year, and the holding cost for the year. That is,

$$TC = IC + SC + HC.$$

The item cost for the year will be the item cost per unit times the yearly demand in units, or

$$IC = C_i(D).$$

The set-up cost for the year will be the cost per set-up times the number of set-ups per year, or

$$SC = C_s(N).$$

But since N is the yearly demand divided by the production quantity,

$$SC = \frac{C_s(D)}{Q}.$$

When items are added to inventory at the rate of R units per year and are taken from inventory at a rate D units per year, where R is greater than D, the net rate of accumulation is $(R - D)$ units per year. The time required to produce D units at the rate R units per year is D/R years. If D units are made in a single lot, the maximum acccumulation in inventory will be $(R - D)$ D/R. Since no units will be in storage at the end of the year, the average number in inventory will be

$$\frac{(R - D)\frac{D}{R} + 0}{2} = (R - D)\frac{D}{2R}.$$

If N lots are produced per year, the average number of units in storage will be

$$(R - D)\frac{D}{2RN}.$$

But, since $N = D/Q$, the average number of units in storage may be expressed as

$$(R - D)\frac{Q}{2R}.$$

The holding cost for the year will be the holding cost per unit times the average number of units in storage for the year, or

$$HC = C_h(R - D)\frac{Q}{2R}.$$

The total yearly cost of producing the required item is the sum of the item cost, set-up cost, and holding cost, or

$$TC = C_i(D) + \frac{C_s(D)}{Q} + C_h(R - D)\frac{Q}{2R}.$$

The production quantity resulting in a minimum yearly cost may be found by differentiating with respect to Q, setting the result equal to zero, and solving for Q as follows:

$$\frac{dTC}{dQ} = -\frac{C_s(D)}{Q^2} + \frac{C_h(R - D)}{2R} = 0$$

$$Q^2 = \frac{C_s(D)2R}{C_h(R - D)} = \frac{C_s(D)2}{C_h\left(1 - \dfrac{D}{R}\right)}$$

$$Q = \sqrt{\frac{2C_s(D)}{C_h\left(1 - \dfrac{D}{R}\right)}}$$

As an example of the use of this model assume that the annual demand for a certain item is 1,000 units. The cost of production is $5.90 per unit and includes the usual cost elements of direct labor, direct material, and factory overhead. The set-up cost per lot is $50 and the item can be produced at the rate of 6,000 units per year. The cost of holding one unit in inventory for one year is estimated to be $1.30.

The economic production quantity may be found from the derived relationship by substituting the appropriate values as follows:

$$Q = \sqrt{\frac{2(\$50)(1,000)}{\$1.30\left(1 - \dfrac{1,000}{6,000}\right)}}$$

$$= 302 \text{ units.}$$

Total cost may be expressed as a function of Q by substituting costs and various values of Q into the total cost equation. The result is shown in Table 19.6. The tabulated total cost value for $Q = 302$ is the minimum cost production quantity for the condition specified. Total cost as a function of Q is illustrated in Figure 19.11.

The "make or buy" decision. The question of whether to manufacture or purchase a needed item may be resolved by the application of minimum cost analysis for multiple alternatives. The alternative of producing may be compared with the alternative of purchasing if the minimum cost procure-

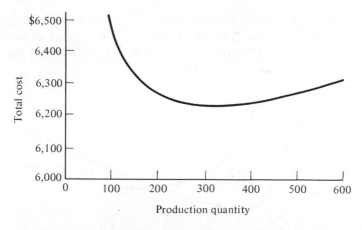

Figure 19.11. Total cost as a function of production quantity.

Table 19.6. TABULATED VALUES OF TOTAL COST AS A FUNCTION OF PRODUCTION QUANTITY

Production Quantity	Total Cost
100	$6,454
150	$6,314
200	$6,258
300	$6,229
302	$6,228
400	$6,241
500	$6,270
600	$6,307

ment quantity for each is computed and used to find the respective total cost values. Choice of the total cost value that is a minimum identifies the best of the two alternatives.

For example, suppose that an item will have a yearly demand of 1,000 units and that the costs associated with purchasing and producing are the same costs which were assumed in the previous two sections, that is,

	Purchase	*Produce*
Item cost	$ 6.00	$ 5.90
Purchase cost	$10.00	—
Set-up cost	—	$50.00
Holding cost	$ 1.32	$ 1.30

For the conditions assumed, total cost as a function of the purchase quantity was given in Table 19.5 and total cost as a function of production quantity was given in Table 19.6. Total cost as a function of purchase quantity was graphed in Figure 19.9 and total cost as a function of production quantity was graphed in Figure 19.11. If Figures 19.9 and 19.11 are superimposed, the result is as shown in Figure 19.12.

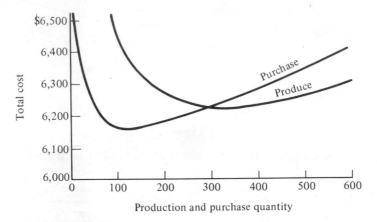

Figure 19.12. Total cost as a function of production and purchase quantities.

The decision of whether to produce or purchase may be made by examining and comparing the minimum cost for each alternative. In this case, the decision to purchase will be the least cost alternative and will result in a saving of $6,228 less $6,162, or $66 per year. If the decision were made on the basis of item cost alone, the needed item would have been supplied by producing with a resultant loss of $66 per year.

The optimal procurement policy may be formally stated by specifying that the item will be procured when the available stock falls to zero units on hand, for a procurement quantity of 123 units, from the purchasing source. In essence, the policy states when, how much, and from what source. The policy, if the decision has been made to purchase, states when and how much, the source being fixed by restriction. Therefore, the decision of "make or buy" is essentially a release of the restriction that the source is fixed. This analysis may be extended to any number of sources; for example, it may be used to compare re-manufacturing with purchasing, to compare alternate manufacturing facilities, or to evaluate alternate vendors.

19.5. Analysis of Waiting Line Operations

A general waiting line or queuing system is shown in Figure 19.13. The system exists because the population shown demands service. To satisfy the demand, a decision maker must specify the level of service capacity at the service facility. This is a problem in operations economy which will involve altering the service rate at existing channels or by adding or deleting channels. One approach to the problem is illustrated by the example of this section.

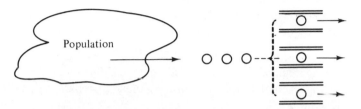

Figure 19.13. A multiple-channel waiting line system.

Failures of equipment do not occur with regularity. In addition, the amount of repair work needed for any given failure may be expected to vary above and below the average amount needed over a long period of time. Each of these events, the occurrence of a failure and the amount of repair needed, are a function of many unpredictable chance causes.

If the number of repair crews provided is just sufficient to take care of the average amount of repair work needed, there will be a considerable backlog of work waiting to be done and the cost of equipment down time will be high. When down time is costly, it may be wise to maintain repair crew capacity in excess of that needed for the average repair load even though this will result in some idleness on the part of the repair crews.

As an example of the analysis required in finding the economical number of repair crews, consider an illustration from the petroleum industry. In petroleum production, heavy equipment is used to pump oil to the surface. When this equipment fails, it is necessary to remove it from the well for repairs. The required repairs are made by crews of three to five men who are equipped with heavy, portable machinery to pull the pumping equipment from the well.

Between the time the pumping equipment fails and the time it is repaired, the well is idle. This results in a loss, known as *lost production*, equal to the amount of oil that would have been produced if there had been no failure of equipment. In some cases lost production may be only production that is deferred to a later date. In other cases lost production is partially lost because of drainage to competitors' wells. This loss may be quite large. Two day's down time of a well producing 200 barrels of oil per day at $10 per barrel, for instance, when drainage to a competitor's well is judged to be half of lost production, results in a loss of $1,000.

The down time of a well is made up of the time that the well is idle before the repair crew gets to it and the time it is idle while repair is in process.

Since the rate at which repairs can be made is substantially controlled by the repair equipment, reduction in loss is brought about by reducing the time a well is idle awaiting repair crews. This is done at the expense of having excess repair crews. Thus, the problem is to balance the number of repair crews with the lost production associated with delays of repairs to wells. Consider the following data adapted from an actual situation in which the well are operated 24 hours a day, 7 days per week:

N = number of oil wells in field = 30;
U = average interval at which individual wells fail = 15 days;
T = average actual time for company crew to repair well failure = 12 hours;
C_1 = hourly cost for company-operated repair unit and repair crew = $60 per hour;
C_2 = lost production per well when "down" = $1,000 per day.

Under the present situation, analysis has revealed that operation of the equivalent of one repair unit 24 hours per day, 7 days per week can keep up with the repair of failures in the long run. This means that one channel is sufficient to service the 30 wells. However, serious losses are arising from delays in getting to wells after notification of their failure. Delay is caused by the chance bunching of well failures. For instance, one period during which no wells fail may be followed by a period of an above-average number of failures. Since repairs cannot be made before failures occur, it is clear that crews sufficient to take care of the average number of failures will have a backlog of failures awaiting them most of the time.

The first step in an analysis to determine the most economical number of repair crews is to determine the number of wells that may be expected to fail during each and every day of a period in the future. The number of failures to expect in the future may be estimated on the basis of past records or by mathematical analysis. The former method is used in the example because it is more revealing.

In the solution of this example the pattern of well failures of a previous 30-day period selected at random will be considered to be representative of future periods. A 30-day period will be taken in the interest of simplicity, even though experience has shown that longer periods are advantageous.

Since the company's unit can repair two wells per day when operating three 8-hour shifts per day, unrepaired wells will be carried over one day each day that the number of failures plus the carry-over from the previous day exceeds two.

Line A of Figure 19.14 shows the number of wells that failed during the 30-day period selected at random and considered to be typical. The number of wells failing during any one day ranged between 0 and 5, and their total was 57.

Day of period

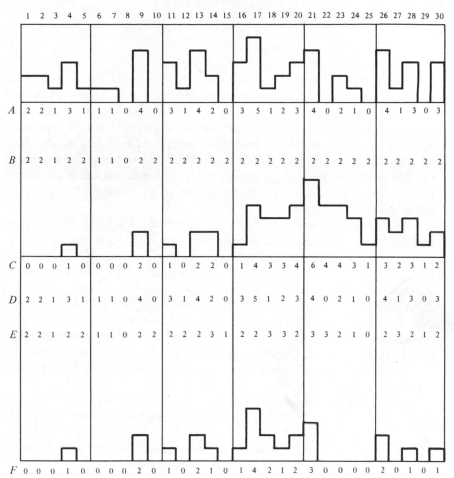

Figure 19.14. Pattern of oil-well failure and two repair methods.

Line *B* gives the number of wells repaired each day with the company's unit and crews. During the 30-day period, 57 wells failed and 55 were repaired. Thus, in spite of the periodic backlogs, there was idle time of the unit and crews on the days numbered 3, 6, 7, and 8—sufficient to repair 5 wells.

The carry-over of unrepaired wells is given in line *C*. The total carry-over for the 30-day period is 52 well-days.

On the present basis of operation, employing the equivalent of one unit 24 hours per day, the total cost incident to well failures and repair during the 30-day period is calculated as follows:

$$TC = C_1(24)(30) + C_2(52)$$
$$= \$60(24)(30) + \$1,000(52)$$
$$= \$95,200.$$

Analysis of the pattern of occurrence of unrepaired wells carried over reveals that, once a backlog has accumulated, wells may remain unrepaired and unproductive for a long period. As a remedy the supervisor in charge considers the feasibility of hiring an additional unit and a crew (operating with two channels) whenever a backlog of unrepaired wells has accumulated. He finds that a repair unit and crew can be hired on short notice when needed for $80 per hour. Because of greater travel distance and other reasons, a hired unit and crew has been found to require an average of 16 hours to repair a well.

The supervisor wishes to determine the effect of a policy of hiring an additional unit and crew for a 16-hour period on days when there is a carry-over of two or more unrepaired wells from the previous day. He assumes that the hired unit and crew will be paid for a minimum of 16 hours each time it is asked to report, whether or not it is used.

If this policy had been in effect during the 30-day period under consideration, the additional unit and repair crew would have been hired for 16 hours on the days numbered 10, 14, 18, 19, 21, 22, and 27; the total number of wells repaired during each day would have been as given in Line E of Figure 19.14.

Under this plan of operation, the total carry-over would have been 24 well-days as is shown in line F.

Line E indicates that there was idle time of units and crew during the days numbered 3, 6, 7, 8, 10, 15, 24, 25, and 29—sufficient for the repair of 11 wells.

On the basis of the analysis above, the total cost incident to well failures and repair for the 30-day period is calculated as follows:

$$TC = \$60(24)(30) + \$80(16)(7) + \$1,000(24)$$
$$= \$76,160.$$

The decision to hire an additional unit and crew on days when there is a carry-over of two or more unrepaired wells from a previous day would result in a reduction of the total cost of operations by $95,200 less $76,160, or $19,040 per month if future months' experience the same pattern of failures. Actually, this is an average value which will be achieved in the long run if the data used exhibit an average pattern.

PROBLEMS

1. The fixed cost of a machine (depreciation, interest, space charges, maintenance, indirect labor, supervision, insurance, and taxes) is F dollars per year. The variable cost of operating the machine (power, supplies, and similar items, excluding direct labor) is V dollars per hour of operation. N is the number of hours the machine is operated per year, TC the annual total cost of operating the machine, TC_h the hourly cost of operating the machine, t the time in hours to process one unit of product, M the machine cost of processing a unit of product, and n the number of units of product processed per year. In terms of these symbols, write expressions for (a) TC, (b) TC_h, (c) M.

2. In Problem 1, $F = \$600$ per year, $t = 0.2$ hour, $V = \$0.50$ per hour, and n varies from 0 to 10,000 in increments of 1,000.
   (a) Plot values of M as a function of n.
   (b) Write the expression for TC_u the total cost of direct labor and machine cost per unit using the symbols given in Problem 1 and letting W equal the hourly cost of direct labor.

3. A semiautomatic arc welding machine that is used for a certain joining process costs $10,000. The machine has a life of 5 years and a salvage value of $1,000. Maintenance, taxes, interest, and other fixed costs amount to $500 per year. The cost of power and supplies is $3.90 per hour, and the operator receives $4.60 per hour. The cycle time per unit of product is 60 minutes. If interest is 9% calculate the cost per unit of processing the product if (a) 200, (b) 600, (c) 1,200, (d) 2,500 units of product are made per year.

4. An analysis of the current operations of a firm results in the conclusion that the production cost is $\$10,900 + 65x + 1500x^{1/2}$ where x is the number of units produced. The selling cost is $\$500\,y^{1/2}$ where y is the number of units sold during the year. The selling price is $180 per unit. Plot the production cost, the selling cost, the total production and selling cost, the gross income from sales, and profit for production levels ranging from 0 to 200 units per year. Determine the level of production for maximum profit.

5. A manufacturing concern estimates that its expenses per year for different levels of operation would be as follows:

Output in units of product	0	10	20	30	40	50
Administrative and sales	$4,900	$ 5,700	$ 6,200	$ 6,700	$ 7,100	$ 7,500
Direct labor and materials	0	2,500	4,600	6,400	8,100	9,800
Overhead expense	4,120	4,190	4,270	4,350	4,400	4,550
Total	$9,020	$12,390	$15,070	$17,450	$19,640	$21,850

(a) What is the incremental cost of maintaining the plant ready to operate (the incremental cost of making zero units of product)?
(b) What is the average incremental cost per unit of manufacturing the first increment of 10 units of product per year?
(c) What is the average incremental cost per unit of manufacturing the increment of 31 to 40 units per year?
(d) What is the average total cost per unit when manufacturing at the rate of 30 units per year?
(e) At a time when the rate of manufacture is 20 units per year, a salesman reports that he can sell 10 additional units at $310 per unit without disturbing the market in which the company sells. Would it be profitable for the company to undertake the production of the 10 additional units?

6. The product of an enterprise has a fixed selling price of $76. An analysis of production and sales costs and the market in which the product is sold has produced the following results:

Level of Operation, Units of Product	Total Production and Selling Cost, Dollars	Level of Operation, Units of Product	Total Production and Selling Cost, Dollars
0	$13,200	600	$35,000
100	17,900	700	38,600
200	21,400	800	47,100
300	24,600	900	55,600
400	27,200	1,000	65,400
500	31,500		

(a) Determine the profit for each level of operation.
(b) Plot production and selling cost, income from sales, profit, average incremental production and selling cost per unit, and average incremental income per unit for each level of operation.

7. A manufacturer of radio receivers has an annual fixed cost of $300,000 and a variable cost of $6.60 per unit produced. An engineering proposal under consideration involves redesign of the present AM unit to include FM. The production engineering staff estimates that the redesigned product will increase the annual fixed cost by $200,000 and the variable cost by $2.60 per unit. A market survey produced the following information:

Unit Selling Price	Sales with AM Only	Sales with AM and FM
$35.00	0	6,200
30.00	3,000	21,000
25.00	11,000	39,500
20.00	25,500	64,000
15.00	48,000	100,000
12.50	74,000	—
11.00	98,000	—

(a) Plot the total fixed cost of each plan, and plot the total income from each plan for the various selling prices against output as the abscissa.

(b) Determine the selling price for maximum annual profit if only AM receivers are manufactured; if AM and FM receivers are manufactured.

(c) Should the design change be adopted?

8. A company has priced its product at $1 per pound and is operating at a loss. Sales at this price total 850,000 pounds per year, which is equivalent to 64% of the total production capacity. The company's fixed cost of manufacture and selling is $480,000 per year and the variable cost is $0.46 per pound. It appears, from information obtained by a market survey, that price reductions of $0.05, $0.10, $0.15, and $0.16 per pound from the present selling price will result in total annual sales of 70%, 75%, 85% and 97% of the total production capacity per year, respectively.

(a) Calculate the annual profit that will result from each of the selling prices given, assuming variable cost per unit will be the same for all production levels.

(b) Determine graphically the annual profit that will result from each of the selling prices by the use of a break-even type chart.

(c) Solve algebraically for the profit that will result from each of the selling prices.

9. A certain firm has the capacity to produce 650,000 units of product per year. At present, it is operating at 64% of capacity. The firm's annual income is $416,000. Annual fixed costs are $192,000 and the variable costs are equal to $0.356 per unit of product.

(a) What is the firm's annual profit or loss?

(b) At what volume of sales does the firm break even?

(c) What will be the profit or loss at 70, 80, and 90% of capacity on the basis of constant income per unit and constant variable cost per unit?

10. A company is operating at capacity for one shift per day. Annual sales are 4,100 units per year and the income per year and the income per units is $190. Fixed costs are $280,000 and total variable costs are $490,000 per year at the present rate of operation.

If output can be increased, 400 additional units can be sold at $200 per unit during the coming year. These additional units can be produced through overtime operation at the expense of a 20% increase in the unit variable cost of the additional units. The elasticity of demand for the product in question is believed to be such that an increase in selling price to $208 will result in curtailing demand to 3,600 units. For greatest profit in the coming year should output be increased or should selling price be increased?

11. A manufacturing company owns two Plants, A and B, that produce an identical product. The capacity of Plant A is 65,000 units annually while that of Plant B is 85,000 units. The annual fixed cost of Plant A is $260,000 per year and the variable cost is $3.20 per unit. The corresponding values for Plant B are $280,000 and $3.90 per unit. At present, Plant A is being operated at 35% of capacity and Plant B is being operated at 40% of capacity.

(a) What are the unit costs of production of Plant A and Plant B?
(b) What are the total cost and the average cost of the total output of both plants?
(c) What would be the total cost to the company and the unit cost if all production were transferred to Plant A?
(d) What would be the total cost to the company and the unit cost if all production were transferred to Plant B?

12. Solve graphically for the values of A and B that maximize total profit expressed as

$$TP = \$0.30A + \$0.42B$$

subject to the constraints

$$A \leq 42$$
$$B \leq 30$$
$$A + B \leq 62$$
$$A \geq 0 \quad \text{and} \quad B \geq 0.$$

13. A small machine shop has capability in turning, milling, drilling, and welding. The machine capacity is 16 hours per day in turning, 16 hours per day in milling, 8 hours per day in drilling, and 8 hours per day in welding. Two products, designated A and B, are under consideration. Each will yield a net profit of $0.25 per unit and will require the following hours of machine time:

	Product A	Product B
Turning	0.064	0.106
Milling	0.106	0.053
Drilling	0.000	0.080

Solve graphically for the number of units of each product that should be scheduled to maximize profit.

14. A refinery produces three grades of gasoline; premium, regular, and economy. Each grade requires straight gasoline, octane, and additives which are available in the amount of 3,200,000, 2,400,000, and 1,100,000 gal per week, respectively. A gallon of premium requires 0.22 gal of straight gasoline, 0.50 gal of octane, and 0.28 gal of additives. One gal of regular requires 0.55 gal of straight gasoline, 0.32 gal of octane, and 0.13 gal of additives. A gallon of economy requires 0.72 gal of straight gasoline, 0.20 gal of octane, and 0.08 gal of additives. A profit of $0.048, $0.040, and $0.029 per gallon is received for premium, regular, and economy, respectively. Formulate the effectiveness function and constraints which may be used to find the number of gallons of each grade to produce which will lead to a maximum profit.

15. A contractor has a requirement for cement in the amount of 45,000 bags per year. Cement costs $1.40 per bag and it costs $25 to process a purchase order.

Holding cost is $2.30 per bag per year because of the high rate of spoilage. Find the minimum cost purchase quantity.

16. A foundry uses 3,600 tons of pig iron per year at a constant rate. The cost per ton delivered to the foundry is $70. It costs $42 to place an order and $8 per ton per year for storage. Find the minimum cost purchase quantity.

17. The demand for 1,000 units of a part to be used at a uniform rate throughout the year may be met by manufacturing. The part can be produced at the rate of 3 per hour in a department which works 1,880 hours per year. The set-up-cost per lot is estimated to be $40 and the manufacturing cost has been established at $5.20 per unit. Interest, insurance, taxes, space, and other holding costs are $3.10 per unit per year. Calculate the economic manufacturing quantity.

18. An aircraft manufacturer uses 1,000,000 special rivets per year at a uniform rate. The rivets are made on a single-spindle automatic screw machine at the rate of 3,000 per hour. The manufacturing cost is $0.0052 per rivet and the storage cost is estimated to be $0.0008 per rivet per year. The set-up cost is $40 per set-up. Calculate the economic production quantity and the allowable variation in this quantity for a maximum deviation from the minimum cost of 5%. Assume the machine operates 1,700 hours per year.

19. Rederive the economic purchase quantity model to reflect holding cost for interest, insurance, taxes, and handling based on the average inventory level and the holding cost for space based on the maximum level.

20. Show that the economic production quantity model reduces to the economic purchase quantity model as R approaches infinity.

21. The annual demand for an item that can be either purchased or produced is 3,600 units. Costs associated with each alternative are as follows:

	Purchase	*Produce*
Item cost	$ 3.90	$ 3.75
Purchase cost per purchase	15.00	—
Set-up cost per set-up	—	175.00
Holding cost per item per year	1.25	1.25
Production rate per year	—	15,000

Should the item be manufactured or purchased? What is the economic lot size for the least cost alternative?

22. The annual demand for a certain item is 1,000 units per year. Holding cost is $0.85 per unit per year. Demand can be met by either purchasing or producing the item with each source described by the following data:

	Purchase	*Produce*
Item cost	$ 8.00	$ 7.40
Procurement cost	20.00	80.00
Production rate	—	2,500 per year

Find the minimum cost procurement source and calculate the economic advantage over the alternate source. What is the minimum cost procurement quantity?

23. For a certain group of machines, a repair crew requires one day to repair a machine that breaks down. A loss of $180 is incurred for each day that a machine is carried-over unrepaired. Each crew costs $75 per day to equip and maintain. The pattern of machine failures on succeeding days of a 30-day period is 7, 5, 8, 0, 3, 4, 5, 0, 2, 3, 8, 4, 5, 6, 1, 3, 4, 0, 2, 8, 6, 5, 5, 3, 4, 6, 7, 1, 3, 2. Assume that all breakdowns occur at the first of each day.

   (a) Determine the idle time of crews when four, five, or six crews are employed.

   (b) What is the optimum number of crews to employ?

24. A company has 24 identical machines that are operated 24 hours per day, 360 days per year. On the average, each machine breaks down at intervals of 6 working days. The time required to repair each machine is 8 man-hours. Repairmen receive $4 per hour and work 8 hours per day. A loss of $100 is sustained for each day that a machine is carried-over unrepaired.

   (a) Determine the number of machines that may be expected by chance to go "down" during each day of a month by tossing a die 24 times to represent the 24 machines. Let the downs for each day be represented by the aces that come up.

   (b) Assuming that all downs as found in part (a) occur at 12 midnight, determine the most economical number of regular repairmen to employ, if no extra repairmen or overtime wages are to be allowed and regular repairmen are not used except to repair the machines in question. Assume that the sequence for the 30-day month will be repeated 12 times during the year.

APPENDICES

APPENDIX A
INTEREST FACTORS FOR ANNUAL
COMPOUNDING INTEREST

Table A.1. $\frac{1}{2}$% INTEREST FACTORS FOR ANNUAL COMPOUNDING INTEREST

	Single Payment		Equal Payment Series				Uniform
	Compound-amount factor	Present-worth factor	Compound-amount factor	Sinking-fund factor	Present-worth factor	Capital-recovery factor	gradient-series factor
n	To find F Given P F/P i, n	To find P Given F P/F i, n	To find F Given A F/A i, n	To find A Given F A/F i, n	To find P Given A P/A i, n	To find A Given P A/P i, n	To find A Given G A/G i, n
1	1.005	0.9950	1.000	1.0000	0.9950	1.0050	0.0000
2	1.010	0.9901	2.005	0.4988	1.9851	0.5038	0.4988
3	1.015	0.9852	3.015	0.3317	2.9703	0.3367	0.9967
4	1.020	0.9803	4.030	0.2481	3.9505	0.2531	1.4938
5	1.025	0.9754	5.050	0.1980	4.9259	0.2030	1.9900
6	1.030	0.9705	6.076	0.1646	5.8964	0.1696	2.4855
7	1.036	0.9657	7.106	0.1407	6.8621	0.1457	2.9801
8	1.041	0.9609	8.141	0.1228	7.8230	0.1278	3.4738
9	1.046	0.9561	9.182	0.1089	8.7791	0.1139	3.9668
10	1.051	0.9514	10.228	0.0978	9.7304	0.1028	4.4589
11	1.056	0.9466	11.279	0.0887	10.6770	0.0937	4.9501
12	1.062	0.9419	12.336	0.0811	11.6189	0.0861	5.4406
13	1.067	0.9372	13.397	0.0747	12.5562	0.0797	5.9302
14	1.072	0.9326	14.464	0.0691	13.4887	0.0741	6.4190
15	1.078	0.9279	15.537	0.0644	14.4166	0.0694	6.9069
16	1.083	0.9233	16.614	0.0602	15.3399	0.0652	7.3940
17	1.088	0.9187	17.697	0.0565	16.2586	0.0615	7.8803
18	1.094	0.9141	18.786	0.0532	17.1728	0.0582	8.3658
19	1.099	0.9096	19.880	0.0503	18.0824	0.0553	8.8504
20	1.105	0.9051	20.979	0.0477	18.9874	0.0527	9.3342
21	1.110	0.9006	22.084	0.0453	19.8880	0.0503	9.8172
22	1.116	0.8961	23.194	0.0431	20.7841	0.0481	10.2993
23	1.122	0.8916	24.310	0.0411	21.6757	0.0461	10.7806
24	1.127	0.8872	25.432	0.0393	22.5629	0.0443	11.2611
25	1.133	0.8828	26.559	0.0377	23.4456	0.0427	11.7407
26	1.138	0.8784	27.692	0.0361	24.3240	0.0411	12.2195
27	1.144	0.8740	28.830	0.0347	25.1980	0.0397	12.6975
28	1.150	0.8697	29.975	0.0334	26.0677	0.0384	13.1747
29	1.156	0.8653	31.124	0.0321	26.9330	0.0371	13.6510
30	1.161	0.8610	32.280	0.0310	27.7941	0.0360	14.1265
31	1.167	0.8568	33.441	0.0299	28.6508	0.0349	14.6012
32	1.173	0.8525	34.609	0.0289	29.5033	0.0339	15.0750
33	1.179	0.8483	35.782	0.0280	30.3515	0.0330	15.5480
34	1.185	0.8440	36.961	0.0271	31.1956	0.0321	16.0202
35	1.191	0.8398	38.145	0.0262	32.0354	0.0312	16.4915
40	1.221	0.8191	44.159	0.0227	36.1722	0.0277	18.8358
45	1.252	0.7990	50.324	0.0199	40.2072	0.0249	21.1595
50	1.283	0.7793	56.645	0.0177	44.1428	0.0227	23.4624
55	1.316	0.7601	63.126	0.0159	47.9815	0.0209	25.7447
60	1.349	0.7414	69.770	0.0143	51.7256	0.0193	28.0064
65	1.383	0.7231	76.582	0.0131	55.3775	0.0181	30.2475
70	1.418	0.7053	83.566	0.0120	58.9394	0.0170	32.4680
75	1.454	0.6879	90.727	0.0110	62.4137	0.0160	34.6679
80	1.490	0.6710	98.068	0.0102	65.8023	0.0152	36.8474
85	1.528	0.6545	105.594	0.0095	69.1075	0.0145	39.0065
90	1.567	0.6384	113.311	0.0088	72.3313	0.0138	41.1451
95	1.606	0.6226	121.222	0.0083	75.4757	0.0133	43.2633
100	1.647	0.6073	129.334	0.0077	78.5427	0.0127	45.3613

Table A.2. $\frac{3}{4}\%$ INTEREST FACTORS FOR ANNUAL COMPOUNDING INTEREST

	Single Payment		Equal Payment Series				Uniform gradient-series factor
	Compound-amount factor	Present-worth factor	Compound-amount factor	Sinking-fund factor	Present-worth factor	Capital-recovery factor	
n	To find F Given P F/P i, n	To find P Given F P/F i, n	To find F Given A F/A i, n	To find A Given F A/F i, n	To find P Given A P/A i, n	To find A Given P A/P i, n	To find A Given G A/G i, n
1	1.008	0.9926	1.000	1.0000	0.9926	1.0075	0.0000
2	1.015	0.9852	2.008	0.4981	1.9777	0.5056	0.4981
3	1.023	0.9778	3.023	0.3309	2.9556	0.3384	0.9950
4	1.030	0.9706	4.045	0.2472	3.9261	0.2547	1.4907
5	1.038	0.9633	5.076	0.1970	4.8894	0.2045	1.9851
6	1.046	0.9562	6.114	0.1636	5.8456	0.1711	2.4782
7	1.054	0.9491	7.159	0.1397	6.7946	0.1472	2.9701
8	1.062	0.9420	8.213	0.1218	7.7366	0.1293	3.4608
9	1.070	0.9350	9.275	0.1078	8.6716	0.1153	3.9502
10	1.078	0.9280	10.344	0.0967	9.5996	0.1042	4.4384
11	1.086	0.9211	11.422	0.0876	10.5207	0.0951	4.9253
12	1.094	0.9142	12.508	0.0800	11.4349	0.0875	5.4110
13	1.102	0.9074	13.601	0.0735	12.3424	0.0810	5.8954
14	1.110	0.9007	14.703	0.0680	13.2430	0.0755	6.3786
15	1.119	0.8940	15.814	0.0632	14.1370	0.0707	6.8606
16	1.127	0.8873	16.932	0.0591	15.0243	0.0666	7.3413
17	1.135	0.8807	18.059	0.0554	15.9050	0.0629	7.8207
18	1.144	0.8742	19.195	0.0521	16.7792	0.0596	8.2989
19	1.153	0.8677	20.339	0.0492	17.6468	0.0567	8.7759
20	1.161	0.8612	21.491	0.0465	18.5080	0.0540	9.2517
21	1.170	0.8548	22.652	0.0442	19.3628	0.0517	9.7261
22	1.179	0.8484	23.822	0.0420	20.2112	0.0495	10.1994
23	1.188	0.8421	25.001	0.0400	21.0533	0.0475	10.6714
24	1.196	0.8358	26.188	0.0382	21.8892	0.0457	11.1422
25	1.205	0.8296	27.385	0.0365	22.7188	0.0440	11.6117
26	1.214	0.8234	28.590	0.0350	23.5422	0.0425	12.0800
27	1.224	0.8173	29.805	0.0336	24.3595	0.0411	12.5470
28	1.233	0.8112	31.028	0.0322	25.1707	0.0397	13.0128
29	1.242	0.8052	32.261	0.0310	25.9759	0.0385	13.4774
30	1.251	0.7992	33.503	0.0299	26.7751	0.0374	13.9407
31	1.261	0.7932	34.754	0.0288	27.5683	0.0363	14.4028
32	1.270	0.7873	36.015	0.0278	28.3557	0.0353	14.8636
33	1.280	0.7815	37.285	0.0268	29.1371	0.0343	15.3232
34	1.289	0.7757	38.565	0.0259	29.9128	0.0334	15.7816
35	1.299	0.7699	39.854	0.0251	30.6827	0.0326	16.2387
40	1.348	0.7417	46.446	0.0215	34.4469	0.0290	18.5058
45	1.400	0.7145	53.290	0.0188	38.0732	0.0263	20.7421
50	1.453	0.6883	60.394	0.0166	41.5665	0.0241	22.9476
55	1.508	0.6630	67.769	0.0148	44.9316	0.0223	25.1223
60	1.566	0.6387	75.424	0.0133	48.1734	0.0208	27.2665
65	1.625	0.6153	83.371	0.0120	51.2963	0.0195	29.3801
70	1.687	0.5927	91.620	0.0109	54.3046	0.0184	31.4634
75	1.751	0.5710	100.183	0.0100	57.2027	0.0175	33.5163
80	1.818	0.5501	109.073	0.0092	59.9945	0.0167	35.5391
85	1.887	0.5299	118.300	0.0085	62.6838	0.0160	37.5318
90	1.959	0.5105	127.879	0.0078	65.2746	0.0153	39.4946
95	2.034	0.4917	137.823	0.0073	67.7704	0.0148	41.4277
100	2.111	0.4737	148.145	0.0068	70.1746	0.0143	43.3311

Table A.3. 1 % INTEREST FACTORS FOR ANNUAL COMPOUNDING INTEREST

	Single Payment		Equal Payment Series				Uniform gradient-series factor
	Compound-amount factor	Present-worth factor	Compound-amount factor	Sinking-fund factor	Present-worth factor	Capital-recovery factor	
n	To find F Given P F/P *i, n*	To find P Given F P/F *i, n*	To find F Given A F/A *i, n*	To find A Given F A/F *i, n*	To find P Given A P/A *i, n*	To find A Given P A/P *i, n*	To find A Given G A/G *i, n*
1	1.010	0.9901	1.000	1.0000	0.9901	1.0100	0.0000
2	1.020	0.9803	2.010	0.4975	1.9704	0.5075	0.4975
3	1.030	0.9706	3.030	0.3300	2.9410	0.3400	0.9934
4	1.041	0.9610	4.060	0.2463	3.9020	0.2563	1.4876
5	1.051	0.9515	5.101	0.1960	4.8534	0.2060	1.9801
6	1.062	0.9421	6.152	0.1626	5.7955	0.1726	2.4710
7	1.072	0.9327	7.214	0.1386	6.7282	0.1486	2.9602
8	1.083	0.9235	8.286	0.1207	7.6517	0.1307	3.4478
9	1.094	0.9143	9.369	0.1068	8.5660	0.1168	3.9337
10	1.105	0.9053	10.462	0.0956	9.4713	0.1056	4.4179
11	1.116	0.8963	11.567	0.0865	10.3676	0.0965	4.9005
12	1.127	0.8875	12.683	0.0789	11.2551	0.0889	5.3815
13	1.138	0.8787	13.809	0.0724	12.1338	0.0824	5.8607
14	1.149	0.8700	14.947	0.0669	13.0037	0.0769	6.3384
15	1.161	0.8614	16.097	0.0621	13.8651	0.0721	6.8143
16	1.173	0.8528	17.258	0.0580	14.7179	0.0680	7.2887
17	1.184	0.8444	18.430	0.0543	15.5623	0.0643	7.7613
18	1.196	0.8360	19.615	0.0510	16.3983	0.0610	8.2323
19	1.208	0.8277	20.811	0.0481	17.2260	0.0581	8.7017
20	1.220	0.8196	22.019	0.0454	18.0456	0.0554	9.1694
21	1.232	0.8114	23.239	0.0430	18.8570	0.0530	9.6354
22	1.245	0.8034	24.472	0.0409	19.6604	0.0509	10.0998
23	1.257	0.7955	25.716	0.0389	20.4558	0.0489	10.5626
24	1.270	0.7876	26.973	0.0371	21.2434	0.0471	11.0237
25	1.282	0.7798	28.243	0.0354	22.0232	0.0454	11.4831
26	1.295	0.7721	29.526	0.0339	22.7952	0.0439	11.9409
27	1.308	0.7644	30.821	0.0325	23.5596	0.0425	12.3971
28	1.321	0.7568	32.129	0.0311	24.3165	0.0411	12.8516
29	1.335	0.7494	33.450	0.0299	25.0658	0.0399	13.3045
30	1.348	0.7419	34.785	0.0288	25.8077	0.0388	13.7557
31	1.361	0.7346	36.133	0.0277	26.5423	0.0377	14.2052
32	1.375	0.7273	37.494	0.0267	27.2696	0.0367	14.6532
33	1.389	0.7201	38.869	0.0257	27.9897	0.0357	15.0995
34	1.403	0.7130	40.258	0.0248	28.7027	0.0348	15.5441
35	1.417	0.7059	41.660	0.0240	29.4086	0.0340	15.9871
40	1.489	0.6717	48.886	0.0205	32.8347	0.0305	18.1776
45	1.565	0.6391	56.481	0.0177	36.0945	0.0277	20.3273
50	1.645	0.6080	64.463	0.0155	39.1961	0.0255	22.4363
55	1.729	0.5785	72.852	0.0137	42.1472	0.0237	24.5049
60	1.817	0.5505	81.670	0.0123	44.9550	0.0223	26.5333
65	1.909	0.5237	90.937	0.0110	47.6266	0.0210	28.5217
70	2.007	0.4983	100.676	0.0099	50.1685	0.0199	30.4703
75	2.109	0.4741	110.913	0.0090	52.5871	0.0190	32.3793
80	2.217	0.4511	121.672	0.0082	54.8882	0.0182	34.2492
85	2.330	0.4292	132.979	0.0075	57.0777	0.0175	36.0801
90	2.449	0.4084	144.863	0.0069	59.1609	0.0169	37.8725
95	2.574	0.3886	157.354	0.0064	61.1430	0.0164	39.6265
100	2.705	0.3697	170.481	0.0059	63.0289	0.0159	41.3426

Table A.4. $1\frac{1}{4}\%$ INTEREST FACTORS FOR ANNUAL COMPOUNDING INTEREST

	Single Payment		Equal Payment Series				Uniform gradient-series factor
	Compound-amount factor	Present-worth factor	Compound-amount factor	Sinking-fund factor	Present-worth factor	Capital-recovery factor	
n	To find F Given P $F/P\ i, n$	To find P Given F $P/F\ i, n$	To find F Given A $F/A\ i, n$	To find A Given F $A/F\ i, n$	To find P Given A $P/A\ i, n$	To find A Given P $A/P\ i, n$	To find A Given G $A/G\ i, n$
1	1.013	0.9877	1.000	1.0001	0.9877	1.0126	0.0000
2	1.025	0.9755	2.013	0.4970	1.9631	0.5095	0.4932
3	1.038	0.9635	3.038	0.3293	2.9265	0.3418	0.9895
4	1.051	0.9516	4.076	0.2454	3.8780	0.2579	1.4830
5	1.064	0.9398	5.127	0.1951	4.8177	0.2076	1.9729
6	1.077	0.9282	6.191	0.1616	5.7459	0.1741	2.4618
7	1.091	0.9168	7.268	0.1376	6.6627	0.1501	2.9491
8	1.105	0.9055	8.359	0.1197	7.5680	0.1322	3.4330
9	1.118	0.8943	9.463	0.1057	8.4623	0.1182	3.9158
10	1.132	0.8832	10.582	0.0946	9.3454	0.1071	4.3960
11	1.147	0.8723	11.714	0.0854	10.2177	0.0979	4.8744
12	1.161	0.8616	12.860	0.0778	11.0792	0.0903	5.3506
13	1.175	0.8509	14.021	0.0714	11.9300	0.0839	5.8248
14	1.190	0.8404	15.196	0.0659	12.7704	0.0784	6.2968
15	1.205	0.8300	16.386	0.0611	13.6004	0.0736	6.7669
16	1.220	0.8198	17.591	0.0569	14.4201	0.0694	7.2350
17	1.235	0.8097	18.811	0.0532	15.2298	0.0657	7.7009
18	1.251	0.7997	20.046	0.0499	16.0293	0.0624	8.1645
19	1.266	0.7898	21.296	0.0470	16.8191	0.0595	8.6264
20	1.282	0.7801	22.563	0.0444	17.5991	0.0569	9.0861
21	1.298	0.7704	23.845	0.0420	18.3695	0.0545	9.5439
22	1.314	0.7609	25.143	0.0398	19.1303	0.0523	9.9993
23	1.331	0.7515	26.457	0.0378	19.8818	0.0503	10.4528
24	1.347	0.7423	27.788	0.0360	20.6240	0.0485	10.9044
25	1.364	0.7331	29.135	0.0344	21.3570	0.0469	11.3539
26	1.381	0.7240	30.499	0.0328	22.0810	0.0453	11.8012
27	1.399	0.7151	31.880	0.0314	22.7960	0.0439	12.2465
28	1.416	0.7063	33.279	0.0301	23.5022	0.0426	12.6898
29	1.434	0.6976	34.695	0.0289	24.1998	0.0414	13.1311
30	1.452	0.6889	36.128	0.0277	24.8886	0.0402	13.5703
31	1.470	0.6804	37.580	0.0267	25.5690	0.0392	14.0074
32	1.488	0.6720	39.050	0.0257	26.2410	0.0382	14.4425
33	1.507	0.6637	40.538	0.0247	26.9047	0.0372	14.8756
34	1.526	0.6555	42.045	0.0238	27.5601	0.0363	15.3066
35	1.545	0.6475	43.570	0.0230	28.2075	0.0355	15.7357
40	1.644	0.6085	51.489	0.0195	31.3266	0.0320	17.8503
45	1.749	0.5718	59.915	0.0167	34.2578	0.0292	19.9144
50	1.861	0.5374	68.880	0.0146	37.0125	0.0271	21.9284
55	1.980	0.5050	78.421	0.0128	39.6013	0.0253	23.8925
60	2.107	0.4746	88.573	0.0113	42.0342	0.0238	25.8072
65	2.242	0.4460	99.375	0.0101	44.3206	0.0226	27.6730
70	2.386	0.4192	110.870	0.0091	46.4693	0.0216	29.4902
75	2.539	0.3939	123.101	0.0082	48.4886	0.0207	31.2594
80	2.702	0.3702	136.116	0.0074	50.3862	0.0199	32.9812
85	2.875	0.3479	149.965	0.0067	52.1696	0.0192	34.6560
90	3.059	0.3270	164.701	0.0061	53.8456	0.0186	36.2844
95	3.255	0.3073	180.382	0.0056	55.4207	0.0181	37.8671
100	3.463	0.2888	197.067	0.0051	56.9009	0.0176	39.4048

Table A.5. $1\frac{1}{2}\%$ INTEREST FACTORS FOR ANNUAL COMPOUNDING INTEREST

	Single Payment		Equal Payment Series				Uniform gradient-series factor
	Compound-amount factor	Present-worth factor	Compound-amount factor	Sinking-fund factor	Present-worth factor	Capital-recovery factor	
n	To find F Given P F/P i,n	To find P Given F P/F i,n	To find F Given A F/A i,n	To find A Given F A/F i,n	To find P Given A P/A i,n	To find A Given P A/P i,n	To find A Given G A/G i,n
1	1.015	0.9852	1.000	1.0000	0.9852	1.0150	0.0000
2	1.030	0.9707	2.015	0.4963	1.9559	0.5113	0.4963
3	1.046	0.9563	3.045	0.3284	2.9122	0.3434	0.9901
4	1.061	0.9422	4.091	0.2445	3.8544	0.2595	1.4814
5	1.077	0.9283	5.152	0.1941	4.7827	0.2091	1.9702
6	1.093	0.9146	6.230	0.1605	5.6972	0.1755	2.4566
7	1.110	0.9010	7.323	0.1366	6.5982	0.1516	2.9405
8	1.127	0.8877	8.433	0.1186	7.4859	0.1336	3.4219
9	1.143	0.8746	9.559	0.1046	8.3605	0.1196	3.9008
10	1.161	0.8617	10.703	0.0934	9.2222	0.1084	4.3772
11	1.178	0.8489	11.863	0.0843	10.0711	0.0993	4.8512
12	1.196	0.8364	13.041	0.0767	10.9075	0.0917	5.3227
13	1.214	0.8240	14.237	0.0703	11.7315	0.0853	5.7917
14	1.232	0.8119	15.450	0.0647	12.5434	0.0797	6.2582
15	1.250	0.7999	16.682	0.0600	13.3432	0.0750	6.7223
16	1.269	0.7880	17.932	0.0558	14.1313	0.0708	7.1839
17	1.288	0.7764	19.201	0.0521	14.9077	0.0671	7.6431
18	1.307	0.7649	20.489	0.0488	15.6726	0.0638	8.0997
19	1.327	0.7536	21.797	0.0459	16.4262	0.0609	8.5539
20	1.347	0.7425	23.124	0.0433	17.1686	0.0583	9.0057
21	1.367	0.7315	24.471	0.0409	17.9001	0.0559	9.4550
22	1.388	0.7207	25.838	0.0387	18.6208	0.0537	9.9018
23	1.408	0.7100	27.225	0.0367	19.3309	0.0517	10.3462
24	1.430	0.6996	28.634	0.0349	20.0304	0.0499	10.7881
25	1.451	0.6892	30.063	0.0333	20.7196	0.0483	11.2276
26	1.473	0.6790	31.514	0.0317	21.3986	0.0467	11.6646
27	1.495	0.6690	32.987	0.0303	22.0676	0.0453	12.0992
28	1.517	0.6591	34.481	0.0290	22.7267	0.0440	12.5313
29	1.540	0.6494	35.999	0.0278	23.3761	0.0428	12.9610
30	1.563	0.6398	37.539	0.0266	24.0158	0.0416	13.3883
31	1.587	0.6303	39.102	0.0256	24.6462	0.0406	13.8131
32	1.610	0.6210	40.688	0.0246	25.2671	0.0396	14.2355
33	1.634	0.6118	42.299	0.0237	25.8790	0.0387	14.6555
34	1.659	0.6028	43.933	0.0228	26.4817	0.0378	15.0731
35	1.684	0.5939	45.592	0.0219	27.0756	0.0369	15.4882
40	1.814	0.5513	54.268	0.0184	29.9159	0.0334	17.5277
45	1.954	0.5117	63.614	0.0157	32.5523	0.0307	19.5074
50	2.105	0.4750	73.683	0.0136	34.9997	0.0286	21.4277
55	2.268	0.4409	84.530	0.0118	37.2715	0.0268	23.2894
60	2.443	0.4093	96.215	0.0104	39.3803	0.0254	25.0930
65	2.632	0.3799	108.803	0.0092	41.3378	0.0242	26.8392
70	2.835	0.3527	122.364	0.0082	43.1549	0.0232	28.5290
75	3.055	0.3274	136.973	0.0073	44.8416	0.0223	30.1631
80	3.291	0.3039	152.711	0.0066	46.4073	0.0216	31.7423
85	3.545	0.2821	169.665	0.0059	47.8607	0.0209	33.2676
90	3.819	0.2619	187.930	0.0053	49.2099	0.0203	34.7399
95	4.114	0.2431	207.606	0.0048	50.4622	0.0198	36.1602
100	4.432	0.2256	228.803	0.0044	51.6247	0.0194	37.5295

Table A.6. 2% INTEREST FACTORS FOR ANNUAL COMPOUNDING INTEREST

	Single Payment		Equal Payment Series				Uniform
	Compound-amount factor	Present-worth factor	Compound-amount factor	Sinking-fund factor	Present-worth factor	Capital-recovery factor	gradient-series factor
n	To find F Given P F/P i, n	To find P Given F P/F i, n	To find F Given A F/A i, n	To find A Given F A/F i, n	To find P Given A P/A i, n	To find A Given P A/P i, n	To find A Given G A/G i, n
1	1.020	0.9804	1.000	1.0000	0.9804	1.0200	0.0000
2	1.040	0.9612	2.020	0.4951	1.9416	0.5151	0.4951
3	1.061	0.9423	3.060	0.3268	2.8839	0.3468	0.9868
4	1.082	0.9239	4.122	0.2426	3.8077	0.2626	1.4753
5	1.104	0.9057	5.204	0.1922	4.7135	0.2122	1.9604
6	1.126	0.8880	6.308	0.1585	5.6014	0.1785	2.4423
7	1.149	0.8706	7.434	0.1345	6.4720	0.1545	2.9208
8	1.172	0.8535	8.583	0.1165	7.3255	0.1365	3.3961
9	1.195	0.8368	9.755	0.1025	8.1622	0.1225	3.8681
10	1.219	0.8204	10.950	0.0913	8.9826	0.1113	4.3367
11	1.243	0.8043	12.169	0.0822	9.7869	0.1022	4.8021
12	1.268	0.7885	13.412	0.0746	10.5754	0.0946	5.2643
13	1.294	0.7730	14.680	0.0681	11.3484	0.0881	5.7231
14	1.319	0.7579	15.974	0.0626	12.1063	0.0826	6.1786
15	1.346	0.7430	17.293	0.0578	12.8493	0.0778	6.6309
16	1.373	0.7285	18.639	0.0537	13.5777	0.0737	7.0799
17	1.400	0.7142	20.012	0.0500	14.2919	0.0700	7.5256
18	1.428	0.7002	21.412	0.0467	14.9920	0.0667	7.9681
19	1.457	0.6864	22.841	0.0438	15.6785	0.0638	8.4073
20	1.486	0.6730	24.297	0.0412	16.3514	0.0612	8.8433
21	1.516	0.6598	25.783	0.0388	17.0112	0.0588	9.2760
22	1.546	0.6468	27.299	0.0366	17.6581	0.0566	9.7055
23	1.577	0.6342	28.845	0.0347	18.2922	0.0547	10.1317
24	1.608	0.6217	30.422	0.0329	18.9139	0.0529	10.5547
25	1.641	0.6095	32.030	0.0312	19.5235	0.0512	10.9745
26	1.673	0.5976	33.671	0.0297	20.1210	0.0497	11.3910
27	1.707	0.5859	35.344	0.0283	20.7069	0.0483	11.8043
28	1.741	0.5744	37.051	0.0270	21.2813	0.0470	12.2145
29	1.776	0.5631	38.792	0.0258	21.8444	0.0458	12.6214
30	1.811	0.5521	40.568	0.0247	22.3965	0.0447	13.0251
31	1.848	0.5413	42.379	0.0236	22.9377	0.0436	13.4257
32	1.885	0.5306	44.227	0.0226	23.4683	0.0426	13.8230
33	1.922	0.5202	46.112	0.0217	23.9886	0.0417	14.2172
34	1.961	0.5100	48.034	0.0208	24.4986	0.0408	14.6083
35	2.000	0.5000	49.994	0.0200	24.9986	0.0400	14.9961
40	2.208	0.4529	60.402	0.0166	27.3555	0.0366	16.8885
45	2.438	0.4102	71.893	0.0139	29.4902	0.0339	18.7034
50	2.692	0.3715	84.579	0.0118	31.4236	0.0318	20.4420
55	2.972	0.3365	98.587	0.0102	33.1748	0.0302	22.1057
60	3.281	0.3048	114.052	0.0088	34.7609	0.0288	23.6961
65	3.623	0.2761	131.126	0.0076	36.1975	0.0276	25.2147
70	4.000	0.2500	149.978	0.0067	37.4986	0.0267	26.6632
75	4.416	0.2265	170.792	0.0059	38.6771	0.0259	28.0434
80	4.875	0.2051	193.772	0.0052	39.7445	0.0252	29.3572
85	5.383	0.1858	219.144	0.0046	40.7113	0.0246	30.6064
90	5.943	0.1683	247.157	0.0041	41.5869	0.0241	31.7929
95	6.562	0.1524	278.085	0.0036	42.3800	0.0236	32.9189
100	7.245	0.1380	312.232	0.0032	43.0984	0.0232	33.9863

Table A.7. 3% INTEREST FACTORS FOR ANNUAL COMPOUNDING INTEREST

	Single Payment		Equal Payment Series				Uniform gradient-series factor
	Compound-amount factor	Present-worth factor	Compound-amount factor	Sinking-fund factor	Present-worth factor	Capital-recovery factor	
n	To find F Given P F/P i, n	To find P Given F P/F i, n	To find F Given A F/A i, n	To find A Given F A/F i, n	To find P Given A P/A i, n	To find A Given P A/P i, n	To find A Given G A/G i, n
1	1.030	0.9709	1.000	1.0000	0.9709	1.0300	0.0000
2	1.061	0.9426	2.030	0.4926	1.9135	0.5226	0.4926
3	1.093	0.9152	3.091	0.3235	2.8286	0.3535	0.9803
4	1.126	0.8885	4.184	0.2390	3.7171	0.2690	1.4631
5	1.159	0.8626	5.309	0.1884	4.5797	0.2184	1.9409
6	1.194	0.8375	6.468	0.1546	5.4172	0.1846	2.4138
7	1.230	0.8131	7.662	0.1305	6.2303	0.1605	2.8819
8	1.267	0.7894	8.892	0.1125	7.0197	0.1425	3.3450
9	1.305	0.7664	10.159	0.0984	7.7861	0.1284	3.8032
10	1.344	0.7441	11.464	0.0872	8.5302	0.1172	4.2565
11	1.384	0.7224	12.808	0.0781	9.2526	0.1081	4.7049
12	1.426	0.7014	14.192	0.0705	9.9540	0.1005	5.1485
13	1.469	0.6810	15.618	0.0640	10.6350	0.0940	5.5872
14	1.513	0.6611	17.086	0.0585	11.2961	0.0885	6.0211
15	1.558	0.6419	18.599	0.0538	11.9379	0.0838	6.4501
16	1.605	0.6232	20.157	0.0496	12.5611	0.0796	6.8742
17	1.653	0.6050	21.762	0.0460	13.1661	0.0760	7.2936
18	1.702	0.5874	23.414	0.0427	13.7535	0.0727	7.7081
19	1.754	0.5703	25.117	0.0398	14.3238	0.0698	8.1179
20	1.806	0.5537	26.870	0.0372	14.8775	0.0672	8.5229
21	1.860	0.5376	28.676	0.0349	15.4150	0.0649	8.9231
22	1.916	0.5219	30.537	0.0328	15.9369	0.0628	9.3186
23	1.974	0.5067	32.453	0.0308	16.4436	0.0608	9.7094
24	2.033	0.4919	34.426	0.0291	16.9356	0.0591	10.0954
25	2.094	0.4776	36.459	0.0274	17.4132	0.0574	10.4768
26	2.157	0.4637	38.553	0.0259	17.8769	0.0559	10.8535
27	2.221	0.4502	40.710	0.0246	18.3270	0.0546	11.2256
28	2.288	0.4371	42.931	0.0233	18.7641	0.0533	11.5930
29	2.357	0.4244	45.219	0.0221	19.1885	0.0521	11.9558
30	2.427	0.4120	47.575	0.0210	19.6005	0.0510	12.3141
31	2.500	0.4000	50.003	0.0200	20.0004	0.0500	12.6678
32	2.575	0.3883	52.503	0.0191	20.3888	0.0491	13.0169
33	2.652	0.3770	55.078	0.0182	20.7658	0.0482	13.3616
34	2.732	0.3661	57.730	0.0173	21.1318	0.0473	13.7018
35	2.814	0.3554	60.462	0.0165	21.4872	0.0465	14.0375
40	3.262	0.3066	75.401	0.0133	23.1148	0.0433	15.6502
►45	3.782	0.2644	92.720	0.0108	24.5187	0.0408	17.1556
50	4.384	0.2281	112.797	0.0089	25.7298	0.0389	18.5575
55	5.082	0.1968	136.072	0.0074	26.7744	0.0374	19.8600
60	5.892	0.1697	163.053	0.0061	27.6756	0.0361	21.0674
65	6.830	0.1464	194.333	0.0052	28.4529	0.0352	22.1841
70	7.918	0.1263	230.594	0.0043	29.1234	0.0343	23.2145
75	9.179	0.1090	272.631	0.0037	29.7018	0.0337	24.1634
80	10.641	0.0940	321.363	0.0031	30.2008	0.0331	25.0354
85	12.336	0.0811	377.857	0.0027	30.6312	0.0327	25.8349
90	14.300	0.0699	443.349	0.0023	31.0024	0.0323	26.5667
95	16.578	0.0603	519.272	0.0019	31.3227	0.0319	27.2351
100	19.219	0.0520	607.288	0.0017	31.5989	0.0317	27.8445

Table A.8. 4% INTEREST FACTORS FOR ANNUAL COMPOUNDING INTEREST

	Single Payment		Equal Payment Series				Uniform gradient-series factor
	Compound-amount factor	Present-worth factor	Compound-amount factor	Sinking-fund factor	Present-worth factor	Capital-recovery factor	
n	To find F Given P F/P i, n	To find P Given F P/F i, n	To find F Given A F/A i, n	To find A Given F A/F i, n	To find P Given A P/A i, n	To find A Given P A/P i, n	To find A Given G A/G i, n
1	1.040	0.9615	1.000	1.0000	0.9615	1.0400	0.0000
2	1.082	0.9246	2.040	0.4902	1.8861	0.5302	0.4902
3	1.125	0.8890	3.122	0.3204	2.7751	0.3604	0.9739
4	1.170	0.8548	4.246	0.2355	3.6299	0.2755	1.4510
5	1.217	0.8219	5.416	0.1846	4.4518	0.2246	1.9216
6	1.265	0.7903	6.633	0.1508	5.2421	0.1908	2.3857
7	1.316	0.7599	7.898	0.1266	6.0021	0.1666	2.8433
8	1.369	0.7307	9.214	0.1085	6.7328	0.1485	3.2944
9	1.423	0.7026	10.583	0.0945	7.4353	0.1345	3.7391
10	1.480	0.6756	12.006	0.0833	8.1109	0.1233	4.1773
11	1.539	0.6496	13.486	0.0742	8.7605	0.1142	4.6090
12	1.601	0.6246	15.026	0.0666	9.3851	0.1066	5.0344
13	1.665	0.6006	16.627	0.0602	9.9857	0.1002	5.4533
14	1.732	0.5775	18.292	0.0547	10.5631	0.0947	5.8659
15	1.801	0.5553	20.024	0.0500	11.1184	0.0900	6.2721
16	1.873	0.5339	21.825	0.0458	11.6523	0.0858	6.6720
17	1.948	0.5134	23.698	0.0422	12.1657	0.0822	7.0656
18	2.026	0.4936	25.645	0.0390	12.6593	0.0790	7.4530
19	2.107	0.4747	27.671	0.0361	13.1339	0.0761	7.8342
20	2.191	0.4564	29.778	0.0336	13.5903	0.0736	8.2091
21	2.279	0.4388	31.969	0.0313	14.0292	0.0713	8.5780
22	2.370	0.4220	34.248	0.0292	14.4511	0.0692	8.9407
23	2.465	0.4057	36.618	0.0273	14.8569	0.0673	9.2973
24	2.563	0.3901	39.083	0.0256	15.2470	0.0656	9.6479
25	2.666	0.3751	41.646	0.0240	15.6221	0.0640	9.9925
26	2.772	0.3607	44.312	0.0226	15.9828	0.0626	10.3312
27	2.883	0.3468	47.084	0.0212	16.3296	0.0612	10.6640
28	2.999	0.3335	49.968	0.0200	16.6631	0.0600	10.9909
29	3.119	0.3207	52.966	0.0189	16.9837	0.0589	11.3121
30	3.243	0.3083	56.085	0.0178	17.2920	0.0578	11.6274
31	3.373	0.2965	59.328	0.0169	17.5885	0.0569	11.9371
32	3.508	0.2851	62.701	0.0160	17.8736	0.0560	12.2411
33	3.648	0.2741	66.210	0.0151	18.1477	0.0551	12.5396
34	3.794	0.2636	69.858	0.0143	18.4112	0.0543	12.8325
35	3.946	0.2534	73.652	0.0136	18.6646	0.0536	13.1199
40	4.801	0.2083	95.026	0.0105	19.7928	0.0505	14.4765
45	5.841	0.1712	121.029	0.0083	20.7200	0.0483	15.7047
50	7.107	0.1407	152.667	0.0066	21.4822	0.0466	16.8123
55	8.646	0.1157	191.159	0.0052	22.1086	0.0452	17.8070
60	10.520	0.0951	237.991	0.0042	22.6235	0.0442	18.6972
65	12.799	0.0781	294.968	0.0034	23.0467	0.0434	19.4909
70	15.572	0.0642	364.290	0.0028	23.3945	0.0428	20.1961
75	18.945	0.0528	448.631	0.0022	23.6804	0.0422	20.8206
80	23.050	0.0434	551.245	0.0018	23.9154	0.0418	21.3719
85	28.044	0.0357	676.090	0.0015	24.1085	0.0415	21.8569
90	34.119	0.0293	817.983	0.0012	24.2673	0.0412	22.2826
95	41.511	0.0241	1012.785	0.0010	24.3978	0.0410	22.6550
100	50.505	0.0198	1237.624	0.0008	24.5050	0.0408	22.9800

Table A.9. 5% INTEREST FACTORS FOR ANNUAL COMPOUNDING INTEREST

	Single Payment		Equal Payment Series				Uniform gradient-series factor
	Compound-amount factor	Present-worth factor	Compound-amount factor	Sinking-fund factor	Present-worth factor	Capital-recovery factor	
n	To find F Given P F/P i, n	To find P Given F P/F i, n	To find F Given A F/A i, n	To find A Given F A/F i, n	To find P Given A P/A i, n	To find A Given P A/P i, n	To find A Given G A/G i, n
1	1.050	0.9524	1.000	1.0000	0.9524	1.0500	0.0000
2	1.103	0.9070	2.050	0.4878	1.8594	0.5378	0.4878
3	1.158	0.8638	3.153	0.3172	2.7233	0.3672	0.9675
4	1.216	0.8227	4.310	0.2320	3.5460	0.2820	1.4391
5	1.276	0.7835	5.526	0.1810	4.3295	0.2310	1.9025
6	1.340	0.7462	6.802	0.1470	5.0757	0.1970	2.3579
7	1.407	0.7107	8.142	0.1228	5.7864	0.1728	2.8052
8	1.477	0.6768	9.549	0.1047	6.4632	0.1547	3.2445
9	1.551	0.6446	11.027	0.0907	7.1078	0.1407	3.6758
10	1.629	0.6139	12.587	0.0795	7.7217	0.1295	4.0991
11	1.710	0.5847	14.207	0.0704	8.3064	0.1204	4.5145
12	1.796	0.5568	15.917	0.0628	8.8633	0.1128	4.9219
13	1.886	0.5303	17.713	0.0565	9.3936	0.1065	5.3215
14	1.980	0.5051	19.599	0.0510	9.8987	0.1010	5.7133
15	2.079	0.4810	21.579	0.0464	10.3797	0.0964	6.0973
16	2.183	0.4581	23.658	0.0423	10.8378	0.0923	6.4736
17	2.292	0.4363	25.840	0.0387	11.2741	0.0887	6.8423
18	2.407	0.4155	28.132	0.0356	11.6896	0.0856	7.2034
19	2.527	0.3957	30.539	0.0328	12.0853	0.0828	7.5569
20	2.653	0.3769	33.066	0.0303	12.4622	0.0803	7.9030
21	2.786	0.3590	35.719	0.0280	12.8212	0.0780	8.2416
22	2.925	0.3419	38.505	0.0260	13.1630	0.0760	8.5730
23	3.072	0.3256	41.430	0.0241	13.4886	0.0741	8.8971
24	3.225	0.3101	44.502	0.0225	13.7987	0.0725	9.2140
25	3.386	0.2953	47.727	0.0210	14.0940	0.0710	9.5238
26	3.556	0.2813	51.113	0.0196	14.3752	0.0696	9.8266
27	3.733	0.2679	54.669	0.0183	14.6430	0.0683	10.1224
28	3.920	0.2551	58.403	0.0171	14.8981	0.0671	10.4114
29	4.116	0.2430	62.323	0.0161	15.1411	0.0661	10.6936
30	4.322	0.2314	66.439	0.0151	15.3725	0.0651	10.9691
31	4.538	0.2204	70.761	0.0141	15.5928	0.0641	11.2381
32	4.765	0.2099	75.299	0.0133	15.8027	0.0633	11.5005
33	5.003	0.1999	80.064	0.0125	16.0026	0.0625	11.7566
34	5.253	0.1904	85.067	0.0118	16.1929	0.0618	12.0063
35	5.516	0.1813	90.320	0.0111	16.3742	0.0611	12.2498
40	7.040	0.1421	120.800	0.0083	17.1591	0.0583	13.3775
45	8.985	0.1113	159.700	0.0063	17.7741	0.0563	14.3644
50	11.467	0.0872	209.348	0.0048	18.2559	0.0548	15.2233
55	14.636	0.0683	272.713	0.0037	18.6335	0.0537	15.9665
60	18.679	0.0535	353.584	0.0028	18.9293	0.0528	16.6062
65	23.840	0.0420	456.798	0.0022	19.1611	0.0522	17.1541
70	30.426	0.0329	588.529	0.0017	19.3427	0.0517	17.6212
75	38.833	0.0258	756.654	0.0013	19.4850	0.0513	18.0176
80	49.561	0.0202	971.229	0.0010	19.5965	0.0510	18.3526
85	63.254	0.0158	1245.087	0.0008	19.6838	0.0508	18.6346
90	80.730	0.0124	1594.607	0.0006	19.7523	0.0506	18.8712
95	103.035	0.0097	2040.694	0.0005	19.8059	0.0505	19.0689
100	131.501	0.0076	2610.025	0.0004	19.8479	0.0504	19.2337

Table A.10. 6% INTEREST FACTORS FOR ANNUAL COMPOUNDING INTEREST

	Single Payment		Equal Payment Series				Uniform gradient-series factor
	Compound-amount factor	Present-worth factor	Compound-amount factor	Sinking-fund factor	Present-worth factor	Capital-recovery factor	
n	To find F Given P F/P i, n	To find P Given F P/F i, n	To find F Given A F/A i, n	To find A Given F A/F i, n	To find P Given A P/A i, n	To find A Given P A/P i, n	To find A Given G A/G i, n
1	1.060	0.9434	1.000	1.0000	0.9434	1.0600	0.0000
2	1.124	0.8900	2.060	0.4854	1.8334	0.5454	0.4854
3	1.191	0.8396	3.184	0.3141	2.6730	0.3741	0.9612
4	1.262	0.7921	4.375	0.2286	3.4651	0.2886	1.4272
5	1.338	0.7473	5.637	0.1774	4.2124	0.2374	1.8836
6	1.419	0.7050	6.975	0.1434	4.9173	0.2034	2.3304
7	1.504	0.6651	8.394	0.1191	5.5824	0.1791	2.7676
8	1.594	0.6274	9.897	0.1010	6.2098	0.1610	3.1952
9	1.689	0.5919	11.491	0.0870	6.8017	0.1470	3.6133
10	1.791	0.5584	13.181	0.0759	7.3601	0.1359	4.0220
11	1.898	0.5268	14.972	0.0668	7.8869	0.1268	4.4213
12	2.012	0.4970	16.870	0.0593	8.3839	0.1193	4.8113
13	2.133	0.4688	18.882	0.0530	8.8527	0.1130	5.1920
14	2.261	0.4423	21.015	0.0476	9.2950	0.1076	5.5635
15	2.397	0.4173	23.276	0.0430	9.7123	0.1030	5.9260
16	2.540	0.3937	25.673	0.0390	10.1059	0.0990	6.2794
17	2.693	0.3714	28.213	0.0355	10.4773	0.0955	6.6240
18	2.854	0.3504	30.906	0.0324	10.8276	0.0924	6.9597
19	3.026	0.3305	33.760	0.0296	11.1581	0.0896	7.2867
20	3.207	0.3118	36.786	0.0272	11.4699	0.0872	7.6052
21	3.400	0.2942	39.993	0.0250	11.7641	0.0850	7.9151
22	3.604	0.2775	43.392	0.0231	12.0416	0.0831	8.2166
23	3.820	0.2618	46.996	0.0213	12.3034	0.0813	8.5099
24	4.049	0.2470	50.816	0.0197	12.5504	0.0797	8.7951
25	4.292	0.2330	54.865	0.0182	12.7834	0.0782	9.0722
26	4.549	0.2198	59.156	0.0169	13.0032	0.0769	9.3415
27	4.822	0.2074	63.706	0.0157	13.2105	0.0757	9.6030
28	5.112	0.1956	68.528	0.0146	13.4062	0.0746	9.8568
29	5.418	0.1846	73.640	0.0136	13.5907	0.0736	10.1032
30	5.744	0.1741	79.058	0.0127	13.7648	0.0727	10.3422
31	6.088	0.1643	84.802	0.0118	13.9291	0.0718	10.5740
32	6.453	0.1550	90.890	0.0110	14.0841	0.0710	10.7988
33	6.841	0.1462	97.343	0.0103	14.2302	0.0703	11.0166
34	7.251	0.1379	104.184	0.0096	14.3682	0.0696	11.2276
35	7.686	0.1301	111.435	0.0090	14.4983	0.0690	11.4319
40	10.286	0.0972	154.762	0.0065	15.0463	0.0665	12.3590
45	13.765	0.0727	212.744	0.0047	15.4558	0.0647	13.1413
50	18.420	0.0543	290.336	0.0035	15.7619	0.0635	13.7964
55	24.650	0.0406	394.172	0.0025	15.9906	0.0625	14.3411
60	32.988	0.0303	533.128	0.0019	16.1614	0.0619	14.7910
65	44.145	0.0227	719.083	0.0014	16.2891	0.0614	15.1601
70	59.076	0.0169	967.932	0.0010	16.3846	0.0610	15.4614
75	79.057	0.0127	1300.949	0.0008	16.4559	0.0608	15.7058
80	105.796	0.0095	1746.600	0.0006	16.5091	0.0606	15.9033
85	141.579	0.0071	2342.982	0.0004	16.5490	0.0604	16.0620
90	189.465	0.0053	3141.075	0.0003	16.5787	0.0603	16.1891
95	253.546	0.0040	4209.104	0.0002	16.6009	0.0602	16.2905
100	339.302	0.0030	5638.368	0.0002	16.6176	0.0602	16.3711

Table A.11. 7% INTEREST FACTORS FOR ANNUAL COMPOUNDING INTEREST

	Single Payment		Equal Payment Series				Uniform
	Compound-amount factor	Present-worth factor	Compound-amount factor	Sinking-fund factor	Present-worth factor	Capital-recovery factor	gradient-series factor
n	To find F Given P F/P i,n	To find P Given F P/F i,n	To find F Given A F/A i,n	To find A Given F A/F i,n	To find P Given A P/A i,n	To find A Given P A/P i,n	To find A Given G A/G i,n
1	1.070	0.9346	1.000	1.0000	0.9346	1.0700	0.0000
2	1.145	0.8734	2.070	0.4831	1.8080	0.5531	0.4831
3	1.225	0.8163	3.215	0.3111	2.6243	0.3811	0.9549
4	1.311	0.7629	4.440	0.2252	3.3872	0.2952	1.4155
5	1.403	0.7130	5.751	0.1739	4.1002	0.2439	1.8650
6	1.501	0.6664	7.153	0.1398	4.7665	0.2098	2.3032
7	1.606	0.6228	8.654	0.1156	5.3893	0.1856	2.7304
8	1.718	0.5820	10.260	0.0975	5.9713	0.1675	3.1466
9	1.838	0.5439	11.978	0.0835	6.5152	0.1535	3.5517
10	1.967	0.5084	13.816	0.0724	7.0236	0.1424	3.9461
11	2.105	0.4751	15.784	0.0634	7.4987	0.1334	4.3296
12	2.252	0.4440	17.888	0.0559	7.9427	0.1259	4.7025
13	2.410	0.4150	20.141	0.0497	8.3577	0.1197	5.0649
14	2.579	0.3878	22.550	0.0444	8.7455	0.1144	5.4167
15	2.759	0.3625	25.129	0.0398	9.1079	0.1098	5.7583
16	2.952	0.3387	27.888	0.0359	9.4467	0.1059	6.0897
17	3.159	0.3166	30.840	0.0324	9.7632	0.1024	6.4110
18	3.380	0.2959	33.999	0.0294	10.0591	0.0994	6.7225
19	3.617	0.2765	37.379	0.0268	10.3356	0.0968	7.0242
20	3.870	0.2584	40.996	0.0244	10.5940	0.0944	7.3163
21	4.141	0.2415	44.865	0.0223	10.8355	0.0923	7.5990
22	4.430	0.2257	49.006	0.0204	11.0613	0.0904	7.8725
23	4.741	0.2110	53.436	0.0187	11.2722	0.0887	8.1369
24	5.072	0.1972	58.177	0.0172	11.4693	o.0872	8.3923
25	5.427	0.1843	63.249	0.0158	11.6536	0.0858	8.6391
26	5.807	0.1722	68.676	0.0146	11.8258	0.0846	8.8773
27	6.214	0.1609	74.484	0.0134	11.9867	0.0834	9.1072
28	6.649	0.1504	80.698	0.0124	12.1371	0.0824	9.3290
29	7.114	0.1406	87.347	0.0115	12.2777	0.0815	9.5427
30	7.612	0.1314	94.461	0.0106	12.4091	0.0806	9.7487
31	8.145	0.1228	102.073	0.0098	12.5318	0.0798	9.9471
32	8.715	0.1148	110.218	0.0091	12.6466	0.0791	10.1381
33	9.325	0.1072	118.933	0.0084	12.7538	0.0784	10.3219
34	9.978	0.1002	128.259	0.0078	12.8540	0.0778	10.4987
35	10.677	0.0937	138.237	0.0072	12.9477	0.0772	10.6687
40	14.974	0.0668	199.635	0.0050	13.3317	0.0750	11.4234
45	21.002	0.0476	285.749	0.0035	13.6055	0.0735	12.0360
50	29.457	0.0340	406.529	0.0025	13.8008	0.0725	12.5287
55	41.315	0.0242	575.929	0.0017	13.9399	0.0717	12.9215
60	57.946	0.0173	813.520	0.0012	14.0392	0.0712	13.2321
65	81.273	0.0123	1146.755	0.0009	14.1099	0.0709	13.4760
70	113.989	0.0088	1614.134	0.0006	14.1604	0.0706	13.6662
75	159.876	0.0063	2269.657	0.0005	14.1964	0.0705	13.8137
80	224.234	0.0045	3189.063	0.0003	14.2220	0.0703	13.9274
85	314.500	0.0032	4478.576	0.0002	14.2403	0.0702	14.0146
90	441.103	0.0023	6287.185	0.0002	14.2533	0.0702	14.0812
95	618.670	0.0016	8823.854	0.0001	14.2626	0.0701	14.1319
100	867.716	0.0012	12381.662	0.0001	14.2693	0.0701	14.1703

Table A.12. 8% INTEREST FACTORS FOR ANNUAL COMPOUNDING INTEREST

	Single Payment		Equal Payment Series				Uniform gradient-series factor
	Compound-amount factor	Present-worth factor	Compound-amount factor	Sinking-fund factor	Present-worth factor	Capital-recovery factor	
n	To find F Given P F/P i,n	To find P Given F P/F i,n	To find F Given A F/A i,n	To find A Given F A/F i,n	To find P Given A P/A i,n	To find A Given P A/P i,n	To find A Given G A/G i,n
1	1.080	0.9259	1.000	1.0000	0.9259	1.0800	0.0000
2	1.166	0.8573	2.080	0.4808	1.7833	0.5608	0.4808
3	1.260	0.7938	3.246	0.3080	2.5771	0.3880	0.9488
4	1.360	0.7350	4.506	0.2219	3.3121	0.3019	1.4040
5	1.469	0.6806	5.867	0.1705	3.9927	0.2505	1.8465
6	1.587	0.6302	7.336	0.1363	4.6229	0.2163	2.2764
7	1.714	0.5835	8.923	0.1121	5.2064	0.1921	2.6937
8	1.851	0.5403	10.637	0.0940	5.7466	0.1740	3.0985
9	1.999	0.5003	12.488	0.0801	6.2469	0.1601	3.4910
10	2.159	0.4632	14.487	0.0690	6.7101	0.1490	3.8713
11	2.332	0.4289	16.645	0.0601	7.1390	0.1401	4.2395
12	2.518	0.3971	18.977	0.0527	7.5361	0.1327	4.5958
13	2.720	0.3677	21.495	0.0465	7.9038	0.1265	4.9402
14	2.937	0.3405	24.215	0.0413	8.2442	0.1213	5.2731
15	3.172	0.3153	27.152	0.0368	8.5595	0.1168	5.5945
16	3.426	0.2919	30.324	0.0330	8.8514	0.1130	5.9046
17	3.700	0.2703	33.750	0.0296	9.1216	0.1096	6.2038
18	3.996	0.2503	37.450	0.0267	9.3719	0.1067	6.4920
19	4.316	0.2317	41.446	0.0241	9.6036	0.1041	6.7697
20	4.661	0.2146	45.762	0.0219	9.8182	0.1019	7.0370
21	5.034	0.1987	50.423	0.0198	10.0168	0.0998	7.2940
22	5.437	0.1840	55.457	0.0180	10.2008	0.0980	7.5412
23	5.871	0.1703	60.893	0.0164	10.3711	0.0964	7.7786
24	6.341	0.1577	66.765	0.0150	10.5288	0.0950	8.0066
25	6.848	0.1460	73.106	0.0137	10.6748	0.0937	8.2254
26	7.396	0.1352	79.954	0.0125	10.8100	0.0925	8.4352
27	7.988	0.1252	87.351	0.0115	10.9352	0.0915	8.6363
28	8.627	0.1159	95.339	0.0105	11.0511	0.0905	8.8289
29	9.317	0.1073	103.966	0.0096	11.1584	0.0896	9.0133
30	10.063	0.0994	113.283	0.0088	11.2578	0.0888	9.1897
31	10.868	0.0920	123.346	0.0081	11.3498	0.0881	9.3584
32	11.737	0.0852	134.214	0.0075	11.4350	0.0875	9.5197
33	12.676	0.0789	145.951	0.0069	11.5139	0.0869	9.6737
34	13.690	0.0731	158.627	0.0063	11.5869	0.0863	9.8208
35	14.785	0.0676	172.317	0.0058	11.6546	0.0858	9.9611
40	21.725	0.0460	259.057	0.0039	11.9246	0.0839	10.5699
45	31.920	0.0313	386.506	0.0026	12.1084	0.0826	11.0447
50	46.902	0.0213	573.770	0.0018	12.2335	0.0818	11.4107
55	68.914	0.0145	848.923	0.0012	12.3186	0.0812	11.6902
60	101.257	0.0099	1253.213	0.0008	12.3766	0.0808	11.9015
65	148.780	0.0067	1847.248	0.0006	12.4160	0.0806	12.0602
70	218.606	0.0046	2720.080	0.0004	12.4428	0.0804	12.1783
75	321.205	0.0031	4002.557	0.0003	12.4611	0.0803	12.2658
80	471.955	0.0021	5886.935	0.0002	12.4735	0.0802	12.3301
85	693.456	0.0015	8655.706	0.0001	12.4820	0.0801	12.3773
90	1018.915	0.0010	12723.939	0.0001	12.4877	0.0801	12.4116
95	1497.121	0.0007	18701.507	0.0001	12.4917	0.0801	12.4365
100	2199.761	0.0005	27484.516	0.0001	12.4943	0.0800	12.4545

Table A.13. 9% INTEREST FACTORS FOR ANNUAL COMPOUNDING INTEREST

	Single Payment		Equal Payment Series				Uniform gradient-series factor
	Compound-amount factor	Present-worth factor	Compound-amount factor	Sinking-fund factor	Present-worth factor	Capital-recovery factor	
n	To find F Given P F/P i, n	To find P Given F P/F i, n	To find F Given A F/A i, n	To find A Given F A/F i, n	To find P Given A P/A i, n	To find A Given P A/P i, n	To find A Given G A/G i, n
1	1.090	0.9174	1.000	1.0000	0.9174	1.0900	0.0000
2	1.188	0.8417	2.090	0.4785	1.7591	0.5685	0.4785
3	1.295	0.7722	3.278	0.3051	2.5313	0.3951	0.9426
4	1.412	0.7084	4.573	0.2187	3.2397	0.3087	1.3925
5	1.539	0.6499	5.985	0.1671	3.8897	0.2571	1.8282
6	1.677	0.5963	7.523	0.1329	4.4859	0.2229	2.2498
7	1.828	0.5470	9.200	0.1087	5.0330	0.1987	2.6574
8	1.993	0.5019	11.028	0.0907	5.5348	0.1807	3.0512
9	2.172	0.4604	13.021	0.0768	5.9953	0.1668	3.4312
10	2.367	0.4224	15.193	0.0658	6.4177	0.1558	3.7978
11	2.580	0.3875	17.560	0.0570	6.8052	0.1470	4.1510
12	2.813	0.3555	20.141	0.0497	7.1607	0.1397	4.4910
13	3.066	0.3262	22.953	0.0436	7.4869	0.1336	4.8182
14	3.342	0.2993	26.019	0.0384	7.7862	0.1284	5.1326
15	3.642	0.2745	29.361	0.0341	8.0607	0.1241	5.4346
16	3.970	0.2519	33.003	0.0303	8.3126	0.1203	5.7245
17	4.328	0.2311	36.974	0.0271	8.5436	0.1171	6.0024
18	4.717	0.2120	41.301	0.0242	8.7556	0.1142	6.2687
19	5.142	0.1945	46.018	0.0217	8.9501	0.1117	6.5236
20	5.604	0.1784	51.160	0.0196	9.1286	0.1096	6.7675
21	6.109	0.1637	56.765	0.0176	9.2923	0.1076	7.0006
22	6.659	0.1502	62.873	0.0159	9.4424	0.1059	7.2232
23	7.258	0.1378	69.532	0.0144	9.5802	0.1044	7.4358
24	7.911	0.1264	76.790	0.0130	9.7066	0.1030	7.6384
25	8.623	0.1160	84.701	0.0118	9.8226	0.1018	7.8316
26	9.399	0.1064	93.324	0.0107	9.9290	0.1007	8.0156
27	10.245	0.0976	102.723	0.0097	10.0266	0.0997	8.1906
28	11.167	0.0896	112.968	0.0089	10.1161	0.0989	8.3572
29	12.172	0.0822	124.135	0.0081	10.1983	0.0981	8.5154
30	13.268	0.0754	136.308	0.0073	10.2737	0.0973	8.6657
31	14.462	0.0692	149.575	0.0067	10.3428	0.0967	8.8083
32	15.763	0.0634	164.037	0.0061	10.4063	0.0961	8.9436
33	17.182	0.0582	179.800	0.0056	10.4645	0.0956	9.0718
34	18.728	0.0534	196.982	0.0051	10.5178	0.0951	9.1933
35	20.414	0.0490	215.711	0.0046	10.5668	0.0946	9.3083
40	31.409	0.0318	337.882	0.0030	10.7574	0.0930	9.7957
45	48.327	0.0207	525.859	0.0019	10.8812	0.0919	10.1603
50	74.358	0.0135	815.084	0.0012	10.9617	0.0912	10.4295
55	114.408	0.0088	1260.092	0.0008	11.0140	0.0908	10.6261
60	176.031	0.0057	1944.792	0.0005	11.0480	0.0905	10.7683
65	270.846	0.0037	2998.288	0.0003	11.0701	0.0903	10.8702
70	416.730	0.0024	4619.223	0.0002	11.0845	0.0902	10.9427
75	641.191	0.0016	7113.232	0.0002	11.0938	0.0902	10.9940
80	986.552	0.0010	10950.574	0.0001	11.0999	0.0901	11.0299
85	1517.932	0.0007	16854.800	0.0001	11.1038	0.0901	11.0551
90	2335.527	0.0004	25939.184	0.0001	11.1064	0.0900	11.0726
95	3593.497	0.0003	39916.635	0.0000	11.1080	0.0900	11.0847
100	5529.041	0.0002	61422.675	0.0000	11.1091	0.0900	11.0930

Table A.14. 10% INTEREST FACTORS FOR ANNUAL COMPOUNDING INTEREST

	Single Payment		Equal Payment Series				Uniform gradient-series factor
	Compound-amount factor	Present-worth factor	Compound-amount factor	Sinking-fund factor	Present-worth factor	Capital-recovery factor	
n	To find F Given P F/P i, n	To find P Given F P/F i, n	To find F Given A F/A i, n	To find A Given F A/F i, n	To find P Given A P/A i, n	To find A Given P A/P i, n	To find A Given G A/G i, n
1	1.100	0.9091	1.000	1.0000	0.9091	1.1000	0.0000
2	1.210	0.8265	2.100	0.4762	1.7355	0.5762	0.4762
3	1.331	0.7513	3.310	0.3021	2.4869	0.4021	0.9366
4	1.464	0.6830	4.641	0.2155	3.1699	0.3155	1.3812
5	1.611	0.6209	6.105	0.1638	3.7908	0.2638	1.8101
6	1.772	0.5645	7.716	0.1296	4.3553	0.2296	2.2236
7	1.949	0.5132	9.487	0.1054	4.8684	0.2054	2.6216
8	2.144	0.4665	11.436	0.0875	5.3349	0.1875	3.0045
9	2.358	0.4241	13.579	0.0737	5.7590	0.1737	3.3724
10	2.594	0.3856	15.937	0.0628	6.1446	0.1628	3.7255
11	2.853	0.3505	18.531	0.0540	6.4951	0.1540	4.0641
12	3.138	0.3186	21.384	0.0468	6.8137	0.1468	4.3884
13	3.452	0.2897	24.523	0.0408	7.1034	0.1408	4.6988
14	3.798	0.2633	27.975	0.0358	7.3667	0.1358	4.9955
15	4.177	0.2394	31.772	0.0315	7.6061	0.1315	5.2789
16	4.595	0.2176	35.950	0.0278	7.8237	0.1278	5.5493
17	5.054	0.1979	40.545	0.0247	8.0216	0.1247	5.8071
18	5.560	0.1799	45.599	0.0219	8.2014	0.1219	6.0526
19	6.116	0.1635	51.159	0.0196	8.3649	0.1196	6.2861
20	6.728	0.1487	57.275	0.0175	8.5136	0.1175	6.5081
21	7.400	0.1351	64.003	0.0156	8.6487	0.1156	6.7189
22	8.140	0.1229	71.403	0.0140	8.7716	0.1140	6.9189
23	8.954	0.1117	79.543	0.0126	8.8832	0.1126	7.1085
24	9.850	0.1015	88.497	0.0113	8.9848	0.1113	7.2881
25	10.835	0.0923	98.347	0.0102	9.0771	0.1102	7.4580
26	11.918	0.0839	109.182	0.0092	9.1610	0.1092	7.6187
27	13.110	0.0763	121.100	0.0083	9.2372	0.1083	7.7704
28	14.421	0.0694	134.210	0.0075	9.3066	0.1075	7.9137
29	15.863	0.0630	148.631	0.0067	9.3696	0.1067	8.0489
30	17.449	0.0573	164.494	0.0061	9.4269	0.1061	8.1762
31	19.194	0.0521	181.943	0.0055	9.4790	0.1055	8.2962
32	21.114	0.0474	201.138	0.0050	9.5264	0.1050	8.4091
33	23.225	0.0431	222.252	0.0045	9.5694	0.1045	8.5152
34	25.548	0.0392	245.477	0.0041	9.6086	0.1041	8.6149
35	28.102	0.0356	271.024	0.0037	9.6442	0.1037	8.7086
40	45.259	0.0221	442.593	0.0023	9.7791	0:1023	9.0962
45	72.890	0.0137	718.905	0.0014	9.8628	0.1014	9.3741
50	117.391	0.0085	1163.909	0.0009	9.9148	0.1009	9.5704
55	189.059	0.0053	1880.591	0.0005	9.9471	0.1005	9.7075
60	304.482	0.0033	3034.816	0.0003	9.9672	0.1003	9.8023
65	490.371	0.0020	4893.707	0.0002	9.9796	0.1002	9.8672
70	789.747	0.0013	7887.470	0.0001	9.9873	0.1001	9.9113
75	1271.895	0.0008	12708.954	0.0001	9.9921	0.1001	9.9410
80	2048.400	0.0005	20474.002	0.0001	9.9951	0.1001	9.9609
85	3298.969	0.0003	32979.690	0.0000	9.9970	0.1000	9.9742
90	5313.023	0.0002	53120.226	0.0000	9.9981	0.1000	9.9831
95	8556.676	0.0001	85556.760	0.0000	9.9988	0.1000	9.9889
100	13780.612	0.0001	137796.123	0.0000	9.9993	0.1000	9.9928

Table A.15. 12% INTEREST FACTORS FOR ANNUAL COMPOUNDING INTEREST

	Single Payment		Equal Payment Series				Uniform gradient-series factor
	Compound-amount factor	Present-worth factor	Compound-amount factor	Sinking-fund factor	Present-worth factor	Capital-recovery factor	
n	To find F Given P F/P i,n	To find P Given F P/F i,n	To find F Given A F/A i,n	To find A Given F A/F i,n	To find P Given A P/A i,n	To find A Given P A/P i,n	To find A Given G A/G i,n
1	1.120	0.8929	1.000	1.0000	0.8929	1.1200	0.0000
2	1.254	0.7972	2.120	0.4717	1.6901	0.5917	0.4717
3	1.405	0.7118	3.374	0.2964	2.4018	0.4164	0.9246
4	1.574	0.6355	4.779	0.2092	3.0374	0.3292	1.3589
5	1.762	0.5674	6.353	0.1574	3.6048	0.2774	1.7746
6	1.974	0.5066	8.115	0.1232	4.1114	0.2432	2.1721
7	2.211	0.4524	10.089	0.0991	4.5638	0.2191	2.5515
8	2.476	0.4039	12.300	0.0813	4.9676	0.2013	2.9132
9	2.773	0.3606	14.776	0.0677	5.3283	0.1877	3.2574
10	3.106	0.3220	17.549	0.0570	5.6502	0.1770	3.5847
11	3.479	0.2875	20.655	0.0484	5.9377	0.1684	3.8953
12	3.896	0.2567	24.133	0.0414	6.1944	0.1614	4.1897
13	4.364	0.2292	28.029	0.0357	6.4236	0.1557	4.4683
14	4.887	0.2046	32.393	0.0309	6.6282	0.1509	4.7317
15	5.474	0.1827	37.280	0.0268	6.8109	0.1468	4.9803
16	6.130	0.1631	42.753	0.0234	6.9740	0.1434	5.2147
17	6.866	0.1457	48.884	0.0205	7.1196	0.1405	5.4353
18	7.690	0.1300	55.750	0.0179	7.2497	0.1379	5.6427
19	8.613	0.1161	63.440	0.0158	7.3658	0.1358	5.8375
20	9.646	0.1037	72.052	0.0139	7.4695	0.1339	6.0202
21	10.804	0.0926	81.699	0.0123	7.5620	0.1323	6.1913
22	12.100	0.0827	92.503	0.0108	7.6447	0.1308	6.3514
23	13.552	0.0738	104.603	0.0096	7.7184	0.1296	6.5010
24	15.179	0.0659	118.155	0.0085	7.7843	0.1285	6.6407
25	17.000	0.0588	133.334	0.0075	7.8431	0.1275	6.7708
26	19.040	0.0525	150.334	0.0067	7.8957	0.1267	6.8921
27	21.325	0.0469	169.374	0.0059	7.9426	0.1259	7.0049
28	23.884	0.0419	190.699	0.0053	7.9844	0.1253	7.1098
29	26.750	0.0374	214.583	0.0047	8.0218	0.1247	7.2071
30	29.960	0.0334	241.333	0.0042	8.0552	0.1242	7.2974
31	33.555	0.0298	271.293	0.0037	8.0850	0.1237	7.3811
32	37.582	0.0266	304.848	0.0033	8.1116	0.1233	7.4586
33	42.092	0.0238	342.429	0.0029	8.1354	0.1229	7.5303
34	47.143	0.0212	384.521	0.0026	8.1566	0.1226	7.5965
35	52.800	0.0189	431.664	0.0023	8.1755	0.1223	7.6577
40	93.051	0.0108	767.091	0.0013	8.2438	0.1213	7.8988
45	163.988	0.0061	1358.230	0.0007	8.2825	0.1207	8.0572
50	289.002	0.0035	2400.018	0.0004	8.3045	0.1204	8.1597

Table A.16. 15% INTEREST FACTORS FOR ANNUAL COMPOUNDING INTEREST

	Single Payment		Equal Payment Series				Uniform gradient-series factor
	Compound-amount factor	Present-worth factor	Compound-amount factor	Sinking-fund factor	Present-worth factor	Capital-recovery factor	
n	To find F Given P F/P i, n	To find P Given F P/F i, n	To find F Given A F/A i, n	To find A Given F A/F i, n	To find P Given A P/A i, n	To find A Given P A/P i, n	To find A Given G A/G i, n
1	1.150	0.8696	1.000	1.0000	0.8696	1.1500	0.0000
2	1.323	0.7562	2.150	0.4651	1.6257	0.6151	0.4651
3	1.521	0.6575	3.473	0.2880	2.2832	0.4380	0.9071
4	1.749	0.5718	4.993	0.2003	2.8550	0.3503	1.3263
5	2.011	0.4972	6.742	0.1483	3.3522	0.2983	1.7228
6	2.313	0.4323	8.754	0.1142	3.7845	0.2642	2.0972
7	2.660	0.3759	11.067	0.0904	4.1604	0.2404	2.4499
8	3.059	0.3269	13.727	0.0729	4.4873	0.2229	2.7813
9	3.518	0.2843	16.786	0.0596	4.7716	0.2096	3.0922
10	4.046	0.2472	20.304	0.0493	5.0188	0.1993	3.3832
11	4.652	0.2150	24.349	0.0411	5.2337	0.1911	3.6550
12	5.350	0.1869	29.002	0.0345	5.4206	0.1845	3.9082
13	6.153	0.1625	34.352	0.0291	5.5832	0.1791	4.1438
14	7.076	0.1413	40.505	0.0247	5.7245	0.1747	4.3624
15	8.137	0.1229	47.580	0.0210	5.8474	0.1710	4.5650
16	9.358	0.1069	55.717	0.0180	5.9542	0.1680	4.7523
17	10.761	0.0929	65.075	0.0154	6.0472	0.1654	4.9251
18	12.375	0.0808	75.836	0.0132	6.1280	0.1632	5.0843
19	14.232	0.0703	88.212	0.0113	6.1982	0.1613	5.2307
20	16.367	0.0611	102.444	0.0098	6.2593	0.1598	5.3651
21	18.822	0.0531	118.810	0.0084	6.3125	0.1584	5.4883
22	21.645	0.0462	137.632	0.0073	6.3587	0.1573	5.6010
23	24.891	0.0402	159.276	0.0063	6.3988	0.1563	5.7040
24	28.625	0.0349	184.168	0.0054	6.4338	0.1554	5.7979
25	32.919	0.0304	212.793	0.0047	6.4642	0.1547	5.8834
26	37.857	0.0264	245.712	0.0041	6.4906	0.1541	5.9612
27	43.535	0.0230	283.569	0.0035	6.5135	0.1535	6.0319
28	50.066	0.0200	327.104	0.0031	6.5335	0.1531	6.0960
29	57.575	0.0174	377.170	0.0027	6.5509	0.1527	6.1541
30	66.212	0.0151	434.745	0.0023	6.5660	0.1523	6.2066
31	76.144	0.0131	500.957	0.0020	6.5791	0.1520	6.2541
32	87.565	0.0114	577.100	0.0017	6.5905	0.1517	6.2970
33	100.700	0.0099	664.666	0.0015	6.6005	0.1515	6.3357
34	115.805	0.0086	765.365	0.0013	6.6091	0.1513	6.3705
35	133.176	0.0075	881.170	0.0011	6.6166	0.1511	6.4019
40	267.864	0.0037	1779.090	0.0006	6.6418	0.1506	6.5168
45	538.769	0.0019	3585.128	0.0003	6.6543	0.1503	6.5830
50	1083.657	0.0009	7217.716	0.0002	6.6605	0.1501	6.6205

Table A.17. 20% INTEREST FACTORS FOR ANNUAL COMPOUNDING INTEREST

	Single Payment		Equal Payment Series				Uniform gradient-series factor
	Compound-amount factor	Present-worth factor	Compound-amount factor	Sinking-fund factor	Present-worth factor	Capital-recovery factor	
n	To find F Given P F/P i, n	To find P Given F P/F i, n	To find F Given A F/A i, n	To find A Given F A/F i, n	To find P Given A P/A i, n	To find A Given P A/P i, n	To find A Given G A/G i, n
1	1.200	0.8333	1.000	1.0000	0.8333	1.2000	0.0000
2	1.440	0.6945	2.200	0.4546	1.5278	0.6546	0.4546
3	1.728	0.5787	3.640	0.2747	2.1065	0.4747	0.8791
4	2.074	0.4823	5.368	0.1863	2.5887	0.3863	1.2742
5	2.488	0.4019	7.442	0.1344	2.9906	0.3344	1.6405
6	2.986	0.3349	9.930	0.1007	3.3255	0.3007	1.9788
7	3.583	0.2791	12.916	0.0774	3.6046	0.2774	2.2902
8	4.300	0.2326	16.499	0.0606	3.8372	0.2606	2.5756
9	5.160	0.1938	20.799	0.0481	4.0310	0.2481	2.8364
10	6.192	0.1615	25.959	0.0385	4.1925	0.2385	3.0739
11	7.430	0.1346	32.150	0.0311	4.3271	0.2311	3.2893
12	8.916	0.1122	39.581	0.0253	4.4392	0.2253	3.4841
13	10.699	0.0935	48.497	0.0206	4.5327	0.2206	3.6597
14	12.839	0.0779	59.196	0.0169	4.6106	0.2169	3.8175
15	15.407	0.0649	72.035	0.0139	4.6755	0.2139	3.9589
16	18.488	0.0541	87.442	0.0114	4.7296	0.2114	4.0851
17	22.186	0.0451	105.931	0.0095	4.7746	0.2095	4.1976
18	26.623	0.0376	128.117	0.0078	4.8122	0.2078	4.2975
19	31.948	0.0313	154.740	0.0065	4.8435	0.2065	4.3861
20	38.338	0.0261	186.688	0.0054	4.8696	0.2054	4.4644
21	46.005	0.0217	225.026	0.0045	4.8913	0.2045	4.5334
22	55.206	0.0181	271.031	0.0037	4.9094	0.2037	4.5942
23	66.247	0.0151	326.237	0.0031	4.9245	0.2031	4.6475
24	79.497	0.0126	392.484	0.0026	4.9371	0.2026	4.6943
25	95.396	0.0105	471.981	0.0021	4.9476	0.2021	4.7352
26	114.475	0.0087	567.377	0.0018	4.9563	0.2018	4.7709
27	137.371	0.0073	681.853	0.0015	4.9636	0.2015	4.8020
28	164.845	0.0061	819.223	0.0012	4.9697	0.2012	4.8291
29	197.814	0.0051	984.068	0.0010	4.9747	0.2010	4.8527
30	237.376	0.0042	1181.882	0.0009	4.9789	0.2009	4.8731
31	284.852	0.0035	1419.258	0.0007	4.9825	0.2007	4.8908
32	341.822	0.0029	1704.109	0.0006	4.9854	0.2006	4.9061
33	410.186	0.0024	2045.931	0.0005	4.9878	0.2005	4.9194
34	492.224	0.0020	2456.118	0.0004	4.9899	0.2004	4.9308
35	590.668	0.0017	2948.341	0.0003	4.9915	0.2003	4.9407
40	1469.772	0.0007	7343.858	0.0002	4.9966	0.2001	4.9728
45	3657.262	0.0003	18281.310	0.0001	4.9986	0.2001	4.9877
50	9100.438	0.0001	45497.191	0.0000	4.9995	0.2000	4.9945

Table A.18. 25% INTEREST FACTORS FOR ANNUAL COMPOUNDING INTEREST

	Single Payment		Equal Payment Series				Uniform gradient-series factor
	Compound-amount factor	Present-worth factor	Compound-amount factor	Sinking-fund factor	Present-worth factor	Capital-recovery factor	
n	To find F Given P F/P i,n	To find P Given F P/F i,n	To find F Given A F/A i,n	To find A Given F A/F i,n	To find P Given A P/A i,n	To find A Given P A/P i,n	To find A Given G A/G i,n
1	1.250	0.8000	1.000	1.0000	0.8000	1.2500	0.0000
2	1.563	0.6400	2.250	0.4445	1.4400	0.6945	0.4445
3	1.953	0.5120	3.813	0.2623	1.9520	0.5123	0.8525
4	2.441	0.4096	5.766	0.1735	2.3616	0.4235	1.2249
5	3.052	0.3277	8.207	0.1219	2.6893	0.3719	1.5631
6	3.815	0.2622	11.259	0.0888	2.9514	0.3388	1.8683
7	4.768	0.2097	15.073	0.0664	3.1611	0.3164	2.1424
8	5.960	0.1678	19.842	0.0504	3.3289	0.3004	2.3873
9	7.451	0.1342	25.802	0.0388	3.4631	0.2888	2.6048
10	9.313	0.1074	33.253	0.0301	3.5705	0.2801	2.7971
11	11.642	0.0859	42.566	0.0235	3.6564	0.2735	2.9663
12	14.552	0.0687	54.208	0.0185	3.7251	0.2685	3.1145
13	18.190	0.0550	68.760	0.0146	3.7801	0.2646	3.2438
14	22.737	0.0440	86.949	0.0115	3.8241	0.2615	3.3560
15	28.422	0.0352	109.687	0.0091	3.8593	0.2591	3.4530
16	35.527	0.0282	138.109	0.0073	3.8874	0.2573	3.5366
17	44.409	0.0225	173.636	0.0058	3.9099	0.2558	3.6084
18	55.511	0.0180	218.045	0.0046	3.9280	0.2546	3.6698
19	69.389	0.0144	273.556	0.0037	3.9424	0.2537	3.7222
20	86.736	0.0115	342.945	0.0029	3.9539	0.2529	3.7667
21	108.420	0.0092	429.681	0.0023	3.9631	0.2523	3.8045
22	135.525	0.0074	538.101	0.0019	3.9705	0.2519	3.8365
23	169.407	0.0059	673.626	0.0015	3.9764	0.2515	3.8634
24	211.758	0.0047	843.033	0.0012	3.9811	0.2512	3.8861
25	264.698	0.0038	1054.791	0.0010	3.9849	0.2510	3.9052
26	330.872	0.0030	1319.489	0.0008	3.9879	0.2508	3.9212
27	413.590	0.0024	1650.361	0.0006	3.9903	0.2506	3.9346
28	516.988	0.0019	2063.952	0.0005	3.9923	0.2505	3.9457
29	646.235	0.0016	2580.939	0.0004	3.9938	0.2504	3.9551
30	807.794	0.0012	3227.174	0.0003	3.9951	0.2503	3.9628
31	1009.742	0.0010	4034.968	0.0003	3.9960	0.2503	3.9693
32	1262.177	0.0008	5044.710	0.0002	3.9968	0.2502	3.9746
33	1577.722	0.0006	6306.887	0.0002	3.9975	0.2502	3.9791
34	1972.152	0.0005	7884.609	0.0001	3.9980	0.2501	3.9828
35	2465.190	0.0004	9856.761	0.0001	3.9984	0.2501	3.9858

Table A.19. 30% INTEREST FACTORS FOR ANNUAL COMPOUNDING INTEREST

n	Single Payment		Equal Payment Series				Uniform
	Compound-amount factor	Present-worth factor	Compound-amount factor	Sinking-fund factor	Present-worth factor	Capital-recovery factor	gradient-series factor
	To find F Given P F/P i, n	To find P Given F P/F i, n	To find F Given A F/A i, n	To find A Given F A/F i, n	To find P Given A P/A i, n	To find A Given P A/P i, n	To find A Given G A/G i, n
1	1.300	0.7692	1.000	1.0000	0.7692	1.3000	0.0000
2	1.690	0.5917	2.300	0.4348	1.3610	0.7348	0.4348
3	2.197	0.4552	3.990	0.2506	1.8161	0.5506	0.8271
4	2.856	0.3501	6.187	0.1616	2.1663	0.4616	1.1783
5	3.713	0.2693	9.043	0.1106	2.4356	0.4106	1.4903
6	4.827	0.2072	12.756	0.0784	2.6428	0.3784	1.7655
7	6.275	0.1594	17.583	0.0569	2.8021	0.3569	2.0063
8	8.157	0.1226	23.858	0.0419	2.9247	0.3419	2.2156
9	10.605	0.0943	32.015	0.0312	3.0190	0.3312	2.3963
10	13.786	0.0725	42.620	0.0235	3.0915	0.3235	2.5512
11	17.922	0.0558	56.405	0.0177	3.1473	0.3177	2.6833
12	23.298	0.0429	74.327	0.0135	3.1903	0.3135	2.7952
13	30.288	0.0330	97.625	0.0103	3.2233	0.3103	2.8895
14	39.374	0.0254	127.913	0.0078	3.2487	0.3078	2.9685
15	51.186	0.0195	167.286	0.0060	3.2682	0.3060	3.0345
16	66.542	0.0150	218.472	0.0046	3.2832	0.3046	3.0892
17	86.504	0.0116	285.014	0.0035	3.2948	0.3035	3.1345
18	112.455	0.0089	371.518	0.0027	3.3037	0.3027	3.1718
19	146.192	0.0069	483.973	0.0021	3.3105	0.3021	3.2025
20	190.050	0.0053	630.165	0.0016	3.3158	0.3016	3.2276
21	247.065	0.0041	820.215	0.0012	3.3199	0.3012	3.2480
22	321.184	0.0031	1067.280	0.0009	3.3230	0.3009	3.2646
23	417.539	0.0024	1388.464	0.0007	3 3254	0.3007	3.2781
24	542.801	0.0019	1806.003	0.0006	3.3272	0.3006	3.2890
25	705.641	0.0014	2348.803	0.0004	3.3286	0.3004	3.2979
26	917.333	0.0011	3054.444	0.0003	3.3297	0.3003	3.3050
27	1192.533	0.0008	3971.778	0.0003	3.3305	0.3003	3.3107
28	1550.293	0.0007	5164.311	0.0002	3.3312	0.3002	3.3153
29	2015.381	0.0005	6714.604	0.0002	3.3317	0.3002	3.3189
30	2619.996	0.0004	8729.985	0.0001	3.3321	0.3001	3.3219
31	3405.994	0.0003	11349.981	0.0001	3.3324	0.3001	3.3242
32	4427.793	0.0002	14755.975	0.0001	3.3326	0.3001	3.3261
33	5756.130	0.0002	19183.768	0.0001	3.3328	0.3001	3.3276
34	7482.970	0.0001	24939.899	0.0001	3.3329	0.3001	3.3288
35	9727.860	0.0001	32422.868	0.0000	3.3330	0.3000	3.3297

Table A.20. 40% INTEREST FACTORS FOR ANNUAL COMPOUNDING INTEREST

	Single Payment		Equal Payment Series				Uniform gradient-series factor
	Compound-amount factor	Present-worth factor	Compound-amount factor	Sinking-fund factor	Present-worth factor	Capital-recovery factor	
n	To find F Given P $F/P\ i, n$	To find P Given F $P/F\ i, n$	To find F Given A $F/A\ i, n$	To find A Given F $A/F\ i, n$	To find P Given A $P/A\ i, n$	To find A Given P $A/P\ i, n$	To find A Given G $A/G\ i, n$
1	1.400	0.7143	1.000	1.0001	0.7143	1.4001	0.0000
2	1.960	0.5103	2.400	0.4167	1.2245	0.8167	0.4167
3	2.744	0.3645	4.360	0.2294	1.5890	0.6294	0.7799
4	3.842	0.2604	7.104	0.1408	1.8493	0.5408	1.0924
5	5.378	0.1860	10.946	0.0914	2.0352	0.4914	1.3580
6	7.530	0.1329	16.324	0.0613	2.1680	0.4613	1.5811
7	10.541	0.0949	23.853	0.0420	2.2629	0.4420	1.7664
8	14.758	0.0678	34.395	0.0291	2.3306	0.4291	1.9186
9	20.661	0.0485	49.153	0.0204	2.3790	0.4204	2.0423
10	28.925	0.0346	69.814	0.0144	2.4136	0.4144	2.1420
11	40.496	0.0247	98.739	0.0102	2.4383	0.4102	2.2215
12	56.694	0.0177	139.234	0.0072	2.4560	0.4072	2.2846
13	79.371	0.0126	195.928	0.0052	2.4686	0.4052	2.3342
14	111.120	0.0090	275.299	0.0037	2.4775	0.4037	2.3729
15	155.568	0.0065	386.419	0.0026	2.4840	0.4026	2.4030
16	217.794	0.0046	541.986	0.0019	2.4886	0.4019	2.4262
17	304.912	0.0033	759.780	0.0014	2.4918	0.4014	2.4441
18	426.877	0.0024	1064.691	0.0010	2.4942	0.4010	2.4578
19	597.627	0.0017	1491.567	0.0007	2.4959	0.4007	2.4682
20	836.678	0.0012	2089.195	0.0005	2.4971	0.4005	2.4761
21	1171.348	0.0009	2925.871	0.0004	2.4979	0.4004	2.4821
22	1639.887	0.0007	4097.218	0.0003	2.4985	0.4003	2.4866
23	2295.842	0.0005	5737.105	0.0002	2.4990	0.4002	2.4900
24	3214.178	0.0004	8032.945	0.0002	2.4993	0.4002	2.4926
25	4499.847	0.0003	11247.110	0.0001	2.4995	0.4001	2.4945
26	6299.785	0.0002	15746.960	0.0001	2.4997	0.4001	2.4959
27	8819.695	0.0002	22046.730	0.0001	2.4998	0.4001	2.4970
28	12347.570	0.0001	30866.430	0.0001	2.4998	0.4001	2.4978
29	17286.590	0.0001	43213.990	0.0001	2.4999	0.4001	2.4984
30	24201.230	0.0001	60500.580	0.0001	2.4999	0.4001	2.4988

Table A.21. 50% INTEREST FACTORS FOR ANNUAL COMPOUNDING INTEREST

	Single Payment		Equal Payment Series				Uniform gradient-series factor
	Compound-amount factor	Present-worth factor	Compound-amount factor	Sinking-fund factor	Present-worth factor	Capital-recovery factor	
n	To find *F* Given *P* *F/P i, n*	To find *P* Given *F* *P/F i, n*	To find *F* Given *A* *F/A i, n*	To find *A* Given *F* *A/F i, n*	To find *P* Given *A* *P/A i, n*	To find *A* Given *P* *A/P i, n*	To find *A* Given *G* *A/G i, n*
1	1.500	0.6667	1.000	1.0000	0.6667	1.5000	0.0001
2	2.250	0.4445	2.500	0.4000	1.1112	0.9001	0.4001
3	3.375	0.2963	4.750	0.2106	1.4075	0.7106	0.7369
4	5.063	0.1976	8.125	0.1231	1.6050	0.6231	1.0154
5	7.594	0.1317	13.188	0.0759	1.7367	0.5759	1.2418
6	11.391	0.0878	20.781	0.0482	1.8245	0.5482	1.4226
7	17.086	0.0586	32.172	0.0311	1.8830	0.5311	1.5649
8	25.629	0.0391	49.258	0.0204	1.9220	0.5204	1.6752
9	38.443	0.0261	74.887	0.0134	1.9480	0.5134	1.7597
10	57.665	0.0174	113.330	0.0089	1.9654	0.5089	1.8236
11	86.498	0.0116	170.995	0.0059	1.9769	0.5059	1.8714
12	129.746	0.0078	257.493	0.0039	1.9846	0.5039	1.9068
13	194.620	0.0052	387.239	0.0026	1.9898	0.5026	1.9329
14	291.929	0.0035	581.858	0.0018	1.9932	0.5018	1.9519
15	437.894	0.0023	873.788	0.0012	1.9955	0.5012	1.9657
16	656.841	0.0016	1311.681	0.0008	1.9970	0.5008	1.9757
17	985.261	0.0011	1968.522	0.0006	1.9980	0.5006	1.9828
18	1477.891	0.0007	2953.783	0.0004	1.9987	0.5004	1.9879
19	2216.837	0.0005	4431.671	0.0003	1.9991	0.5003	1.9915
20	3325.256	0.0004	6648.511	0.0002	1.9994	0.5002	1.9940
21	4987.882	0.0003	9973.765	0.0002	1.9996	0.5002	1.9958
22	7481.824	0.0002	14961.640	0.0001	1.9998	0.5001	1.9971
23	11222.730	0.0001	22443.470	0.0001	1.9999	0.5001	1.9980
24	16834.100	0.0001	33666.210	0.0001	1.9999	0.5001	1.9986
25	25251.160	0.0001	50500.330	0.0001	2.0000	0.5001	1.9991

Table B.1. EFFECTIVE INTEREST RATES CORRESPONDING TO NOMINAL RATE r

r	Compounding Frequency					
	Semi-annually $\left(1 + \frac{r}{2}\right)^2 - 1$	Quarterly $\left(1 + \frac{r}{4}\right)^4 - 1$	Monthly $\left(1 + \frac{r}{12}\right)^{12} - 1$	Weekly $\left(1 + \frac{r}{52}\right)^{52} - 1$	Daily $\left(1 + \frac{r}{365}\right)^{365} - 1$	Continuously $\left(1 + \frac{r}{\infty}\right)^{\infty} - 1$
.01	.010025	.010038	.010046	.010049	.010050	.010050
.02	.020100	.020151	.020184	.020197	.020200	.020201
.03	.030225	.030339	.030416	.030444	.030451	.030455
.04	.040400	.040604	.040741	.040793	.040805	.040811
.05	.050625	.050945	.051161	.051244	.051261	.051271
.06	.060900	.061364	.061678	.061797	.061799	.061837
.07	.071225	.071859	.072290	.072455	.072469	.072508
.08	.081600	.082432	.082999	.083217	.083246	.083287
.09	.092025	.093083	.093807	.094085	.094132	.094174
.10	.102500	.103813	.104713	.105060	.105126	.105171
.11	.113025	.114621	.115718	.116144	.116231	.116278
.12	.123600	.125509	.126825	.127336	.127447	.127497
.13	.134225	.136476	.138032	.138644	.138775	.138828
.14	.144900	.147523	.149341	.150057	.150217	.150274
.15	.155625	.158650	.160755	.161582	.161773	.161834
.16	.166400	.169859	.172270	.173221	.173446	.173511
.17	.177225	.181148	.183891	.184974	.185235	.185305
.18	.188100	.192517	.195618	.196843	.197142	.197217
.19	.199025	.203971	.207451	.208828	.209169	.209250
.20	.210000	.215506	.219390	.220931	.221316	.221403
.21	.221025	.227124	.231439	.233153	.233584	.233678
.22	.232100	.238825	.243596	.245494	.245976	.246077
.23	.243225	.250609	.255863	.257957	.258492	.258600
.24	.254400	.262477	.268242	.270542	.271133	.271249
.25	.265625	.274429	.280731	.283250	.283901	.284025
.26	.276900	.286466	.293333	.296090	.296796	.296930
.27	.288225	.298588	.306050	.309049	.309821	.309964
.28	.299600	.310796	.318880	.322135	.322976	.323130
.29	.311025	.323089	.331826	.335350	.336264	.336428
.30	.322500	.335469	.344889	.348693	.349684	.349859
.31	.334025	.347936	.358068	.362168	.363238	.363425
.32	.345600	.360489	.371366	.375775	.376928	.377128
.33	.357225	.373130	.384784	.389515	.390756	.390968
.34	.368900	.385859	.398321	.403389	.404722	.404948
.35	.380625	.398676	.411979	.417399	.418827	.419068

Table C.1. 1% INTEREST FACTORS FOR CONTINUOUS COMPOUNDING INTEREST

	Single Payment		Equal Payment Series				Uniform gradient-series factor
	Compound-amount factor	Present-worth factor	Compound-amount factor	Sinking-fund factor	Present-worth factor	Capital-recovery factor	
n	To find F Given P $F/P \quad r, n$	To find P Given F $P/F \quad r, n$	To find F Given A $F/A \quad r, n$	To find A Given F $A/F \quad r, n$	To find P Given A $P/A \quad r, n$	To find A Given P $A/P \quad r, n$	To find A Given G $A/G \quad r, n$
1	1.010	0.9901	1.000	1.0000	0.9901	1.0101	0.0000
2	1.020	0.9802	2.010	0.4975	1.9703	0.5076	0.4975
3	1.030	0.9705	3.030	0.3300	2.9407	0.3401	0.9933
4	1.041	0.9608	4.061	0.2463	3.9015	0.2563	1.4875
5	1.051	0.9512	5.102	0.1960	4.8527	0.2061	1.9800
6	1.062	0.9418	6.153	0.1625	5.7945	0.1726	2.4708
7	1.073	0.9324	7.215	0.1386	6.7269	0.1487	2.9600
8	1.083	0.9231	8.287	0.1207	7.6500	0.1307	3.4475
9	1.094	0.9139	9.370	0.1067	8.5639	0.1168	3.9334
10	1.105	0.9048	10.465	0.0956	9.4688	0.1056	4.4175
11	1.116	0.8958	11.570	0.0864	10.3646	0.0965	4.9000
12	1.128	0.8869	12.686	0.0788	11.2515	0.0889	5.3809
13	1.139	0.8781	13.814	0.0724	12.1296	0.0825	5.8600
14	1.150	0.8694	14.952	0.0669	12.9990	0.0769	6.3376
15	1.162	0.8607	16.103	0.0621	13.8597	0.0722	6.8134
16	1.174	0.8522	17.264	0.0579	14.7118	0.0680	7.2876
17	1.185	0.8437	18.438	0.0542	15.5555	0.0643	7.7601
18	1.197	0.8353	19.623	0.0510	16.3908	0.0610	8.2310
19	1.209	0.8270	20.821	0.0480	17.2177	0.0581	8.7002
20	1.221	0.8187	22.030	0.0454	18.0365	0.0555	9.1677
21	1.234	0.8106	23.251	0.0430	18.8470	0.0531	9.6336
22	1.246	0.8025	24.485	0.0409	19.6496	0.0509	10.0978
23	1.259	0.7945	25.731	0.0389	20.4441	0.0489	10.5604
24	1.271	0.7866	26.990	0.0371	21.2307	0.0471	11.0213
25	1.284	0.7788	28.261	0.0354	22.0095	0.0454	11.4806
26	1.297	0.7711	29.545	0.0339	22.7806	0.0439	11.9381
27	1.310	0.7634	30.842	0.0324	23.5439	0.0425	12.3941
28	1.323	0.7558	32.152	0.0311	24.2997	0.0412	12.8484
29	1.336	0.7483	33.475	0.0299	25.0480	0.0399	13.3010
30	1.350	0.7408	34.811	0.0287	25.7888	0.0388	13.7520
31	1.363	0.7335	36.161	0.0277	26.5223	0.0377	14.2013
32	1.377	0.7262	37.525	0.0267	27.2484	0.0367	14.6490
33	1.391	0.7189	38.902	0.0257	27.9673	0.0358	15.0950
34	1.405	0.7118	40.293	0.0248	28.6791	0.0349	15.5394
35	1.419	0.7047	41.698	0.0240	29.3838	0.0340	15.9821
40	1.492	0.6703	48.937	0.0204	32.8034	0.0305	18.1711
45	1.568	0.6376	56.548	0.0177	36.0563	0.0277	20.3190
50	1.649	0.6065	64.548	0.0155	39.1505	0.0256	22.4261
55	1.733	0.5770	72.959	0.0137	42.0939	0.0238	24.4926
60	1.822	0.5488	81.802	0.0122	44.8936	0.0223	26.5187
65	1.916	0.5221	91.097	0.0110	47.5569	0.0210	28.5045
70	2.014	0.4966	100.869	0.0099	50.0902	0.0200	30.4505
75	2.117	0.4724	111.142	0.0090	52.5000	0.0191	32.3567
80	2.226	0.4493	121.942	0.0082	54.7922	0.0183	34.2235
85	2.340	0.4274	133.296	0.0075	56.9727	0.0176	36.0513
90	2.460	0.4066	145.232	0.0069	59.0468	0.0169	37.8402
95	2.586	0.3868	157.779	0.0063	61.0198	0.0164	39.5907
100	2.718	0.3679	170.970	0.0059	62.8965	0.0159	41.3032

Table C.2. 2% INTEREST FACTORS FOR CONTINUOUS COMPOUNDING INTEREST

	Single Payment		Equal Payment Series				Uniform gradient-series factor
	Compound-amount factor	Present-worth factor	Compound-amount factor	Sinking-fund factor	Present-worth factor	Capital-recovery factor	
n	To find F Given P F/P r,n	To find P Given F P/F r,n	To find F Given A F/A r,n	To find A Given F A/F r,n	To find P Given A P/A r,n	To find A Given P A/P r,n	To find A Given G A/G r,n
1	1.020	0.9802	1.000	1.0000	0.9802	1.0202	0.0000
2	1.041	0.9608	2.020	0.4950	1.9410	0.5152	0.4950
3	1.062	0.9418	3.061	0.3267	2.8828	0.3469	0.9867
4	1.083	0.9231	4.123	0.2426	3.8059	0.2628	1.4750
5	1.105	0.9048	5.206	0.1921	4.7107	0.2123	1.9600
6	1.128	0.8869	6.311	0.1585	5.5976	0.1787	2.4417
7	1.150	0.8694	7.439	0.1344	6.4670	0.1546	2.9200
8	1.174	0.8522	8.589	0.1164	7.3191	0.1366	3.3951
9	1.197	0.8353	9.763	0.1024	8.1544	0.1226	3.8667
10	1.221	0.8187	10.960	0.0913	8.9731	0.1115	4.3351
11	1.246	0.8025	12.181	0.0821	9.7757	0.1023	4.8002
12	1.271	0.7866	13.427	0.0745	10.5623	0.0947	5.2619
13	1.297	0.7711	14.699	0.0680	11.3333	0.0882	5.7203
14	1.323	0.7558	15.995	0.0625	12.0891	0.0827	6.1754
15	1.350	0.7408	17.319	0.0578	12.8299	0.0780	6.6272
16	1.377	0.7262	18.668	0.0536	13.5561	0.0738	7.0757
17	1.405	0.7118	20.046	0.0499	14.2679	0.0701	7.5209
18	1.433	0.6977	21.451	0.0466	14.9655	0.0668	7.9628
19	1.462	0.6839	22.884	0.0437	15.6494	0.0639	8.4015
20	1.492	0.6703	24.346	0.0411	16.3197	0.0613	8.8368
21	1.522	0.6571	25.838	0.0387	16.9768	0.0589	9.2688
22	1.553	0.6440	27.360	0.0366	17.6208	0.0568	9.6976
23	1.584	0.6313	28.913	0.0346	18.2521	0.0548	10.1231
24	1.616	0.6188	30.497	0.0328	18.8709	0.0530	10.5453
25	1.649	0.6065	32.113	0.0312	19.4774	0.0514	10.9643
26	1.682	0.5945	33.762	0.0296	20.0719	0.0498	11.3801
27	1.716	0.5828	35.444	0.0282	20.6547	0.0484	11.7925
28	1.751	0.5712	37.160	0.0269	21.2259	0.0471	12.2018
29	1.786	0.5599	38.910	0.0257	21.7858	0.0459	12.6078
30	1.822	0.5488	40.696	0.0246	22.3346	0.0448	13.0106
31	1.859	0.5380	42.518	0.0235	22.8725	0.0437	13.4102
32	1.896	0.5273	44.377	0.0225	23.3998	0.0427	13.8065
33	1.935	0.5169	46.274	0.0216	23.9167	0.0418	14.1997
34	1.974	0.5066	48.209	0.0208	24.4233	0.0410	14.5897
35	2.014	0.4966	50.182	0.0199	24.9199	0.0401	14.9765
40	2.226	0.4493	60.666	0.0165	27.2591	0.0367	16.8630
45	2.460	0.4066	72.253	0.0139	29.3758	0.0341	18.6714
50	2.718	0.3679	85.058	0.0118	31.2910	0.0320	20.4028
55	3.004	0.3329	99.210	0.0101	33.0240	0.0303	22.0588
60	3.320	0.3012	114.850	0.0087	34.5921	0.0289	23.6409
65	3.669	0.2725	132.135	0.0076	36.0109	0.0278	25.1507
70	4.055	0.2466	151.238	0.0066	37.2947	0.0268	26.5899
75	4.482	0.2231	172.349	0.0058	38.4564	0.0260	27.9604
80	4.953	0.2019	195.682	0.0051	39.5075	0.0253	29.2640
85	5.474	0.1827	221.468	0.0045	40.4585	0.0247	30.5028
90	6.050	0.1653	249.966	0.0040	41.3191	0.0242	31.6786
95	6.686	0.1496	281.461	0.0036	42.0978	0.0238	32.7937
100	7.389	0.1353	316.269	0.0032	42.8024	0.0234	33.8499

Table C.3. 3% INTEREST FACTORS FOR CONTINUOUS COMPOUNDING INTEREST

	Single Payment		Equal Payment Series				Uniform gradient-series factor
	Compound-amount factor	Present-worth factor	Compound-amount factor	Sinking-fund factor	Present-worth factor	Capital-recovery factor	
n	To find F Given P $F/P \quad r, n$	To find P Given F $P/F \quad r, n$	To find F Given A $F/A \quad r, n$	To find A Given F $A/F \quad r, n$	To find P Given A $P/A \quad r, n$	To find A Given P $A/P \quad r, n$	To find A Given G $A/G \quad r, n$
1	1.030	0.9705	1.000	1.0000	0.9705	1.0305	0.0000
2	1.062	0.9418	2.030	0.4925	1.9122	0.5230	0.4925
3	1.094	0.9139	3.092	0.3234	2.8262	0.3538	0.9800
4	1.128	0.8869	4.186	0.2389	3.7131	0.2693	1.4625
5	1.162	0.8607	5.314	0.1882	4.5738	0.2186	1.9400
6	1.197	0.8353	6.476	0.1544	5.4090	0.1849	2.4126
7	1.234	0.8106	7.673	0.1303	6.2196	0.1608	2.8801
8	1.271	0.7866	8.907	0.1123	7.0063	0.1427	3.3427
9	1.310	0.7634	10.178	0.0983	7.7696	0.1287	3.8003
10	1.350	0.7408	11.488	0.0871	8.5105	0.1175	4.2529
11	1.391	0.7189	12.838	0.0779	9.2294	0.1084	4.7006
12	1.433	0.6977	14.229	0.0703	9.9271	0.1007	5.1433
13	1.477	0.6771	15.662	0.0639	10.6041	0.0943	5.5811
14	1.522	0.6571	17.139	0.0584	11.2612	0.0888	6.0139
15	1.568	0.6376	18.661	0.0536	11.8988	0.0841	6.4419
16	1.616	0.6188	20.229	0.0494	12.5176	0.0799	6.8650
17	1.665	0.6005	21.845	0.0458	13.1181	0.0762	7.2831
18	1.716	0.5828	23.511	0.0425	13.7008	0.0730	7.6964
19	1.768	0.5655	25.227	0.0397	14.2663	0.0701	8.1049
20	1.822	0.5488	26.995	0.0371	14.8152	0.0675	8.5085
21	1.878	0.5326	28.817	0.0347	15.3477	0.0652	8.9072
22	1.935	0.5169	30.695	0.0326	15.8646	0.0630	9.3012
23	1.994	0.5016	32.629	0.0307	16.3662	0.0611	9.6904
24	2.054	0.4868	34.623	0.0289	16.8529	0.0593	10.0748
25	2.117	0.4724	36.678	0.0273	17.3253	0.0577	10.4545
26	2.181	0.4584	38.795	0.0258	17.7837	0.0562	10.8294
27	2.248	0.4449	40.976	0.0244	18.2286	0.0549	11.1996
28	2.316	0.4317	43.224	0.0231	18.6603	0.0536	11.5652
29	2.387	0.4190	45.540	0.0220	19.0792	0.0524	11.9261
30	2.460	0.4066	47.927	0.0209	19.4858	0.0513	12.2823
31	2.535	0.3946	50.387	0.0199	19.8803	0.0503	12.6339
32	2.612	0.3829	52.921	0.0189	20.2632	0.0494	12.9810
33	2.691	0.3716	55.533	0.0180	20.6348	0.0485	13.3235
34	2.773	0.3606	58.224	0.0172	20.9954	0.0476	13.6614
35	2.858	0.3499	60.998	0.0164	21.3453	0.0469	13.9948
40	3.320	0.3012	76.183	0.0131	22.9459	0.0436	15.5953
45	3.857	0.2593	93.826	0.0107	24.3235	0.0411	17.0874
50	4.482	0.2231	114.324	0.0088	25.5092	0.0392	18.4750
55	5.207	0.1921	138.140	0.0072	26.5297	0.0377	19.7623
60	6.050	0.1653	165.809	0.0060	27.4081	0.0365	20.9538
65	7.029	0.1423	197.957	0.0051	28.1642	0.0355	22.0540
70	8.166	0.1225	235.307	0.0043	28.8149	0.0347	23.0677
75	9.488	0.1054	278.702	0.0036	29.3750	0.0341	23.9996
80	11.023	0.0907	329.119	0.0030	29.8570	0.0335	24.8543
85	12.807	0.0781	387.696	0.0026	30.2720	0.0330	25.6368
90	14.880	0.0672	455.753	0.0022	30.6291	0.0327	26.3516
95	17.288	0.0579	534.823	0.0019	30.9365	0.0323	27.0033
100	20.086	0.0498	626.690	0.0016	31.2010	0.0321	27.5963

Table C.4. 4% INTEREST FACTORS FOR CONTINUOUS COMPOUNDING INTEREST

	Single Payment		Equal Payment Series				Uniform gradient-series factor
	Compound-amount factor	Present-worth factor	Compound-amount factor	Sinking-fund factor	Present-worth factor	Capital-recovery factor	
n	To find F Given P F/P r, n	To find P Given F P/F r, n	To find F Given A F/A r, n	To find A Given F A/F r, n	To find P Given A P/A r, n	To find A Given P A/P r, n	To find A Given G A/G r, n
1	1.041	0.9608	1.000	1.0000	0.9608	1.0408	0.0000
2	1.083	0.9231	2.041	0.4900	1.8839	0.5308	0.4900
3	1.128	0.8869	3.124	0.3201	2.7708	0.3609	0.9734
4	1.174	0.8522	4.252	0.2352	3.6230	0.2760	1.4500
5	1.221	0.8187	5.425	0.1843	4.4417	0.2251	1.9201
6	1.271	0.7866	6.647	0.1505	5.2283	0.1913	2.3835
7	1.323	0.7558	7.918	0.1263	5.9841	0.1671	2.8402
8	1.377	0.7262	9.241	0.1082	6.7103	0.1490	3.2904
9	1.433	0.6977	10.618	0.0942	7.4079	0.1350	3.7339
10	1.492	0.6703	12.051	0.0830	8.0783	0.1238	4.1709
11	1.553	0.6440	13.543	0.0738	8.7223	0.1147	4.6013
12	1.616	0.6188	15.096	0.0663	9.3411	0.1071	5.0252
13	1.682	0.5945	16.712	0.0598	9.9356	0.1007	5.4425
14	1.751	0.5712	18.394	0.0544	10.5068	0.0952	5.8534
15	1.822	0.5488	20.145	0.0497	11.0556	0.0905	6.2578
16	1.896	0.5273	21.967	0.0455	11.5829	0.0863	6.6558
17	1.974	0.5066	23.863	0.0419	12.0895	0.0827	7.0474
18	2.054	0.4868	25.837	0.0387	12.5763	0.0795	7.4326
19	2.138	0.4677	27.892	0.0359	13.0440	0.0767	7.8114
20	2.226	0.4493	30.030	0.0333	13.4933	0.0741	8.1840
21	2.316	0.4317	32.255	0.0310	13.9250	0.0718	8.5503
22	2.411	0.4148	34.572	0.0289	14.3398	0.0697	8.9105
23	2.509	0.3985	36.983	0.0270	14.7383	0.0679	9.2644
24	2.612	0.3829	39.492	0.0253	15.1212	0.0661	9.6122
25	2.718	0.3679	42.104	0.0238	15.4891	0.0646	9.9539
26	2.829	0.3535	44.822	0.0223	15.8425	0.0631	10.2896
27	2.945	0.3396	47.651	0.0210	16.1821	0.0618	10.6193
28	3.065	0.3263	50.596	0.0198	16.5084	0.0606	10.9431
29	3.190	0.3135	53.661	0.0186	16.8219	0.0595	11.2609
30	3.320	0.3012	56.851	0.0176	17.1231	0.0584	11.5730
31	3.456	0.2894	60.171	0.0166	17.4125	0.0574	11.8792
32	3.597	0.2780	63.626	0.0157	17.6905	0.0565	12.1797
33	3.743	0.2671	67.223	0.0149	17.9576	0.0557	12.4746
34	3.896	0.2567	70.966	0.0141	18.2143	0.0549	12.7638
35	4.055	0.2466	74.863	0.0134	18.4609	0.0542	13.0475
40	4.953	0.2019	96.862	0.0103	19.5562	0.0511	14.3845
45	6.050	0.1653	123.733	0.0081	20.4530	0.0489	15.5918
50	7.389	0.1353	156.553	0.0064	21.1872	0.0472	16.6775
55	9.025	0.1108	196.640	0.0051	21.7883	0.0459	17.6498
60	11.023	0.0907	245.601	0.0041	22.2805	0.0449	18.5172
65	13.464	0.0743	305.403	0.0033	22.6834	0.0441	19.2882
70	16.445	0.0608	378.445	0.0027	23.0133	0.0435	19.9710
75	20.086	0.0498	467.659	0.0021	23.2834	0.0430	20.5737
80	24.533	0.0408	576.625	0.0017	23.5045	0.0426	21.1038
85	29.964	0.0334	709.717	0.0014	23.6856	0.0422	21.5687
90	36.598	0.0273	872.275	0.0012	23.8338	0.0420	21.9751
95	44.701	0.0224	1070.825	0.0009	23.9552	0.0418	22.3295
100	54.598	0.0183	1313.333	0.0008	24.0545	0.0416	22.6376

Table C.5. 5% INTEREST FACTORS FOR CONTINUOUS COMPOUNDING INTEREST

	Single Payment		Equal Payment Series				Uniform gradient-series factor
	Compound-amount factor	Present-worth factor	Compound-amount factor	Sinking-fund factor	Present-worth factor	Capital-recovery factor	
n	To find F Given P F/P r,n	To find P Given F P/F r,n	To find F Given A F/A r,n	To find A Given F A/F r,n	To find P Given A P/A r,n	To find A Given P A/P r,n	To find A Given G A/G r,n
1	1.051	0.9512	1.000	1.0000	0.9512	1.0513	0.0000
2	1.105	0.9048	2.051	0.4875	1.8561	0.5388	0.4875
3	1.162	0.8607	3.156	0.3168	2.7168	0.3681	0.9667
4	1.221	0.8187	4.318	0.2316	3.5355	0.2829	1.4376
5	1.284	0.7788	5.540	0.1805	4.3143	0.2318	1.9001
6	1.350	0.7408	6.824	0.1466	5.0551	0.1978	2.3544
7	1.419	0.7047	8.174	0.1224	5.7598	0.1736	2.8004
8	1.492	0.6703	9.593	0.1043	6.4301	0.1555	3.2382
9	1.568	0.6376	11.084	0.0902	7.0678	0.1415	3.6678
10	1.649	0.6065	12.653	0.0790	7.6743	0.1303	4.0892
11	1.733	0.5770	14.301	0.0699	8.2513	0.1212	4.5025
12	1.822	0.5488	16.035	0.0624	8.8001	0.1136	4.9077
13	1.916	0.5221	17.857	0.0560	9.3221	0.1073	5.3049
14	2.014	0.4966	19.772	0.0506	9.8187	0.1019	5.6941
15	2.117	0.4724	21.786	0.0459	10.2911	0.0972	6.0753
16	2.226	0.4493	23.903	0.0418	10.7404	0.0931	6.4487
17	2.340	0.4274	26.129	0.0383	11.1678	0.0896	6.8143
18	2.460	0.4066	28.468	0.0351	11.5744	0.0864	7.1721
19	2.586	0.3868	30.928	0.0323	11.9611	0.0836	7.5222
20	2.718	0.3679	33.514	0.0298	12.3290	0.0811	7.8646
21	2.858	0.3499	36.232	0.0276	12.6789	0.0789	8.1996
22	3.004	0.3329	39.090	0.0256	13.0118	0.0769	8.5270
23	3.158	0.3166	42.094	0.0238	13.3284	0.0750	8.8471
24	3.320	0.3012	45.252	0.0221	13.6296	0.0734	9.1599
25	3.490	0.2865	48.572	0.0206	13.9161	0.0719	9.4654
26	3.669	0.2725	52.062	0.0192	14.1887	0.0705	9.7638
27	3.857	0.2593	55.732	0.0180	14.4479	0.0692	10.0551
28	4.055	0.2466	59.589	0.0168	14.6945	0.0681	10.3395
29	4.263	0.2346	63.644	0.0157	14.9291	0.0670	10.6170
30	4.482	0.2231	67.907	0.0147	15.1522	0.0660	10.8877
31	4.711	0.2123	72.389	0.0138	15.3645	0.0651	11.1517
32	4.953	0.2019	77.101	0.0130	15.5664	0.0643	11.4091
33	5.207	0.1921	82.054	0.0122	15.7584	0.0635	11.6601
34	5.474	0.1827	87.261	0.0115	15.9411	0.0627	11.9046
35	5.755	0.1738	92.735	0.0108	16.1149	0.0621	12.1429
40	7.389	0.1353	124.613	0.0080	16.8646	0.0593	13.2435
45	9.488	0.1054	165.546	0.0061	17.4485	0.0573	14.2024
50	12.183	0.0821	218.105	0.0046	17.9032	0.0559	15.0329
55	15.643	0.0639	285.592	0.0035	18.2573	0.0548	15.7480
60	20.086	0.0498	372.247	0.0027	18.5331	0.0540	16.3604
65	25.790	0.0388	483.515	0.0021	18.7479	0.0533	16.8822
70	33.115	0.0302	626.385	0.0016	18.9152	0.0529	17.3245
75	42.521	0.0235	809.834	0.0012	19.0455	0.0525	17.6979
80	54.598	0.0183	1045.387	0.0010	19.1469	0.0522	18.0116
85	70.105	0.0143	1347.843	0.0008	19.2260	0.0520	18.2742
90	90.017	0.0111	1736.205	0.0006	19.2875	0.0519	18.4931
95	115.584	0.0087	2234.871	0.0005	19.3354	0.0517	18.6751
100	148.413	0.0067	2875.171	0.0004	19.3728	0.0516	18.8258

Table C.6. 6% INTEREST FACTORS FOR CONTINUOUS COMPOUNDING INTEREST

	Single Payment		Equal Payment Series				Uniform gradient-series factor
	Compound-amount factor	Present-worth factor	Compound-amount factor	Sinking-fund factor	Present-worth factor	Capital-recovery factor	
n	To find F Given P F/P r, n	To find P Given F P/F r, n	To find F Given A F/A r, n	To find A Given F A/F r, n	To find P Given A P/A r, n	To find A Given P A/P r, n	To find A Given G A/G r, n
1	1.062	0.9418	1.000	1.0000	0.9418	1.0618	0.0000
2	1.128	0.8869	2.062	0.4850	1.8287	0.5469	0.4850
3	1.197	0.8353	3.189	0.3136	2.6640	0.3754	0.9600
4	1.271	0.7866	4.387	0.2280	3.4506	0.2898	1.4251
5	1.350	0.7408	5.658	0.1768	4.1914	0.2386	1.8802
6	1.433	0.6977	7.008	0.1427	4.8891	0.2045	2.3254
7	1.522	0.6571	8.441	0.1185	5.5461	0.1803	2.7607
8	1.616	0.6188	9.963	0.1004	6.1649	0.1622	3.1862
9	1.716	0.5828	11.579	0.0864	6.7477	0.1482	3.6020
10	1.822	0.5488	13.295	0.0752	7.2965	0.1371	4.0080
11	1.935	0.5169	15.117	0.0662	7.8133	0.1280	4.4044
12	2.054	0.4868	17.052	0.0587	8.3001	0.1205	4.7912
13	2.181	0.4584	19.106	0.0523	8.7585	0.1142	5.1685
14	2.316	0.4317	21.288	0.0470	9.1902	0.1088	5.5363
15	2.460	0.4066	23.604	0.0424	9.5968	0.1042	5.8949
16	2.612	0.3829	26.064	0.0384	9.9797	0.1002	6.2442
17	2.773	0.3606	28.676	0.0349	10.3403	0.0967	6.5845
18	2.945	0.3396	31.449	0.0318	10.6799	0.0936	6.9157
19	3.127	0.3198	34.393	0.0291	10.9997	0.0909	7.2379
20	3.320	0.3012	37.520	0.0267	11.3009	0.0885	7.5514
21	3.525	0.2837	40.840	0.0245	11.5845	0.0863	7.8562
22	3.743	0.2671	44.366	0.0225	11.8517	0.0844	8.1525
23	3.975	0.2516	48.109	0.0208	12.1032	0.0826	8.4403
24	4.221	0.2369	52.084	0.0192	12.3402	0.0810	8.7199
25	4.482	0.2231	56.305	0.0178	12.5633	0.0796	8.9913
26	4.759	0.2101	60.786	0.0165	12.7734	0.0783	9.2546
27	5.053	0.1979	65.545	0.0153	12.9713	0.0771	9.5101
28	5.366	0.1864	70.598	0.0142	13.1577	0.0760	9.7578
29	5.697	0.1755	75.964	0.0132	13.3332	0.0750	9.9980
30	6.050	0.1653	81.661	0.0123	13.4985	0.0741	10.2307
31	6.424	0.1557	87.711	0.0114	13.6542	0.0732	10.4561
32	6.821	0.1466	94.135	0.0106	13.8008	0.0725	10.6743
33	7.243	0.1381	100.956	0.0099	13.9389	0.0718	10.8855
34	7.691	0.1300	108.198	0.0093	14.0689	0.0711	11.0899
35	8.166	0.1225	115.889	0.0086	14.1914	0.0705	11.2876
40	11.023	0.0907	162.091	0.0062	14.7046	0.0680	12.1809
45	14.880	0.0672	224.458	0.0045	15.0849	0.0663	12.9295
50	20.086	0.0498	308.645	0.0032	15.3665	0.0651	13.5519
55	27.113	0.0369	422.285	0.0024	15.5752	0.0642	14.0654
60	36.598	0.0273	575.683	0.0017	15.7298	0.0636	14.4862
65	49.402	0.0203	782.748	0.0013	15.8443	0.0631	14.8288
70	66.686	0.0150	1062.257	0.0010	15.9292	0.0628	15.1060
75	90.017	0.0111	1439.555	0.0007	15.9920	0.0625	15.3291
80	121.510	0.0082	1948.854	0.0005	16.0386	0.0624	15.5078
85	164.022	0.0061	2636.336	0.0004	16.0731	0.0622	15.6503
90	221.406	0.0045	3564.339	0.0003	16.0986	0.0621	15.7633
95	298.867	0.0034	4817.012	0.0002	16.1176	0.0621	15.8527
100	403.429	0.0025	6507.944	0.0002	16.1316	0.0620	15.9232

Table C.7. 7% INTEREST FACTORS FOR CONTINUOUS COMPOUNDING INTEREST

	Single Payment		Equal Payment Series				Uniform gradient-series factor
	Compound-amount factor	Present-worth factor	Compound-amount factor	Sinking-fund factor	Present-worth factor	Capital-recovery factor	
n	To find F Given P F/P r, n	To find P Given F P/F r, n	To find F Given A F/A r, n	To find A Given F A/F r, n	To find P Given A P/A r, n	To find A Given P A/P r, n	To find A Given G A/G r, n
1	1.073	0.9324	1.000	1.0000	0.9324	1.0725	0.0000
2	1.150	0.8694	2.073	0.4825	1.8018	0.5550	0.4825
3	1.234	0.8106	3.223	0.3103	2.6123	0.3828	0.9534
4	1.323	0.7558	4.456	0.2244	3.3681	0.2969	1.4126
5	1.419	0.7047	5.780	0.1730	4.0728	0.2455	1.8603
6	1.522	0.6571	7.199	0.1389	4.7299	0.2114	2.2965
7	1.632	0.6126	8.721	0.1147	5.3425	0.1872	2.7211
8	1.751	0.5712	10.353	0.0966	5.9137	0.1691	3.1344
9	1.878	0.5326	12.104	0.0826	6.4463	0.1551	3.5364
10	2.014	0.4966	13.981	0.0715	6.9429	0.1440	3.9272
11	2.160	0.4630	15.995	0.0625	7.4059	0.1350	4.3069
12	2.316	0.4317	18.155	0.0551	7.8376	0.1276	4.6756
13	2.484	0.4025	20.471	0.0489	8.2401	0.1214	5.0334
14	2.664	0.3753	22.955	0.0436	8.6154	0.1161	5.3804
15	2.858	0.3499	25.620	0.0390	8.9654	0.1161	5.7168
16	3.065	0.3263	28.478	0.0351	9.2917	0.1076	6.0428
17	3.287	0.3042	31.542	0.0317	9.5959	0.1042	6.3585
18	3.525	0.2837	34.829	0.0287	9.8795	0.1012	6.6640
19	3.781	0.2645	38.355	0.0261	10.1440	0.0986	6.9596
20	4.055	0.2466	42.136	0.0237	10.3906	0.0963	7.2453
21	4.349	0.2299	46.191	0.0217	10.6205	0.0942	7.5215
22	4.665	0.2144	50.540	0.0198	10.8349	0.0923	7.7882
23	5.003	0.1999	55.205	0.0181	11.0348	0.0906	8.0456
24	5.366	0.1864	60.208	0.0166	11.2212	0.0891	8.2940
25	5.755	0.1738	65.573	0.0153	13.3949	0.0878	8.5335
26	6.172	0.1620	71.328	0.0140	11.5570	0.0865	8.7643
27	6.619	0.1511	77.500	0.0129	11.7080	0.0854	8.9867
28	7.099	0.1409	84.119	0.0119	11.8489	0.0844	9.2009
29	7.614	0.1313	91.218	0.0110	11.9802	0.0835	9.4070
30	8.166	0.1225	98.833	0.0101	12.1027	0.0826	9.6052
31	8.758	0.1142	106.999	0.0094	12.2169	0.0819	9.7958
32	9.393	0.1065	115.757	0.0086	12.3233	0.0812	9.9790
33	10.047	0.0993	125.150	0.0080	12.4226	0.0805	10.1550
34	10.805	0.0926	135.225	0.0074	12.5151	0.0799	10.3239
35	11.588	0.0863	146.030	0.0069	12.6014	0.0794	10.4860
40	16.445	0.0608	213.006	0.0047	12.9529	0.0772	11.2017
45	23.336	0.0429	308.049	0.0033	13.2006	0.0758	11.7769
50	33.115	0.0302	442.922	0.0023	13.3751	0.0748	12.2347
55	46.993	0.0213	634.316	0.0016	13.4981	0.0741	12.5957
60	66.686	0.0150	905.916	0.0011	13.5847	0.0736	12.8781
65	94.632	0.0106	1291.336	0.0008	13.6458	0.0733	13.0974
70	134.290	0.0075	1838.272	0.0006	13.6889	0.0731	13.2664
75	190.566	0.0053	2614.412	0.0004	13.7192	0.0729	13.3959
80	270.426	0.0037	3715.807	0.0003	13.7406	0.0728	13.4946
85	383.753	0.0026	5278.761	0.0002	13.7556	0.0727	13.5695
90	544.572	0.0019	7496.698	0.0001	13.7662	0.0727	13.6260
95	772.784	0.0013	10644.100	0.0001	13.7737	0.0726	13.6685
100	1096.633	0.0009	15110.476	0.0001	13.7790	0.0726	13.7003

Table C.8. 8% INTEREST FACTORS FOR CONTINUOUS COMPOUNDING INTEREST

	Single Payment		Equal Payment Series				Uniform gradient-series factor
	Compound-amount factor	Present-worth factor	Compound-amount factor	Sinking-fund factor	Present-worth factor	Capital-recovery factor	
n	To find F Given P F/P r, n	To find P Given F P/F r, n	To find F Given A F/A r, n	To find A Given F A/F r, n	To find P Given A P/A r, n	To find A Given P A/P r, n	To find A Given G A/G r, n
1	1.083	0.9231	1.000	1.0000	0.9231	1.0833	0.0000
2	1.174	0.8522	2.083	0.4800	1.7753	0.5633	0.4800
3	1.271	0.7866	3.257	0.3071	2.5619	0.3903	0.9467
4	1.377	0.7262	4.528	0.2209	3.2880	0.3041	1.4002
5	1.492	0.6703	5.905	0.1694	3.9584	0.2526	1.8405
6	1.616	0.6188	7.397	0.1352	4.5772	0.2185	2.2676
7	1.751	0.5712	9.013	0.1110	5.1484	0.1942	2.6817
8	1.896	0.5273	10.764	0.0929	5.6757	0.1762	3.0829
9	2.054	0.4868	12.660	0.0790	6.1624	0.1623	3.4713
10	2.226	0.4493	14.715	0.0680	6.6117	0.1513	3.8470
11	2.411	0.4148	16.940	0.0590	7.0265	0.1423	4.2102
12	2.612	0.3829	19.351	0.0517	7.4094	0.1350	4.5611
13	2.829	0.3535	21.963	0.0455	7.7629	0.1288	4.8998
14	3.065	0.3263	24.792	0.0403	8.0891	0.1236	5.2265
15	3.320	0.3012	27.857	0.0359	8.3903	0.1192	5.5415
16	3.597	0.2780	31.177	0.0321	8.6684	0.1154	5.8449
17	3.896	0.2567	34.774	0.0288	8.9250	0.1121	6.1369
18	4.221	0.2369	38.670	0.0259	9.1620	0.1092	6.4178
19	4.572	0.2187	42.891	0.0233	9.3807	0.1066	6.6879
20	4.953	0.2019	47.463	0.0211	9.5826	0.1044	6.9473
21	5.366	0.1864	52.416	0.0191	9.7689	0.1024	7.1963
22	5.812	0.1721	57.781	0.0173	9.9410	0.1006	7.4352
23	6.297	0.1588	63.594	0.0157	10.0998	0.0990	7.6642
24	6.821	0.1466	69.890	0.0143	10.2464	0.0976	7.8836
25	7.389	0.1353	76.711	0.0130	10.3818	0.0963	8.0937
26	8.004	0.1249	84.100	0.0119	10.5067	0.0952	8.2948
27	8.671	0.1153	92.105	0.0109	10.6220	0.0942	8.4870
28	9.393	0.1065	100.776	0.0099	10.7285	0.0932	8.6707
29	10.176	0.0983	110.169	0.0091	10.8267	0.0924	8.8461
30	11.023	0.0907	120.345	0.0083	10.9175	0.0916	9.0136
31	11.941	0.0838	131.368	0.0076	11.0012	0.0909	9.1734
32	12.936	0.0773	143.309	0.0070	11.0785	0.0903	9.3257
33	14.013	0.0714	156.245	0.0064	11.1499	0.0897	9.4708
34	15.180	0.0659	170.258	0.0059	11.2157	0.0892	9.6090
35	16.445	0.0608	185.439	0.0054	11.2765	0.0887	9.7405
40	24.533	0.0408	282.547	0.0035	11.5173	0.0868	10.3069
45	36.598	0.0273	427.416	0.0023	11.6786	0.0856	10.7426
50	54.598	0.0183	643.535	0.0016	11.7868	0.0849	11.0738
55	81.451	0.0123	965.947	0.0010	11.8593	0.0843	11.3230
60	121.510	0.0082	1446.928	0.0007	11.9079	0.0840	11.5088
65	181.272	0.0055	2164.469	0.0005	11.9404	0.0838	11.6461
70	270.426	0.0037	3234.913	0.0003	11.9623	0.0836	11.7469
75	403.429	0.0025	4831.828	0.0002	11.9769	0.0835	11.8203
80	601.845	0.0017	7214.146	0.0002	11.9867	0.0834	11.8735
85	897.847	0.0011	10768.146	0.0001	11.9933	0.0834	11.9119
90	1339.431	0.0008	16070.091	0.0001	11.9977	0.0834	11.9394
95	1998.196	0.0005	23979.664	0.0001	12.0007	0.0833	11.9591
100	2980.958	0.0004	35779.360	0.0000	12.0026	0.0833	11.9731

Table C.9. 9% INTEREST FACTORS FOR CONTINUOUS COMPOUNDING INTEREST

	Single Payment		Equal Payment Series				Uniform gradient-series factor
	Compound-amount factor	Present-worth factor	Compound-amount factor	Sinking-fund factor	Present-worth factor	Capital-recovery factor	
n	To find *F* Given *P* *F/P r, n*	To find *P* Given *F* *P/F r, n*	To find *F* Given *A* *F/A r, n*	To find *A* Given *F* *A/F r, n*	To find *P* Given *A* *P/A r, n*	To find *A* Given *P* *A/P r, n*	To find *A* Given *G* *A/G r, n*
1	1.094	0.9139	1.000	1.0000	0.9139	1.0942	0.0000
2	1.197	0.8353	2.094	0.4775	1.7492	0.5717	0.4775
3	1.310	0.7634	3.291	0.3038	2.5126	0.3980	0.9401
4	1.433	0.6977	4.601	0.2173	3.2103	0.3115	1.3878
5	1.568	0.6376	6.035	0.1657	3.8479	0.2599	1.8206
6	1.716	0.5828	7.603	0.1315	4.4306	0.2257	2.2388
7	1.878	0.5326	9.319	0.1073	4.9632	0.2015	2.6424
8	2.054	0.4868	11.197	0.0893	5.4500	0.1835	3.0316
9	2.248	0.4449	13.251	0.0755	5.8948	0.1697	3.4065
10	2.460	0.4066	15.499	0.0645	6.3014	0.1587	3.7674
11	2.691	0.3716	17.959	0.0557	6.6730	0.1499	4.1145
12	2.945	0.3396	20.650	0.0484	7.0126	0.1426	4.4479
13	3.222	0.3104	23.594	0.0424	7.3230	0.1366	4.7680
14	3.525	0.2837	26.816	0.0373	7.6066	0.1315	5.0750
15	3.857	0.2593	30.342	0.0330	7.8658	0.1271	5.3691
16	4.221	0.2369	34.199	0.0293	8.1028	0.1234	5.6507
17	4.618	0.2165	38.420	0.0260	8.3193	0.1202	5.9201
18	5.053	0.1979	43.038	0.0232	8.5172	0.1174	6.1776
19	5.529	0.1809	48.091	0.0208	8.6981	0.1150	6.4234
20	6.050	0.1653	53.620	0.0187	8.8634	0.1128	6.6579
21	6.619	0.1511	59.670	0.0168	9.0144	0.1109	6.8815
22	7.243	0.1381	66.289	0.0151	9.1525	0.1093	7.0945
23	7.925	0.1262	73.532	0.0136	9.2787	0.1078	7.2972
24	8.671	0.1153	81.457	0.0123	9.3940	0.1065	7.4900
25	9.488	0.1054	90.128	0.0111	9.4994	0.1053	7.6732
26	10.381	0.0963	99.616	0.0100	9.5958	0.1042	7.8471
27	11.359	0.0880	109.997	0.0091	9.6838	0.1033	8.0122
28	12.429	0.0805	121.356	0.0083	9.7643	0.1024	8.1686
29	13.599	0.0735	133.784	0.0075	9.8378	0.1017	8.3169
30	14.880	0.0672	147.383	0.0068	9.9050	0.1010	8.4572
31	16.281	0.0614	162.263	0.0062	9.9664	0.1003	8.5900
32	17.814	0.0561	178.544	0.0056	10.0225	0.0998	8.7155
33	19.492	0.0513	196.358	0.0051	10.0739	0.0993	8.8341
34	21.328	0.0469	215.850	0.0046	10.1207	0.0988	8.9460
35	23.336	0.0429	237.178	0.0042	10.1636	0.0984	9.0516
40	36.598	0.0273	378.004	0.0027	10.3285	0.0968	9.4950
45	57.397	0.0174	598.863	0.0017	10.4336	0.0959	9.8207
50	90.017	0.0111	945.238	0.0011	10.5007	0.0952	10.0569
55	141.175	0.0071	1488.463	0.0007	10.5434	0.0949	10.2263
60	221.406	0.0045	2340.410	0.0004	10.5707	0.0946	10.3464
65	347.234	0.0029	3676.528	0.0003	10.5880	0.0945	10.4309
70	544.572	0.0019	5771.978	0.0002	10.5991	0.0944	10.4898
75	854.059	0.0012	9058.298	0.0001	10.6062	0.0943	10.5307
80	1339.431	0.0008	14212.274	0.0001	10.6107	0.0943	10.5588
85	2100.646	0.0005	22295.318	0.0001	10.6136	0.0942	10.5781
90	3294.468	0.0003	34972.053	0.0000	10.6154	0.0942	10.5913
95	5166.754	0.0002	54853.132	0.0000	10.6166	0.0942	10.6002
100	8103.084	0.0001	86032.870	0.0000	10.6173	0.0942	10.6063

Table C.10. 10% INTEREST FACTORS FOR CONTINUOUS COMPOUNDING
INTEREST

	Single Payment		Equal Payment Series				Uniform gradient-series factor
	Compound-amount factor	Present-worth factor	Compound-amount factor	Sinking-fund factor	Present-worth factor	Capital-recovery factor	
n	To find F Given P F/P r, n	To find P Given F P/F r, n	To find F Given A F/A r, n	To find A Given F A/F r, n	To find P Given A P/A r, n	To find A Given P A/P r, n	To find A Given G A/G r, n
1	1.105	0.9048	1.000	1.0000	0.9048	1.1052	0.0000
2	1.221	0.8187	2.105	0.4750	1.7236	0.5802	0.4750
3	1.350	0.7408	3.327	0.3006	2.4644	0.4058	0.9335
4	1.492	0.6703	4.676	0.2138	3.1347	0.3190	1.3754
5	1.649	0.6065	6.168	0.1621	3.7412	0.2673	1.8009
6	1.822	0.5488	7.817	0.1279	4.2901	0.2331	2.2101
7	2.014	0.4966	9.639	0.1038	4.7866	0.2089	2.6033
8	2.226	0.4493	11.653	0.0858	5.2360	0.1910	2.9806
9	2.460	0.4066	13.878	0.0721	5.6425	0.1772	3.3423
10	2.718	0.3679	16.338	0.0612	6.0104	0.1664	3.6886
11	3.004	0.3329	19.056	0.0525	6.3433	0.1577	4.0198
12	3.320	0.3012	22.060	0.0453	6.6445	0.1505	4.3362
13	3.669	0.2725	25.381	0.0394	6.9170	0.1446	4.6381
14	4.055	0.2466	29.050	0.0344	7.1636	0.1396	4.9260
15	4.482	0.2231	33.105	0.0302	7.3867	0.1354	5.2001
16	4.953	0.2019	37.587	0.0266	7.5886	0.1318	5.4608
17	5.474	0.1827	42.540	0.0235	7.7713	0.1287	5.7086
18	6.050	0.1653	48.014	0.0208	7.9366	0.1260	5.9437
19	6.686	0.1496	54.063	0.0185	8.0862	0.1237	6.1667
20	7.389	0.1353	60.749	0.0165	8.2215	0.1216	6.3780
21	8.166	0.1225	68.138	0.0147	8.3440	0.1199	6.5779
22	9.025	0.1108	76.305	0.0131	8.4548	0.1183	6.7669
23	9.974	0.1003	85.330	0.0117	8.5550	0.1169	6.9454
24	11.023	0.0907	95.304	0.0105	8.6458	0.1157	7.1139
25	12.183	0.0821	106.327	0.0094	8.7279	0.1146	7.2727
26	13.464	0.0743	118.509	0.0084	8.8021	0.1136	7.4223
27	14.880	0.0672	131.973	0.0076	8.8693	0.1128	7.5631
28	16.445	0.0608	146.853	0.0068	8.9301	0.1120	7.6954
29	18.174	0.0550	163.297	0.0061	8.9852	0.1113	7.8198
30	20.086	0.0498	181.472	0.0055	9.0349	0.1107	7.9365
31	22.198	0.0451	201.557	0.0050	9.0800	0.1101	8.0459
32	24.533	0.0408	223.755	0.0045	9.1208	0.1097	8.1485
33	27.113	0.0369	248.288	0.0040	9.1576	0.1092	8.2446
34	29.964	0.0334	275.400	0.0036	9.1910	0.1088	8.3345
35	33.115	0.0302	305.364	0.0033	9.2212	0.1085	8.4185
40	54.598	0.0183	509.629	0.0020	9.3342	0.1071	8.7620
45	90.017	0.0111	846.404	0.0012	9.4027	0.1064	9.0028
50	148.413	0.0067	1401.653	0.0007	9.4443	0.1059	9.1692
55	244.692	0.0041	2317.104	0.0004	9.4695	0.1056	9.2826
60	403.429	0.0025	3826.427	0.0003	9.4848	0.1054	9.3592
65	665.142	0.0015	6314.879	0.0002	9.4940	0.1053	9.4105
70	1096.633	0.0009	10417.644	0.0001	9.4997	0.1053	9.4445
75	1808.042	0.0006	17181.959	0.0001	9.5031	0.1052	9.4668
80	2980.958	0.0004	28334.430	0.0001	9.5052	0.1052	9.4815
85	4914.769	0.0002	46721.745	0.0000	9.5064	0.1052	9.4910
90	8103.084	0.0001	77037.303	0.0000	9.5072	0.1052	9.4972
95	13359.727	0.0001	127019.209	0.0000	9.5076	0.1052	9.5012
100	22026.466	0.0001	209425.440	0.0000	9.5079	0.1052	9.5038

Table C.11. 12% INTEREST FACTORS FOR CONTINUOUS COMPOUNDING INTEREST

	Single Payment		Equal Payment Series				Uniform gradient-series factor
	Compound-amount factor	Present-worth factor	Compound-amount factor	Sinking-fund factor	Present-worth factor	Capital-recovery factor	
n	To find *F* Given *P* F/P r, n	To find *P* Given *F* P/F r, n	To find *F* Given *A* F/A r, n	To find *A* Given *F* A/F r, n	To find *P* Given *A* P/A r, n	To find *A* Given *P* A/P r, n	To find *A* Given *G* A/G r, n
1	1.128	0.8869	1.000	1.0000	0.8869	1.1275	0.0000
2	1.271	0.7866	2.128	0.4700	1.6736	0.5975	0.4700
3	1.433	0.6977	3.399	0.2942	2.3712	0.4217	0.9202
4	1.616	0.6188	4.832	0.2070	2.9900	0.3345	1.3506
5	1.822	0.5488	6.448	0.1551	3.5388	0.2826	1.7615
6	2.054	0.4868	8.270	0.1209	4.0256	0.2484	2.1531
7	2.316	0.4317	10.325	0.0969	4.4573	0.2244	2.5257
8	2.612	0.3829	12.641	0.0791	4.8402	0.2066	2.8796
9	2.945	0.3396	15.253	0.0656	5.1798	0.1931	3.2153
10	3.320	0.3012	18.197	0.0550	5.4810	0.1825	3.5332
11	3.743	0.2671	21.518	0.0465	5.7481	0.1740	3.8337
12	4.221	0.2369	25.261	0.0396	5.9850	0.1671	4.1174
13	4.759	0.2101	29.482	0.0339	6.1952	0.1614	4.3848
14	5.366	0.1864	34.241	0.0292	6.3815	0.1567	4.6364
15	6.050	0.1653	39.606	0.0253	6.5468	0.1528	4.8728
16	6.821	0.1466	45.656	0.0219	6.6935	0.1494	5.0947
17	7.691	0.1300	52.477	0.0191	6.8235	0.1466	5.3025
18	8.671	0.1153	60.167	0.0166	6.9388	0.1441	5.4969
19	9.777	0.1023	68.838	0.0145	7.0411	0.1420	5.6785
20	11.023	0.0907	78.615	0.0127	7.1318	0.1402	5.8480
21	12.429	0.0805	89.638	0.0112	7.2123	0.1387	6.0058
22	14.013	0.0714	102.067	0.0098	7.2836	0.1373	6.1528
23	15.800	0.0633	116.080	0.0086	7.3469	0.1361	6.2893
24	17.814	0.0561	131.880	0.0076	7.4031	0.1351	6.4160
25	20.086	0.0498	149.694	0.0067	7.4528	0.1342	6.5334
26	22.646	0.0442	169.780	0.0059	7.4970	0.1334	6.6422
27	25.534	0.0392	192.426	0.0052	7.5362	0.1327	6.7428
28	28.789	0.0347	217.960	0.0046	7.5709	0.1321	6.8358
29	32.460	0.0308	246.749	0.0041	7.6017	0.1316	6.9215
30	36.598	0.0273	279.209	0.0036	7.6290	0.1311	7.0006
31	41.264	0.0242	315.807	0.0032	7.6533	0.1307	7.0734
32	46.525	0.0215	357.071	0.0028	7.6748	0.1303	7.1404
33	52.457	0.0191	403.597	0.0025	7.6938	0.1300	7.2020
34	59.145	0.0169	456.054	0.0022	7.7107	0.1297	7.2586
35	66.686	0.0150	515.200	0.0020	7.7257	0.1294	7.3105
40	121.510	0.0082	945.203	0.0011	7.7788	0.1286	7.5114
45	221.406	0.0045	1728.720	0.0006	7.8079	0.1281	7.6392
50	403.429	0.0025	3156.382	0.0003	7.8239	0.1278	7.7191

Table C.12. 15% INTEREST FACTORS FOR CONTINUOUS COMPOUNDING
INTEREST

	Single Payment		Equal Payment Series				Uniform gradient-series factor
	Compound-amount factor	Present-worth factor	Compound-amount factor	Sinking-fund factor	Present-worth factor	Capital-recovery factor	
n	To find F Given P F/P r,n	To find P Given F P/F r,n	To find F Given A F/A r,n	To find A Given F A/F r,n	To find P Given A P/A r,n	To find A Given P A/P r,n	To find A Given G A/G r,n
1	1.162	0.8607	1.000	1.0000	0.8607	1.1618	0.0000
2	1.350	0.7408	2.162	0.4626	1.6015	0.6244	0.4626
3	1.568	0.6376	3.512	0.2848	2.2392	0.4466	0.9004
4	1.822	0.5488	5.080	0.1969	2.7880	0.3587	1.3137
5	2.117	0.4724	6.902	0.1449	3.2603	0.3067	1.7029
6	2.460	0.4066	9.019	0.1109	3.6669	0.2727	2.0685
7	2.858	0.3499	11.479	0.0871	4.0168	0.2490	2.4110
8	3.320	0.3012	14.336	0.0698	4.3180	0.2316	2.7311
9	3.857	0.2593	17.657	0.0566	4.5773	0.2185	3.0295
10	4.482	0.2231	21.514	0.0465	4.8004	0.2083	3.3070
11	5.207	0.1921	25.996	0.0385	4.9925	0.2003	3.5645
12	6.050	0.1653	31.203	0.0321	5.1578	0.1939	3.8028
13	7.029	0.1423	37.252	0.0269	5.3000	0.1887	4.0228
14	8.166	0.1225	44.281	0.0226	5.4225	0.1844	4.2255
15	9.488	0.1054	52.447	0.0191	5.5279	0.1809	4.4119
16	11.023	0.0907	61.935	0.0162	5.6186	0.1780	4.5829
17	12.807	0.0781	72.958	0.0137	5.6967	0.1756	4.7394
18	14.880	0.0672	85.765	0.0117	5.7639	0.1735	4.8823
19	17.288	0.0579	100.645	0.0099	5.8217	0.1718	5.0127
20	20.086	0.0498	117.933	0.0085	5.8715	0.1703	5.1313
21	23.336	0.0429	138.018	0.0073	5.9144	0.1691	5.2390
22	27.113	0.0369	161.354	0.0062	5.9513	0.1680	5.3367
23	31.500	0.0318	188.467	0.0053	5.9830	0.1672	5.4251
24	36.598	0.0273	219.967	0.0046	6.0103	0.1664	5.5050
25	42.521	0.0235	256.566	0.0039	6.0339	0.1657	5.5771
26	49.402	0.0203	299.087	0.0034	6.0541	0.1652	5.6420
27	57.397	0.0174	348.489	0.0029	6.0715	0.1647	5.7004
28	66.686	0.0150	405.886	0.0025	6.0865	0.1643	5.7529
29	77.478	0.0129	472.573	0.0021	6.0994	0.1640	5.8000
30	90.017	0.0111	550.051	0.0018	6.1105	0.1637	5.8422
31	104.585	0.0096	640.068	0.0016	6.1201	0.1634	5.8799
32	121.510	0.0082	744.653	0.0014	6.1283	0.1632	5.9136
33	141.175	0.0071	866.164	0.0012	6.1354	0.1630	5.9438
34	164.022	0.0061	1007.339	0.0010	6.1415	0.1628	5.9706
35	190.566	0.0053	1171.361	0.0009	6.1467	0.1627	5.9945
40	403.429	0.0025	2486.673	0.0004	6.1639	0.1622	6.0798
45	854.059	0.0012	5271.188	0.0002	6.1719	0.1620	6.1264
50	1808.042	0.0006	11166.008	0.0001	6.1758	0.1619	6.1515

Table C.13. 20% INTEREST FACTORS FOR CONTINUOUS COMPOUNDING
INTEREST

	Single Payment		Equal Payment Series				Uniform gradient-series factor
	Compound-amount factor	Present-worth factor	Compound-amount factor	Sinking-fund factor	Present-worth factor	Capital-recovery factor	
n	To find F Given P F/P r, n	To find P Given F P/F r, n	To find F Given A F/A r, n	To find A Given F A/F r, n	To find P Given A P/A r, n	To find A Given P A/P r, n	To find A Given G A/G r, n
1	1.221	0.8187	1.000	1.0000	0.8187	1.2214	0.0000
2	1.492	0.6703	2.221	0.4502	1.4891	0.6716	0.4502
3	1.822	0.5488	3.713	0.2693	2.0379	0.4907	0.8676
4	2.226	0.4493	5.535	0.1807	2.4872	0.4021	1.2528
5	2.718	0.3679	7.761	0.1289	2.8551	0.3503	1.6068
6	3.320	0.3012	10.479	0.0954	3.1563	0.3168	1.9306
7	4.055	0.2466	13.799	0.0725	3.4029	0.2939	2.2255
8	4.953	0.2019	17.854	0.0560	3.6048	0.2774	2.4929
9	6.050	0.1653	22.808	0.0439	3.7701	0.2653	2.7344
10	7.389	0.1353	28.857	0.0347	3.9054	0.2561	2.9515
11	9.025	0.1108	36.246	0.0276	4.0162	0.2490	3.1460
12	11.023	0.0907	45.271	0.0221	4.1069	0.2435	3.3194
13	13.464	0.0743	56.294	0.0178	4.1812	0.2392	3.4736
14	16.445	0.0608	69.758	0.0143	4.2420	0.2357	3.6102
15	20.086	0.0498	86.203	0.0116	4.2918	0.2330	3.7307
16	24.533	0.0408	106.288	0.0094	4.3326	0.2308	3.8368
17	29.964	0.0334	130.821	0.0077	4.3659	0.2291	3.9297
18	36.598	0.0273	160.785	0.0062	4.3933	0.2276	4.0110
19	44.701	0.0224	197.383	0.0051	4.4156	0.2265	4.0819
20	54.598	0.0183	242.084	0.0041	4.4339	0.2255	4.1435
21	66.686	0.0150	296.683	0.0034	4.4489	0.2248	4.1970
22	81.451	0.0123	363.369	0.0028	4.4612	0.2242	4.2432
23	99.484	0.0101	444.820	0.0023	4.4713	0.2237	4.2831
24	121.510	0.0082	544.304	0.0018	4.4795	0.2232	4.3175
25	148.413	0.0067	665.814	0.0015	4.4862	0.2229	4.3471
26	181.272	0.0055	814.228	0.0012	4.4917	0.2226	4.3724
27	221.406	0.0045	995.500	0.0010	4.4963	0.2224	4.3942
28	270.426	0.0037	1216.906	0.0008	4.5000	0.2222	4.4127
29	330.300	0.0030	1487.333	0.0007	4.5030	0.2221	4.4286
30	403.429	0.0025	1817.632	0.0006	4.5055	0.2220	4.4421
31	492.749	0.0020	2221.061	0.0005	4.5075	0.2219	4.4536
32	601.845	0.0017	2713.810	0.0004	4.5092	0.2218	4.4634
33	735.095	0.0014	3315.655	0.0003	4.5105	0.2217	4.4717
34	897.847	0.0011	4050.750	0.0003	4.5116	0.2217	4.4788
35	1096.633	0.0009	4948.598	0.0002	4.5125	0.2216	4.4847
40	2980.958	0.0004	13459.444	0.0001	4.5152	0.2215	4.5032
45	8103.084	0.0001	36594.322	0.0000	4.5161	0.2214	4.5111
50	22026.466	0.0001	99481.443	0.0000	4.5165	0.2214	4.5144

Table C.14. 25% INTEREST FACTORS FOR CONTINUOUS COMPOUNDING
INTEREST

	Single Payment		Equal Payment Series				Uniform gradient-series factor
	Compound-amount factor	Present-worth factor	Compound-amount factor	Sinking-fund factor	Present-worth factor	Capital-recovery factor	
n	To find F Given P F/P r, n	To find P Given F P/F r, n	To find F Given A F/A r, n	To find A Given F A/F r, n	To find P Given A P/A r, n	To find A Given P A/P r, n	To find A Given G A/G r, n
1	1.284	0.7788	1.000	1.0000	0.7788	1.2840	0.0000
2	1.649	0.6065	2.284	0.4378	1.3853	0.7219	0.4378
3	2.117	0.4724	3.933	0.2543	1.8577	0.5383	0.8351
4	2.718	0.3679	6.050	0.1653	2.2256	0.4493	1.1929
5	3.490	0.2865	8.768	0.1141	2.5121	0.3981	1.5131
6	4.482	0.2231	12.258	0.0816	2.7352	0.3656	1.7975
7	5.755	0.1738	16.740	0.0597	2.9090	0.3438	2.0486
8	7.389	0.1353	22.495	0.0445	3.0443	0.3285	2.2687
9	9.488	0.1054	29.884	0.0335	3.1497	0.3175	2.4605
10	12.183	0.0821	39.371	0.0254	3.2318	0.3094	2.6266
11	15.643	0.0639	51.554	0.0194	3.2957	0.3034	2.7696
12	20.086	0.0498	67.197	0.0149	3.3455	0.2989	2.8921
13	25.790	0.0388	87.282	0.0115	3.3843	0.2955	2.9964
14	33.115	0.0302	113.072	0.0089	3.4145	0.2929	3.0849
15	42.521	0.0235	146.188	0.0069	3.4380	0.2909	3.1596
16	54.598	0.0183	188.709	0.0053	3.4563	0.2893	3.2223
17	70.105	0.0143	243.307	0.0041	3.4706	0.2881	3.2748
18	90.017	0.0111	313.413	0.0032	3.4817	0.2872	3.3186
19	115.584	0.0087	403.430	0.0025	3.4904	0.2865	3.3550
20	148.413	0.0067	519.014	0.0019	3.4971	0.2860	3.3851
21	190.566	0.0053	667.427	0.0015	3.5023	0.2855	3.4100
22	244.692	0.0041	857.993	0.0012	3.5064	0.2852	3.4305
23	314.191	0.0032	1102.685	0.0009	3.5096	0.2849	3.4474
24	403.429	0.0025	1416.876	0.0007	3.5121	0.2847	3.4612
25	518.013	0.0019	1820.305	0.0006	3.5140	0.2846	3.4725
26	665.142	0.0015	2338.318	0.0004	3.5155	0.2845	3.4817
27	854.059	0.0012	3003.459	0.0003	3.5167	0.2844	3.4892
28	1096.633	0.0009	3857.518	0.0003	3.5176	0.2843	3.4953
29	1408.105	0.0007	4954.151	0.0002	3.5183	0.2842	3.5002
30	1808.042	0.0006	6362.256	0.0002	3.5189	0.2842	3.5042
31	2321.572	0.0004	8170.298	0.0001	3.5193	0.2842	3.5075
32	2980.958	0.0004	10491.871	0.0001	3.5196	0.2841	3.5101
33	3827.626	0.0003	13472.829	0.0001	3.5199	0.2841	3.5122
34	4914.769	0.0002	17300.455	0.0001	3.5201	0.2841	3.5139
35	6310.688	0.0002	22215.223	0.0001	3.5203	0.2841	3.5153

Table C.15. 30% INTEREST FACTORS FOR CONTINUOUS COMPOUNDING
INTEREST

	Single Payment		Equal Payment Series				Uniform
	Compound-amount factor	Present-worth factor	Compound-amount factor	Sinking-fund factor	Present-worth factor	Capital-recovery factor	gradient-series factor
n	To find F Given P F/P r, n	To find P Given F P/F r, n	To find F Given A F/A r, n	To find A Given F A/F r, n	To find P Given A P/A r, n	To find A Given P A/P r, n	To find A Given G A/G r, n
1	1.350	0.7408	1.000	1.0000	0.7408	1.3499	0.0000
2	1.822	0.5488	2.350	0.4256	1.2896	0.7754	0.4256
3	2.460	0.4066	4.172	0.2397	1.6962	0.5896	0.8030
4	3.320	0.3012	6.632	0.1508	1.9974	0.5007	1.1343
5	4.482	0.2231	9.952	0.1005	2.2205	0.4504	1.4222
6	6.050	0.1653	14.433	0.0693	2.3858	0.4192	1.6701
7	8.166	0.1225	20.483	0.0488	2.5083	0.3987	1.8815
8	11.023	0.0907	28.649	0.0349	2.5990	0.3848	2.0602
9	14.880	0.0672	39.672	0.0252	2.6662	0.3751	2.2099
10	20.086	0.0498	54.552	0.0183	2.7160	0.3682	2.3343
11	27.113	0.0369	74.638	0.0134	2.7529	0.3633	2.4371
12	36.598	0.0273	101.750	0.0098	2.7802	0.3597	2.5212
13	49.402	0.0203	138.349	0.0072	2.8004	0.3571	2.5897
14	66.686	0.0150	187.751	0.0053	2.8154	0.3552	2.6452
15	90.017	0.0111	254.437	0.0039	2.8266	0.3538	2.6898
16	121.510	0.0082	344.454	0.0029	2.8348	0.3528	2.7255
17	164.022	0.0061	465.965	0.0022	2.8409	0.3520	2.7540
18	221.406	0.0045	629.987	0.0016	2.8454	0.3515	2.7766
19	298.867	0.0034	851.393	0.0012	2.8487	0.3510	2.7945
20	403.429	0.0025	1150.261	0.0009	2.8512	0.3507	2.8086
21	544.572	0.0018	1553.689	0.0007	2.8531	0.3505	2.8197
22	735.095	0.0014	2098.261	0.0005	2.8544	0.3503	2.8283
23	992.275	0.0010	2833.356	0.0004	2.8554	0.3502	2.8351
24	1339.431	0.0008	3825.631	0.0003	2.8562	0.3501	2.8404
25	1808.042	0.0006	5165.062	0.0002	2.8567	0.3501	2.8445
26	2440.602	0.0004	6973.104	0.0002	2.8571	0.3500	2.8476
27	3294.468	0.0003	9413.706	0.0001	2.8574	0.3500	2.8501
28	4447.067	0.0002	12708.174	0.0001	2.8577	0.3499	2.8520
29	6002.912	0.0002	17155.241	0.0001	2.8578	0.3499	2.8535
30	8103.084	0.0001	23158.153	0.0001	2.8580	0.3499	2.8546
31	10938.019	0.0001	31261.237	0.0000	2.8580	0.3499	2.8555
32	14764.782	0.0001	42199.257	0.0000	2.8581	0.3499	2.8561
33	19930.370	0.0001	56964.038	0.0000	2.8582	0.3499	2.8566
34	26903.186	0.0001	76894.409	0.0000	2.8582	0.3499	2.8570
35	36315.503	0.0000	103797.595	0.0000	2.8582	0.3499	2.8573

APPENDIX D
FUNDS FLOW CONVERSION FACTORS

Table D.1. FUNDS FLOW
CONVERSION FACTORS

r	$\dfrac{e^r - 1}{r}$ $\left(A/\overline{A} \quad r \right)$
1	1.005020
2	1.010065
3	1.015150
4	1.020270
5	1.025422
6	1.030608
7	1.035831
8	1.041088
9	1.046381
10	1.051709
11	1.057073
12	1.062474
13	1.067910
14	1.073384
15	1.078894
16	1.084443
17	1.090028
18	1.095652
19	1.101313
20	1.107014
21	1.112752
22	1.118530
23	1.124347
24	1.130204
25	1.136101
26	1.142038
27	1.148016
28	1.154035
29	1.160094
30	1.166196
31	1.172339
32	1.178524
33	1.184751
34	1.191022
35	1.197335
36	1.203692
37	1.210093
38	1.216538
39	1.223027
40	1.229561

SELECTED REFERENCES

AMERICAN TELEPHONE AND TELEGRAPH CO., *Engineering Economy* (New York: American Telephone and Telegraph Co., 1963).

BARISH, N. N., *Economic Analysis* (New York: McGraw-Hill Book Company, 1962).

BARNARD, CHESTER I., *The Functions of the Executive* (Cambridge, Mass.: Harvard University Press, 1938).

BIERMAN, H., AND S. SMIDT, *The Capital Budgeting Decision* (New York: The Macmillan Company, 1971).

BUFFA, E. S., *Models for Production and Operations Management* (New York: John Wiley & Sons, Inc., 1963).

CANADA, J. R., *Intermediate Economic Analysis for Management and Engineering* (Englewood Cliffs., N. J.: Prentice-Hall, Inc., 1971).

COUGHLAN, J. D., AND W. K. STRAND, *Depreciation* (New York: The Ronald Press Company, 1969).

DASGUPTA, A. K., AND D. W. PEARCE, *Cost-Benefit Analysis: Theory and Practice* (New York: Barnes and Noble, 1972).

DEAN, J., *Managerial Economics* (Englewood Cliffs, N. J.: Prentice-Hall, Inc., 1951).

DEGARMO, E. P., AND J. R. CANADA, *Engineering Economy* (New York: The Macmillan Company, 1973).

ENGLISH, J. M., ed., *Economics of Engineering and Social Systems* (New York, Wiley-Interscience, 1972).

ENGLISH, J. M., ed., *Cost Effectiveness* (New York: John Wiley and Sons, Inc., 1968).

FABRYCKY, W. J., P. M. GHARE, AND P. E. TORGERSEN, *Industrial Operations Research* (Englewood Cliffs, N. J.: Prentice-Hall, Inc., 1972).

FABRYCKY, W. J., AND G. J. THUESEN, *Economic Decision Analysis* (Englewood Cliffs, N. J.: Prentice-Hall, Inc., 1974).

FLEISCHER, G. A., *Capital Allocation Theory* (New York: Appleton-Century-Crofts, 1969).

GRANT, E. L., and L. F. BELL, *Basic Accounting and Cost Accounting* (New York: McGraw-Hill Book Company, 1964).

GRANT, E. L., W. G. IRESON, AND R. S. LEAVENWORTH, *Principles of Engineering Economy* (New York: The Ronald Press Company, 1976).

JELEN, F. C., *Cost and Optimization Engineering* (New York: McGraw-Hill Book Company, 1970).

JEYNES, P. H., *Profitability and Economic Choice* (Ames, Iowa: Iowa State University Press, 1968).

MARSTON, A., R. WINFREY, AND J. C. HEMPSTEAD, *Engineering Valuation and Depreciation* (Ames, Iowa: Iowa State University Press, 1963).

579

McNEIL, T. F., AND D. S. CLARK, *Cost Estimating and Contract Pricing* (New York: American Elsevier Publishing Co., 1966).

NEWNAN, DONALD G., *Engineering Economic Analysis* (San Jose, California: Engineering Press, 1976).

OSTWALD, P. F., *Cost Estimating for Engineering and Management* (Englewood Cliffs, N. J.: Prentice-Hall, Inc., 1974).

PARK, W. R., *Cost Engineering Analysis: A Guide to the Economic Evaluation of Engineering Projects* (New York: John Wiley and Sons, Inc., 1973).

PEURIFOY, R. L., *Estimating Construction Costs* (New York: McGraw-Hill Book Company, 1975).

REISMAN, A., *Managerial and Engineering Economics* (Boston, Mass.: Allyn and Bacon, 1971).

RIGGS, J. L., *Economic Decision Models* (New York: McGraw-Hill Book Company, 1968).

SCHWEYER, H. E., *Analytical Models for Managerial and Engineering Economics* (New York: Reinhold Publishing Corporation, 1964).

SMITH, G. W., *Engineering Economy* (Ames, Iowa: Iowa State University Press, 1973).

SPECTHRIE, S. W., *Industrial Accounting* (Englewood Cliffs, N.J.: Prentice-Hall, Inc., 1959).

SPORN, P., *Technology, Engineering, and Economics* (Cambridge, Massachusetts: MIT Press, 1969).

TARQUIN, A. J., AND L. BLANK, *Engineering Economy: A Behavioral Approach* (New York: McGraw-Hill Book Company, 1976).

TAYLOR, G. A., *Managerial and Engineering Economy* (Princeton, N.J.: D. Van Nostrand Company, Inc., 1964).

TERBORGH, G. *Business Investment Management* (Washington, D.C.: Machinery and Allied Products Institute, 1967).

TERBORGH, G., *Dynamic Equipment Policy* (New York: McGraw-Hill Book Company, 1949).

TYLER, C., AND C. H. WINTER, JR., *Chemical Engineering Economics* (New York: McGraw-Hill Book Company, 1959).

WINFREY, R., *Economic Analysis for Highways* (Scranton, Pennsylvania: International Textbook Company, 1969).

INDEX